Methods in Molecular Biology™

Series Editor
John M. Walker
School of Life Sciences
University of Hertfordshire
Hatfield, Hertfordshire, AL10 9AB, UK

For further volumes:
http://www.springer.com/series/7651

Lipid-Protein Interactions

Methods and Protocols

Edited by

Jörg H. Kleinschmidt

Institut für Biologie, Abteilung Biophysik
Universität Kassel, Heinrich-Plett-Straße 40, Kassel, Germany

Humana Press

Editor
Jörg H. Kleinschmidt
Institut für Biologie, Abteilung Biophysik
Universität Kassel
Kassel, Germany

ISSN 1064-3745 ISSN 1940-6029 (electronic)
ISBN 978-1-4939-6263-1 ISBN 978-1-62703-275-9 (eBook)
DOI 10.1007/978-1-62703-275-9
Springer New York Heidelberg Dordrecht London

Printed on acid-free paper

Humana Press is a brand of Springer
Springer is part of Springer Science+Business Media (www.springer.com)

Preface

Biological membranes are the essential structuring elements of all living cells. The core elements of a membrane are a lipid bilayer and integral and peripheral proteins. While the lipid bilayer constitutes the hydrophobic barrier that prevents arbitrary exchange of solutes, the transmembrane proteins allow the regulated exchange of solutes or they transduce signals from one side of the membrane to the other. Many enzymatic reactions take place at the membrane–water interface. Specific lipid–protein interactions are important for the stable integration and activity of integral and peripheral membrane proteins. The unique structure of the lipid bilayer requires specific surface properties of integral and peripheral proteins so that the proteins can function. Protein surfaces exposed to the fatty core of the membrane are typically hydrophobic, while protein surfaces exposed to the aqueous space usually are composed of polar amino acid residues. The polar–apolar interface of the lipid bilayer is formed by the glycerol backbone and by the polar head group of the various phospholipid species and therefore an important region for lipid–protein interactions. The fatty acyl chains of the lipids can vary a lot in length and degree of unsaturation and the membranes may contain cholesterol, sphingolipids, etc. This has consequences, e.g., for membrane thickness and fluidity. Furthermore, the complex membrane composition often leads to the formation of micro-domains with distinct physicochemical properties.

To gain detailed insight into membrane properties, it is therefore of great importance to understand the complex nature of the interactions of membrane proteins with lipids. This volume provides a selection of protocols to examine protein–lipid interactions, membrane and membrane protein structure, how membrane proteins affect lipids, and how they are in turn affected by the lipid bilayer and lipid properties.

Numerous methodologies have been developed in the past, each with its own advantages and limitations. The methods described here are all actively used, complementary, and necessary to obtain comprehensive information about membrane structure and function. The method of choice is determined by the information that is sought, but is dependent on the properties of the sample, the available quantity, and the required sensitivity. Label-free approaches described in this book include methodologies like quartz-crystal microbalances with dissipation, surface plasmon resonance, isothermal titration calorimetry, and differential scanning calorimetry. These are useful methods, e.g., to monitor binding events and to obtain the free energies, enthalpies, and entropies of protein–lipid interactions. Imaging techniques like electron microscopy and atomic force microscopy are used to examine the structure and organization of protein–lipid complexes in membranes. Atomic force spectroscopy allows the probing of mechanical properties of macromolecules, e.g., the force necessary to unfold a single protein in a lipid environment. Neutron scattering is an emerging technique to study the structure of protein–lipid complexes, which in combination with deuteration of either lipids or proteins allows resolving the inner structure of big and dynamic lipid–protein complexes. The secondary structure of native and nonnative proteins in lipid membranes can conveniently be monitored by circular dichroism spectroscopy and synchrotron radiation circular dichroism spectroscopy. The development of the latter greatly extended the recordable wavelength range, strengthening structural investigations. Secondary structure and in addition the orientation and order parameters of membrane proteins in lipid bilayers can be obtained from infrared-spectroscopic methods. In combination with isotope editing methods, these are also used to determine changes in local protein conformation.

The specificity and selectivity of protein interactions with lipid species is efficiently investigated by combining the labeling of lipids and proteins with either fluorescence or electron paramagnetic resonance spectroscopy, which are both very sensitive techniques to examine the protein–lipid interface. These methods are applied with great success to probe the topology of peptides and proteins in membranes. Fluorescence quenching is a fast and reliable technique to determine the location of fluorescent amino acids, especially tryptophan residues in lipid bilayers. Förster energy transfer is a highly sensitive fluorescence method that is useful to measure distances up to 10 nm and able to detect direct binding as well as deviations from homogeneity of the lipid distribution around a protein. Fluorescence methodologies have also been successfully applied, e.g., to track single molecules like transmembrane proteins in planar supported membranes. Single particle tracking allows imaging and tracking single fluorescent molecules with good spatial and temporal resolution. Particle association and dissociation events can be monitored. Fluorescence correlation spectroscopy uses the time correlation of temporal fluctuations of fluorescence, which are detected in a focal volume, to explore dynamic events with high temporal resolution and statistical accuracy. Lipid-spin labeling and electron paramagnetic resonance (EPR also called electron spin resonance, ESR) spectroscopy have been highly successful in the determination of the protein-solvating lipid shell, that is, the stoichiometry of the lipid interactions with integral membrane proteins. Since the technique allows the estimation of mobile and protein-immobilized lipids and their exchange rates, it has also been very successful in probing the lipid-selectivity of transmembrane proteins. In combination with site-directed mutagenesis, EPR of spin-labeled proteins has become a powerful tool to examine protein structure and dynamics, even in complex systems that are not accessible with other approaches. Nuclear magnetic resonance (NMR) spectroscopy, combined with isotopic labeling, is widely used to probe the structure and dynamics of proteins either in solution or in a lipid environment. Solution-phase NMR is performed with detergent-solubilized membrane proteins and usually yields high-resolution structures. Solid-state NMR is performed either as magic angle spinning (MAS) NMR to obtain highly resolved protein structures with spectra resembling fast isotropically tumbling proteins in solution or with oriented samples. When applied to oriented samples, NMR gives valuable information on the dynamics and orientation of lipids and proteins in membranes and allows determining, e.g., peptide orientation, lipid-order parameters, lipid-phases, etc. Lipid–protein interactions have been investigated in small, large, and giant vesicles, in supported lipid bilayers, in lipid monolayers, with short-chain lipid micelles, or even with single lipids. More recently lipid nanodiscs have been developed, which are disc-like fragments of lipid bilayers that are stabilized by two amphipathic helical proteins. Nanodiscs have been shown to be a robust means for stabilizing and investigating protein–lipid interactions and nanodisc-applications are reviewed here. The last chapter in this book is dedicated to a range of molecular dynamics simulation approaches to lipid–protein interactions, including both atomistically detailed and coarse-grain methods. Lipid–protein interactions are heavily investigated and while it is clear that not all methods can be described in a single book, the present volume covers a wide range of the methods typically used in this area of research. I thank all contributing authors for providing interesting and highly valuable methods, insights, and reviews for this volume of Methods in Molecular Biology.

Kassel, Germany ***Jörg H. Kleinschmidt***

Contents

Contributors

CHRISTOPHER AISENBREY • *Institut de chimie, CNRS, Université de Strasbourg, UMR 7177, Strasbourg, France*
GREGOR ANDERLUH • *National Institute of Chemistry, Ljubljana, Slovenia; Department of Biology, Biotechnical Faculty, University of Ljubljana, Ljubljana, Slovenia*
ASHISH ARORA • *Molecular and Structural Biology, CSIR-Central Drug Research Institute, Lucknow, UP, India*
BURKHARD BECHINGER • *Institut de chimie, CNRS, Université de Strasbourg, UMR 7177, Strasbourg, France*
VIKTORIA BETANELI • *Biophysics, BIOTEC, Technische Universitat Dresden, Dresden, Germany*
PETER J. BOND • *Department of Chemistry, Unilever Centre for Molecular Informatics, University of Cambridge, Cambridge, UK*
ENRICA BORDIGNON • *Laboratory for Physical Chemistry, ETH Zurich, Zurich, Switzerland*
JAMES A. BROZIK • *Department of Chemistry, Washington State University, Pullman, WA, USA*
OLGA CAÑADAS • *Department of Biochemistry and Molecular Biology I, Faculty of Biology, CIBER de Enfermedades Respiratorias, Madrid, Spain*
GREGORY A. CAPUTO • *Department of Chemistry and Biochemistry, Rowan University, Glassboro, NJ, USA*
CRISTINA CASALS • *Department of Biochemistry and Molecular Biology I, Faculty of Chemistry, Complutense University of Madrid, Madrid, Spain*
LUKE A. CLIFTON • *Large Scale Structures Group, ISIS Spallation Neutron Source, Harwell Science and Innovation Campus, Rutherford Appleton Laboratory, Didcot, Oxfordshire, UK*
ILIA G. DENISOV • *Departments of Biochemistry, University of Illinois, Urbana, IL, USA*
FÁBIO FERNANDES • *Centro de Química Física Molecular and Institute of Nanosciences and Nanotechnologies, Complexo I, Instituto Superior Técnico, Lisbon, Portugal*
MARK GAREWAL • *Molecular Foundry, Lawrence Berkeley National Laboratory, Berkeley, CA, USA*
DIANA GARZÓN • *Theoretische Molekulare Biophysik, Max Planck Institute of Biophysics, Frankfurt am Main, Germany*
KUNIHIKO GEKKO • *Institute for Sustainable Sciences and Development, Hiroshima University, Higashi-Hiroshima, Japan*
VESNA HODNIK • *Department of Biology, Biotechnical Faculty, University of Ljubljana, Ljubljana, Slovenia*
DANIEL A. HOLDBROOK • *School of Chemistry, University of Southampton, Highfield, Southampton, UK*
JEFFREY P. JONES • *Department of Chemistry, Washington State University, Pullman, WA, USA*
SYMA KHALID • *School of Chemistry, University of Southampton, Highfield, Southampton, UK*

Zoltán Kóta • *Institute of Biophysics, Biological Research Centre, Szeged, Hungary*
Jeremy H. Lakey • *Institute for Cell and Molecular Biosciences, Newcastle University, Newcastle upon Tyne, UK*
Erwin London • *Department of Biochemistry and Cell Biology, Stony Brook University, Stony Brook, NY, USA*
Luís M. S. Loura • *Faculdade de Farmácia, Universidade de Coimbra, Pólo das Ciências da Saúde, Azinhaga de Santa Comba, Coimbra, Portugal; Centro de Química de Coimbra, Universidade de Coimbra, Coimbra, Portugal*
Koichi Matsuo • *Hiroshima Synchrotron Radiation Center, Hiroshima University, Higashi-Hiroshima, Japan*
Matthias Michalek • *Institut de chimie, CNRS, Université de Strasbourg, UMR 7177, Strasbourg, France*
Cameron Neylon • *Science and Technology Facilities Council, ISIS Spallation Neutron Source, Harwell Science and Innovation Campus, Rutherford Appleton Laboratory, Didcot, Oxfordshire, UK*
Søren B. Nielsen • *Department of Molecular Biology and Genetics, Interdisciplinary Nanoscience Center (iNANO), Aarhus University, Aarhusc, Denmark*
Daniel E. Otzen • *Department of Molecular Biology and Genetics, Interdisciplinary Nanoscience Center (iNANO), Aarhus University, Aarhusc, Denmark*
Tibor Páli • *Institute of Biophysics, Biological Research Centre, Szeged, Hungary*
Teresa Paramo • *Department of Chemistry, Unilever Centre for Molecular Informatics, University of Cambridge, Cambridge, UK*
Yevhen Polyhach • *Laboratory for Physical Chemistry, ETH Zurich, Zurich, Switzerland*
Kumud R. Poudel • *Department of Chemistry, Washington State University, Pullman, WA, USA*
Manuel Prieto • *Centro de Química Física Molecular and Institute of Nanosciences and Nanotechnologies, Complexo I, Instituto Superior Técnico, Lisbon, Portugal*
Gang Ren • *Molecular Foundry, Lawrence Berkeley National Laboratory, Berkeley, CA, USA*
Evgeniy S. Salnikov • *Institut de chimie, CNRS, Université de Strasbourg, UMR 7177, Strasbourg, France*
Rajeshwer S. Sankhala • *Department of Biochemistry & Molecular Biology, Thomas Jefferson University, Philadelphia, PA, USA*
K. Tanuj Sapra • *Department of Chemistry, University of Oxford, Oxford, UK*
Mary A. Schuler • *Departments of Biochemistry and Cell and Developmental Biology, University of Illinois, Urbana, IL, USA*
Petra Schwille • *Biophysics, BIOTEC, Technische Universitat Dresden, Dresden, Germany*
Stephen G. Sligar • *Department of Biochemistry and Chemistry, University of Illinois, Urbana, IL, USA*
Musti J. Swamy • *School of Chemistry, University of Hyderabad, Hyderabad, India*
Suren A. Tatulian • *Department of Physics, University of Central Florida, Orlando, FL, USA*
Lei Zhang • *Molecular Foundry, Lawrence Berkeley National Laboratory, Berkeley, CA, USA*

Chapter 1

Quartz Crystal Microbalances as Tools for Probing Protein–Membrane Interactions

Søren B. Nielsen and Daniel E. Otzen

Abstract

Extensive studies on the spontaneous collapse of phospholipid vesicles into supported lipid bilayers (SLBs) have led to procedures which allow SLB formation on a wealth of substrates and lipid compositions. SLBs provide a widely accepted and versatile model system which mimics the natural cell membrane separating the extracellular and intracellular fluids of the living cell. The quartz crystal microbalance with dissipation monitoring (QCM-D) has been central both in the understanding of vesicle collapse into SLBs on various substrates and in probing the kinetics and mechanisms of biomolecular interactions with SLBs in real time. We describe a robust procedure to form SLBs of zwitterionic and charged lipids on SiO_2 sensor crystals which subsequently can be exploited to probe the interaction between proteins and peptides with the SLB.

Key words: Supported lipid bilayer, SiO_2, Quartz crystal microbalance with dissipation, Interaction

1. Introduction

A quartz crystal microbalance consists of a small circular piece of quartz (the sensor crystal), whose piezoelectric properties make it oscillate at a characteristic resonance frequency f in response to an oscillating electric field. The QCM-D technique simultaneously measures changes in the resonance frequency (Δf) and the energy dissipation (ΔD). These values depend not only on the sensor crystal itself but also on the molecules (e.g., water, ions, lipid bilayers proteins, and peptides) trapped in the oscillation through contact with the crystal surface (1, 2). Provided this material absorbs as a rigid film, Δf is directly proportional to the mass adsorbed to the sensor surface (the Sauerbrey relation). However, this relation breaks down if the film is not sufficiently rigid. Information about film rigidity is provided by ΔD. Consequently, the combination of Δf and ΔD (described in more detail in Subheading 3.4.2) provides detailed information about the general

Jörg H. Kleinschmidt (ed.), *Lipid-Protein Interactions: Methods and Protocols*, Methods in Molecular Biology, vol. 974,
DOI 10.1007/978-1-62703-275-9_1, © Springer Science+Business Media New York 2013

viscoelastic properties of material bound to the sensor surface, such as those associated with various biological processes. The application of QCM-D has been demonstrated in numerous studies of protein adsorption (1, 3–6), antibody reactions with antigens (7, 8), attachment and spreading of cells (9, 10), and highly hydrated protein films (11).

Another extensively studied yet not completely understood phenomenon is the formation of planar-supported lipid bilayers (SLBs) on various substrate surfaces. SiO_2, Si_3N_4, and mica supports are popular choices for SLB formation, while recent progress has further allowed the formation of SLBs on TiO_2 and Au substrates. QCM-D has been key in providing mechanistic insights leading to the identification and understanding of conditions which determine whether adsorbing vesicles (1) remain intact (e.g., oxidized Au), (2) form SLBs through direct collapse upon interaction with the substrate (e.g., cationic lipids binding on SiO_2), or (3) form SLBs when adsorbed vesicles rupture and fuse at a critical surface coverage level (12–14). This latter process depends on factors such as solvent ions (divalent cations bridge anionic lipids to SiO_2 surfaces, while monovalent ions will have a more modest effect through charge shielding), lipid composition (charge repulsion of anionic lipids binding to the SiO_2 surface), and surface chemistry (e.g., charge and roughness) (12).

Once SLBs are formed on the sensor surface (as confirmed by characteristic Δf and ΔD signals), it is possible to monitor interactions with biomolecules which interact with SLBs or interact with specific phospholipids or functional/reactive groups intentionally added to the SLB studied such as maleimide-activated or biotin-doped bilayers (15–17). By coupling these functionalities to specific biomolecules, such functionalities can be exploited to probe interactions with, e.g., collagen (18), hyaluronan (19), DNA, or specific peptide sequences (16, 17). SLBs thus provide an excellent model system for biological cell membranes which separate and regulate the transport of biomolecules between the exterior and interior environments *in vivo*.

In this chapter, we present a robust protocol to explore interactions of different proteins with SLBs on SiO_2. We will also mention examples of applications including protein interactions with specific lipid headgroups (Ca^{2+}-dependent interaction of annexin A5 with anionic phosphatidylserine headgroups (20)), enzymatic hydrolysis of SLBs by phospholipase A2 (21, 22), binding and membrane perturbation by antimicrobial peptides on SLBs (23–29), and the interaction of protein:lipid complexes with SLBs (30). Nonspecific protein-SLB interactions are only observed to a very small extent (31), making SPBs an attractive platform for experiments in which low nonspecific binding is essential and for further functionalization of SPBs such as biotinylation and glycosylation.

2. Materials

2.1. QCM-D System and Accessories

1. Q-sense E4, E1, or D300 QCM-D system from Q-sense AB (Västra Frölunda, Sweden).
2. 5 MHz silicon dioxide (SiO_2)-coated QCM-D crystal sensors (Cat# QSX 303) from Q-sense AB (Västra Frölunda, Sweden).
3. Q-sense Sensor Holder.
4. Ozone lamp, e.g., Bioforce Nanosciences UV/Ozone Procleaner.
5. 2% SDS in MilliQ H_2O.
6. 20 mM Tris–HCl and 150 mM NaCl.
7. 1 M $CaCl_2$ or $MgCl_2$.

2.2. Lipid Bilayers

1. 1,2-diooleoyl-sn-glycero-3-phosphocholine (DOPC), 1,2-diooleoyl-sn-glycero-3-phosphoserine (DOPS), and/or 1,2-diooleoyl-sn-glycero-3-phosphatidylglycerol (DOPG) from Avanti Polar Lipids (Alabaster, AL).
2. Sonicator probe or Avanti Polar Lipids MiniExtruder (incl. suitable supports and polycarbonate membranes in 30–100 nm range for vesicle sizing).
3. Liquid nitrogen or dry ice bath.
4. Water bath.

2.3. Software

1. Qsoft 401 v2.5.10 for Q-sense E1 and E4 QCM-D systems for data acquisition.
2. Qsoft v1.6.18 for Q-sense D300 QCM-D systems for data acquisition.
3. Qtools v3.0.13 for data analysis.

The software is supplied by the manufacturer, and updates can be downloaded from the user area at http://www.q-sense.com (registration requires instrument serial number and user name).

3. Methods

3.1. Preparations

3.1.1. Cleaning of SiO_2-Coated Sensors

Maintaining clean QCM-D sensor crystals and QCM-D flow system is essential to obtain reliable results, and we here describe a procedure for cleaning of SiO_2 sensor crystals. The reader is referred to Subheading 4 Notes 2 and 3 for guidelines and implications of (improper) system cleaning.

1. Place sensor crystals in sensor holder.
2. Immerse sensor crystal in 2% sodium dodecyl sulfate (SDS) for 30 min at room temperature to remove adsorbed proteins and

lipids. Do not allow the sensor to dry after exposure to SDS until they have been washed in MilliQ H_2O.

3. Rinse thoroughly with MilliQ H_2O.
4. Dry under a stream of nitrogen gas.
5. Subject to UV/ozone treatment for 10 min to oxidize residual organic contaminants.
6. Rinse thoroughly with MilliQ H_2O.
7. Dry under a stream of nitrogen gas.

3.1.2. Vesicle Preparation

Vesicles can be prepared by several methods including but not limited to detergent removal, sonication, and extrusion. Of these methods, extrusion of multilamellar vesicles through polycarbonate membranes with defined pore sizes as described in the text below is preferred in our laboratory. Refer to Note 4 for an alternative procedure to produce small unilamellar vesicles by sonication.

1. Dissolve individual lipids to 10 mg/ml in 1:1 chloroform: methanol in a clean glass vial.
2. Mix a total amount of 5–10 mg lipid to desired molar or weight ratio in a clean glass vial (1.5 mL). Up to 50% anionic lipids such as PS can be tolerated (14, 32). A higher content of anionic lipids will increase electrostatic repulsion between vesicles and increase the activation barrier to collapse and vesicle fusion.
3. Remove solvent in a stream of nitrogen gas while slowly turning the glass vial to create a thin lipid film on the bottom and wall of the glass vial.
4. Continue flushing with nitrogen gas for at least 1 h or incubate lipid film overnight in vacuum desiccator to remove residual solvent.
5. Add 1 mL 10 mM Tris–HCl, pH 7.4, and 150 mM NaCl to a final lipid concentration of 5–10 mg/ml and allow the film to swell for at least 10 min at a temperature 5–10°C above the highest t_m of the lipids employed (see Note 5). Gently vortex the suspension to remove remaining lipid film from the glass vial walls. This procedure will create large multilamellar vesicles (LMVs).
6. Subject lipid suspension to 4–5 freeze–thaw cycles by successive transfer between liquid nitrogen (or dry ice bath) and a water bath (NB! Keep water bath temperature above t_m).
7. Prepare large unilamellar vesicles (LUVs) by at least 11 passages through a 30–100 nm polycarbonate membrane mounted between support discs in the Avanti Polar Lipids MiniExtruder. LUVs will generally be slightly larger than the pore size employed.

8. Vesicles can be stored for up to 3–4 days at 4°C. Avoid freezing after extrusion.

3.1.3. Sensor Mounting and Equilibration

1. After thorough cleaning of the sensor crystals, the dry crystal is mounted in the QCM-D sensor chamber.
2. Flow through at least 1 mL buffer at 250 μl/min. To avoid buffer-dependent changes in F and D in subsequent steps, this buffer should be the same as that intended for use in protein: SLB interaction study.
3. Start the Qsoft program and set to desired temperature (this temperature should be above ambient. The QCM-D apparatus does not allow cooling or heating above 60°C unless an external heating/cooling source is attached). Allow temperature equilibration for 5–15 min (depending on temperature set point; the higher above ambient, the longer equilibration time is necessary to avoid drift in ΔF and ΔD due to temperature drift.
4. Start measurement by sweeping frequencies to locate resonance peaks at multiple harmonics, e.g., 1st to 9th harmonic. Recording data at multiple harmonics allows more detailed data analysis as described in Subheading 3.4.2.
5. Obtain at least 5 min of stable (preferably <1 Hz/h) baseline before proceeding with the experiment.

3.2. Preparation of Supported Lipid Bilayers

Successful preparation of supported lipid bilayers depends on several factors including the support surface and characteristics of lipid and buffers employed. The QCM-D technique has, together with complementary techniques such as AFM, SPR, and ellipsometry, been central in understanding the process of SPB formation (12, 33–37). Depending on conditions, lipids, and substrate, phospholipid vesicles adsorbing onto solid supports may bind as intact vesicles spontaneously rupture, undergo vesicle–vesicle fusion followed by rupture, or intact vesicles may adsorb until a critical surface coverage at which they rupture into a SPB.

SPBs have successfully been prepared on various supports with silicon dioxide (SiO_2), glass and mica being the most popular and successful substrates (13, 35–37). More recently, methods for SPB preparation on titanium dioxide (TiO_2) (38, 39), silicon nitride (Si_3N_4) (40), and gold (Au) have been presented, the latter through the use of an α-helical (AH) peptide which induces vesicle rupture (38). For further details on SPB formation from zwitterionic and charged phospholipids on gold and TiO_2 substrates, we refer to protocols and considerations on SPB formation by (33). The following protocol allows adsorption and rupture of phospholipid vesicles composed by pure zwitterionic (e.g., DOPC) vesicles and mixtures of zwitterionic and up to ~50 w/w% anionic phospholipids (e.g., DOPS and DOPG) on SiO_2 sensor surfaces. SiO_2 surfaces

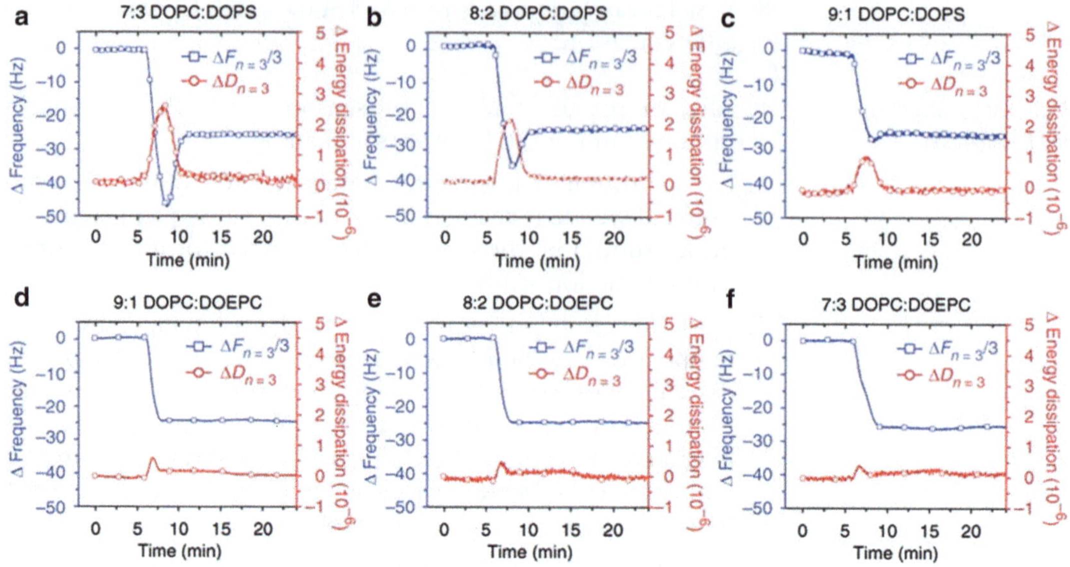

Fig. 1. Examples of formation of SLBs by different lipids on SiO_2-coated QCM-D sensor crystals. Data presented are from the third overtone (Reproduced with permission from (33), Copyright 2006 Elsevier Press).

have previously been shown to allow symmetrical distribution of anionic lipids in the final SLB, whereas TiO_2 and mica lead to substantial asymmetry in the lipid distribution across SLBs (32). The basic protocol may require adjustments such as addition of divalent cations and creation of osmotic gradients to facilitate collapse of charged vesicles. These elements will be discussed in the text. Upon addition of vesicles, a rapid decrease in the resonance frequency is seen along with an increase in dissipation due to the adsorption of the vesicles and coupled water. The maximum resonance frequency shift and dissipation values depend largely on the vesicle size and buffer composition (e.g., salts). At the critical surface coverage, Δf starts to increase to -26 ± 1 Hz, and ΔD decreases to ~0–0.2 × 10^{-6} due to the transformation of adsorbed vesicles into a continuous and rigid SLB and the concurrent release of water entrapped in vesicles. The distinct transition in Δf and ΔD may not always be observed, owing to a (by QCM standards) rapid (i.e., within a ~1 s time frame) vesicle collapse into SLBs upon surface adsorption. For instance, the electrostatic interaction between positively charged DOEPC (1,2-dioleoyl-*sn*-glycero-3-ethylphosphocholine) lipids and the SiO_2 support is sufficient to cause vesicle rupture, and vesicles continue to collapse until saturation around -24 ± 1 Hz, indicating complete surface coverage. Typical collapse time profiles are shown in Fig. 1.

1. Start by flowing buffer over the mounted sensor and obtain at least 5 min of stable baseline.

2. Dilute vesicle preparation to 0.1 mg/ml in 10 mM Tris–HCl, pH 7.4, and 150 mM NaCl. Include 2–5 mM $CaCl_2$ when working with anionic lipids.
3. Inject 0.1 mg/ml lipid suspension at 100 μl/min for 5 min to ensure formation of a defect-free bilayer. This should produce bilayers with a resonance frequency of -26 ± 1 Hz and a dissipation value of less than 0.2×10^{-6}. Washing may be required to remove loosely bound vesicles.
4. Wash with 1 mL buffer at 100 μl/min to remove residual loosely bound vesicles. Proceed only with the experiment at this stage if a stable frequency of -26 ± 1 Hz and dissipation values less than 0.2×10^{-6} are obtained as an indicator of proper vesicle collapse. Deviation from this may arise due to incomplete vesicle collapse or artifacts as discussed in Notes 6–8.

3.3. Using the QCM-D Technique to Monitor Protein Interactions with SLBs

Once an SLB is formed on SiO_2 support and a baseline has been established, the SLB can be exposed to proteins, peptides, or protein complexes in order to study phenomena such as (1) binding events, (2) the effect on membrane structure and integrity, and (3) enzymatic activities. The QCM-D data collected may be affected by both the concentration and size of injected protein, and we thus attempt to provide basic guidelines here.

The impact of antimicrobial peptides (AMPs) and mode of action and binding to SLBs has been investigated in several papers (23–29). The studies have been made using ~1–20 μM of the studied AMPs, and their interaction with SLBs often gives weak *F* and *D* signals in the low end of the range and leads to lipid removal and/or membrane disruption in the larger concentration range. Thus, while the activity of AMPs may depend on ionic strength, peptide, and SLB charge, the 1–20 μM range may be a good starting point for AMP interaction studies with SLBs as a function of, e.g., lipid composition.

The binding and activity of phospholipase A2 (PLA2) enzymes which specifically recognize and cleave the ester bond between the glycerol group and acyl group at the sn-2 position of phospholipids embedded in SLBs has been monitored by QCM-D at 500 nM (22) and 0.37 mg/ml (21).

The Ca^{2+}-dependent interaction of anionic phosphatidylserine headgroups with ~kDa prothrombin (32) and annexins A5 (~36 kDa monomer and ~72 kDa dimer) and A6 (~76 kDa) (41) employed protein concentrations of 1.7–68 μg/ml (~26–1,000 nM) and 20–40 μg/ml (~570 nM), respectively. Further, concentrations from ~0–1 μM were employed to construct a adsorption isotherm of annexin A2 (~39 kDa) to 4:1 POPC:POPS membranes (42). Thus, obtaining significant changes in *F* and *D* signals during adsorption of proteins appears to require less protein compared to small peptides. The higher mass uptake of larger proteins adsorbing to surfaces observed by QCM-D has been attributed primarily to bound water and water coupled to the adsorbed proteins (3). Owing to the variation in protein size and associated bound water, analysis of QCM-D data has to be done in a case-by-case manner and preferentially aided by complementary data. We here present three ways of QCM-D data treatments to describe and obtain information on (1) the adsorbed mass of protein adsorbed on SLBs as a rigid film through the use of the Sauerbrey relation; (2) evaluation of film mass, thickness, viscosity, and shear modules through Voigt-based viscoelastic modeling upon exposure of SLBs to proteins; and (3) the mechanism of interaction through use of frequency-dissipation plots.

The interpretation of QCM-D data may be challenging for both new and more experienced users, and we thus recommend new QCM-D users to attend Q-sense webinars at http://www.q-sense.com to gain insights in basic QCM-D data interpretation and the use of the Qtools software.

1. Ensure proper SLB formation (see previous section).
2. Obtain at least 5 min of stable baseline.
3. Equilibrate sample at measuring temperature for 5 min in heating block or water bath.
4. Inject protein of interest at 250 μl/min.
   (a) The appropriate injection volume depends on the length of the tubing. The internal volume of the QCM-D cell is approximately 150 μl, of which the chamber is 40 μl. To ensure reproducible results, we recommend injection of a volume corresponding to 150 μl + inlet tubing volume + 120 μl (~3× chamber volume).
5. Wash the cell to evaluate reversibility of binding interaction.
   (a) Washing can be done in reverse direction to avoid residual protein in the tubing to enter and potentially interact with the sensor surface.
6. Wash in 2% SDS for 10 min at 250 μl/min to clean the flow system and mounted sensor crystals.
7. Rinse thoroughly with MilliQ H_2O to remove residual SDS before removing sensor crystals.
8. Do not leave crystals mounted in the QCM-D system for prolonged time periods as O-rings tend to stick to the sensor electrodes and reduce the lifespan of sensor crystals.

3.4. Data Interpretation

After finishing the QCM-D experiment, we first plot the raw Δf and ΔD data at two harmonics or more as a function of time to obtain an overview of the experimental data. This can done using the commercial Qtools software package (supplied with the Q-sense instruments) and allow first interpretation of acquired data. As mentioned in the introduction, a decrease in Δf data can under certain simplifying conditions be interpreted using the Sauerbrey relation established in 1959, which links Δf with the absorbed mass Δm (43):

$$\Delta m = -C\frac{\Delta f_n}{n}. \quad (1)$$

where n is an odd overtone number ($n = 1, 3,\ldots,13$) and C is a sensor-dependent mass sensitivity constant. C equals 17.7 ng/cm^2 Hz for the 14 mm 5 MHz sensor crystals commonly employed in Q-sense QCM-D systems. In practice, the fundamental resonance frequency ($n = 1$) is neglected in data analysis since it is prone to instabilities. The Sauerbrey relation assumes that (1) the adsorbed mass is small compared to the mass of the sensor crystal and (2) the resulting film is laterally homogeneous, rigidly adsorbed to the sensor surface, and (3) rigid by itself (2). The Sauerbrey relation is integrated in the Qtools software and can be found in the "data" tab. The added mass can further be used to evaluate film thickness, provided the layer density is known.

Film rigidity is evaluated through the dissipation factor, ΔD:

$$D = \frac{E_{lost}}{2\pi E_{stored}} \quad (2)$$

where E_{lost} is the energy dissipated through viscoelastic losses and E_{stored} is the energy stored in the oscillation within one oscillation cycle. D is recorded by momentarily shutting off the current driving the oscillation of the sensor crystal and measuring the dampening of oscillations. Since soft films dissipate the energy of the oscillation more readily than a rigid film, D provides a quantitative measure of the rigidity/softness of an adsorbed film. It is difficult to provide an exact guideline to evaluate when the Sauerbrey relation is a proper measure of the adsorbed mass and when it is not. However, the Sauerbrey relation generally provides good estimates of the mass or thickness of adlayers when the dissipation per mass is low ($\Delta D/\Delta F \leq 0.1 \times 10^{-6}$ Hz^{-1}) and becomes inaccurate for softer films ($\Delta D/\Delta F > 0.1 \times 10^{-6}$ Hz^{-1}) which require compensation for viscous losses as described in following sections. The $\Delta D/\Delta f$ value of a proper SLB is typically in the order of 0.01 Hz^{-1} or less (30, 33).

Δf and ΔD data for the 3rd to 7th harmonic of a typical SLB formation on SiO_2 surface are shown in Fig. 2. After obtaining a stable baseline of at least 5 min, 0.1 mg/ml SUVs made of POPC were injected, leading to a rapid decrease in Δf and an increase in D at all harmonics, indicative of the adsorption of intact vesicles. Around 8 min of incubation time, a critical surface coverage is reached, and vesicles begin to rupture onto the SiO_2 surface as shown by the maxima in Δf and ΔD. After complete vesicle rupture, the Δf and ΔD values stabilize around −25 Hz and ~0.1×10^{-6}, respectively.

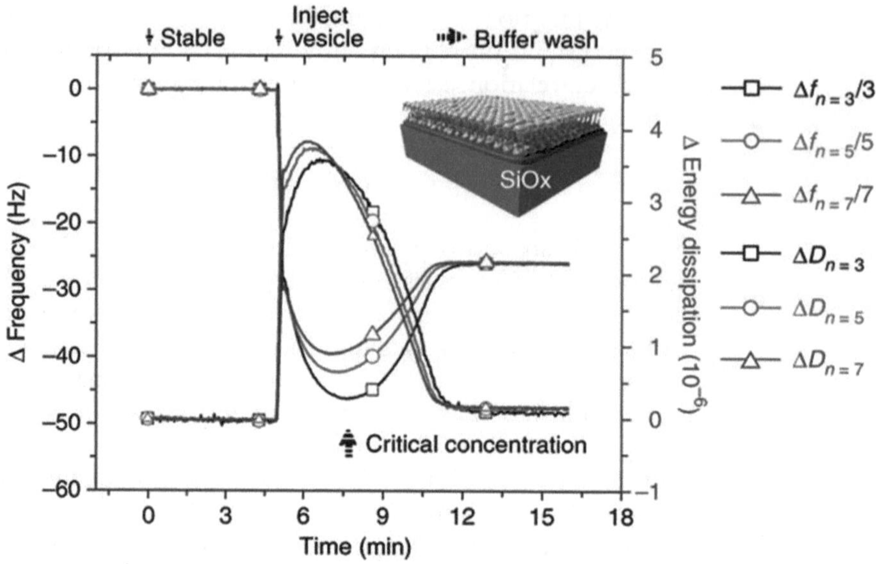

Fig. 2. Formation of an SLB on SiO_2 support by injection of 0.1 mg/ml POPC vesicles. The initial absorption leads to a large drop in frequency and increase in dissipation. Once the critical surface concentration of vesicles has been reached, the vesicles fuse and collapse, leading to a rise in frequency as water is expelled from the SLB, while the dissipation drops to a value close to the initial level (Reproduced with permission from (33), Copyright 2006 Elsevier Press).

Notice that the Sauerbrey relation is sufficient to describe the final SLB but not valid during vesicle adsorption and rupture. Under these conditions, ΔD is high and individual harmonics spread out as an indication of a viscoelastic film. This requires modeling of the viscoelastic properties for accurate mass estimates. The difference between harmonics arises due to two effects: (1) decreasing penetration depths (δ_n) with increasing frequency (harmonic number, n) and (2) films which do not couple fully to the oscillation of the sensor crystal.

The penetration depth (δ_n) at which the amplitude of the shear wave has decreased by a factor e at a given overtone can be estimated by

$$\delta_n = \sqrt{\frac{\eta}{\rho f_n \pi}}$$

where ρ is the density of the layer and η is the shear viscosity. Thus, the penetration depth is inversely proportional to the overtone number and equals ~137 nm for the 3rd overtone and ~90 nm for the 7th overtone in water at 25°C ($\rho = 997$ kg/m^3, $\eta = 8.9 \times 10^{-4}$ kg/ms). The Sauerbrey relation has been shown to provide valid estimate of film thicknesses of up to ~2δ (44) under the limitations previously described. SLBs are around ~4.5–5 nm thick, so a substantial region can be probed beyond the lipid surface. This difference in penetration depths can give rise to some interesting observations. The antimicrobial peptide aurein 1.2

binds to DMPC membranes, giving rise to positive Δf values (mass loss) for the 3rd and 5th overtones but negative Δf values (mass gain) for the 7th and 9th overtones (26). The higher overtones register binding closer to the membrane surface than the lower overtones. Accordingly, the data may be interpreted to indicate that peptide binding to the surface reduces the coupling between the membrane and the bulk solution (26).

3.4.1. Protein Adsorption onto SLBs

SLBs are generally highly resistant toward nonspecific adsorption of proteins. No or weak interactions have been observed for HSA, hIgG, fibrinogen, hemoglobin, cytochrome c, and serum (31) and not many have been shown to adsorb as rigid adlayers. Specific binding is however demonstrated by annexins A1, A5, and A6, which bind in a Ca^{2+}-dependent fashion to phosphatidylserine lipid headgroups, and prothrombin has been used to probe the interleaflet distribution of DOPS lipids (32, 41, 42, 45). Buzhynskyy et al. (41) prepared SLBs from 0.1 mg/ml SUVs consisting of 4:1 DOPC:DOPS in the presence of 2 mM Ca^{2+} to facilitate collapse of anionic lipids on SiO_2 and monitored the binding of annexin A5 monomer (Fig. 3a) and dimer (Fig. 3b) species at 20 and 40 µg/ml protein, respectively (~570 nM). The use of SiO_2 support has been shown to produce SLBs with an equal distribution of anionic DOPS between bilayer leaflets (32). With $\Delta f = 16 \pm 1$ Hz and low ΔD at $0.1 \times 10^{-6} \pm 0.05 \times 10^{-6}$ ($\Delta D/\Delta f$~0.006×10^{-6} Hz^{-1}), the adsorbed annexin A5 monomer film is rigid and may be described by the Sauerbrey relation Eq. 1, yielding an adsorbed mass of approximately 280 ± 18 ng/cm^2 (41). Adsorption of the dimer yielded larger Δf and ΔD values at 38 ± 1 Hz and $1 \times 10^{-6} \pm 0.1 \times 10^{-6}$ ($\Delta D/\Delta f$ ~ 0.026×10^{-6} Hz^{-1}), respectively. This corresponds to an adsorbed dimer mass of ~670 ± 18 ng/cm^2 which is more than twice that of the monomer, and the excess mass may be ascribed to the fact that QCM-D in addition to the protein also measures coupled water trapped in the cavities between protein molecules as also emphasized by the larger ΔD. However, the film remains sufficiently rigid to obtain appropriate mass estimates by the Sauerbrey relation for comparison. After protein adsorption, the authors further demonstrate that SLB-bound annexin A5 dimers but not monomers are able to bind to injected SUVs and that annexin A5 binding to the SLB is Ca^{2+} dependent and reversible upon injection of EGTA-containing buffer (41). Note that while the SLB-annexin A5 dimer film is sufficiently rigid to be described by Eq. 1, the relation is not valid when intact SUVs adsorb onto annexin A5 dimers (Fig. 3b).

3.4.2. Viscoelastic Films

For soft films, which are also formed in the initial stages of SLB formation, the Sauerbrey relation underestimates the mass, since the oscillation of the film is not fully coupled to the sensor surface, i.e., the film "slips." Further data analysis using Voigt-based viscoelastic

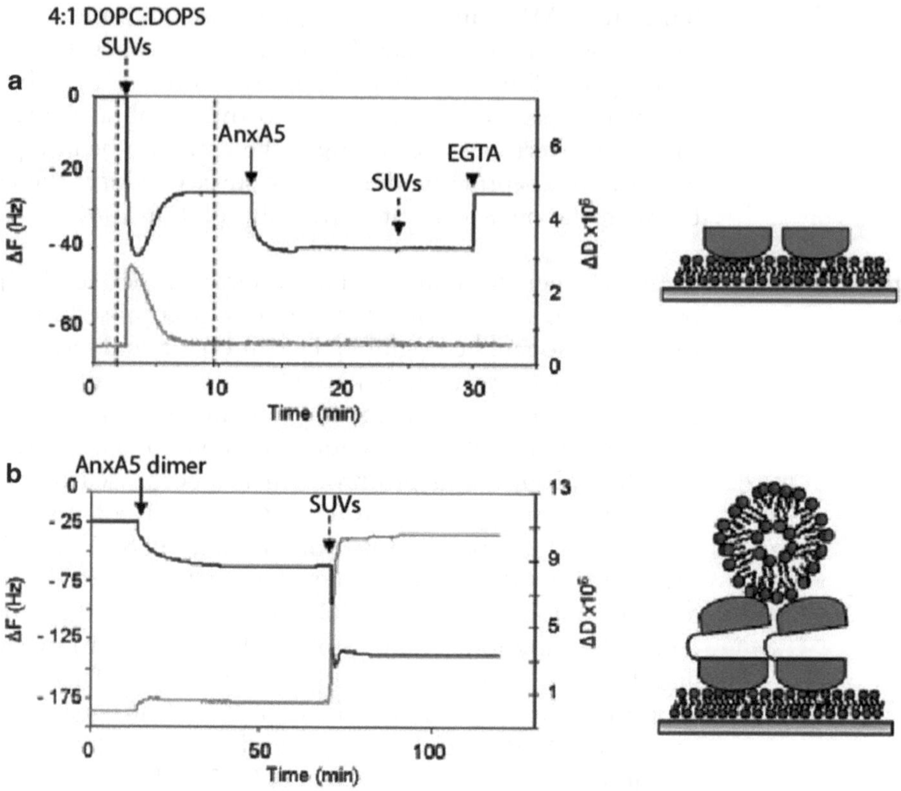

Fig. 3. Interaction of AnxA5 and AnxA5 dimers with 4:1 DOPC:DOPS SLBs on SiO_2 support. (**a**) The characteristic overshoot of ΔF (*black*) and ΔD (*orange*) during SUV collapse into SLBs is followed by the injection of 20 μg/ml annexin A5 (AnxA5) monomer, leading to adsorption of 280 ± 18 ng/cm^2 AnxA5. This layer of adsorbed monomeric protein is resistant toward nonspecific interaction with SUVs. However, rinsing with EGTA eliminates the Ca^{2+}-dependent AnxA5:SLB interaction. (**b**) The profile starts with preformed SLB, after which 40 μg/ml AnxA5 dimer is injected, leading to adsorption of ~670 ± 18 ng/cm^2 AnxA5 dimer. The dimer allows subsequent binding of SUVs. The models illustrate the proposed adsorption of AnxA5 monomer and dimers and the specific association of 4:1 DOPC:DOPS SUVs to AnxA5 dimers (Modified with permission from (41), Copyright 2009 Elsevier Press).

models is in such cases necessary and requires use of the change in dissipation ΔD accompanying film formation (46). As a guideline, a rigid non-slipping film shows (1) low dissipation (D~0–0.2 × 10^{-6}) and (2) a low increase in dissipation at increasing frequency shifts (low slope in frequency-dissipation plots as discussed later), generally in the order of $\Delta D/\Delta f < 1 \times 10^{-8}\ \text{Hz}^{-1}$. If the layer is sufficiently rigid (SLBs are rigid in the QCM-D regime), mass estimates from the Sauerbrey relation are generally in good agreement with viscoelastic models described in the following but do not provide estimates of film density, viscosity, and shear modulus.

The Voigt model is included in the Qtools software package provided by Q-sense and provides means to quantify not only the adsorbed mass but also film thickness, shear, and viscous moduli if the effective film density is known *a priori*. The value for the effective density of the film lies between the density of proteins (~1.2–1.35 g/cm^3) and water (~1.0 g/cm^3), taking into account that coupled water is detected by the QCM-D (3, 47). For SLBs, a density of 1.1 g/cm^3 can be assumed, whereas the density of a densely packed DNA film is ≈1.7 g/cm^3 (47). More precise estimates of the effective film density can be obtained using optical techniques such as SPR and OWLS in combination with QCM-D (3, 48). However, a good estimate of the adsorbed mass can still be obtained even if good estimates of the film density is not known *a priori* since the mass estimate is independent of the choice of density (Δm = density × thickness) within the biologically relevant range. However, estimates of viscosity and shear modulus are dependent on the choice of film density (47). In practice, QCM-D provides estimates of the adsorbed mass and coupled water ("wet mass"), whereas optical techniques such as SPR and OWLS provide estimates of the adsorbed protein alone ("dry mass").

In Voigt viscoelastic modeling, it is essential to verify that estimates of shear modulus and shear viscosities are within a biologically relevant range. The shear modulus of Mefp-1 protein films was previously reported to be 0.066 MPa, increasing to 0.3 MPa upon cross-linking (1). Concurrently, estimates of shear viscosity increased from 0.002 to 0.006 kg/m s for the cross-linked Mefp-1 film (1). Similarly, cross-linking of polylysine films by glutaraldehyde increased shear modulus from 0.25 to 2.5 MPa and shear viscosity from 0.003 to 0.0175 kg/m s (49). BSA adsorption onto stainless steel and alumina led to shear elastic moduli of 0.5 ± 0.1 and 0.3 ± 0.1 MPa, respectively, and shear viscosity estimates of 0.0028 ± 0.001 kg/m s and 0.0022 ± 0.0003 kg/m s, respectively (50). Lastly, β-lactoglobulin adsorption onto polyethersulfone films led to shear viscosity estimates of ~0.0023–0.01 kg/m s and shear modulus of ~0.1–0.92 MPa (51). Thus, biologically relevant estimates of shear modulus and shear viscosity (for both protein and lipid films) should be in the range of ~0.01–2.5 MPa and ~0.001–0.02 kg/m s, respectively. Very few examples of viscoelastic modeling applied to protein films on SLBs are available in the literature, presumably due to the complexity of data treatment and verification. However, the Δf and ΔD data during adsorption and cross-linking of Mefp-1 onto a hydrophobic QCM-D sensor surface in Fig. 4 serve to illustrate the relevant biological range of shear viscosity and shear modulus estimates and can be used as a guideline for experiments in which protein adsorbs to SLBs.

3.4.3. Membrane Modulation/Disintegration

It is challenging to make a qualitative description of the interaction between SPBs and proteins/peptides from Δf and ΔD values alone. However, useful insight can often be provided by frequency-dissipation plots ($\Delta f - \Delta D$ plots). These provide a "phase diagram" describing how the conformation or structure of

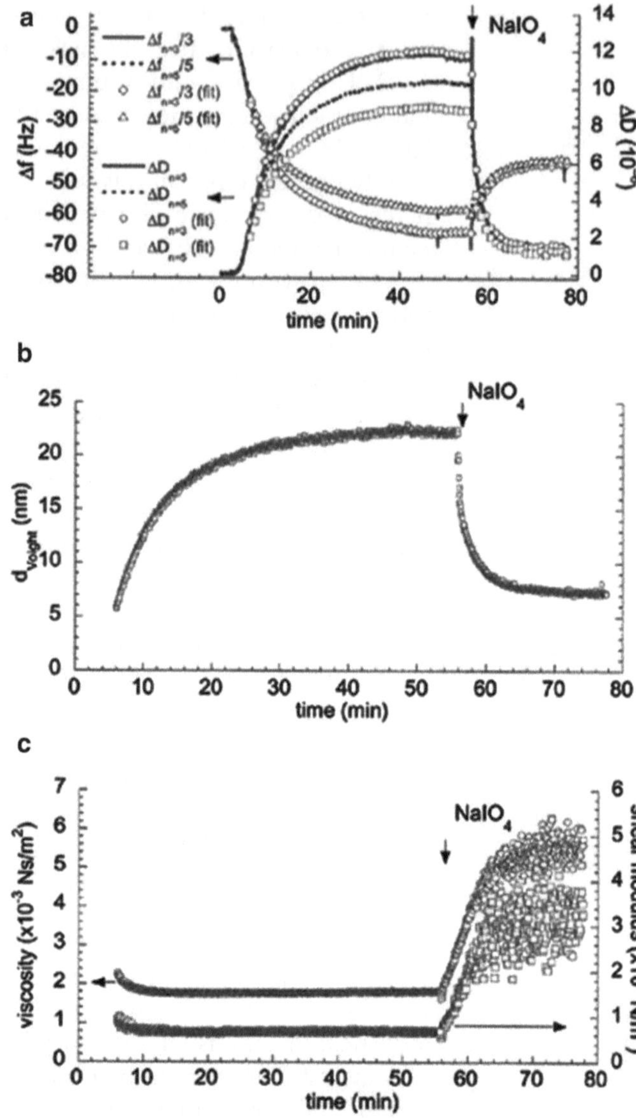

Fig. 4. (**a**) Raw f-*D* data for the adsorption and cross-linking (by $NaIO_4$) of a Mefp-1 film on an alkane-terminated QCM-D sensor surface. $NaIO_4$ is added around 56 min (*arrows*). Subsequently Voigt viscoelastic modeling is used to obtain estimates of (**b**) film thickness and (**c**) shear viscosity and shear modules (Reproduced with permission from (1), Copyright 2001 American Chemical Society).

proteins (and lipids) changes for a mass unit adsorbing to the film, eliminating time as an explicit parameter.

$\Delta f - \Delta D$ plots consist of discrete points each representing a single point in time. Thus, a linear relation in the $\Delta f - \Delta D$ plot suggests a simple adsorption process in which added mass leads to a constant change in dissipation (i.e., film rigidity) per mass unit. In these plots, dissipation may either be constant or increase,

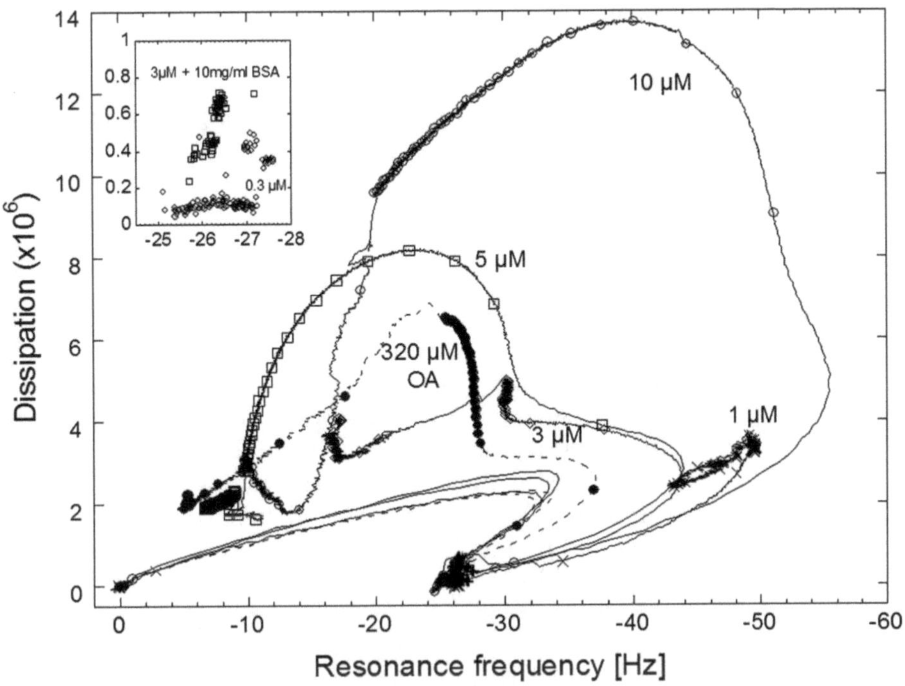

Fig. 5. $\Delta f - \Delta D$ plot of equine lysozyme:oleic acid complexes with oleic acid and 4:1 DOPC:DOPG SLBs (Reproduced with permission from (30), Copyright 2010 Elsevier Press).

depending on whether the adsorbed film is essentially rigid throughout or softens upon adsorption of a given mass, respectively. A larger increase in dissipation per mass unit suggests a more flexible or unfolded conformation of the adsorbed proteins, leading to a more hydrated film. In contrast, a low increase in dissipation per mass unit indicates the adsorption of a rigid film, which may involve a more compact protein state. Such information may shed more light on the mechanism of interaction and can be exploited to examine how factors such as pH, temperature, lipid composition, salts, or mutations in the polypeptide chain affect protein adsorption (4, 7, 52, 53). Further, deviations from linear relationships (not easily picked up from raw Δf and ΔD time plots) usually indicate one or more secondary processes occurring at the surface and provide essential mechanistic information about the adsorption and/or film rearrangement process(es) (23, 30). Deviations from linear relations may be observed as the protein concentration and/or lipid composition changes.

We have found $\Delta f - \Delta D$ plots particularly helpful in determining the mechanism of interaction between equine lysozyme complexes with oleic acid (ELOA) and SLBs. As shown in Fig. 5, $\Delta f - \Delta D$ plots reveal the typical vesicle adsorption-rupture process resulting in SLBs with Δf around 25 Hz and $\Delta D < 0.2 \times 10^{-6}$ (30). On exposure to low concentrations of ELOA (<1 μM), an essentially linear trajectory suggests a simple binding of the complex to the SLB. Higher concentrations of ELOA displayed a similar adsorption but revealed much more complex trajectories indicative of secondary processes following initial adsorption.

4. Notes

4.1. Choice of module type

1. *Use of open modules.* In addition to the closed QCM-D chamber which is the main focus of this chapter, Q-sense provides open modules which allow direct pipetting of the protein to the sensor surface. Although open modules are more prone to temperature variation and drift due to sample evaporation or variations in bulk volume, they allow precise control of the amount of protein added and may (depending on the bulk volume) require less sample owing to decreased loss of unused protein within the flow system.

4.2. QCM-D System Maintenance and Guidelines

2. *General system cleaning.* Proper maintenance of the QCM-D system is a prerequisite for good experimental results. Many problems encountered during the use of QCM-D equipment fall back on improper cleaning of the flow system and/or sensor crystals. It is thus essential that sensor crystals and the flow system of the QCM-D system are cleaned on a regular basis to avoid artifacts, baseline drift, or other unexplained behavior due to adsorption of contaminants. Between experiments, the system can be cleaned by washing in 2% SDS, 2% Hellmanex, or other compatible cleaning agent followed by extensive flushing with MilliQ water. If washing with 2% SDS and rinsing with MilliQ is insufficient to obtain stable baselines, the measuring cells should be disassembled and cleaned according to manufacturer's description (see http://www.q-sense.com user area for instructions and chemical compatibility charts). Sensor crystals are cleaned as described in Subheading 3.1.1. Noisy or drifting frequency and/or dissipation readings are thus often caused by several factors including but not limited to (a) a dirty sensor crystal, cell, or flow system resulting in slow adsorption to (sensor surface is unintentionally blocked with contaminants) or desorption from sensor surfaces; (b) leaks and/or residual humidity on the sensor surfaces or O-rings in the QCM-D cell may result in drifting baselines; (c) a broken crystal as a result of nonplanar O-rings during QCM-D cell assembly; and (d) sensor crystals become worn due to removal of surface and/or electrode material during cleaning, mounting, and disassembly and are typically only reusable up to 7–8 times.

3. *Ensure proper temperature equilibration.* The oscillation of the crystal is highly sensitive to temperature changes, and excessive temperature fluctuations around the equipment should thus be avoided. Drifting baselines may thus be reduced or avoided by allowing at least 5–15 min of temperature equilibration before starting experiments. Further, equilibration of the buffer and sample at or close to the measuring temperature is a prerequisite. Working within ~5°C above ambient temperature usually does not lead to temperature-induced changes in ΔD and Δf readings after initial equilibration, whereas higher temperature gradients require temperature equilibration of buffer and samples. Reducing the flow rate allows longer temperature equilibration within the measuring cell and may thus improve results. Further, it is essential that the peristaltic pump delivers equal volumes/flow rates to each channel in order to obtain reliable and reproducible results. This can easily be checked by monitoring the onset time of vesicle adsorption and by collecting and weighing outlet liquid (e.g., in Eppendorf tubes). Unequal flow rates and/or volumes may be caused by worn pump tubing which must be replaced. Small differences in flow rates may result in variations in the amount of protein each crystal is subjected to upon injection, making it difficult to compare individual time profiles.

4.3. Working with Lipids

4. *Preparation of small unilamellar vesicles.* As an alternative to LUV preparation described in Subheading 3.1.2, small unilamellar vesicles (SUVs) of ~25 nm in diameter can be prepared by sonication of LMVs obtained from step 1 to 5 in Subheading 3.1.2 using a probe or bath sonicator for 5 min or until the suspension becomes clear. SUVs prepared using a probe sonicator should be centrifuged for 5 min at 14,000 rfc to remove larger vesicles and potential titanium particles which may potentially break off the sonicator probe during sonication. The stability of SUVs is lower than LUVs due to high surface tension, and they should therefore be used immediately after preparation. However, SUVs may be preferred since they more readily collapse on SiO_2 (14, 33).
5. *Work above the transition temperature of lipids.* We recommend always to work at least 5°C above the phase transition temperature (t_m) of the lipids employed to ensure proper mixing. This is easily accomplished by working at room temperature for lipids containing oleoyl or dioleoyl chains. Dimyristoyl lipids with phosphatidylcholine, phosphatidylglycerol, and phosphatidylserine headgroups have transition temperatures around 23–35°C. Consequently, experiments performed at or below ambient temperatures will involve lipids in the gel phase,

making vesicle collapse more challenging and complicating data interpretation.

4.4. Troubleshooting Bilayer Collapses

6. Vesicle collapse into supported lipid bilayers generally leads to stable frequencies of -26 ± 1 Hz and dissipation values of less than 0.2×10^{-6} (compared to baseline). However, deviations may be observed due to several factors such as varying lipid chain lengths and headgroups, QCM-D system contamination (see Notes sections above), or buffer changes.
7. *Frequency shifts and dissipation readings are too large.* Larger frequency shifts and dissipation readings suggest residual non-ruptured vesicles which cannot fuse with the bilayer. This can be remedied as follows: (a) washing with a buffer of higher ionic strength changes the osmotic pressure across membranes and thus increases the likelihood of vesicle rupture. Wash thoroughly with initial buffer afterward until a stable baseline is obtained. (b) Preparation of vesicles in MilliQ water followed by dilution into the buffer of choice also creates an osmotic gradient which may aid vesicles in collapse. (c) Inclusion of 2–5 mM Ca^{2+} decreases the critical surface coverage required for vesicle collapse and thus aids bilayer formation. Note that changes in buffer composition may have an influence on QCM-D results.
8. *Frequency readings are lower than expected.* Lower end-level frequencies may suggest the presence of membrane defects or a change in buffer composition. Remedy by (a) increasing lipid concentration or increasing incubation time to allow complete vesicle collapse or (b) by ensuring that no changes in buffer composition have been made since this may alter the solution viscosity which affects QCM-D readings. (c) Electrostatic attraction of positively charged lipids such as DOEPC may reduce hydration layer between the SLB and support, causing lower frequency readings of -24 ± 1 Hz (33). Likewise, different lipid compositions may alter the final frequency readings. Thus, the formation of partially positively charged DOEPC SLBs on SiO_2 appears to create SLBs with lower $\Delta f \sim -24 \pm 1$ Hz, presumably due to stronger lipid-substrate interactions and/or thinner hydration layer between the SLB and the SiO_2 support (33). In general, increasing the proportion of anionic lipids increases the electrostatic repulsion between vesicles and increases the activation barrier to collapse and vesicle fusion. Thus, decreasing in anionic lipid content may improve the success rate of bilayer formation.

References

1. Höök F, Kasemo B, Nylander T, Fant C, Sott K, Elwing H (2001) Variations in coupled water, viscoelastic properties, and film thickness of a Mefp-1 protein film during adsorption and cross-linking: a quartz crystal microbalance with dissipation monitoring, ellipsometry, and surface plasmon resonance study. Anal Chem 73:5796–5804
2. Dixon MC (2008) Quartz crystal microbalance with dissipation monitoring: enabling real-time characterization of biological materials and their interactions. J Biomol Tech 19:151
3. Höök F, Vörös J, Rodahl M, Kurrat R, Böni P, Ramsden JJ, Textor M, Spencer ND, Tengvall P, Gold J (2002) A comparative study of protein adsorption on titanium oxide surfaces using in situ ellipsometry, optical waveguide lightmode spectroscopy, and quartz crystal microbalance/dissipation. Colloids Surf B Biointerfaces 24:155–170
4. Höök F, Rodahl M, Kasemo B, Brzezinski P (1998) Structural changes in hemoglobin during adsorption to solid surfaces: effects of pH, ionic strength, and ligand binding. Proc Nat Acad Sci U S A 95:12271
5. Rodahl M, Höök F, Fredriksson C, Keller CA, Krozer A, Brzezinski P, Voinova M, Kasemo B (1997) Simultaneous frequency and dissipation factor QCM measurements of biomolecular adsorption and cell adhesion. Faraday Discuss 107:229–246
6. Holst J, Watson S, Lord MS, Eamegdool SS, Bax DV, Nivison-Smith LB, Kondyurin A, Ma L, Oberhauser AF, Weiss AS, Rasko JEJ (2010) Substrate elasticity provides mechanical signals for the expansion of hemopoietic stem and progenitor cells. Nat Biotech 28:1123–1128
7. Höök F, Rodahl M, Brzezinski P, Kasemo B (1998) Energy dissipation kinetics for protein and antibody-antigen adsorption under shear oscillation on a quartz crystal microbalance. Langmuir 14:729–734
8. Ayela C, Roquet F, Valera L, Granier C, Nicu L, Pugnière M (2007) Antibody–antigenic peptide interactions monitored by SPR and QCM-D: a model for SPR detection of IA-2 autoantibodies in human serum. Biosens Bioelectron 22:3113–3119
9. Modin C, Stranne AL, Foss M, Duch M, Justesen J, Chevallier J, Andersen LK, Hemmersam AG, Pedersen FS (2006) QCM-D studies of attachment and differential spreading of preosteoblastic cells on Ta and Cr surfaces. Biomaterials 27:1346–1354
10. Svedhem S, Dahlborg D, Ekeroth J, Kelly J, Hook F, Gold J (2003) In situ peptide-modified supported lipid bilayers for controlled cell attachment. Langmuir 19:6730–6736
11. Malmström J, Agheli H, Kingshott P, Sutherland DS (2007) Viscoelastic modeling of highly hydrated Laminin layers at homogeneous and nanostructured surfaces: quantification of protein layer properties using QCM-D and SPR. Langmuir 23:9760–9768
12. Richter RP, Berat R, Brisson AR (2006) Formation of solid-supported lipid bilayers: an integrated view. Langmuir 22:3497–3505
13. Richter R, Mukhopadhyay A, Brisson A (2003) Pathways of lipid vesicle deposition on solid surfaces: a combined QCM-D and AFM study. Biophys J 85:3035–3047
14. Reimhult E, Hook F, Kasemo B (2003) Intact vesicle adsorption and supported biomembrane formation from vesicles in solution: influence of surface chemistry, vesicle size, temperature, and osmotic pressure. Langmuir 19:1681–1691
15. Höök F, Ray A, Nordén B, Kasemo B (2001) Characterization of PNA and DNA immobilization and subsequent hybridization with DNA using acoustic-shear-wave attenuation measurements. Langmuir 17:8305–8312
16. Edvardsson M, Svedhem S, Wang G, Richter R, Rodahl M, Kasemo B (2008) QCM-D and reflectometry instrument: applications to supported lipid structures and their biomolecular interactions. Anal Chem 81:349–361
17. Thid D, Holm K, Eriksson PS, Ekeroth J, Kasemo B, Gold J (2008) Supported phospholipid bilayers as a platform for neural progenitor cell culture. J Biomed Mater Res Part A 84A:940–953
18. Huang C-J, Cho N-J, Hsu C-J, Tseng P-Y, Frank CW, Chang Y-C (2010) Type I collagen-functionalized supported lipid bilayer as a cell culture platform. Biomacromolecules 11:1231–1240
19. Richter RP, Hock KK, Burkhartsmeyer J, Boehm H, Bingen P, Wang G, Steinmetz NF, Evans DJ, Spatz JP (2007) Membrane-grafted hyaluronan films: a well-defined model system of glycoconjugate cell coats. J Am Chem Soc 129:5306–5307
20. Fezoua-Boubegtiten Z, Desbat B, Brisson A, Gounou C, Laguerre M, Lecomte S (2011) Effect of Mg^{2+} versus Ca^{2+} on the behavior of Annexin A5 in a membrane-bound state. Eur Biophys J 40:641–649

21. Justesen PH, Kristensen T, Ebdrup T, Otzen D (2004) Investigating porcine pancreatic phospholipase A2 action on vesicles and supported planar bilayers using a quartz crystal microbalance with dissipation. J Colloid Interface Sci 279:399–409
22. Jackman JA, Cho N-J, Duran RS, Frank CW (2009) Interfacial binding dynamics of bee venom phospholipase A2 investigated by dynamic light scattering and quartz crystal microbalance. Langmuir 26:4103–4112
23. Nielsen SB, Otzen DE (2010) Impact of the antimicrobial peptide Novicidin on membrane structure and integrity. J Colloid Interface Sci 345:248–256
24. Piantavigna S, McCubbin GA, Boehnke S, Graham B, Spiccia L, Martin LL (2011) A mechanistic investigation of cell-penetrating Tat peptides with supported lipid membranes. Biochimica et Biophysica Acta (BBA) Biomembranes 1808:1811–1817
25. Piantavigna S, Czihal P, Mechler A, Richter M, Hoffmann R, Martin LL (2009) Cell penetrating apidaecin peptide interactions with biomimetic phospholipid membranes. Int J Pept Res Therapeut 15:139–146
26. McCubbin GA, Praporski S, Piantavigna S, Knappe D, Hoffmann R, Bowie JH, Separovic F, Martin LL (2010) QCM-D fingerprinting of membrane-active peptides. Eur Biophys J 40:437–446.
27. Sherman PJ, Jackway RJ, Gehman JD, Praporski S, McCubbin GA, Mechler A, Martin LL, Separovic F, Bowie JH (2009) Solution structure and membrane interactions of the antimicrobial peptide fallaxidin 4.1 a: an NMR and QCM study. Biochemistry 48:11892–11901
28. Knappe D, Piantavigna S, Hansen A, Mechler A, Binas A, Nolte O, Martin LL, Hoffmann R (2010) Oncocin (VDKPPYLPRPRPPRRIYNR-NH2): a novel antibacterial peptide optimized against gram-negative human pathogens. J Med Chem 53:5240–5247
29. Mechler A, Praporski S, Atmuri K, Boland M, Separovic F, Martin LL (2007) Specific and selective peptide-membrane interactions revealed using quartz crystal microbalance. Biophys J 93:3907–3916
30. Nielsen SB, Wilhelm K, Vad B, Schleucher J, Morozova-Roche LA, Otzen D (2010) The interaction of equine lysozyme: oleic acid complexes with lipid membranes suggests a cargo off-loading mechanism. J Mol Biol 398:351–361
31. Glasmästar K, Larsson C, Höök F, Kasemo B (2002) Protein adsorption on supported phospholipid bilayers. J Colloid Interface Sci 246:40–47
32. Richter RP, Maury N, Brisson AR (2005) On the effect of the solid support on the interleaflet distribution of lipids in supported lipid bilayers. Langmuir 21:299–304
33. Cho NJ, Frank CW, Kasemo B, Höök F (2010) Quartz crystal microbalance with dissipation monitoring of supported lipid bilayers on various substrates. Nat Protocol 5:1096–1106
34. Anderson TH, Min Y, Weirich KL, Zeng H, Fygenson D, Israelachvili JN (2009) Formation of supported bilayers on silica substrates. Langmuir 25:6997–7005
35. Richter RP, Brisson AR (2005) Following the formation of supported lipid bilayers on mica: a study combining AFM, QCM-D, and ellipsometry. Biophys J 88:3422–3433
36. Keller CA, Glasmästar K, Zhdanov VP, Kasemo B (2000) Formation of supported membranes from vesicles. Phys Rev Lett 84:5443–5446
37. Keller CA, Kasemo B (1998) Surface specific kinetics of lipid vesicle adsorption measured with a quartz crystal microbalance. Biophys J 75:1397–1402
38. Cho NJ, Cho SJ, Cheong KH, Glenn JS, Frank CW (2007) Employing an amphipathic viral peptide to create a lipid bilayer on Au and TiO_2. J Am Chem Soc 129:10050–10051
39. Rossetti FF, Bally M, Michel R, Textor M, Reviakine I (2005) Interactions between titanium dioxide and phosphatidyl serine-containing liposomes: formation and patterning of supported phospholipid bilayers on the surface of a medically relevant material. Langmuir 21:6443–6450
40. Mashaghi A, Swann M, Popplewell J, Textor M, Reimhult E (2008) Optical anisotropy of supported lipid structures probed by waveguide spectroscopy and its application to study of supported lipid bilayer formation kinetics. Anal Chem 80:3666–3676
41. Buzhynskyy N, Golczak M, Lai-Kee-Him J, Lambert O, Tessier B, Gounou C, Bérat R, Simon A, Granier T, Chevalier JM (2009) Annexin-A6 presents two modes of association with phospholipid membranes. A combined QCM-D, AFM and cryo-TEM study. J Struct Biol 168:107–116
42. Kastl K, Ross M, Gerke V, Steinem C (2002) Kinetics and thermodynamics of annexin A1 binding to solid-supported membranes: a QCM study. Biochemistry 41:10087–10094

43. Sauerbrey G (1959) The use of quartz oscillators for weighing thin layers and for microweighing. Z Phys 155:206–222
44. Rodahl M, Kasemo B (1996) On the measurement of thin liquid overlayers with the quartz-crystal microbalance. Sens Actuators A Phys 54:448–456
45. Richter RP, Lai Kee Him J, Tessier B, Tessier C, Brisson AR (2005) On the kinetics of adsorption and two-dimensional self-assembly of annexin A5 on supported lipid bilayers. Biophys J 89:3372–3385
46. Voinova MV, Jonson M, Kasemo B (2002) Missing mass effect in biosensor's QCM applications. Biosens Bioelectron 17:835–841
47. Larsson C, Rodahl M, Höök F (2003) Characterization of DNA immobilization and subsequent hybridization on a 2D arrangement of streptavidin on a biotin-modified lipid bilayer supported on SiO_2. Anal Chem 75:5080–5087
48. Voros J (2004) The density and refractive index of adsorbing protein layers. Biophys J 87:553–561
49. Dutta AK, Nayak A, Belfort G (2008) Viscoelastic properties of adsorbed and cross-linked polypeptide and protein layers at a solid–liquid interface. J Colloid Interface Sci 324:55–60
50. Gispert M, Serro A, Colaço R, Saramago B (2008) Bovine serum albumin adsorption onto 316 L stainless steel and alumina: a comparative study using depletion, protein radiolabeling, quartz crystal microbalance and atomic force microscopy. Surf Interface Anal 40:1529–1537
51. Kim JT, Weber N, Shin GH, Huang Q, Liu SX (2007) The study of-lactoglobulin adsorption on polyethersulfone thin film surface using QCM-D and AFM. J Food Sci 72: E214–E221
52. Otzen DE, Oliveberg M, Höök F (2003) Adsorption of a small protein to a methyl-terminated hydrophobic surfaces: effect of protein-folding thermodynamics and kinetics. Colloids Surf B: Biointerfaces 29:67–73
53. Otzen D (2008) Differential adsorption of variants of the *Thermomyces lanuginosus* lipase on a hydrophobic surface suggests a role for local flexibility. Colloids Surf B Biointerfaces 64:223–228

Chapter 2

Surface Plasmon Resonance for Measuring Interactions of Proteins with Lipid Membranes

Vesna Hodnik and Gregor Anderluh

Abstract

Surface plasmon resonance (SPR) is an established method for studying molecular interactions in real time. It allows obtaining qualitative and quantitative data on interactions of proteins with lipid membranes. In most of the approaches, a lipid membrane or a membrane-mimetic surface is prepared on the surface of Biacore (GE Healthcare) sensor chips HPA or L1, and the studied protein is then injected across the surface. Here, we provide an overview of SPR in protein–membrane interactions, different approaches described in the literature, and a general protocol for conducting an SPR experiment including lipid membranes, together with some experimental considerations.

Key words: Protein–membrane interactions, Lipid membrane, Sensor chip L1, Sensor chip HPA, Biacore

1. Introduction

One of the most important approaches to study molecular interactions is the use of optical biosensors that employ the surface plasmon resonance (SPR) phenomenon. This label-free method allows monitoring molecular interactions in real time. The use of commercial biosensors enables facile determination of kinetic parameters of binding. Basically, any molecular interaction can be studied, i.e., protein–protein, protein–DNA, protein–small ligand, virus–antibody, and also protein–lipid membrane interactions. The greatest advantages of the SPR approach are label-free detection, real-time monitoring, and low sample consumption. During the years, it became a strong experimental tool that can easily provide qualitative or quantitative data on molecular interactions. The technique, however, has certain particularities that can challenge the inexperienced researcher. The interested reader should therefore also consult yearly overviews of the rich SPR literature (1, 2) and recent books on SPR (3, 4), to obtain the insights into how

Jörg H. Kleinschmidt (ed.), *Lipid-Protein Interactions: Methods and Protocols*, Methods in Molecular Biology, vol. 974, DOI 10.1007/978-1-62703-275-9_2, © Springer Science+Business Media New York 2013

the SPR experiment is properly conducted and the data analyzed and interpreted. SPR was successfully used to study protein–lipid membrane interactions, e.g., membrane binding of proteins that participate in cell signaling, pore-forming proteins and peptides, binding of coagulation factors, enzymes, and amyloidogenic proteins (5). Additionally, some very good reviews were published on the use of SPR in protein–membrane interactions that highlight advantages over other biophysical approaches and are complementary to this review (5–7). Here, we aim to provide the status of SPR in protein–membrane interactions in the last few years by providing examples of qualitative and quantitative data that can be obtained. We will particularly focus on SPR experiments performed on Biacore (GE Healthcare) machines, since they are still the most commonly used.

1.1. SPR Basics

Any biosensor based on SPR is composed of SPR detector, fluidic system that brings interacting molecules together and gold-coated glass slides, so-called »sensor chips«, where interactions occur. A P-polarized laser light is directed through a prism, a medium of high refractive index, to a gold layer on the border of a flow cell with the sample, providing a medium of low refractive index. Laser light is reflected on the sensor chip surface and detected by the detector. At a critical angle of an incident light, the SPR phenomenon occurs and reduces the intensity of the reflected light. Several factors affect the optical properties of the system, the most important for the approach being the refractive index of the medium in the measurement cell. In the Biacore systems, one of the molecular partners is attached to the surface of the sensor chip (in the SPR biosensors terminology named ligand), while the second one (analyte) is injected across the surface of the sensor chip by employing a microfluidic system. Molecular interactions close to the surface of the sensor chip change the refractive index of the solution and consequently the angle at which SPR occurs. This is viewed on-line as an increase in the signal (Fig. 1). The so-called sensorgram is thus a plot that shows the change of the angle at which SPR occurs against time. The preferred units to describe the rise of the signal are so-called resonance units (RU). There is a linear relationship between the amount of the analyte on the surface of the sensor chip and the increase in the signal, i.e., 1 RU equals to approximately 1 pg protein/mm^2 (8). Since SPR detects changes of the mass concentration at the sensor chip surface, it is a label-free method, and no additional labeling of ligand or analyte is needed.

The surface of the sensor chip is composed of chemical groups that allow the capture of the ligand. The study of protein–lipid interaction can be performed by attaching a protein to the surface of the sensor chip and lipid vesicles injected over the protein (9). However, more researchers use an approach in which a lipid membrane or membrane-mimetic surface is formed on the surface of the

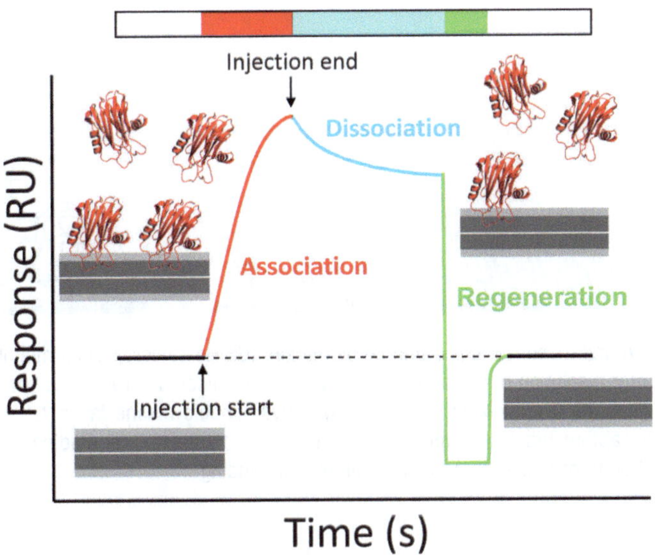

Fig. 1. Sensorgram is a plot that shows SPR response as a function of time. Different phases of an experiment are here labeled with different colors. During the association phase, the surface of the sensor chip with immobilized ligand (here a lipid membrane) is in the contact with the analyte (a protein molecule) dissolved in the running buffer. In the dissociation phase, the sensor chip is flushed with the running buffer only. Regeneration is used to remove the remaining analyte from the ligand surface, so that the new cycle of binding analysis can be performed.

sensor chip and protein then injected across such a surface. Over the years, many different approaches on how to prepare a lipid membrane on the surface of the sensor chip were described and reviewed (5, 10). However, the most used sensor chips for membrane-related work offered by Biacore are HPA and L1 sensor chips (Fig. 2). HPA sensor chip harbors a layer of alkanes on the gold surface. When small unilamellar liposomes are injected across such a surface, they attach and fuse to generate a hybrid bilayer membranes (11). An L1 sensor chip possesses lipophilic alkane groups on the dextran matrix, which efficiently captures intact liposomes (Fig. 2). Since capturing of intact liposomes is possible, this sensor chip is the most preferred of all. The use of L1 sensor chip is described below in a more detail. Captured liposomes were characterized in a more detail, and these papers should be also consulted for more information (12, 13). The experiment is typically performed in several steps: surface preparation, binding experiment, and regeneration of the sensor chip surface (Fig. 3). Both sensor chips can be regenerated easily by injecting a detergent solution, and thus, it is possible to use them many times.

Other ways to attach intact liposomes are by using liposomes that possess traces of biotinylated lipids and sensor chip with immobilized streptavidin (SA sensor chip), liposomes that contain trace amounts of lipopolysaccharide and sensor chip with immobilized

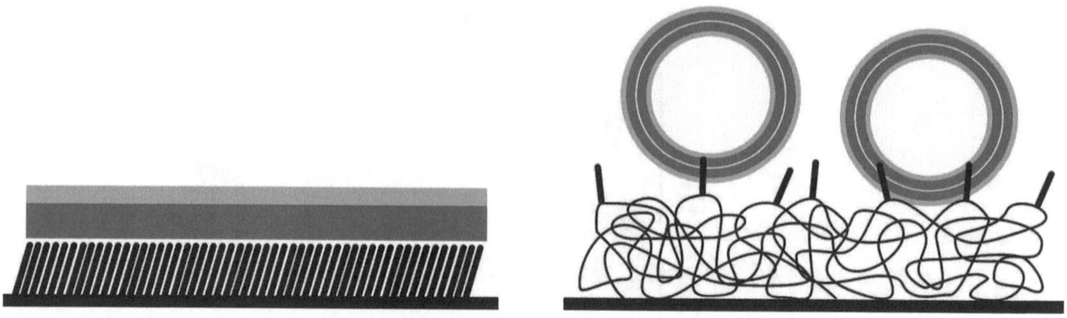

Fig. 2. HPA and L1 sensor chips offered by Biacore (GE Healthcare). Lipid monolayer, also termed hybrid lipid bilayer, is formed after a solution of small unilamellar vesicles is injected across the surface of the HPA sensor chip. Vesicles then fuse to generate a monolayer supported by a hydrophobic alkane layer. L1 chip possess long aliphatic anchors that efficiently capture intact liposomes. This is the preferred approach for studies of integral membrane proteins and proteins that need both monolayers for efficient membrane binding.

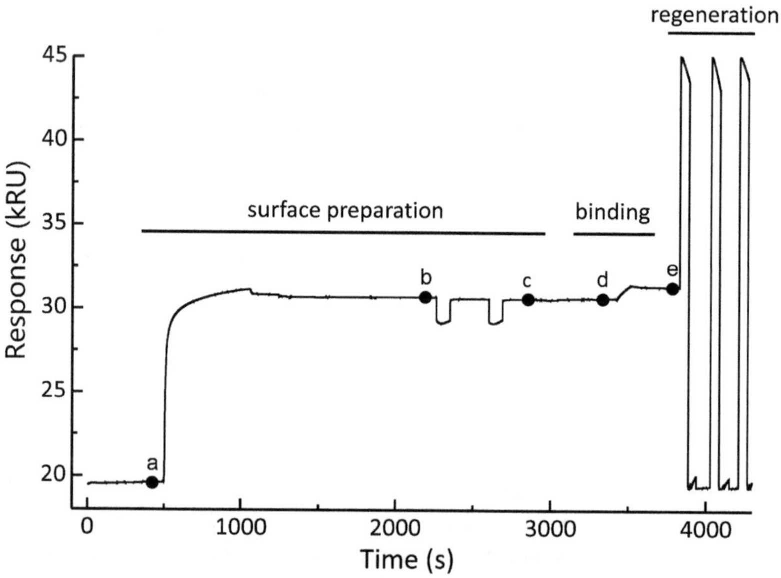

Fig. 3. A single cycle in a protein–membrane interaction consists of surface preparation, binding experiment, and regeneration of the sensor chip. In the first step, liposomes are injected (**a**) over an L1 sensor chip surface at a low flow rate (typically 2 μl/min) and then conditioned with several injections of 100 mM NaOH (**b**) to remove the loosely bound liposomes. A single injection of bovine serum albumin is then used (**c**) to assess the degree of sensor chip exposure. It is possible to immobilize enough liposomes so that the whole surface of the chip is covered and no lipophilic anchors are exposed, as shown here. In the binding step, the protein of interest is injected across the so-prepared surface (**d**). Regeneration is the last step (**e**) and is used to remove the liposomes with the bound protein, and so cleaned chip is ready for another measurement cycle. The best regeneration solution is mixture of isopropanol and 50 mM NaOH (2:3, vol:vol) or detergent solution (40 mM octyl glucoside). Sometimes, protein can be removed from the surface of liposomes by a brief injection of high-salt solution (0.5–2 M NaCl), low-pH (10 mM glycine pH 2–3), or high-pH (10–200 mM NaOH) solution. In such case, many injections of different protein concentration can be performed on a single liposome surface, and sensor chip is regenerated at the end of the experiment.

LPS-specific antibody, or by using DNA-derivatized liposomes that allows hybridization to DNA tethers attached on a gold chip (for overview, see Beseničar Podlesnik et al.(5, 10)). The liposomes used for the interaction studies can be composed of a single synthetic lipid or mixtures of several lipids. Lipid extracts from whole cells, plasma membrane, or some other cell compartment were also employed for the preparation of liposomes. L1 sensor chip also allows capturing of various cellular membrane preparations, such as red blood cell ghosts (14, 15).

1.2. SPR in Protein–Membrane Interactions

In general, SPR gives qualitative and quantitative data of molecular interactions. The most straightforward experiment in protein–membrane interaction is determination of lipid specificity or influence of some other factor on the binding of protein to the lipid membrane, i.e., pH, buffer composition, and salt concentration (Fig. 4) (14). Such qualitative experiments are easy to perform and can be done fast, since it is easy to change lipid composition of the liposomes attached on the sensor chip or change the buffer composition.

However, the most important advantage of SPR over other biophysical approaches is the ability to determine the apparent rate and affinity constants from sensorgrams. This is particularly important when the differences between different conditions, e.g., different variants of studied protein, are small. Typically, such experiments are performed to get an insight into the magnitude of the effects particular amino acid side chain of the protein has on membrane association or dissociation. Here, one needs to perform binding experiment with several different concentrations of the protein, and then binding constants can be determined directly from the

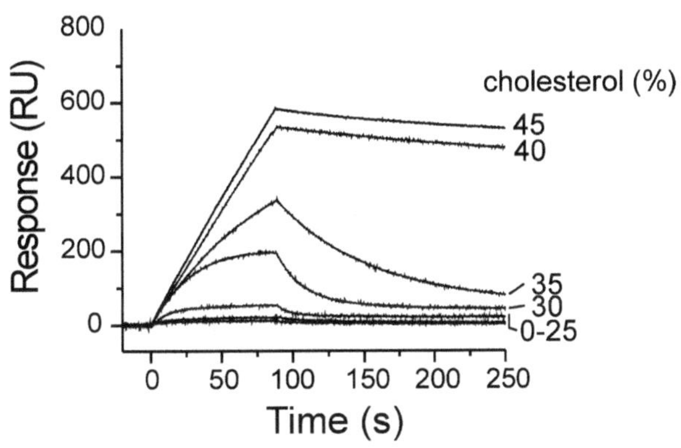

Fig. 4. Qualitative assessment of protein lipid specificity. Here, binding of a protein toxin listeriolysin O at 40 nM concentration to phosphatidylcholine liposomes containing different amounts of cholesterol was monitored. Listeriolysin O activity is dependent on cholesterol content in the membranes. Cholesterol enables initial interaction of the toxin with the membrane, which is evident from the experiment. The amount of cholesterol was 0, 10, 20, 25, 30, 35, 40, and 45% (mol) (from *bottom* to *top*) (Adapted from Bavdek et al. (14) with permission).

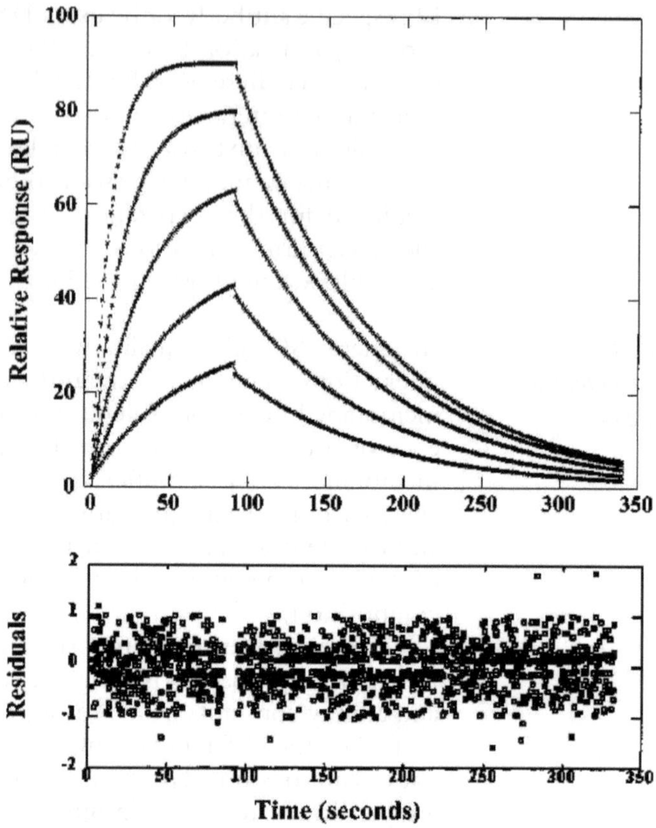

Fig. 5. Binding analysis of *Naja naja atra* phospholipase A_2 to lipid vesicles composed of 1,2-di-*O*-hexadecyl-*sn*-glycero-3-phosphocholine. Different concentrations of the protein (0.1, 0.2, 0.4, 0.8, and 1.6 μM from *bottom* to *top*) were injected over the vesicle surface, and association and dissociation were monitored as shown in the *upper panel*. The data fit well to the 1:1 Langmuir model (see Note 10), as is indicated by low and random residual scatter. The fit is presented with solid lines (Reproduced from Stahelin and Cho (17) with permission).

sensorgrams by numerical integration analysis (Fig. 5) (6, 16). This is conveniently done by Biacore evaluation software or some other dedicated programs, employing the appropriate binding model. The equilibrium affinity constants can also be directly determined from the equilibrium binding responses over a range of protein concentrations by fitting the data to a Langmuir adsorption isotherm (Fig. 6) (17).

In addition, SPR can provide some further insights into the mechanisms of protein action on membranes, as highlighted with some elegant examples of the recent literature. SPR allows assessing the strength of the protein interaction with the membranes. It is easy to perform the screening of conditions that desorb the protein from the surface of the lipid membrane after the binding (Fig. 7a) (19). Molecular interaction of ternary complexes on membranes can also be easily studied. Sometimes, molecular interaction between

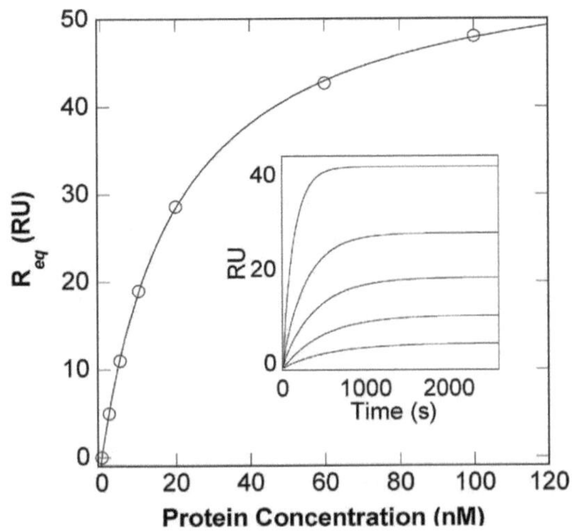

Fig. 6. Equilibrium binding analysis allows determination of equilibrium affinity constant. The PLD1 PX domain was injected at varying concentrations (2, 5, 10, 20, and 60 nM from *bottom* to *top*; inset) over the liposome surface. Here, long association times were used in order to reach equilibrium responses. A binding isotherm was then generated from equilibrium responses (R_{eq}) versus the concentration of the protein. A solid line represents the fit of the data by using the following equation: $R_{eq} = R_{max}/(1 + K_D/C)$, where R_{max} is the maximal response, K_D is equilibrium dissociation constant, and C is the concentration of the protein (Reproduced from Stahelin et al. (18) with permission).

different proteins occurs only after one of the binding partners is first associated with the membrane. Membrane binding may cause conformational change that exposes the binding site for the other partner (Fig. 7b) (20, 21). Some proteins can extract lipids from membranes, and SPR was successfully used to study the kinetics of removal of particular lipid component. Here, the decrease of the signal during the protein injection is indicative of lipid removal from liposomes (Fig. 7c). Some nice examples include extraction of lipids by saposin (22), ceramide by a CERT protein (23), and also cholesterol by methyl-β-cyclodextrin (24). Finally, kinetics of pore formation by human perforin was followed by SPR (Fig. 7d) (25). Perforin is a pore-forming protein from the immune system. Its pore-forming ability was studied by liposomes filled with fluorescent probe calcein. Such vesicles can be attached to the surface of L1 sensor chip without compromising the integrity of the liposomes. In fact, this is a useful control for proteins that bind only to the outer vesicle leaflet, since in this case liposomes are not ruptured or damaged during the binding process, since no fluorescence can be detected during the binding process (17). However, in the case of perforin, the SPR signal dropped considerably during the association phase (Fig. 7d). Additional controls were done to show that this decrease of the signal is due to the released calcein, i.e., the eluted solution was strongly fluorescent (25). Finally, the use of both L1 and HPA sensor chips for a particular protein–membrane

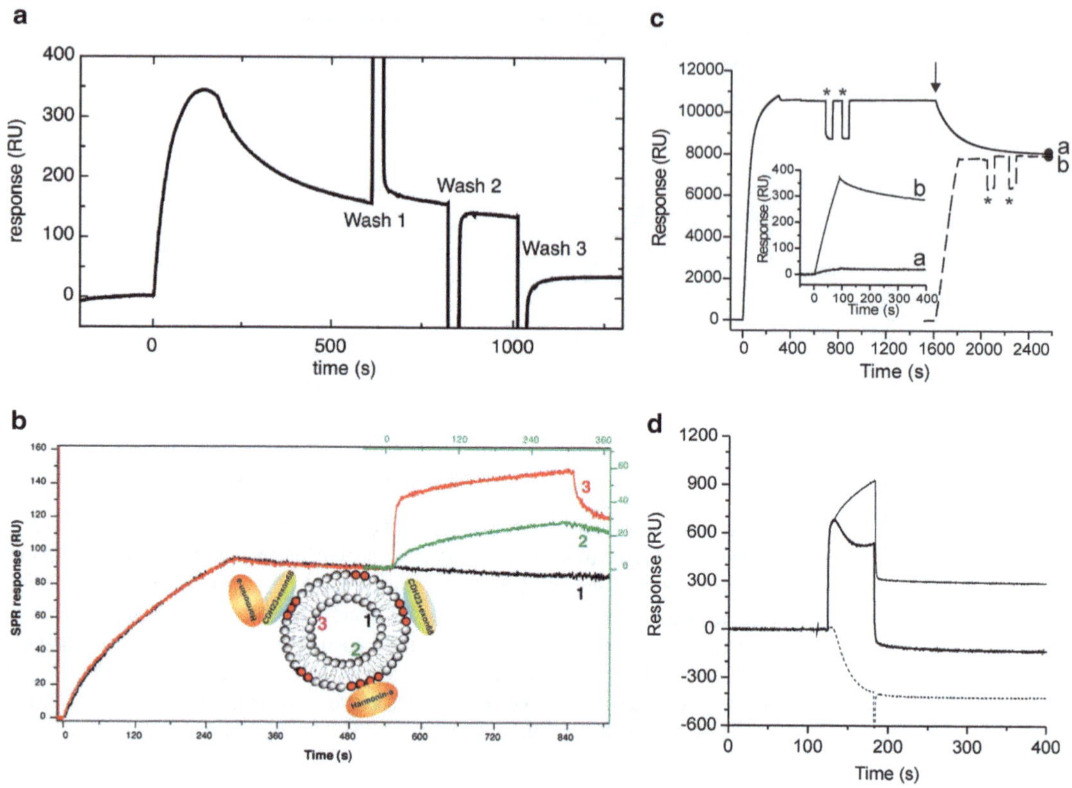

Fig. 7. Some additional examples from the literature on the certain aspects of protein–membrane interactions. (**a**) The stability of protein–membrane complexes may be assessed on-line by washing the surfaces with buffers commonly used to disrupt protein–membrane interactions. TorA–GFP fusion was bound to the liposome and subsequently washed with 300 mM NaCl (wash 1), 100 mM Na_2CO_3 (wash 2), and 100 mM NaOH (wash 3) (19). (**b**) Assessment of molecular complexes attached to the lipid vesicle. In such assay, binding of one of the partners to the lipid membrane is checked in the absence or presence of the other partner (20). (**c**) Selective extraction of cholesterol from membranes by using methyl-β-cyclodextrin. The decrease of the signal is observed after injection of methyl-β-cyclodextrin across the sensor chip surface covered by the liposomes composed of phosphatidylcholine and cholesterol at 40 mol%. Two injections of 100 mM NaOH are denoted by asterisks. The arrow denotes the injection of methyl-β-cyclodextrin. The surface after depletion was tested for the binding of cholesterol-dependent cytolysin streptolysin, which was negligible (*inset*). The control was surface that contains approximately the same amount of original liposomes (trace b and dashed trace of the inset) (24). (**d**) Interaction of pore-forming protein perforin with the liposomes (25). The interaction with liposomes (*thin curve*) is compared to the liposomes filled with the fluorescent probe calcein (*thick curve*). In the latter case, the signal decreases in the association phase, which is indicative of the pore formation in the membrane of the liposome and release of calcein from the vesicle. The difference between the responses (*dashed curve*) shows the kinetics of the probe release from the liposome (Adapted from refs. (19, 20, 24, 25) with permission).

interaction can provide information about the depth of the protein insertion. If protein attaches only superficially to the lipid membrane, then kinetic constants of the binding should not differ in both systems, as indeed observed in the case of coagulation factor VII (26). However, if protein needs both monolayers for insertion, then weaker binding is observed in the case of HPA chip (27).

All these examples show the capability and versatility of the SPR approach in studying protein interactions with membranes. Other examples of using SPR and various membrane preparations include the reconstitution of receptors and assessment of their functionality (28), attachment and characterization of membrane systems prepared from cellular membranes, i.e., nanosomes with functional proteins (29), and transport of solutes across biological membranes by membrane protein (30). Some new approaches to prepare model membranes, such as nanosized bilayer disks, were also reported recently (31). We will next describe the most commonly used approach to study protein–membrane interactions by employing L1 and HPA sensor chips (the following protocol describes the binding experiment as presented on Fig. 3). Some variations of this protocol, other different approaches on preparing membrane surfaces for protein interactions studies, and some additional experimental considerations may be found in some recent reviews (32, 33).

2. Materials

2.1. Preparation of Lipid Vesicles

1. Lipid stocks in organic solvents (Avanti Polar Lipids, USA).
2. Acid-washed glass beads (Sigma-Aldrich, USA).
3. Cryogen vials (Pierce, USA).
4. Vesicle buffer: 20 mM Tris–HCl, 140 mM NaCl, 1 mM EDTA, pH 7.4 in ultrapure water. Pass through cellulose acetate filters with 0.22 μm pores (Sartorius, Germany) and store at room temperature.

2.2. Immobilization of Vesicles on the Surface of L1 and HPA Sensor Chips

1. Solutions for conditioning and regeneration of the L1 sensor chip: isopropanol:50 mM NaOH 2:3 (vol:vol), 100 mM NaOH, 0.1 mg/ml bovine serum albumin (BSA; Sigma-Aldrich, USA). 40 mM octyl glucoside is used instead of isopropanol:50 mM NaOH 2:3 (vol:vol) when regenerating HPA sensor chip.
2. L1 or HPA sensor chips, T100 optical biosensor (Biacore, GE Healthcare, Sweden).

2.3. Binding Experiment

1. Stock solution of protein in the vesicle buffer. Usually micromolar concentrations of proteins should be enough.

3. Methods

3.1. Preparation of Lipid Vesicles

1. Add 5 mg of desired lipids dissolved in the appropriate organic solvent to a round-bottom flask and dry under vacuum using the rotary evaporator for at least 3 h (see Note 1).
2. Add 1 ml of vesicle buffer and one-third of teaspoon of glass beads. Agitate vigorously on vortex approximately 1 min or

until all lipids are removed from the walls of the flask. The temperature of the vesicle buffer should be above the gel–liquid crystal transition temperature.

3. Transfer the suspension of large multilamellar vesicles to the cryogen vial and freeze it by using liquid nitrogen. Repeat the freeze–thaw cycles 6 times.
4. Use extruder equipped with polycarbonate filters of the defined size (Avestin, Germany) to obtain large unilamellar vesicles. Pass the suspension through filters at temperature that is above the gel–liquid crystal transition temperature until translucent solution is obtained. Store the vesicles at 4°C and use them within 2 days. Do not freeze (see Note 2).

3.2. Immobilization of Vesicles on the Surface of L1 Sensor Chips

1. Equilibrate sensor chip at room temperature, dock it into the apparatus, and prime the system twice with the vesicle buffer.
2. Set the flow rate to 10 μl/min and precondition the surface with two 1-min injections of isopropanol:50 mM NaOH 2:3 (see Note 3).
3. Prepare 200 μl of 1 mM lipids. Use slow flow rate (2 μl/min) and long injection time (10 min) to immobilize the vesicles in the desired flow cells (see Notes 4 and 5).
4. Increase the flow rate to 100 μl/min for few minutes to rinse the loosely bound vesicles from the surface.
5. Stabilize the lipid surface with two 1-min injections of 100 mM NaOH at 10 μl/min. To cover the unbound area on the chip, inject 0.1 mg/ml BSA for 1 min. Allow the surface to stabilize (baseline drift should be lower than 1 RU/min) before performing the analysis.

3.3. Immobilization of Vesicles on the Surface of HPA Sensor Chips

1. Clean the instrument with desorb and sanitize procedures. Run the ultrapure water with low flow rate over the surface overnight to remove all traces of detergent (see Note 6).
2. Equilibrate sensor chip at room temperature, dock it into the apparatus, and prime the system twice with the vesicle buffer.
3. Set the flow rate to 10 μl/min and precondition the surface with 5-min injection of 40 mM octyl glucoside.
4. Prepare 200 μl of 1 mM lipids. Use slow flow rate (2 μl/min) and long injection time (30 min–3 h) to immobilize the vesicles in the desired flow cells (see Notes 4 and 5).
5. Continue with the procedures 4 and 5 in paragraph 3.2.

3.4. Binding Experiment

1. Set flow rate to 10 μl/min. Inject the protein at appropriate concentration for several minutes and follow the dissociation for several minutes to half an hour (see Notes 7 and 8).

2. Regenerate surface with brief injection (1–2 min) of one of the following solutions: 0.5–2 mM NaCl or up to 200 mM NaOH (see Note 9). If protein cannot be removed from the surface of liposomes, proceed to step 4.
3. Repeat the binding of protein by injecting different concentration.
4. Regenerate the surface of the L1 sensor chip by three 1-min injections of isopropanol:50 mM NaOH 2:3 (Fig. 3), whereas for the regeneration of the HPA chip use 5-min injection of OG.
5. Fit the obtained sensorgrams using the evaluation program and the appropriate binding model (see Notes 10 and 11).

4. Notes

1. L1 sensor chip allows capture of liposomes of different composition or size. It is also possible to deposit membrane preparations from cells, such as red blood cell ghosts, plasma membrane remnants, and cellular organelles (5).
2. Small unilamellar vesicles prepared by sonication may also be effectively used.
3. It is important to clean the surface of the sensor chip before the deposition of the liposomes. This is conveniently done by regeneration solutions. Apart from isopropanol:50 mM NaOH 2:3, also some detergent solutions may be used, i.e., 0.5% SDS or 40 mM octyl glucoside.
4. The maximum immobilization level depends on the lipids used. It is higher (11,000–12,000 RU) for the noncharged lipids, such as phosphatidylcholine, and lower (up to 8,000 RU) for negatively charged phospholipids, such as phosphatidylglycerol or phosphatidylserine (13).
5. In general, L1 sensor chip allows capture of intact liposomes (12, 13, 34), although some reports indicate that vesicles may fuse to form the bilayer (7, 35).
6. The surface of HPA chip is composed of long alkanethiol chains that form hydrophobic layer which is very sticky for various hydrophobic molecules. Extra care should be taken when preparing solutions. Be sure that no traces of detergents are present in buffers.
7. The concentrations that should be used in the analysis cover the range from the lowest, where there is hardly any binding seen, to the highest concentration, reaching the saturation. In other words, concentrations used should be $0.1 \times K_D$–$10 \times K_D$.

Use at least five different concentrations to cover this range, do at least one repetition, and include the buffer injection.

8. The association time should be optimized for each interaction separately. For equilibrium analysis, the sensorgrams should reach the equilibrium level during the injection. The duration of dissociation phase is thus not crucial, since the equilibrium response levels are used for the evaluation of the interaction.
9. The level of lipids on the surface should remain the same during the whole experiment. If the analyte could not be effectively removed from the lipid vesicles (see Fig. 7a for the procedure that is used to determine the most effective way in how to remove the analyte from the membrane), then liposomes with bound analytes should be removed with three 1-min injections of isopropanol:NaOH 2:3, and the lipids should be applied for each concentration of protein separately.
10. Special care should be taken to perform experiments at conditions where interaction is not affected by mass transport effect, rebinding of the analyte during the dissociation phase, etc. (16). The evaluation programs allow data to be fitted to several models. The appropriate model should be carefully chosen, possibly also by the use of some supportive data from other experiments.
11. The simple 1:1 interaction model (also termed Langmuir model) implies that molecules bind without other interactions. The dynamic equilibrium is given by

$$A + B \underset{k_d}{\overset{k_a}{\rightleftarrows}} AB$$

where A represents the analyte and B is the ligand. k_a and k_d are association and dissociation rate constants, respectively. The association and dissociation rate constants thus determine the formation and breakdown of the complex at the surface of the sensor chip. The net rate equation is expressed as

$$\frac{d[AB]}{dt} = k_a \cdot [A] \cdot [B] - k_d \cdot [AB]$$

In SPR experiments, the response, R, scales linearly with the complex concentration, [AB], so the rate equation is expressed as

$$\frac{dR}{dt} = k_a \cdot C \cdot (R_{\max} - R) - k_d \cdot R$$

where C is the concentration of the analyte and $R_{\max}$ is the response signal at the saturation. This equation is used to fit the data, as presented on Fig. 5, to obtain k_a, k_d, and $R_{\max}$. The equilibrium dissociation constant, K_D, is expressed by the rate constants:

$$K_D = \frac{k_d}{k_a}$$

Acknowledgments

We would like to thank to the Slovenian Research Agency for the continuing support of the Infrastructural Centre for Molecular Interaction Analysis at the Department of Biology, Biotechnical Faculty, University of Ljubljana, Slovenia.

References

1. Rich RL, Myszka DG (2008) Survey of the year 2007 commercial optical biosensor literature. J Mol Recognit 21:355–400
2. Rich RL, Myszka DG (2010) Grading the commercial optical biosensor literature-class of 2008: "The Mighty Binders". J Mol Recognit 23:1–64
3. Schasfoort RBM, Tudos AJ (eds) (2008) Handbook of surface plasmon resonance. The Royal Society of Chemistry, Cambridge, UK
4. de Mol NJ, Fischer MJE (eds) (2010) Surface plasmon resonance: methods and protocols. Humana Press, New York, USA
5. Beseničar M, Maček P, Lakey JH et al (2006) Surface plasmon resonance in protein-membrane interactions. Chem Phys Lipids 141:169–178
6. Cho W, Bittova L, Stahelin RV (2001) Membrane binding assays for peripheral proteins. Anal Biochem 296:153–161
7. Cooper MA (2004) Advances in membrane receptor screening and analysis. J Mol Recognit 17:286–315
8. Stenberg E, Persson B, Roos H et al (1990) Quantitative determination of surface concentration of protein with surface plasmon resonance using radiolabeled proteins. J Colloid Interface Sci 143:513–526
9. Mohri I, Taniike M, Okazaki I et al (2006) Lipocalin-type prostaglandin D synthase is upregulated in oligodendrocytes in lysosomal storage diseases and binds gangliosides. J Neurochem 97:641–651
10. Beseničar Podlesnik M, Anderluh G (2010) Preparation of lipid membrane surfaces for molecular interaction studies by surface plasmon resonance biosensors. Methods Mol Biol 627:191–200
11. Cooper MA, Try AC, Carroll J et al (1998) Surface plasmon resonance analysis at a supported lipid monolayer. Biochim Biophys Acta 1373:101–111
12. Cooper MA, Hansson A, Lofas S et al (2000) A vesicle capture sensor chip for kinetic analysis of interactions with membrane-bound receptors. Anal Biochem 277:196–205
13. Anderluh G, Beseničar M, Kladnik A et al (2005) Properties of nonfused liposomes immobilized on an L1 Biacore chip and their permeabilization by a eukaryotic pore-forming toxin. Anal Biochem 344:43–52
14. Bavdek A, Gekara NO, Priselac D et al (2007) Sterol and pH interdependence in the binding, oligomerization, and pore formation of listeriolysin O. Biochemistry 46:4425–4437
15. Bakrač B, Gutierrez-Aguirre I, Podlesek Z et al (2008) Molecular determinants of sphingomyelin specificity of a eukaryotic pore-forming toxin. J Biol Chem 283:18665–18677
16. Myszka DG (1997) Kinetic analysis of macromolecular interactions using surface plasmon resonance biosensors. Curr Opin Biotechnol 8:50–57
17. Stahelin RV, Cho W (2001) Differential roles of ionic, aliphatic, and aromatic residues in membrane-protein interactions: a surface plasmon resonance study on phospholipases A2. Biochemistry 40:4672–4678
18. Stahelin RV, Ananthanarayanan B, Blatner NR et al (2004) Mechanism of membrane binding

of the phospholipase D1 PX domain. J Biol Chem 279:54918–54926

19. Shanmugham A, WongFongSang HW, Bollen YJM et al (2006) Membrane binding of twin arginine preproteins as an early step in translocation. Biochemistry 45:2243–2249
20. Bahloul A, Michel V, Hardelin JP et al (2010) Cadherin-23, myosin VIIa and harmonin, encoded by usher syndrome type I genes, form a ternary complex and interact with membrane phospholipids. Hum Mol Gen 19:3557–3565
21. Hekman M, Albert S, Galmiche A et al (2006) Reversible membrane interaction of BAD requires two C-terminal lipid binding domains in conjunction with 14-3-3 protein binding. J Biol Chem 281:17321–17336
22. Locatelli-Hoops S, Remmel N, Klingenstein R et al (2006) Saposin a mobilizes lipids from low cholesterol and high bmp containing membranes. patient variant saposin a lacks lipid extraction capacity. J Biol Chem 281:32451–32460
23. Sugiki T, Takahashi H, Nagasu M et al (2010) Real-time assay method of lipid extraction activity. Anal Biochem 399:162–167
24. Beseničar MP, Bavdek A, Kladnik A et al (2008) Kinetics of cholesterol extraction from lipid membranes by methyl-beta-cyclodextrin—a surface plasmon resonance approach. Biochim Biophys Acta 1778:175–184
25. Praper T, Beseničar MP, Istinič H et al (2010) Human perforin permeabilizing activity, but not binding to lipid membranes, is affected by pH. Mol Immunol 47:2492–2504
26. Saenko E, Sarafanov A, Ananyeva N et al (2001) Comparison of the properties of phospholipid surfaces formed on HPA and L1 biosensor chips for the binding of the coagulation factor VIII. J Chromatogr A 921:49–56
27. Papo N, Shai Y (2003) Exploring peptide membrane interaction using surface plasmon resonance: differentiation between pore formation versus membrane disruption by lytic peptides. Biochemistry 42:458–466
28. Karlsson OP, Lofas S (2002) Flow-mediated on-surface reconstitution of G-protein coupled receptors for applications in surface plasmon resonance biosensors. Anal Biochem 300:132–138
29. Vidic JM, Grosclaude J, Persuy MA et al (2006) Quantitative assessment of olfactory receptors activity in immobilized nanosomes: a novel concept for bioelectronic nose. Lab Chip 6:1026–1032
30. Branden M, Tabaei SR, Fischer G et al (2010) Refractive-index-based screening of membrane-protein-mediated transfer across biological membranes. Biophys J 99:124–133
31. Lundquist A, Hansen SB, Nordström H et al (2010) Biotinylated lipid bilayer disks as model membranes for biosensor analyses. Anal Biochem 405:153–159
32. Hodnik V, Anderluh G (2010) Capture of intact liposomes on biacore sensor chips for protein-membrane interaction studies. Methods Mol Biol 627:201–211
33. Hall K, Aguilar MI (2010) Surface plasmon resonance spectroscopy for studying the membrane binding of antimicrobial peptides. Methods Mol Biol 627:213–223
34. Höning S, Ricotta D, Krauss M et al (2005) Phosphatidylinositol-(4,5)-bisphosphate regulates sorting signal recognition by the clathrin-associated adaptor complex AP2. Mol Cell 19:577–577
35. Erb EM, Chen X, Allen S et al (2000) Characterization of the surfaces generated by liposome binding to the modified dextran matrix of a surface plasmon resonance sensor chip. Anal Biochem 280:29–35

Chapter 3

Probing the Thermodynamics of Protein–Lipid Interactions by Isothermal Titration Calorimetry

Musti J. Swamy and Rajeshwer S. Sankhala

Abstract

Isothermal titration calorimetry is a highly sensitive technique for the study of molecular interactions. This method has been applied quite extensively to investigate the interaction of proteins with small ligands, other proteins, and nucleic acids as well as with drugs and metal ions. In this chapter, we describe the application of ITC for the investigation of thermodynamics of protein–lipid interaction. A number of parameters such as enthalpy of binding (ΔH), entropy of binding (ΔS), association constant (K_a), binding stoichiometry (n), and free energy of binding (ΔG) can be obtained from a single calorimetric titration, providing a complete thermodynamic characterization of the interaction. The method is described in detail taking the major protein of the bovine seminal plasma, PDC-109, which exhibits a high preference for interaction with choline-containing lipids, as an example. The method can be applied to investigate the thermodynamics of the interaction of other soluble proteins with lipid membranes.

Key words: Isothermal titration calorimetry, Protein–lipid interaction, Binding stoichiometry, Association constant, Enthalpy of binding, Entropy of binding

1. Introduction

Isothermal titration calorimetry is a highly sensitive physical technique which is being used extensively to determine the thermodynamic parameters associated with biomolecular interactions in the solution phase. Before going into more details of this technique, a simple question that one might ask is "why ITC?" since a number of approaches such as various spectroscopic techniques and other physical methods such as equilibrium dialysis, are available and have been used quite extensively in the past to investigate biomolecular interactions. The major advantages of ITC over other methods are:

(a) All the thermodynamic parameters such as binding enthalpy (ΔH), binding entropy (ΔS), association constant (K_a), and

Jörg H. Kleinschmidt (ed.), *Lipid-Protein Interactions: Methods and Protocols*, Methods in Molecular Biology, vol. 974, DOI 10.1007/978-1-62703-275-9_3, © Springer Science+Business Media New York 2013

stoichiometry of binding (n) can be determined in a single experiment.

(b) Calorimetric measurements do not require the presence of a chromophore or fluorophore in the molecule; hence spectroscopically silent molecules can also be studied.

(c) Although in general the sample requirement for ITC studies is significant, with the advancement in the technology, in favorable cases, it is possible to perform the binding reactions at micromolar concentrations of biomolecules.

1.1. Basic Principle and Instrumentation

An ITC instrument consists of two identical cells (a sample and a reference cell), usually made up of an alloy with high thermal conductivity (e.g., Hastelloy) and surrounded by an adiabatic jacket (Fig. 1). In some instruments, the cell is made up of gold. Reference cell is filled up with water/buffer, and the biomolecule under investigation taken in the same medium is placed in the sample cell which is then titrated by adding the ligand/titrant through an injector syringe. ITC output depends on the temperature difference between the sample and reference cell. A measure of the heat taken up from the surroundings (endothermic process) or heat released into the surroundings (exothermic process) can be correlated to the amount of the reaction that has occurred. The temperature differences between the two cells are measured and calibrated in terms of power units. Therefore, the output signal in an ITC experiment is always differential power and is represented by the acronym *DP* (for *differential power*). This signal is also referred to as "feedback," that is, power used to maintain the two cells at the

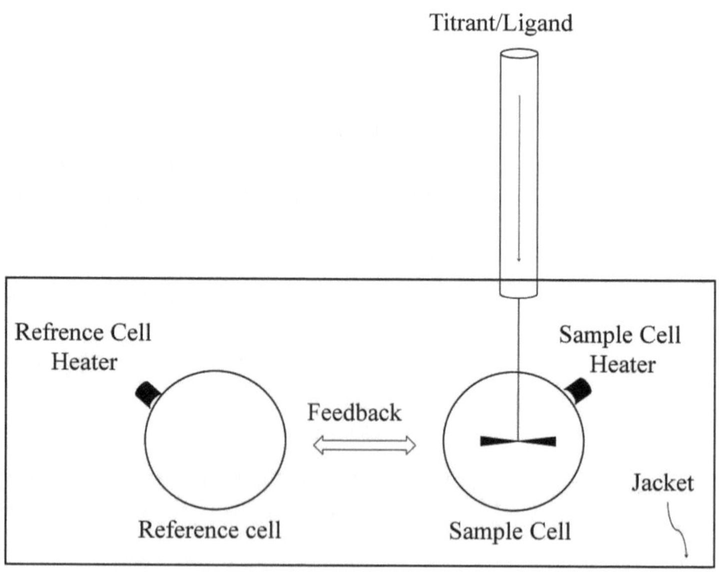

Fig. 1. A schematic representation of ITC.

same temperature. The temperature difference between the two cells is constantly monitored using a thermopile/thermocouple, and heaters located on both cells are activated when necessary to maintain them at identical temperatures. In the absence of any reaction, feedback power (*DP*) will be at baseline level. Before the experiment, a constant power is supplied to the reference cell which then directs the feedback circuit to activate the heater connected to the sample cell, and this yields a baseline. When a reaction occurs, the temperature of the sample cell will change due to uptake or release of heat. Since the sample cell temperature will increase in an exothermic reaction, the feedback circuit will decrease power supply to the sample cell; in contrast, in an endothermic reaction, the feedback circuit will increase the power to sample cell. The heat evolved or absorbed by the reaction is then obtained by integration of these deflections from baseline, with respect to time. The data obtained can be analyzed using different fitting routines available in the ITC data analysis software (supplied along with the instrument).

A typical ITC experiment can be divided into the following steps:

1. Sample preparation and relative concentrations of interacting partners
2. Filling of the cells and injector syringe
3. Setting up experimental parameters
4. Blank/control titration
5. Data analysis

Appropriate care must be taken at each of the above steps. Otherwise the data obtained could be erroneous and misleading.

1.2. Major Applications

In principle any binding reaction or physical change that is accompanied by a change in enthalpy (release or absorption of heat) can be measured using ITC. Some of the major scientific problems that have been addressed using ITC are the investigation of protein–ligand interaction, protein–protein interaction, protein–nucleic acid interaction, nucleic acid–drug interaction, protein folding/unfolding, and reaction rates (see, for reviews, (1–7)). The interaction of a number of proteins with metal ions and metal nanostructures has also been investigated in a number of studies, and some examples can be found in (8–11).

1.2.1. Protein–Lipid Interaction

In this chapter, we will discuss the application of isothermal titration calorimetry to characterize thermodynamics of protein/phospholipid interactions. It has been estimated that about 30% of the proteins in human are associated with membranes (12). These membrane-associated proteins can be categorized into the following two types.

1. Peripheral proteins: The peripheral proteins are usually soluble in aqueous buffers without the need for detergents, but bind to phospholipid membranes, usually by a combination of electrostatic and hydrophobic interactions or by the specific interaction with the head group of a particular membrane lipid. Such proteins are generally dislodged from the membrane by treatment with high ionic strength buffers containing chelating agents. A classic example of this type is the myelin basic protein, which binds to membranes containing acidic lipids such as phosphatidylglycerol, gangliosides, and sulfatide (13–16). Some peripheral proteins bind to lipid membranes by their specific interaction with the head groups of particular lipids. Examples of the latter type include PDC-109, the major protein of the bovine seminal plasma, which specifically interacts with the zwitterionic phosphatidylcholine and other choline-containing phospholipids such as sphingomyelin, lysophosphatidylcholine, and platelet-activating factor (17–19) as well as lectins and toxins that bind specifically to the glycosyl moieties of neutral glycolipids and gangliosides (20–22). Some peripheral proteins adhere to the membranes by interacting with the integral membrane proteins. However, such proteins are not relevant to the present discussion as the main focus of this chapter is to discuss how ITC can be used to investigate the thermodynamics of protein–lipid interactions.
2. Integral membrane proteins: These proteins are permanently attached to the membrane. They contain hydrophobic amino acid-rich segments through which they interact with the hydrophobic tails of phospholipids, whereas their hydrophilic segments could interact with the aqueous medium on either side of the membrane. Although the interaction of detergent-solubilized integral proteins with amphipols was investigated by ITC (23, 24), the characterization was restricted to determining the enthalpy of the exchange of the detergent-solubilized proteins into amphipol-solubilized form, and no information on the thermodynamic parameters such as binding stoichiometry, enthalpy of binding, and entropy of binding could be obtained. Further, despite the amphiphilic nature of the amphipols, this interaction is different from protein–lipid interaction as the amphipols used in the above studies are not lipid-like, but are soluble polymers carrying cationic or anionic groups in order to solubilize them in water or aqueous buffers. Therefore, these studies are not further considered in this chapter.

In addition to the above, the interaction of a number of small peptides with phospholipid membranes has been investigated by ITC. An excellent review, though not very recent, on this topic can be found in (25). These studies will not be considered further here, as a discussion on this aspect is beyond the scope of this chapter. However, the method that will be described below for studying protein–membrane interaction can also be applied to the study of the interaction of peptides that are soluble in aqueous media with lipid aggregates/membranes.

2. Materials

2.1. Reagents

The following chemical/biochemicals or reagents are required to perform an ITC experiment:

1. The protein of interest, whose interaction with lipids is to be characterized in terms of phospholipid binding, in pure, soluble form.
2. The lipids to be used in the study. A variety of phospholipids, some glycolipids, and sterols can be purchased from different commercial suppliers. A large variety of phospholipids, sterols, and glycolipids are available from Avanti Polar Lipids (Alabaster, AL, USA) which specializes in lipids. Suppliers of biochemicals such as Sigma-Aldrich, Fluka, and Serva also have a number of lipids in their product list.
3. Buffering system: The buffering system to be used depends on the optimal conditions (pH, ionic strength, etc.) suitable to the protein and required for the preparation of stable phospholipid vesicles.
4. Organic solvents (chloroform, dichloromethane, ethanol, methanol, etc.) to dissolve the lipid can be purchased from locally available distributors. Dry solvents of high purity should be used.

2.2. Equipment

1. Several commercial isothermal titration calorimeters are available in the market. The most widely used models are from MicroCal (which is now a part of GE Healthcare) and TA Instruments (a part of Waters Corporation). MicroCal has two models, VP-ITC and iTC_{200}. The iTC_{200} is also available in an automated version, Auto-iTC_{200}, which can handle up to 75 titrations per day with a capacity to run more than 350 samples unattended. The VP-ITC has a cell volume of ~1.45 ml, whereas in the iTC_{200} the cell volume is about 200 μl and therefore requires considerably less sample; however, the sensitivity of these two instruments is comparable, and since smaller volume would translate into smaller heat changes during the titration, it would often be necessary to use somewhat higher concentrations of the protein and the lipid while using the iTC_{200}. The titration calorimeters from TA Instruments are called Nano ITC and are available with two different

cell volumes, 1.0 ml and 190 μl. ITC instruments from both MicroCal and TA Instruments can be used over a wide range of temperatures between 2 and 80°C. Results from the authors' laboratory presented in this chapter were obtained using a MicroCal VP-ITC equipment.

2. Vacuum desiccator: A vacuum desiccator is required to remove the traces of organic solvent from thin phospholipid layers before they are hydrated.
3. Vacuum pump: A vacuum pump is required for degassing of the samples and automated cleaning of the cells.
4. Syringes: Graduated microsyringes (typically 50–250 μl total volume) are used for injecting the titrant into the sample cell. Syringes of larger capacity (~0.5–2 ml) are used for washing and loading samples into the calorimeter cells. Both types of syringes are normally supplied by the manufacturer along with the ITC equipment.
5. Sonicator: Typically, a probe sonicator is used to prepare small unilamellar vesicles (SUVs). Bath sonicators are also sometimes used, but the liposomes obtained from bath sonication may not be of uniform size. A large number of sonicators of these two types are commercially available.
6. Extruder: If the experimental design requires unilamellar vesicles of different sizes, then these can be prepared using a simple syringe-based, small-volume extruder (mini-extruder). Originally designed by MacDonald et al. (26), the extruder is available from Avanti Polar Lipids (Alabaster, AL, USA). Typically, a suspension of multilamellar vesicles (MLVs) made up of the desired lipid (or lipid mixture) is prepared by hydrating a thin lipid film with the appropriate buffer. The MLVs are then taken in a syringe (syringe 1) and passed through a membrane filter such that the liposomes that pass through the filter enter another syringe (syringe 2) connected to the filter holder on the other side. The liposomal suspension is then passed back into syringe 1. After several (odd number of) passages (ca. 11, 13, or 15) through the membrane, the liposomal suspension is collected from syringe 2. The mini-extruder is suitable for processing small volumes (~0.25–1 ml) of liposomal suspensions. Liposomes of various sizes can be prepared by using filters with different pore sizes. Filters of pore sizes ranging between 30 and 1000 nm are commercially available. Further details regarding the use of the extruder can be obtained from the above reference or from the literature provided by the supplier. Extruders with larger capacity are also commercially available, for example, from Northern Lipids, Inc. (Vancouver, Canada).
7. Dialysis tubing: To dialyze the protein in order to obtain a fresh sample.

3. Methods

Isothermal titration calorimeters are available with single and dual mode of injection facilities (1). The present discussion will focus on a calorimeter with single-injection mode, which is most widely used for investigating the thermodynamics of biological interactions. In such a calorimeter, the titrant is added only to the sample cell. Therefore, a control experiment has to be performed separately where the titrant is added to the sample cell containing the buffer alone, and the blank/control titration data is subtracted from the data obtained from the main experiment. In contrast, in a dual-injection mode calorimeter, the titrant is added to both the cells (sample and reference) simultaneously, obviating the need to perform a separate blank (control) titration.

3.1. Sample Preparation and Relative Concentrations of Interacting Partners

Sample preparation is the most critical step for a successful ITC experiment and can be divided into the following subsections for better understanding.

3.1.1. Preparation of Protein Sample

It is very important to check the purity and exact concentration of the protein prior to the ITC experiment (see Note 1). Protein concentration can be determined using extinction coefficient or standard protein estimation assays such as Lowry assay (27), Bradford assay (28), or other dye-based methods (29). If the amino acid sequence of the protein is known, then the extinction coefficient of the protein can be estimated by the method of Edelhoch (30) as described by Gill and Von Hippel (31). Protein sample must be dialyzed against freshly prepared buffer, followed by centrifugation and degassing prior to run. The last dialysate is retained and used to prepare the lipid vesicles. Both the protein and ligand must be in the same buffering system in order to minimize the heat of dilution (see Notes 2, 3).

3.1.2. Preparation of Lipid Vesicles

The desired quantity of lipid (usually a few milligrams) is taken in a glass test tube and dissolved in an appropriate amount of a suitable organic solvent. Dichloromethane or dichloromethane/methanol mixture is most commonly employed as most of the lipids used are soluble in these solvents. The solvent is then dried under a gentle stream of nitrogen gas followed by vacuum desiccation for a minimum of 3 h (see Note 4). The dry lipid film thus obtained is hydrated with the appropriate buffer to obtain a desired final concentration. This suspension is then either subjected to sonication in a sonicator or extruded through a filter of desired pore size in order to obtain clear, homogenous, and uniform-sized lipid vesicles.

3.1.3. Relative Concentrations of the Interacting Partners

Prior to the ITC experiment, it is important to choose appropriate concentrations of protein and ligand to be used. This can be done with the help of Eq. 1, which provides a good approximation of the protein concentration to start with:

$$K_a \times C = 5 \text{ to } 10 \tag{1}$$

Here K_a is the association constant, and C is the concentration of protein in the sample cell (32). Association constant measured from other techniques, if available, can be used in the above equation to calculate the approximate concentration of the protein required to start the experiment. If K_a is unknown, a best guess can be used for trial experiments. A titrant/ligand can be used at ca. 20-fold higher concentration than that of the protein in sample cell. However, the above hints should be taken as "rules of thumb," and not as absolute values. The ideal protein/ligand ratio can only be decided after a few trial runs for each system.

3.2. Filling of the Cells and Injector Syringe

Reference cell is filled with the final dialysis buffer, whereas the macromolecule under investigation is usually filled in the sample cell. However, for the study of lipid–protein interaction, calorimetric experiments are often conducted by taking the liposomal suspension in the calorimeter cell and titrating it with the protein (see Note 5). Both cells will have the same sample volume of ~1.5 ml; therefore, a glass syringe containing ~2.0 ml of the respective solution should be used to fill the cells. Utmost care should be taken while filling to avoid bubbling which otherwise can hinder the experiment. Injector syringe also should be filled up with extreme care to avoid the bubbling and physical damage to the syringe (see Note 6). Any sort of bubbling or bending (physical damage) can cause ambiguity in the output data. Commercially available calorimeters such as the VP-ITC from MicroCal are supplied with a facility to purge and refill which is helpful in eliminating small bubbles that may form while filling. It must be noted that all the samples should be in the same buffering system and must be degassed properly before loading. It is important to clean the cells thoroughly after the experiment, especially if the reaction system is being changed (see Note 7).

3.3. Setting Up Experimental Parameters

The experimental parameters are entered into the software program supplied along with the instrument. Total volume and number of injections, duration of the injection, temperature of the cells, and concentrations of the protein and ligand are important parameters and should be filled in properly. Baseline setting is also an important factor in an ITC run in which a small constant power is supplied to the reference cell heater. This activates the DP feedback system which then supplies a compensating power to the sample cell. In ITC data display window baseline starts from a value, fed prior to the run. Adequate spacing should be given between successive injections so that output signal (DP) can reach the baseline level after each injection (see Note 8). To ensure proper mixing of the titrant/ligand with the sample in the cell, the injector syringe must be given an adequate stirring speed (~300 rpm is a good default value for most titrations; however, see Note 9 for more details). Another important factor is "feedback mode." For normal operation, a high feedback is recommended which means fast response time. However, monitoring of the slow thermal processes may require low feedback. The first injection is often inaccurate; therefore, a 1–2 μl injection is added first, and the resultant data point is deleted before analyzing the remaining data (see Note 10).

3.4. Blank/Control Titration

As mentioned above in a *single-injection mode* calorimeter, a blank titration must be performed separately in order to eliminate the heat of dilution from the actual experimental data. The heat of binding measured in an ITC experiment also includes contributions from heats of dilution of ligand/titrant and protein. Heat of dilution of ligand is more critical and must be subtracted from the actual data to obtain a meaningful data. Control experiment is performed by titrating the ligand/titrant solution into a buffer solution under identical conditions, and the resulting heat changes are subtracted from the measured heats of binding (see Note 11). Although heat of dilution of the macromolecule (sample in the cell) is generally not crucial, it can also be eliminated to obtain a more precise data.

3.5. Data Analysis

The titration data obtained from the calorimetric experiment is analyzed using ITC data analysis software supplied along with the instrument. Since the mechanism of protein–ligand interaction can vary from system to system, a range of fitting tools such as *one set of sites*, *two sets of sites*, and *sequential binding* are usually provided in the software. In this chapter, we describe the *one set of sites* and *two sets of sites* binding models, which have been used to characterize various protein–lipid interactions (9, 19, 33, 34).

3.5.1. One Set of Sites Binding Model

For a system of one set of identical binding sites, the total heat evolved (or absorbed) during the binding process at the end of the *i*th injection, $Q(i)$, is given by Eq. 2 (35):

$$Q(i) = \frac{nP_t\Delta HV\left\{1 + \frac{X_t}{nP_t} + \frac{1}{nKP_t} - \left[\left(1 + \frac{X_t}{nP_t} + \frac{1}{nKP_t}\right)^2 - \frac{4X_t}{nP_t}\right]^{\frac{1}{2}}\right\}}{2} \quad (2)$$

where n is the number of binding sites, P_t is the total protein concentration, X_t is the total ligand concentration, V is the cell volume, K is the binding constant, and ΔH is the binding enthalpy. The heat corresponding to the *i*th injection only, $\Delta Q(i)$, is equal to the difference between $Q(i)$ and $Q(i-1)$ and is given by Eq. 3, which involves the necessary correction factor for the displaced volume (the injection volume dV_i):

$$\Delta Q(i) = Q(i) + \frac{dV_i}{V_o}\left[\frac{Q(i) + Q(i-1)}{2}\right] - Q(i-1) \quad (3)$$

The ITC unit measures $\Delta Q(i)$ value for every injection. These values are then fitted to Eqs. 2 and 3 by a nonlinear least-squares method using the data analysis program Origin® [MicroCal™]. The fit process involves initial guess of n, K, and ΔH which allows calculation of $\Delta Q(i)$ values as mentioned above for all injections and comparing them with the corresponding experimentally determined values. Based on this comparison, the initial guess of n, K_a, and ΔH is improved, and the process is repeated till no further significant improvement in the fit can be obtained.

3.5.2. Two Sets of Sites Binding Model

For a system with two sets of independent binding sites, the total heat evolved (or absorbed) during the binding process at the end of the ith injection, $Q(i)$, is given by Eq. 4 (32):

$$Q(i) = P_t V(n_1\Theta_1\Delta H_1 + n_2\Theta_2\Delta H_2) \quad (4)$$

where n_1 and n_2 are the number of binding sites of type 1 and type 2 and ΔH_1 and ΔH_2 are the corresponding enthalpies of binding. Θ_1 and Θ_2 are the fractions of the type 1 and type 2 sites that are occupied and are related to the association constants K_1 and K_2, according to Eq. 5:

$$K_1 = \frac{\Theta_1}{(1-\Theta_1)[X]} \qquad K_2 = \frac{\Theta_2}{(1-\Theta_2)[X]} \quad (5)$$

The heat corresponding to the ith injection only, $\Delta Q(i)$, is equal to the difference between $Q(i)$ and $Q(i-1)$ and is given by Eq. 5, which involves the necessary correction factor for the displaced volume (the injection volume dV_i), according to Eq. 3.

The $\Delta Q(i)$ values measured by the ITC from the titration are then fitted to Eqs. 3 and 4 by a nonlinear least-squares method. The fit process involves initial guess of n_1, n_2, K_1, K_2, ΔH_1, and ΔH_2 which allows calculation of $\Delta Q(i)$ values as mentioned above for all injections and comparing them with the corresponding experimentally determined values. Based on this comparison, the initial guess of the values of the above parameters is improved, and the process is repeated till no further significant improvement in the fit can be obtained.

Using the values of ΔH and K_a obtained from the above analysis, the thermodynamic parameters, ΔG and ΔS, are calculated according to the basic thermodynamic Eqs. 6 and 7:

$$\Delta G = -RT \ln K_a \quad (6)$$

$$\Delta G = \Delta H - T\Delta S \quad (7)$$

The ITC data acquisition system is loaded with the Origin software, and the above analysis is quite routinely performed using the options available in it.

3.6. Running a Typical ITC Experiment

Choice of lipid is very critical for protein/lipid interaction studies. The binding affinity of proteins or peptides varies significantly with the lipid composition of the membrane. Therefore, it would be possible to investigate the thermodynamic parameters characterizing the interaction of a given protein with a variety of lipids by performing calorimetric titrations with different protein/lipid combinations. A number of soluble peripheral proteins require the presence of anionic lipids in the membrane in order to bind (see Subheading 1.2.1 above). Studies can be performed at various temperatures for the same protein/lipid combination, and the effect of the lipid phase on the interaction can be investigated. To illustrate some of these aspects, ITC studies on the interaction of PDC-109 with diacyl phospholipid membranes and lysophosphatidylcholine (Lyso-PC) micelles are described below.

3.6.1. Interaction of PDC-109 with Diacyl Phospholipids

The interaction of PDC-109 with various phospholipids such as dimyristoylphosphatidylcholine (DMPC), dimyristoylphosphatidylglycerol (DMPG), and dipalmitoylphosphatidylcholine (DPPC) was investigated by ITC (19). Interaction of PDC-109 with DMPC above its phase transition temperature (24°C) yielded a complex isotherm which could not be analyzed using the fitting tools available in the ITC analysis software. On the other hand, calorimetric titrations performed in the gel phase region of DMPC yielded endothermic isotherms, which could be fitted quite satisfactorily using the *one set of sites* binding model (Fig. 2).

As mentioned above, ITC can provide direct information of the various thermodynamic parameters associated with the binding, viz., n, K, ΔH, and ΔS. The values obtained for the interaction of PDC-109 with DMPC, DPPC, and DMPG are listed in Table 1.

From the temperature dependence of ΔH associated with the binding of each lipid, the corresponding change in heat capacity (ΔC_p) can be obtained. Heat capacity is defined as the ratio of amount of heat transferred to an object and the resulting increase in the temperature (Eq. 8).

$$\Delta C_p = \Delta Q/\Delta T \tag{8}$$

where Q is the amount of heat transferred to the object. In the present case, the energy transferred to or retrieved from the calorimeter cell (sample cell) is defined as enthalpy change (ΔH). Therefore, heat capacity can be represented as shown in Eq. 9:

$$\Delta C_p = \partial \Delta H/\partial T \tag{9}$$

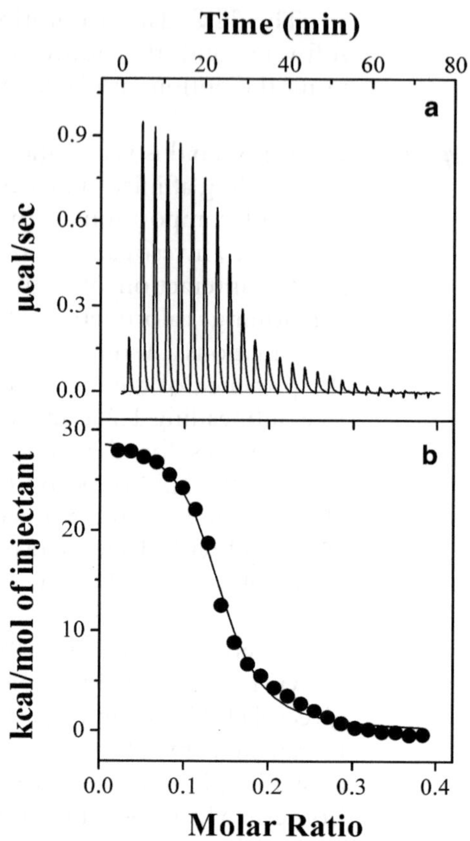

Fig. 2. Calorimetric titrations for the binding of PDC-109 to phosphatidylcholine membranes at 20°C in gel phase (below transition temperature). *Upper panel* (**a**) shows the raw data for the titration of DMPC unilamellar vesicles (in the calorimeter cell) with protein (in the syringe). The *lower panel* (**b**) shows the integrated heats of binding obtained from the raw data, after subtracting the heat of dilution. The solid line in the *bottom panel* represents the best curve fit of the experimental data to the *one set of sites* model from MicroCal Origin (Adapted from ref. (19)).

One can determine the ΔC_p value by measuring the enthalpy change of a binding reaction at different temperatures. Plotting ΔH versus temperature would yield ΔC_p as its slope. In the case of PDC-109 interaction with DMPC and DPPC, although the enthalpy of binding increased with increase in temperature for both the lipids, clear-cut linear dependence was not observed (19). Therefore, the heat capacity changes associated with the binding process could not be estimated.

3.6.2. Interaction of PDC-109 with Lyso-PC Micelles

When Lyso-PC micelles (in the calorimeter cell) were titrated with PDC-109, the binding isotherm (Fig. 3) was found to be exothermic, in contrast to the endothermic isotherms obtained with DMPC and DPPC (19). Further, analysis of the binding data by the *one set of sites* binding model did not yield satisfactory fits,

Table 1
Thermodynamic parameters obtained by isothermal titration calorimetry for PDC-109 binding to phosphatidylcholine and phosphatidylglycerol unilamellar vesicles

T (°C)	*n*	$K_a \times 10^{-5}$ (M^{-1})	ΔG (kcal/mol)	ΔH (kcal/mol)	ΔS (cal/mol K)
DPPC					
20	18.2 (±3.3)	0.8 (±0.3)	−6.57	0.5 (±0.1)	23.7 (±0.8)
25	14.4 (±1.0)	1.3 (±0.4)	−6.98	0.9 (±0.4)	26.5 (±1.0)
30	7.6 (±1.2)	2.2 (±0.4)	−7.41	5.1 (±0.9)	41.1 (±2.5)
36	7.2 (±0.1)	20.7 (±8.9)	−8.93	5.1 (±0.5)	45.2 (±2.5)
39[a]	6.5 (±0.1)	24.5 (±6.5)	−9.12	5.5 (±0.1)	46.7
DMPC					
10	21.6 (±0.2)	3.9 (±0.8)	−7.24	1.2 ± 0.1	29.7 (±0.4)
15	13.4 (±0.9)	3.5 (±0.6)	−7.31	2.3 ± 0.2	33.3 (±0.5)
20	7.2 (±0.3)	4.1 (±1.5)	−7.52	4.5 ± 0.5	40.8 (±1.8)
DMPG					
7.4	55	0.26 (±0.03)	−5.91	2.9 (±0.6)	30.0 (±2.4)

Values reported are averages of results obtained from 2 to 4 independent experiments with standard deviations indicated in parentheses (adapted from ref. (19))
[a]Data from a single measurement

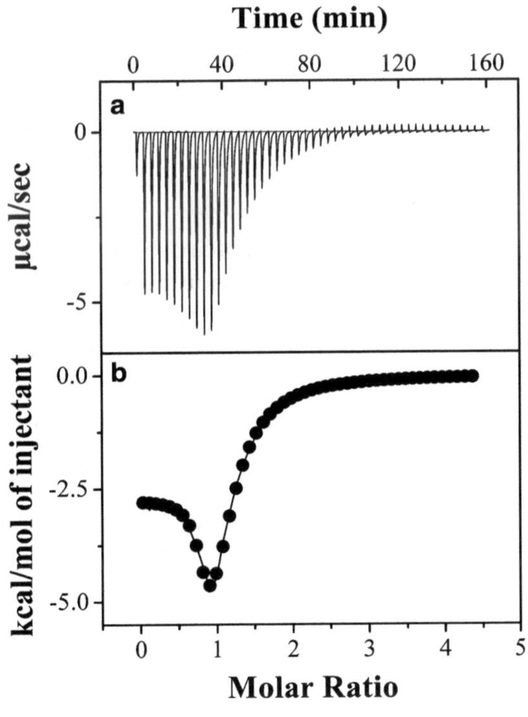

Fig. 3. Binding of PDC-109 to Lyso-PC micelles. (**a**) Raw data for the titration of 450 µM protein with 10.88 mM Lyso-PC at 25 °C. (**b**) Integrated heats of binding obtained from the raw data, after subtracting the heat of dilution. The solid line in (**b**) represents the best curve fit to the experimental data, using the *two sets of sites* model from MicroCal Origin (Adapted from ref. (19)).

whereas successful fitting was achieved with the *two sets of sites* binding model. This suggested two kinds of binding interactions between the protein and Lyso-PC; one characterized by a higher affinity and the other with a lower affinity. The high-affinity interaction was characterized by the following binding parameters: stoichiometry (n = the number of Lyso-PC molecules bound to each molecule of PDC-109) = 0.9 (± 0.1), $K_a = 7.0\ (\pm\ 0.2) \times 10^5\ M^{-1}$, $\Delta H = -2.6$ (±0.1) kcal/mol, and $\Delta S = 17.9$ (±0.4) cal/mol/K. The corresponding values obtained for the low-affinity interaction are $n = 3.5$, $K_a = 1.2\ (\pm 0.1) \times 10^4\ M^{-1}$, $\Delta H = -13.7$ (±0.6) kcal/mol, and $\Delta S = -2.6$ (± 0.1) cal/mol/K. Interestingly, while enthalpy of binding is favorable for both types of interactions, entropic factors are favorable only for the high-affinity interaction (ΔS = +ve), with the low-affinity interaction being characterized by a large unfavorable contribution from the binding entropy (ΔS = −ve).

The two types of binding interactions were assigned to the interaction of PDC-109 with Lyso-PC micelles and monomer. Since the binding of the protein to both DMPC and DPPC vesicles was characterized by a positive enthalpy, and since the entropy of binding for the higher affinity interaction was also positive, it was assigned to the binding of the protein with Lyso-PC micelles, whereas the lower-affinity interaction was assigned to the binding of the monomeric form of the lipid to PDC-109 (19).

4. Notes

1. The protein and ligand (phospholipid) concentrations used in the titration must be estimated/calculated with great care and high accuracy. Errors in the input concentrations of the protein or ligand (lipid) while analyzing the data would lead to errors in the values of the thermodynamic parameters obtained.
2. It is very important to ensure that the buffer in which the ligand is dissolved is an **exact match** (i.e., pH, buffer strength, and salt concentration) to the buffer in which the macromolecule is dissolved, or else large spurious heat effects from buffer mixing may result, which will complicate the data analysis or may even render the data useless. An approach that has been found to work well is to use the last dialysate obtained from dialyzing the protein as the medium for preparing the ligand solution.
3. Sometimes binding depends on the buffering system. Some binding reactions which involve uptake or release of proton may not show enough heat change in a buffer with low H_{ion} (e.g., phosphate buffer). For such systems, if one uses a H_{ion}-rich buffer, it may yield a better signal (see ITC manual for details).

4. Often, due to cooling due to evaporation of the solvent when nitrogen is blown over the lipid solution, the lipid may precipitate out of the solvent. If this happens, it would be useful to place the bottom of the glass tube in lukewarm water and gently shake the tube in it while blowing the nitrogen gas over the surface of the lipid solution.
5. While investigating protein/lipid interaction, adequate care must be taken to see that the phase of the lipid does not change during the titration. This can be achieved either by performing the experiment at ambient temperature or the lipid can be taken in the cell so that it will be in same phase throughout the experiment.
6. Adequate care must be taken while loading the sample. One should avoid both the contamination and bubbling of the cells and injector syringe, which otherwise may yield complex/non-analyzable data. To avoid bubbling in the syringe, the sample taken in the syringe can be purged and refilled a few times using the automated feature available in the ITC equipment. Bubbling in the cells can also be eliminated by performing the purge–refill process manually in which all the solution in the cell is first sucked into the syringe and then dispensed back into the cell. Running several cycles of purge–refill removes unwanted bubbling.
7. It is important to clean the ITC cells carefully after each experiment. A vacuum pump can be used for a thorough cleaning. Sometimes protein or ligand may adhere to the cells rather strongly. In such a case, strong acidic or basic reagents can be used to clean such cells. Suitable cleaning agents are recommended in the ITC manual.
8. Time intervals between successive injections must be sufficient to allow the DP signal to reach the baseline after each injection. This will depend on the system being studied as some reactions are completed faster than others; therefore, the interval period may require to be adjusted after performing one or two trial runs. However, for most systems reported in the literature, an interval of 3–5 min between successive injections was found to be adequate.
9. If the sample under investigation is not homogenous (e.g., a suspension), an increased stirring speed (rotation speed of injector syringe) can be used to in order to obtain more uniform data.
10. Very often it is found that the first injection is inaccurate, possibly because of manual loading of the injection syringe, due to which sample is not always filled to the same position of the needle tip. In order to minimize the error due to this, in the final analysis, usually a small volume (1–2 μl) of the sample is

added in the first injection, and the resulting data point is deleted prior to analyzing the data.

11. In single-injection mode calorimeters, it is necessary to perform a separate control experiment which must be subtracted from the experimental titration prior to analysis. The experimental parameters and samples used for the control experiment must be identical to those of the main experiment. If the lipid suspension (liposomes or micelles) is taken in the calorimeter cell and then titrated with the protein (in the syringe), then the blank titration is done by injecting exactly identical aliquots of the protein of identical concentration into the buffer in the ITC cell.

Acknowledgments

MJS acknowledges the Department of Science and Technology (India) for a research grant which supported the work done in his laboratory and referred to here. RSS thanks the CSIR (India) for a Senior Research Fellowship. We thank Dr. V. Anbazhagan for a fruitful collaboration in the ITC studies on protein–lipid interaction.

References

1. Pierce MM, Raman CS, Nall BT (1999) Isothermal titration calorimetry of protein–protein interactions. Methods 19:213–221
2. Velázquez-Campoy A, Ohtaka H, Nezami A, Muzammil S, Freire E (2004) Isothermal titration calorimetry. Curr Protoc Cell Biol. Chapter 17, Unit 17.8.
3. Weber PC, Salemme FR (2003) Applications of calorimetric methods to drug discovery and the study of protein interactions. Curr Opin Struct Biol 13:115–121
4. Privalov PL, Dragan AI (2007) Microcalorimetry of biological macromolecules. Biophys Chem 126:16–24
5. Buurma NJ, Haq I (2007) Advances in the analysis of isothermal titration calorimetry data for ligand–DNA interactions. Methods 42:162–172
6. Freyer MW, Lewis EA (2008) Isothermal titration calorimetry: experimental design, data analysis, and probing macromolecule/ligand binding and kinetic interactions. Methods Cell Biol 84:79–113
7. Ladbury JE (2010) Calorimetry as a tool for understanding biomolecular interactions and an aid to drug design. Biochem Soc Trans 38:888–893
8. Gopal B, Swaminathan CP, Bhattacharya S, Bhattacharya A, Murthy MR, Surolia A (1997) Thermodynamics of metal ion binding and denaturation of a calcium binding protein from *Entamoeba histolytica*. Biochemistry 36:10910–10916
9. De M, You CC, Srivastava S, Rotello VM (2007) Biomimetic interactions of proteins with functionalized nanoparticles: a thermodynamic study. J Am Chem Soc 129:10747–10753
10. Zhang L, Zheng Y, Xi Z, Luo Z, Xu X, Wang C, Liu Y (2009) Metal ions binding to recA inteins from Mycobacterium tuberculosis. Mol Biosyst 5:644–650
11. Chakraborty S, Joshi P, Shanker V, Ansari ZA, Singh SP, Chakrabarti P (2011) Contrasting effect of gold nanoparticles and nanorods with different surface modifications on the structure and activity of bovine serum albumin. Langmuir 27:7722–7731
12. Cooper MA (2004) Advances in membrane receptor screening and analysis. J Mol Recognit 17:286–315
13. Boggs JM, Stamp D, Moscarello MA (1981) Interaction of myelin basic protein with dipalmitoylphosphatidylglycerol: dependence on

the lipid phase and investigation of a metastable state. Biochemistry 20:6066–6072

14. Boggs JM, Wood DD, Moscarello MA (1981) Hydrophobic and electrostatic interactions of myelin basic proteins with lipid. Participation of N-terminal and C-terminal portions. Biochemistry 20:1065–1073
15. Sankaram MB, Brophy PJ, Marsh D (1989) Selectivity of interaction of phospholipids with bovine spinal cord myelin basic protein studied by spin-label electron spin resonance. Biochemistry 28:9699–9707
16. Maggio B, Sturtevant JM, Yu RK (1987) Effect of myelin basic protein on the thermotropic behavior of aqueous dispersions of neutral and anionic glycosphingolipids and their mixtures with dipalmitoylphosphatidylcholine. J Biol Chem 262:2652–2659
17. Desnoyers L, Manjunath P (1992) Major proteins of bovine seminal plasma exhibit novel interactions with phospholipids. J Biol Chem 267:10149–10155
18. Ramakrishnan M, Anbazhagan V, Pratap TV, Marsh D, Swamy MJ (2001) Membrane insertion and lipid–protein interactions of bovine seminal plasma protein, PDC-109 investigated by spin label electron spin resonance spectroscopy. Biophys J 81:2215–2225
19. Anbazhagan V, Sankhala RS, Singh BP, Swamy MJ (2011) Isothermal titration calorimetric studies on the interaction of the major bovine seminal plasma protein, PDC-109 with phospholipid membranes. PLoS One 6:e25993
20. Surolia A, Bachhawat BK, Podder SK (1975) Interaction between lectin from Ricinus communis and liposomes containing gangliosides. Nature 257:802–804
21. Grant CW, Peters MW (1984) Lectin-membrane interactions. Information from model systems. Biochim Biophys Acta 779:403–422
22. Ramalingam TS, Das PK, Podder SK (1994) Ricin-membrane interaction: membrane penetration depth by fluorescence quenching and resonance energy transfer. Biochemistry 33:12247–12254
23. Diab C, Tribet C, Gohon Y, Popot JL, Winnik FM (2007) Complexation of integral membrane proteins by phosphorylcholine-based amphipols. Biochim Biophys Acta 1768:2737–2747
24. Tribet C, Diab C, Dahmane T, Zoonens M, Popot JL, Winnik FM (2009) Thermodynamic characterization of the exchange of detergents and amphipols at the surfaces of integral membrane proteins. Langmuir 25:12623–12634
25. Seelig J (2004) Thermodynamics of lipid-peptide interactions. Biochim Biophys Acta 1666:40–50
26. MacDonald RC, MacDonald RI, Menco BP, Takeshita K, Subbarao NK, Hu L (1991) Small-volume extrusion apparatus for preparation of large, unilamellar vesicles. Biochim Biophys Acta 1061:297–303
27. Lowry OH, Rosebrough NJ, Farr AL, Randall RJ (1951) Protein measurement with the Folin phenol reagent. J Biol Chem 193:265–275
28. Bradford MM (1976) A rapid and sensitive method for the quantitation of microgram quantities of protein utilizing the principle of protein-dye binding. Anal Biochem 72:248–254
29. Noble JE, Knight AE, Reason AJ, Di Matola A, Bailey MJ (2007) A comparison of protein quantitation assays for biopharmaceutical applications. Mol Biotechnol 37:99–111
30. Edelhoch H (1967) Spectroscopic determination of tryptophan and tyrosine in proteins. Biochemistry 6:1948–1954
31. Gill SC, Von Hippel PH (1989) Calculation of protein extinction coefficients from amino acid sequence data. Anal Biochem 182:319–326
32. (1998) ITC Data Analysis in Origin®, Tutorial Guide Version 5.0. MicroCal, LLC, Northampton, MA, USA
33. Dimitrova MN, Matsumura H, Terezova N, Neytchev V (2002) Binding of globular proteins to lipid membranes studied by isothermal titration calorimetry and fluorescence. Colloids Surf B Biointerfaces 24:53–61
34. Arnulphi C, Jin L, Tricerri MA, Jonas A (2004) Enthalpy-driven apolipoprotein A-I and lipid bilayer interaction indicating protein penetration upon lipid binding. Biochemistry 43:12258–12264
35. Wiseman T, Williston S, Brandts JF, Lin LN (1989) Rapid measurement of binding constants and heats of binding using a new titration calorimeter. Anal Biochem 179:131–137

Chapter 4

Differential Scanning Calorimetry of Protein–Lipid Interactions

Olga Cañadas and Cristina Casals

Abstract

Differential scanning calorimetry (DSC) is a highly sensitive non-perturbing technique for measuring the thermodynamic properties of thermally induced transitions. This technique is particularly useful for the characterization of lipid/protein interactions. This chapter presents an introduction to DSC instrumentation, basic theory, and methods and describes DSC applications for characterizing protein effects on model lipid membranes. Examples of the use of DSC for the evaluation of protein effects on modulation of membrane domains and membrane stability are given.

Key words: Differential scanning calorimetry, Lipids, Proteins, Lipid–protein interaction, Antimicrobial peptides, Data analysis

1. Introduction

Differential scanning calorimetry (DSC) is an effective analytical tool to characterize the thermotropic properties of protein–lipid interactions. This chapter attempts to provide the reader with an understanding of the capabilities of DSC instrumentation and the type of information that can be achieved from DSC studies of lipid–protein interaction. In particular, we describe in detail the analysis of DSC data to assess the effects of proteins on model lipid bilayers.

In DSC experiments, the sample under study and an appropriate reference material (a material which does not undergo a phase transition within the temperature range of interest) are simultaneously heated or cooled at identical, predetermined rates. If the sample undergoes a phase transition induced by temperature, absorption (or release) of heat from the sample will occur. As a result, a temperature differential (ΔT) develops between the sample and reference cells. This temperature differential is minimized by the instrumental control system, which supplies more (or less) heat

Jörg H. Kleinschmidt (ed.), *Lipid-Protein Interactions: Methods and Protocols*, Methods in Molecular Biology, vol. 974,
DOI 10.1007/978-1-62703-275-9_4, © Springer Science+Business Media New York 2013

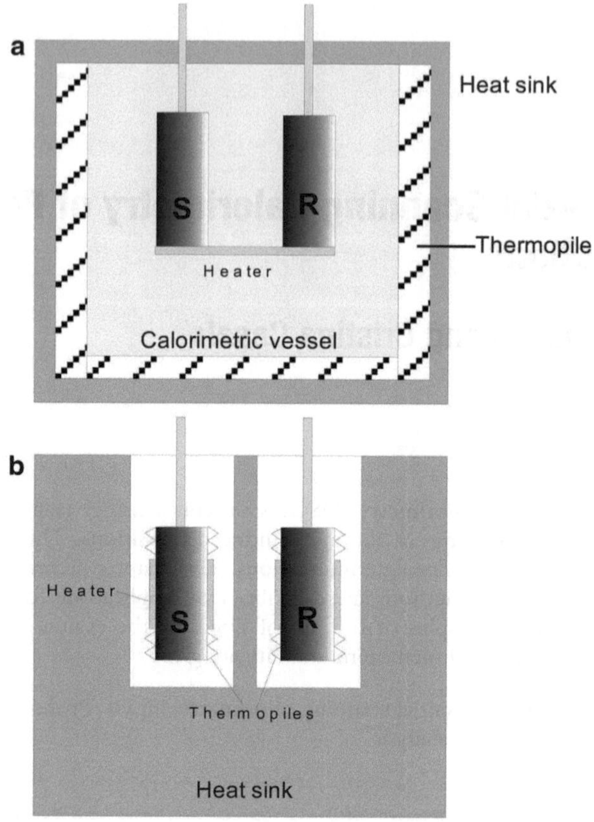

Fig. 1. Different types of differential scanning calorimeters: (**a**) heat conduction calorimeter; (**b**) power compensation calorimeter.

to the sample cell so as to maintain its temperature equal to that of the reference cell. Based on the mechanism of operation, DSC instruments can be classified into two types.

Heat Conduction Calorimeters: The calorimetric vessel is in thermal contact with a heat flow sensor which is placed on a heat sink, usually a constant temperature bath, and both sample and reference cells are heated or cooled indirectly by contact with the heat sink (Fig. 1a). The heat released (or absorbed) during the calorimetric event is allowed to flow to (or from) the surrounding heat sink. Usually, the sensor of the heat flow is a thermopile located between the sample cell and the heat sink. The temperature difference between the sample and reference cells and the heat sink results in an electrical power difference over the thermopile. The output differential power generated by the heat flow sensor is related to the flow of heat from the calorimetric vessel to the heat sink. These calorimeters have a large time constant of the measurement system. Therefore, it is necessary to operate at very slow scan rates to maintain thermal equilibrium.

Power Compensation Calorimeters: Sample and reference cells are surrounded by a cooling jacket that is maintained some 10–20°C below the temperature of the calorimetric vessel. Sample and reference cells are heated (or cooled) by independent heat sources (Fig. 1b). Both cells are maintained at the desired temperature by continuously manipulating the power supplied to an electrical heating element placed within the vessel. At the beginning of the experiment, cooling power and electrical heat are in balance. As the calorimetric event occurs, the electrical power is varied in order to maintain the desired process temperature. The difference between the electrical power at the beginning and at any time during the measurement is proportional to the heat absorbed or liberated by the calorimetric process. These calorimeters have a very short response time and operate at considerably faster scanning rates than those of heat conduction instruments.

DSC instruments are characterized by high sensitivity, a stable instrumental baseline, and a wide operational range, being able to scan aqueous solutions from 0°C up to 100°C. To supercool the sample below 0°C, the instrument must be equipped with an automatic system that heats the cells to avoid the freezing of water. On the other hand, to heat the samples above 100°C, an excess pressure of 2 atm should be applied to instruments equipped with pressure-resistant sample cells. This overpressure allows raising the temperature to 120°C and dissolves bubbles forming in the solution upon heating. Since DSC instruments compare the heat capacities of identical volumes (1), it is important to avoid the formation of bubbles that would affect the accuracy of the volume.

A calorimetric experiment usually consists of three or more scans. The first one corresponds to a buffer scan; successive scans are of the sample (see Note 1). These scans are performed to assess the reversibility of the sample since DSC data analysis is performed on equilibrium data, and samples that undergo irreversible transitions cannot be analyzed by equilibrium thermodynamics-based approaches. The equilibrium criterion usually applied is the reproducibility of the sample DSC trace in a second heating.

1.1. Lipid Thermotropic Phase Behavior

The output of any DSC experiment is the heat flow rate as a function of time. The time integral of the measured differential heat flow provides the energy of the sample (2). Figure 2 shows the typical DSC thermogram for the two-state gel (L_β)-to-liquid crystalline (L_α) phase transition exhibited by an aqueous dispersion of a single phospholipid. This phase transition is characterized by a number of thermodynamic parameters, that is, the phase transition temperature (T_m), the relative cooperativity of the phase transition, and the enthalpy of the calorimetric event (ΔH) that can be directly determined from the thermogram. The phase transition temperature, T_m, for lipid hydrocarbon chain-melting transitions corresponds to the maximum excess specific heat absorbed by the

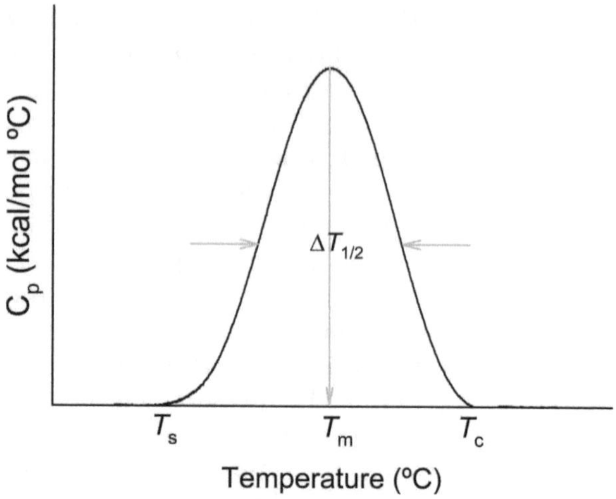

Fig. 2. DSC thermogram for a gel-to-liquid crystalline endothermic phase transition of a pure phospholipid that has reached a thermodynamic equilibrium. T_m, the transition midpoint temperature; $\Delta T_{1/2}$, the transition with at half-height; *Ts*, the onset or lower boundary of the phase transition; *Tc*, the completion or upper boundary temperature.

system (3–5). For lipid samples that have reached a thermodynamic equilibrium and hence show a symmetric DSC trace, T_m represents the temperature at which the gel-to-liquid crystalline phase transition is half complete (6) (see Note 2). At any point under the transition endotherm, the sample comprises a mixture of lipids in the gel and fluid phases. Below the onset temperature (T_s), the membrane is in the gel state, and above the completion temperature (T_L), the sample is in the liquid crystalline phase.

The relative cooperativity of a phase transition can be determined directly from the DSC curve since it is related to the sharpness of the transition peak, which is given by the temperature width at half-height, $\Delta T_{1/2}$, of the DSC trace. Pure single lipids usually have $\Delta T_{1/2}$ values less than 0.1°C. The presence of impurities within the bilayers broadens the transition peak. Thus, for complex lipid mixtures $\Delta T_{1/2}$ reaches values of 10–15°C (3–5). The cooperativity of a pure lipid transition is described by the cooperative unit (CU) that measures the degree of cooperation between lipid molecules. CU reaches a zero value for noncooperative processes, whereas large values of CU indicate a high cooperativity (3).

The enthalpy of the endothermic or exothermic event is evaluated by the integration of the area of the scan rate-normalized DSC peak relative to the experimental baseline (3, 4, 7).

In addition to the gel-to-liquid crystalline lamellar phases, other metastable lipid phases can occur. Examples are the pre-transition from a gel to a rippled gel ($P_{\beta'}$) phase of disaturated phosphatidylcholines and the lamellar to inverted hexagonal (H_{II}) phase transition of unsaturated phospholipids that contain small, strongly interacting polar headgroups such as phosphatidylethanolamines.

These thermotropic interconversions do not proceed under equilibrium conditions but are kinetically controlled, that is, these processes approach thermodynamic equilibrium at a slower rate than the temperature scanning rate. As a consequence, both the shape and the transition temperature of the phase transition vary with the scanning rate, and the only accessible thermodynamic parameter is the enthalpy of the process. Thus, to obtain reliable data from these types of lipid phase transitions, it is necessary to perform the experiments at slow heating rates.

1.2. Theory

Differential scanning calorimetry measures the excess heat capacity, C_p, of the sample solution with respect to the reference material, which usually is the same buffer used in the preparation of the sample solution. The calorimetric enthalpy of the transition, ΔH_{cal}, can be determined by integrating the DSC peak:

$$\Delta H_{cal} = \int C_p \cdot dT \tag{1}$$

and the entropy of the transition is given by

$$\Delta S = \frac{\Delta H_{cal}}{T_m} \tag{2}$$

where T_m is the transition temperature.

The dependence of the equilibrium constant, K, for a simple two-state transition ($A \leftrightarrows B$) on enthalpy is given by the van't Hoff equation:

$$\left(\frac{\partial \ln K}{\partial T}\right)_p = \frac{\Delta H_{vH}}{RT^2} \tag{3}$$

where ΔH_{vH} is the van't Hoff enthalpy and R is the gas constant.

Since the standard free energy change that occurs during the transition, ΔG^0, is related to the standard enthalpy and entropy changes:

$$\Delta G^0 = \Delta H^0 - T\Delta S^0 \tag{4}$$

and

$$\Delta G^0 = -RT \ln K \tag{5}$$

it follows that

$$\ln K = -\frac{\Delta H^0}{RT} + \frac{\Delta S^0}{R} \tag{6}$$

Therefore, a plot of the natural logarithm of K versus $1/T$ gives a straight line, the slope of which is equal to minus the van't Hoff enthalpy divided by the gas constant, $-\Delta H^0/R$, and the intercept corresponds to the standard entropy change divided by the gas constant, $\Delta S^0/R$ (see Note 3).

The van't Hoff enthalpy is the amount of heat required for each cooperative unit to undergo the phase transition. Comparison of ΔH_{vH} with the calorimetric enthalpy provides information about the nature of the transition. For a first-order two-state transition, ΔH_{vH} is equal to the calorimetric enthalpy, whereas for systems that involve intermediate stages, the van't Hoff enthalpy is smaller than the calorimetric enthalpy. Finally, when oligomers are involved in the thermotropic transition, the van't Hoff enthalpy is larger than the calorimetric enthalpy.

The number of molecules (or moles) per cooperative unit, CU, is given by the ratio:

$$CU = \frac{\Delta H_{vH}}{\Delta H_{cal}} \tag{7}$$

Comparison of the CU values for phospholipid bilayers, in the absence and presence of different amounts of reconstituted protein, allows evaluation of the effects of the protein on the cooperativity of the phase transition.

2. Materials

2.1. Equipment

1. DSC instrument (MicroCal VP-DSC, TA NanoDSC, Setaram μDSC7 evo, or equivalent).
2. Degassing equipment (vacuum desiccator, magnetic stirrer).
3. Dialysis tubing or cassettes (e.g., Pierce Slide-A-Lyzer Dialysis Cassettes).
4. Polystyrene beads (e.g., Sigma-Aldrich Amberlite XAD-25).
5. Gel filtration columns (e.g., Sephadex G-50).
6. Centrifuge (Beckman XL-90, or equivalent).

2.2. Reagents

2.2.1. Buffer

A buffer with a small or zero enthalpy of ionization is preferable since this eliminates any ionization contributions in the measurement. Buffers such as phosphate, formate, and acetate are ideal since they have negligible ionization heats and their pH remains constant with changes in temperature. However, phosphate buffer should not be used when the phospholipid concentration is determined by phosphorus assay, since the presence of inorganic phosphorus in the buffer would interfere in the measurement. The pH and ionic strength of the buffer solution should be chosen to avoid sample aggregation (see Note 4). Although the choice of buffer depends on the type of study, the buffer most widely used to study lipid–protein interactions is 10 mM HEPES, 0.1 mM NaCl, pH 7.4.

2.2.2. Organic Solvents

Usually, a chloroform/methanol mixture is used to dissolve the lipids. When the reconstitution of hydrophobic membrane proteins into proteoliposomes is performed by co-solving lipids and protein in organic solvents, the chosen solvents should affect neither the structure nor the function of the protein.

2.2.3. Detergent

The most common detergents used for protein reconstitution into proteoliposomes are octylglucoside, CHAPS, or Triton X-100, since these mild detergents preserve the native biological structure and activity of the proteins.

2.2.4. Sample

The lipid component should exhibit phase transition temperatures below the thermal denaturation temperature of the protein of interest (see Note 5).

3. Methods

3.1. Sample Preparation

Given the variety of lipids and proteins that can be studied, each one with its own thermodynamic characteristics, there is no single protocol optimal for the preparation and characterization of all the possible lipoprotein complexes that can be formed. Therefore, we present some general guidelines for sample preparation.

3.1.1. Preparation of Proteoliposomes

Only the purest lipid and protein samples should be used in lipid–protein interaction studies. Phospholipid vesicles that contain integral membrane proteins are known as proteoliposomes and are an excellent tool for studying the structure and function of membrane proteins as well as protein effects on the physical properties of membranes. Some extremely hydrophobic membrane proteins are reconstituted into proteoliposomes by co-solving both components in an organic solvent that is subsequently evaporated to dryness under a stream of nitrogen or argon, leaving a lipid–protein film. This film is rehydrated for 1 h in an aqueous buffer at a temperature above the gel-to-liquid phase transition temperature of the lipid or lipid mixture to form multilamellar vesicles (MLVs) (see Note 6). However, most integral membrane proteins are isolated and reconstituted using detergents. There are two common approaches for reconstituting proteoliposomes via a detergent-mediated pathway (8) which involve the transfer of an integral membrane protein from a detergent solution to phospholipid vesicles. The first is a simple dilution approach: If a protein–detergent mixture is diluted in a liposome solution, the protein would transfer into the liposome when the concentration of detergent falls below its critical micelle concentration (CMC). The second approach involves introducing detergent to preformed liposomes such that the liposome bilayer becomes saturated with detergent. The detergent disrupts lipid–lipid interactions, which results in a more permeable bilayer. These structures are more receptive to protein uptake, and after protein is introduced, the excess detergent can be removed by several methods (dialysis, column chromatography, or incubation with detergent-adsorbing beads). Dialysis is done against the hydration buffer devoid of detergents, the dialysate is centrifuged, and the pellet is resuspended in the buffer of interest. For detergent adsorption, polystyrene beads such as Amberlite XAD-25 can be used. Finally, for gel filtration the lipoprotein complex is passed through a gel filtration column that has been previously equilibrated with the buffer used for film hydration devoid of detergent. Turbid fractions containing the lipoprotein complex are collected, the sample centrifuged, and the pellet suspended in the buffer of interest.

3.1.2. Proteoliposome Characterization

The mode and extent of interaction of membrane proteins with phospholipid vesicles might depend on the conditions of preparation of lipid–protein samples (9), and care should be taken in the interpretation of findings from reconstituted systems. Prior to the DSC measurement, it is necessary to determine the chemical composition of the reconstituted proteoliposomes and characterize their morphology (9). It is also necessary to check the presence of residual solvents or detergents in the sample (see Note 7).

3.1.3. Sample Degassing

To avoid bubble formation, which would affect the accuracy of the volume and add experimental noise to the thermogram, it is necessary to degas under vacuum with gentle stirring both the sample and reference solutions (see Note 8).

3.2. DSC Procedure

3.2.1. Loading the Calorimeter

Load the degassed buffer solution into the sample and reference cells and scan the calorimeter up and down several times until the baseline stabilizes. Without stopping scanning, replace the solvent in one of the calorimetric cells with the sample solution. This must be done when the instrument is cooling down the sample and the temperature of the cells is close to room temperature.

3.2.2. Application of Excess Pressure

Apply an excess pressure not lower than 2 atm to the sample and reference solutions.

3.3. Data Analysis

3.3.1. Reference Substraction

In most experiments both sample and reference data are obtained, hence the first step in data analysis is to subtract the reference from the sample data.

3.3.2. Baseline Substraction

After normalizing the result by concentration, the following step is to subtract a pre-transition to post-transition baseline from the overall curve. Due to ΔC_p effects, the pre- and post-transition baselines usually differ. This makes it necessary to estimate the behavior of the baseline in the region between the onset and completion of the transition, where a mixture of lipid molecules in different phase states occurs. To that end, the temperature range scanned during a DSC assay should be broad enough for the instrument to establish stable baselines before and after the transition peak. Different approaches can be used to establish a proper baseline, as the progress, step, linear, or quadratic interpolations of pre- and post-transition baselines. Given that each possible baseline somehow affects the apparent shape and area under the transition, which in turn results in errors in ΔH_{cal} and ΔH_{vH}, it is desirable to check several baseline procedures to get an idea of the uncertainties derived from each baseline type.

3.3.3. Data Deconvolution

The final step is to deconvolute the data to obtain the thermotropic parameters T_m, $\Delta T_{1/2}$, and ΔH_{cal}, from which it is possible to calculate the van't Hoff enthalpy, entropy, and Gibbs free energy of the transition.

To properly characterize the effect of a protein on the thermotropic behavior of a lipid bilayer, it is necessary to determine, in the first place, the thermodynamic parameters of the lipids alone, that is, T_m, $\Delta T_{1/2}$, and ΔH_{cal}.

3.4. Data Interpretation

The sensibility of the phase transitions to the presence of exogenously added compounds has converted DSC in a valuable tool to study lipid–protein interactions. Monitoring changes in T_m, $\Delta T_{1/2}$, and ΔH can yield information concerning the ability of proteins to interact with lipid membranes and affect the lipid acyl chain packing, which provides insight into their interaction mechanism.

According to Papahadjopoulos (10), the binding mode of the lipid–protein interaction can be determined from the effect of the protein on the gel-to-liquid crystalline lipid phase transition. Thus, hydrophilic proteins that interact with the bilayer surface through electrostatic forces produce no change or a slight increase in the T_m and the $\Delta T_{1/2}$ of the transition, and a substantial increase in ΔH in a dose-dependent manner. These proteins normally do not expand phospholipid monolayers, nor alter the permeability of phospholipid vesicles into which they are incorporated. These peripheral membrane proteins are thought to interact with the surface of the phospholipid bilayer exclusively by electrostatic forces. Hence, protein interaction induces stronger effects on the phase transitions of negatively charged phospholipids rather than on zwitterionic phospholipids bearing no net charge. An example of this behavior would be the interaction of pulmonary surfactant protein A (SP-A) with surfactant-like membranes (11). The interaction of SP-A with liposomes consisting of 1,2-dipalmitoyl-phosphatidylcholine (DPPC)/1-palmitoyl-2-oleoyl-phosphatidylglycerol (POPG)/palmitic acid (PA) (28:9:5.6, w/w/w) leads to a protein concentration-dependent increase in the T_m (Fig. 3). Moreover, SP-A leads to a narrowing of the phase transition, indicating that this peripheral protein mediates stabilization of DPPC-rich assemblies (11).

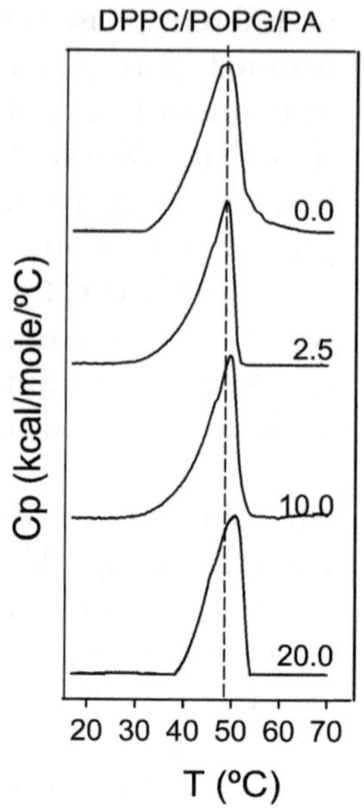

Fig. 3. Effect of SP-A on DSC heating scans of DPPC/POPG/PA vesicles (1 mM). The % SP-A/lipid weight ratio is indicated on each thermogram. Calorimetric scans were performed at a rate of 0.5°C/min. For clarity of presentation, the tick labels of heat capacity (*Cp*) are not shown. For each thermogram, the scale of *Cp* was from 0 to 1.4. One representative experiment of three experiments is shown (Adapted from (11), © Cell Press, 2007).

Another example is the interaction of the basic polypeptide, poly (L-lysine), with phosphatidylglycerol membranes that leads to a substantial increase in ΔH in a dose-dependent manner (4). However, with other anionic phospholipids, the ΔH of the phospholipid gel-to-liquid crystalline phase transition is reduced rather than increased. McElhaney (4) suggests that in the poly(L-lysine)-phospholipid interaction two competing effects occur: an electrostatic, charge-neutralization interaction at the bilayer surface, which condenses and stabilizes the gel-state phospholipid bilayer, and a hydrophobic interaction due to a partial penetration of the amino acid side chains into the lipid bilayer core, which decreases the ΔH as the amount of polypeptide added is increased.

In fact, if the binding of the peripheral membrane protein to the bilayer surface is followed by the partial penetration of the protein into the bilayer, where it interacts with the lipid hydrocarbon chains, a decrease in T_m and/or ΔH is observed, which might be explained by a decrease of van der Waals interactions between lipid acyl chains after protein insertion. These proteins are believed to interact with phospholipid bilayers by a combination of electrostatic and hydrophobic forces. These peripheral proteins initially adsorb to the charged polar head groups of the phospholipids and subsequently partially penetrate the hydrophilic–hydrophobic

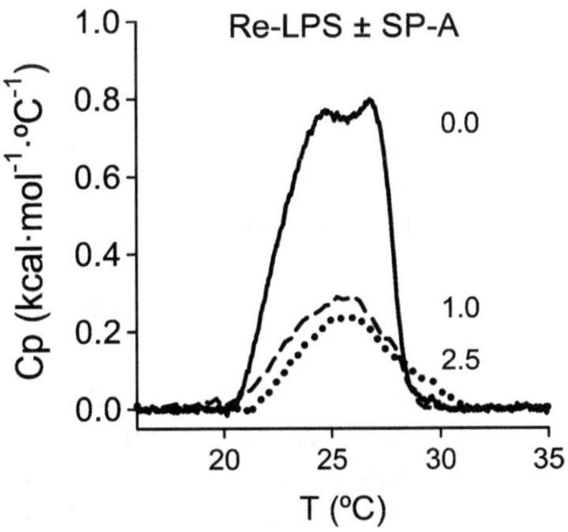

Fig. 4. Effect of SP-A on the DSC heating scans of Re-LPS membranes (0.4 mM) in 5 mM Tris–HCl, 150 mM NaCl, pH 7.4, containing 150 μM $CaCl_2$. Experiments were performed in the presence of different SP-A/Re-LPS weight ratios: 0 % (*solid line*), 1 % (*dotted line*), and 2.5 % (*filled circles*). Calorimetric scans were performed at a rate of 0.5 °C/min. Data shown are the means of four thermograms (Adapted from (13), © Cell Press, 2008).

interface of the bilayer to interact with a portion of the lipid hydrocarbon chains. These proteins normally expand phospholipid monolayers and alter the permeability of phospholipid vesicles into which they are incorporated. A good example of this type of proteins is the C-reactive protein (CRP) (12), which is a plasma protein considered the prototypical acute-phase protein in humans and other animal species. The presence of packing defects produced by coexisting ordered/disordered domains within membranes facilitates the partial insertion of CRP in the membrane (12). Another example is the effect of pulmonary surfactant protein A (SP-A) on model bacterial membranes composed of deep rough lipopolysaccharide (Re-LPS) (13). SP-A favors the penetration of water molecules in Re-LPS membranes and causes a sharp decrease of the overall transition enthalpy of these membranes (Fig. 4). Such SP-A effects are due to protein-induced perturbation of the lipid packing. SP-A induces Re-LPS molecular loss by promoting the formation of calcium-mediated protein aggregates that contain LPS. This would result in decreased van der Waals interactions between LPS hydrocarbon chains, consistent with SP-A-mediated decrease in the overall transition enthalpy of Re-LPS membranes observed by DSC. Other examples of peripheral proteins that reduce the T_m and significantly reduce the ΔH of the chain-melting transition of anionic phospholipids include the myelin basic protein and cytochrome c, indicating that hydrophobic as well as electrostatic lipid–protein interactions are important in these systems (4).

Finally, hydrophobic integral membrane proteins, which deeply penetrate into the hydrophobic core of the lipid bilayers, prevent some lipid molecules from participating in the melting transition. As a consequence, these proteins have little effect on T_m, but reduce the cooperativity of gel-to-liquid crystalline phase transitions and decrease ΔH usually linearly, with increasing protein concentration. In this case, the dependence of the transition enthalpy on the protein/lipid molar ratio is given by (14):

$$\Delta H = \Delta H_0 \cdot \left[1 - m \cdot \frac{P}{L}\right] \quad (8)$$

where ΔH_0 and ΔH are the transition enthalpies in the absence and presence of protein, respectively; P/L is the protein/lipid molar ratio; and m is the number of phospholipid molecules prevented from participating in the gel-to-liquid crystalline phase transition by one molecule of protein.

The number of phospholipid molecules that do not participate in the transition can be determined by extrapolating ΔH to zero in the ΔH versus protein/lipid molar ratio plot (see Note 9).

The decrease of the T_m induced by the protein, known as the van't Hoff freezing point depression, can be estimated by the following equation:

$$\Delta T_m = -\left(\frac{RT_m^2}{\Delta H}\right)X \quad (9)$$

where ΔH is the lipid-melting enthalpy of the lipid bilayers, R the universal gas constant, T_m the lipid-melting temperature, and X the molar fraction of the protein or peptide incorporated into the membrane. In the case of nonideal mixing:

$$\Delta T_m = -\left(\frac{RT_m^2}{\Delta H}\right)X(1 - wX) \quad (10)$$

where w is the nonideal mixing parameter.

There are a number of integral transmembrane proteins that exhibit this behavior (increase in $\Delta T_{1/2}$ and substantial decrease in ΔH). These include the VSV G protein (15), myelin proteolipid protein (3, 16), glycophorin (14), bacteriorhodopsin (17), cytochrome oxidase (18), cytochrome b_5 (19), (Ca^{2++} Mg^{2+})–ATPase (20), the transmembrane pulmonary surfactant peptide SP-C (21), and the monotopic integral membrane surfactant protein SP-B (9).

Information on miscibility of lipids and proteins can also be obtained by studying the effect of proteins on the transition temperature and shape of the DSC peaks. If one protein exhibits ideal mixing with both lipid phases, a symmetric broadening of the DSC curve is observed (22). Conversely, if the protein does not bind well with both lipid phases, it aggregates into clusters at all temperatures, reducing its interface with the lipids and slightly affecting the transition. For proteins that mix well with the gel phase, the thermogram shifts upward, broadening the high-temperature side of

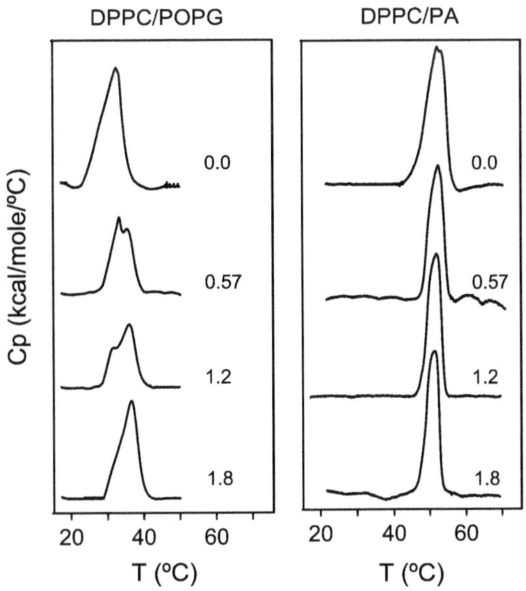

Fig. 5. DSC heating scans of DPPC/POPG and DPPC/PA multilamellar vesicles (1 mM) in the absence and presence of different concentrations of KL4. The mole percentage of KL4 is indicated on each thermogram. Calorimetric scans were performed at a rate of 0.5 °C/min (Adapted from (23), © Wiley, 2006).

the transition. In the case of proteins that show good mixing with the fluid phase lipids, the protein accumulates in the fluid phase, shifting the DSC curve to lower temperatures and broadening the low-temperature side of the transition.

The specific interaction of a protein or peptide with a phospholipid class could induce the formation of domains within lipid membranes. The appearance of shoulders or double peaks in the DSC trace of lipid mixtures indicates that the protein has a higher affinity for one lipid species (23, 24). Figure 5 shows an example of the application of DSC to investigate peptide-induced redistribution of lipids in DPPC/POPG bilayers (23). DSC thermograms indicated that, in the presence of a cationic and hydrophobic 21-residue peptide KLLLLKLLLLKLLLLKLLLLK, KL4, the thermal transition of DPPC/POPG vesicles is characterized by a double peak, which is not observed in DPPC/PA vesicles. It must be generated by electrostatic interactions between the positively charged lysine residues of KL4 and the anionic headgroup of POPG. Another example is that of penetratin, an antimicrobial peptide that induces the appearance of a second peak at around 41°C in DPPC/cardiolipin bilayers (24). Without the peptide, these vesicles show a single, broad phase transition due to lipid intercalation, with a Tm of 28°C. The penetratin-induced second peak corresponds to the temperature at which the phase transition of pure DPPC occurs. This indicates that penetratin preferentially binds and recruits cardiolipin, leaving a domain highly enriched in DPPC.

The stabilizing or destabilizing effect of a peptide or protein is related to the induction of a curvature strain in the membrane. DSC can be used to assess this curvature stress exerted on the membrane. In particular, the lamellar (L_α) to type II hexagonal (H_{II}) transition of suitable model lipids, such as 1-palmitoyl-2-oleoyl-phosphatidylethanolamine (POPE), is usually studied as the lamellar phase has no curvature, whereas the type II hexagonal phase is negatively curved. Compounds that stabilize the membrane by favoring the lamellar phase increase the L_α–H_{II} phase transition temperature, shifting the formation of the H_{II} phase to a higher temperature, whereas a downward shift of the L_α–H_{II} phase transition indicates the promotion of the H_{II} phase (25), that is, the destabilization of the membrane.

4. Conclusions

The diversity of lipid thermotropic phases and the different ways in which proteins and peptides can affect them constitute the main problem in the application of DSC to the study of lipid/protein interactions. In order to optimize the protocols for data acquisition, analysis, and interpretation, the user should have general knowledge of the thermodynamically and kinetically controlled lipid phase transitions as well as the capabilities and limitations of the available instruments.

5. Notes

1. The first scan is less reproducible than the following scans because the DSC instrument has a different thermal history prior to this scan. Therefore, the first scan usually corresponds to a buffer scan, and the next scans correspond to the sample under study. To obtain comparable results during a series of related experiments, each sample solution should be treated exactly the same. In order to improve the repeatability of the thermal scanning, it is necessary to allow the DSC instrument to remain in cycle, programming identical scan parameters: that is, equilibration prior to the run, scan rate, starting and final temperatures, filtering period, gain, cooldown, and re-equilibration. This ensures that the DSC thermal history is the same from the 2nd to the *n*th scan, which increases the repeatability of scans.

   For pure lipids that exhibit a sharp transition, scan rates of around 0.1 K/min or less, filter periods of 1–5 s, and high gain feedback are usually appropriate.

2. When the phase transition does not proceed under equilibrium conditions, as in many lipid mixtures, the DSC curve is asymmetric, and the T_m and the phase transition midpoint temperature do not coincide (3).
3. This plot is linear only in certain simple cases and over relatively narrow ranges of temperature. In general, a curvature of the van't Hoff plot is observed due to the temperature dependence of ΔH. It is necessary to subject the data to nonlinear least squares analysis employing the integrated form of the van't Hoff equation (26):

$$\ln \frac{K}{K_0} = \frac{\Delta H_0 - T_0 \Delta C_p}{R} \left(\frac{1}{T_0} - \frac{1}{T} \right) + \frac{\Delta C_p}{R} \ln \frac{T}{T_0} \qquad (11)$$

where ΔC_p is the temperature-dependent heat capacity change, T_0 is an arbitrarily selected reference temperature, and K_0 and ΔH_0 are the equilibrium constant and the van't Hoff enthalpy at that temperature.
4. Sample aggregation causes a rapid downward shift and/or an increase in baseline noise, hampering data analysis. This can be reduced by decreasing the concentration, changing the pH, and/or using a different buffer.
5. Usually, DSC studies have been performed with model bilayers composed of a single phospholipid class, that is, phosphatidylcholine to model eukaryotic cell membranes and anionic phospholipids to model bacterial cell membranes. However, lipid mixtures should be used to examine the role of different parameters, such as electrostatic interactions, lipid bilayer fluidity, or the preferential interaction of a protein with a certain lipid component, on the lipid–protein interaction. When designing a lipid mixture, it is necessary to consider the miscibility of the lipids since not all the lipids are miscible.
6. Multilamellar vesicles are usually used for DSC studies since they provide the clearest resolution of the phase transitions (7, 27).
7. For proteins that partition between water and the membrane, the concentration of protein that interacts with the membrane decreases depending on the lipid/water ratio. Since the accuracy of the estimated ΔH_{cal} depends critically on the purity of the sample, it is necessary to remove any uncomplexed lipid or protein and determine the lipid/protein ratio of the reconstituted lipoprotein complex. Protein concentration is usually determined by measuring the absorbance at 280 nm, provided a reliable molar extinction coefficient is available, whereas lipid concentration is evaluated through a phosphorus assay or by inclusion of trace quantities of ^{14}C lipids in the reconstitution mixture followed by liquid scintillation counting of the reconstituted proteoliposomes. On the other hand, because of

the preparation procedures of the proteoliposomes, the thermotropic properties of the reconstituted lipoprotein complex can be affected by detergent or solvent traces (3). Therefore, it is necessary to check the proteoliposomes for the presence of residual organic solvent or detergent molecules. Finally, alterations in lipid vesicle size and homogeneity induced by the reconstitution procedure or the presence of the protein can modify the thermodynamic properties of lipid chain-melting transitions (3). Hence, to prevent the failure of the DSC study, it is also necessary to characterize the morphology of the lipoprotein complex.

8. The amount of air that can be dissolved in water decreases with the temperature. Thus, the sample should be equilibrated at room temperature before the degassing is started. If the sample is being stirred, it will take only 5 min of vacuum to completely degas the sample and reference solutions.
9. Sometimes, positive deviations of linearity are observed. This indicates that either the protein aggregates within the bilayers, and hence the number of lipid molecules that are affected by the protein decreases, or the conformation of the hydrophilic regions of the protein has changed, modifying their interaction with the bilayer.

Acknowledgments

This work was supported by the Ministerio de Ciencia e Innovación (SAF2009-07810) and Instituto de Salud Carlos III (cibeRES-CB06/06/0002).

References

1. Privalov PL, Potekhin SA (1986) Scanning microcalorimetry in studying temperature-induced changes in proteins. Methods Enzymol 131:4–51
2. Privalov PL, Dragan AI (2007) Microcalorimetry of biological macromolecules. Biophys Chem 126:16–24
3. McElhaney RN (1982) The use of differential scanning calorimetry and differential thermal analysis in studies of model and biological membranes. Chem Phys Lipids 30:229–259
4. McElhaney RN (1986) Differential scanning calorimetric studies of lipid-protein interactions in model membrane systems. Biochim Biophys Acta 864:361–421
5. Demetzos C (2008) Differential scanning calorimetry (DSC): a tool to study the thermal behavior of lipid bilayers and liposome stability. J Liposome Res 18:159–173
6. Lewis RN, Mannock DA, McElhaney RN (2007) Differential scanning calorimetry in the study of lipid phase transitions in model and biological membranes: practical considerations. Methods Mol Biol 400:171–195
7. Biltonen RL, Lichtenberg D (1993) The use of differential scanning calorimetry as a tool to characterize liposome preparations. Chem Phys Lipids 64:129–142
8. Seddon AM, Curnow P, Booth PJ (2004) Membrane proteins, lipids and detergents:

not just a soap opera. Biochim Biophys Acta 1666:105–117

9. Cruz A, Casals C, Keough KM, Pérez-Gil J (1997) Different modes of interaction of pulmonary surfactant protein SP-B in phosphatidylcholine bilayers. Biochem J 327:133–138
10. Papahadjopoulos D, Moscarello M, Eylar EH, Isac T (1975) Effects of proteins on thermotropic phase transitions of phospholipid membranes. Biochim Biophys Acta 401:317–335
11. Sáenz A, Cañadas O, Bagatolli LA, Sánchez-Barbero F, Johnson ME, Casals C (2007) Effect of surfactant protein A (SP-A) on the physical properties and surface activity of KL4-surfactant. Biophys J 92:482–492
12. Sáenz A, Lopez-Sanchez A, Mojica-Lazaro J, Martinez-Caro L, Nin N, Bagatolli LA, Casals C (2010) Fluidizing effects of C-reactive protein on lung surfactant membranes: protective role of surfactant protein A. FASEB J 24:3662–3673
13. Cañadas O, García-Verdugo I, Keough KMW, Casals C (2008) SP-A permeabilizes lipopolysaccharide membranes by forming aggregates that extract lipids from the membrane. Biophys J 95:3287–3294
14. van Zoelen EJ, van Dijck PW, de Kruijff B, Verkleij AJ, van Deenen LL (1978) Effect of glycophorin incorporation on the physicochemical properties of phospholipid bilayers. Biochim Biophys Acta 514:9–24
15. Petri WA, Estep TN, Pal R, Thompson TE, Biltonen RL, Wagner RR (1980) Thermotropic behavior of dipalmitoylphosphatidylcholine vesicles reconstituted with the glycoprotein of vesicular stomatitis virus. Biochemistry 19:3088–3091
16. Boggs JM, Moscarello MA (1978) Dependence of boundary lipid on fatty acid chain length in phosphatidylcholine vesicles containing a hydrophobic protein from myelin proteolipid. Biochemistry 17:5734–5739
17. Heyn MP, Blume A, Rehorek M, Dencher NA (1981) Calorimetric and fluorescence depolarization studies on the lipid phase transition of bacteriorhodopsin–dimyristoylphosphatidylcholine vesicles. Biochemistry 20:7109–7115
18. Semin BK, Saraste M, Wikström M (1984) Calorimetric studies of cytochrome oxidase-phospholipid interactions. Biochim Biophys Acta 769:15–22
19. Freire E, Markello T, Rigell C, Holloway PW (1983) Calorimetric and fluorescence characterization of interactions between cytochrome b5 and phosphatidylcholine bilayers. Biochemistry 22:1675–1680
20. Gómez-Fernández JC, Goñi FM, Bach D, Restall C, Chapman D (1979) Protein–lipid interactions. A study of (Ca2+–Mg2+)ATPase reconstituted with synthetic phospholipids. FEBS Lett 98:224–228
21. Plasencia I, Cruz A, Lopez-Lacomba JL, Casals C, Perez-Gil J (2001) Selective labeling of pulmonary surfactant protein SP-C in organic solution. Anal Biochem 296:49–56
22. Ivanova VP, Makarov IM, Schaffer TE, Heimburg T (2003) Analyzing heat capacity profiles of peptide-containing membranes: cluster formation of gramicidin A. Biophys J 84:2427–2439
23. Sáenz A, Cañadas O, Bagatolli LA, Johnson ME, Casals C (2006) Physical properties and surface activity of surfactant-like membranes containing the cationic and hydrophobic peptide KL4. FEBS J 273:2515–2527
24. Joanne P, Galanth C, Goasdoue N, Nicolas P, Sagan S, Lavielle S et al (2009) Lipid reorganization induced by membrane-active peptides probed using differential scanning calorimetry. Biochim Biophys Acta 1788:1772–1781
25. Powers JP, Tan A, Ramamoorthy A, Hancock REW (2005) Solution structure and interaction of the antimicrobial polyphemusins with lipid membranes. Biochemistry 44:15504–15513
26. Naghibi H, Tamura A, Sturtevant JM (1995) Significant discrepancies between van't Hoff and calorimetric enthalpies. Proc Natl Acad Sci U S A 92:5597–5599
27. Mason JT, Huang CM, Biltonen RL (1983) Effect of liposomal size on the calorimetric behavior of mixed-chain phosphatidylcholine bilayers dispersions. Biochemistry 22:2013–2018

Chapter 5

Atomic Force Microscopy and Spectroscopy to Probe Single Membrane Proteins in Lipid Bilayers

K. Tanuj Sapra

Abstract

The atomic force microscope (AFM) has opened vast avenues hitherto inaccessible to the biological scientist. The high temporal (millisecond) and spatial (nanometer) resolutions of the AFM are suited for studying many biological processes in their native conditions. The AFM cantilever stylus is aptly termed as a "lab on a tip" owing to its versatility as an imaging tool as well as a handle to manipulate single bonds and proteins. Recent examples assert that the AFM can be used to study the mechanical properties and monitor processes of single proteins and single cells, thus affording insight into important mechanistic details. This chapter specifically focuses on practical and analytical protocols of single-molecule AFM methodologies related to high-resolution imaging and single-molecule force spectroscopy of membrane proteins. Both these techniques are operator oriented, and require specialized working knowledge of the instrument, theoretical, and practical skills.

Key words: Atomic force microscope, Cantilever, Force spectroscopy, High-resolution imaging, Lipid bilayer, Membrane protein, Single molecule

1. Introduction

1.1. Atomic Force Microscope

In the early 1980s, a simple and elegant demonstration by Binnig and Rohrer marked the beginning of a new scientific era through the invention of the scanning tunneling microscope (STM) (1). The ability to manipulate single atoms was there! In 1986, Binnig, Quate, and Gerber built the scanning probe microscope (SPM), which used the short-range van der Waals interactions for probing a surface. This was the advent of the AFM, also known as the scanning force microscope (SFM) (2). Subsequent advances in hardware and software have metamorphosed AFM to its current form.

Of the several powerful techniques available for probing the interaction forces between (bio)surfaces, use of shear flow detachment (3), surface force apparatus (SFA) (4), bioforce probe

Jörg H. Kleinschmidt (ed.), *Lipid-Protein Interactions: Methods and Protocols*, Methods in Molecular Biology, vol. 974,
DOI 10.1007/978-1-62703-275-9_5, © Springer Science+Business Media New York 2013

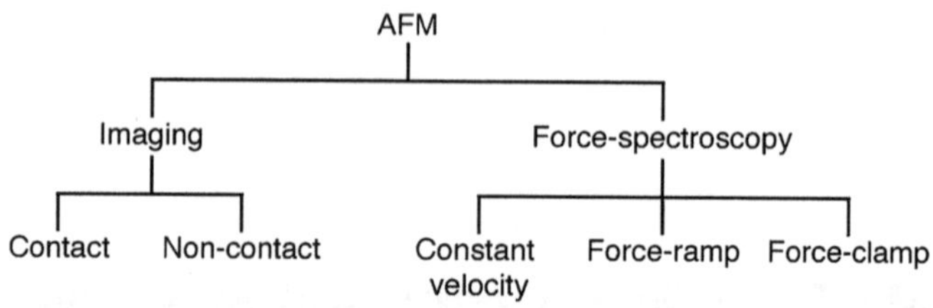

Fig. 1. AFM modes. The AFM can be used for imaging structural features (*high* and *low resolution*) in the contact or noncontact modes. The AFM is routinely used to determine the mechanical characteristics of nanosystems (e.g., proteins, biological and synthetic polymers) with a technique called force spectroscopy. Force spectroscopy can be employed in three different modes, viz., constant velocity, force ramp, and force clamp.

(BFP) (5), laser optical tweezers (LOT) (6), and AFM, the AFM offers the smallest force sensor (tip radii 2–60 nm) providing the highest lateral resolution. The high force sensitivity (≈10 pN) (7) and spatial resolution (lateral ≈0.5 nm and vertical ≈0.1 nm) (8, 9) of the AFM has made it an important tool to study soft biological materials. Since its invention, the AFM (or the conventional STM) has been applied in various modes (Fig. 1), in disciplines ranging-from biology (biochemical, enzymology, cellular biology) (10–12) to (biological) polymer physics (13, 14) to organic chemistry (15, 16). Information that cannot be extracted from the routinely used fluorescence-based techniques (microscopy and spectroscopy) can be obtained from AFM experiments. For example, high-resolution imaging is able to provide a structural map of a membrane protein in its native environment at a reasonably high spatial and, more recently, temporal resolution (17–19). On the other hand, force spectroscopy is used to quantify forces enabling the mechanical characterization of proteins and cells (10, 20). Combining the advantages of the AFM and fluorescence microscopy is a powerful way to gain comprehensive understanding of a system.

1.2. AFM Design and Working Principle

The AFM works on a very basic principle (Fig. 2a). In its simplest form, the AFM incorporates a piezoelectric actuator, a laser diode housed in the so-called head, and very importantly a sharp nano-tip (or the stylus) mounted at the end of a soft cantilever spring (Fig. 2b). Under an applied voltage, the piezo-ceramic elements of the actuator can move with an accuracy of atomic dimensions in the *x*, *y*, and *z* directions. The force exerted by the cantilever stylus on a sample is determined by measuring the deflection (vertical bending) of the cantilever. A laser beam aligned on the dorsal side of the cantilever apex is reflected by a polished mirror onto a quadrant photodiode to detect the cantilever deflection. The photodiode signal also provides the required feedback (force, frequency, amplitude) to maintain the actuator response in *z*. Additionally, for studying biological systems, the AFM is equipped with a fluid cell; for example, a quartz fluid cell with capillary

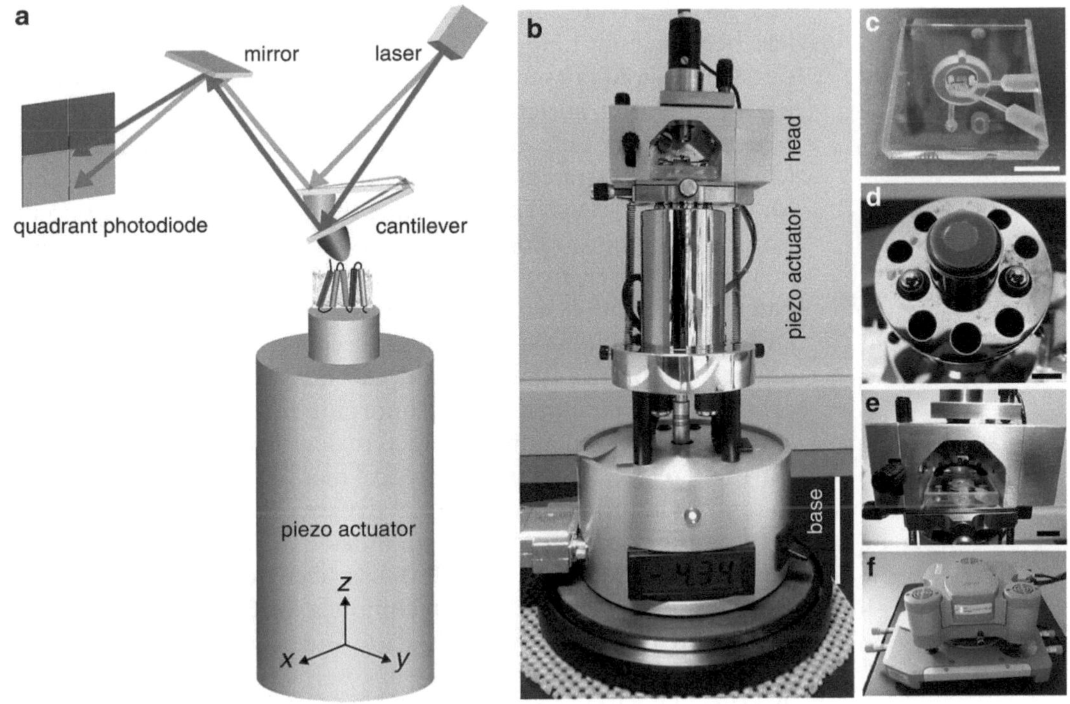

Fig. 2. Principle and setup of AFM. (**a**) A cartoon representation showing the different parts and working of the AFM. Except the piezoelectric actuator, all the other components are housed in a so-called head of the microscope. A laser beam ($\lambda = 625$ nm) is adjusted on the edge of a cantilever spring so as to obtain a maximum sum signal on the photodiode. As a sharp stylus at the end of the cantilever scans the sample, the beam is deflected on the photodiode with the up–down movement of the cantilever. The voltage signal is converted to force (pN) or deflection (nm) with the help of software. The output of the deflection signal varies with the AFM mode. (**b**) A working AFM setup. The whole instrument is <30 cm. The cantilever is fixed in a fluid cell using a spring clip (**c**), which is then clamped inside the head. (**d**) A support assembly (magnetic steel disc, Teflon, mica) fixed magnetically on the piezoelectric actuator. Once the sample has been adsorbed onto the mica, the head is carefully mounted on the actuator (**e**). (**f**) The newly developed automated AFM setup ForceRobot® 300. This instrument can be used as a stand alone or mounted on an optical microscope for simultaneous fluorescence and force measurements. Scale bars, (**b**, **d**) 5 cm, (**c**, **e**) 1 cm.

channels (Fig. 2c). Such a fluid cell serves two important purposes: (1) it holds the cantilever chip, and (2) it facilitates an easy exchange of buffers while maintaining the volume. The fluid cell is mounted inside the head of the AFM such that the cantilever stylus is just above the sample support, which is held magnetically to the piezoelectric actuator (Fig. 2d, e). In other AFM types (Fig. 2f), the fluid-cell may be an open chamber devoid of channels for buffer exchange. The cantilever holder, in this case, is not an integral part of the fluid chamber.

1.3. Studying Membrane Proteins with AFM

Lipid bilayers and the proteins that they host form the basic foundation of a biological cell (21). Membrane proteins are the gateway to a cell's interior forming essential components of cellular dynamics. As elusive as these proteins have been in revealing their structural design, it has not deterred researchers from pursuing them

(22, 23). Structure and function elucidation of membrane proteins has been mainly hindered by their tendency to aggregate once outside the lipid bilayer (24). Without understating the importance of X-ray crystal structures, it is important to stress that static structures offer only partial solution to a very complex puzzle (25). Different strategies are used to stabilize membrane proteins for forming crystals. This may shift the average structural population of a protein as compared to its native structure in the lipid bilayer. Moreover, two different states of a protein should be compared under the exact crystallization conditions, which may not always be possible. It is increasingly accepted that the chemical and physical properties of lipids and the density of the surrounding proteins in a membrane are important in tuning the structural and hence the functional characteristics of a membrane protein (26). Therefore, alternate biophysical methods should be employed to understand the conformational dynamics of membrane proteins without the need to remove them outside their home environment.

A pertinent use of the AFM has been realized in membrane biology (27). AFM-related techniques offer the possibility to image and mechanically manipulate single membrane proteins in different conformational states while in the native lipid bilayer (28, 29) (Fig. 1). AFM-based imaging and force spectroscopy have been used to study the influence of different biological, physical, and chemical parameters on the functional dynamics of membrane proteins, for example, species-to-species variation in membrane protein stoichiometry (30), effect of voltage (31), pH (32, 33), salt concentration (34), temperature (35), ligands and substrates (33), protein partners (36, 37), and mutations (38, 39). It has also been possible to follow the unfolding and folding kinetics of a 12 transmembrane α-helical antiporter, NhaA, in a lipid bilayer (40). In conjunction with (fluorescence) microscopy, the AFM cantilever is a sensitive tool for monitoring cellular processes (41, 42), cell surface receptors, and other membrane proteins in situ (43).

This review will provide a theoretical introduction to imaging and force spectroscopy using AFM, and the basic protocols of single-molecule imaging and force spectroscopy to study membrane proteins. It is suggested that the purple membrane protein bacteriorhodopsin be used during the beginning stages of imaging and force spectroscopy. Bacteriorhodopsin from *Halobacterium salinarum* is structurally and biochemically the most well-characterized membrane protein. Bacteriorhodopsin forms a two-dimensional trigonal lattice in the lipid bilayer, and the assembly is chemically and physically stable. Moreover, the protein is commercially available, is cheap, and can be relatively easily extracted from the archaebacterium. All these features make bacteriorhodopsin a good model system for learning high-resolution imaging and force spectroscopy with the AFM. The well-characterized titin immunoglobulin domain $(Ig)_{27}$ may serve as the system of choice for laboratories interested in the force spectroscopy of water-soluble proteins. These proteins can also be regularly used to check the calibration of the AFM piezoelectric sensors.

1.4. High-Resolution Imaging

Fluorescence-based techniques are routinely employed for noninvasive monitoring of cellular dynamics, intracellular processes, and in vitro biochemical reactions. However, a mandatory need of the method is to tag the system with a fluorescent dye (or use a fluorescent analog of a ligand). Fluorescent labeling of a sample may cause experimental difficulties ranging from a weak fluorescent signal to cell toxicity to altering the system dynamics altogether. Imaging by AFM overcomes some of these problems, at least for studying cell membrane processes on the surface. A spatial lateral resolution of $\approx$0.5 nm and a vertical resolution of $\approx$0.1 nm are unmatched by any of the current optical microscopy techniques. Such superb resolution and a high signal-to-noise ratio afford the visualization of structural features such as loops and individual helices of membrane proteins and their dynamic conformational changes. (9, 32, 44). A number of channels, receptors, and transporters have been characterized by high-resolution imaging under different experimental conditions (Fig. 3).

Surface topography maps are created using the AFM by scanning a sharp stylus in the $x - y$ plane over a sample (contact or noncontact modes). A feedback loop, between the detection system and the piezoelectric drive, controls the vertical movement of the actuator to keep the applied force (frequency or amplitude depending on the mode) constant (45, 46). This locks the vertical movement of the actuator to its lateral movement during scanning and the sample topography is contoured. The vertical information of the image is due to structural height differences but can exhibit additional information based on the specific interactions between the stylus and the sample. These can be of electrostatic, magnetic, or chemical nature.

1.5. Contact or DC Mode

In the contact mode, topographic images are recorded as the stylus makes soft physical contact with the sample and scans the surface to monitor changes in the cantilever deflection, as explained above. Topographic data is generated in one of the two modes: constant-height or constant-force mode. In the constant-height mode, the height of the scanner is kept fixed while scanning, and the spatial variation of the cantilever deflection is used to generate the topographic data set. It is most often used when speed is essential, for example, real-time imaging of dynamic surfaces, or where cantilever deflections are small, for example, atomically flat surfaces. In the constant-force mode, the total applied force on the sample is kept constant by moving the scanner (or the cantilever) up and down in response to the topography. The preferred mode for high-resolution imaging of membrane proteins, the constant-force mode, is limited by the response time of the feedback circuit.

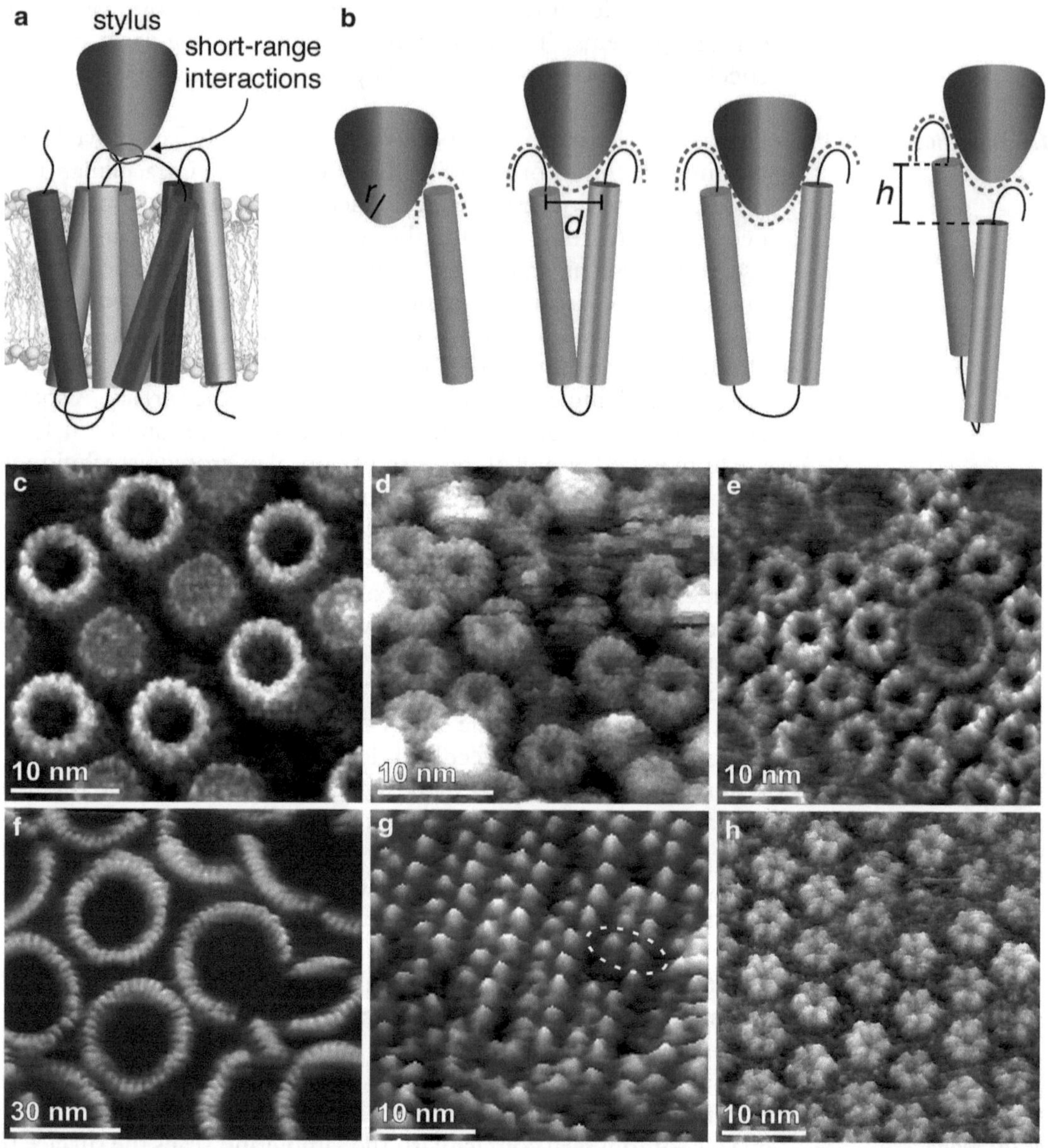

Fig. 3. High-resolution imaging with AFM. (**a**) Long-range electrostatic repulsions (not shown) and short-range van der Waals attractive forces between the stylus and the sample should be balanced for obtaining optimum resolution in the contact mode (47). The long-range forces act over a larger area of the macromolecules in the membrane owing to the large stylus size, whereas the short-range interactions are limited to the stylus apex. With an apparent stylus radius of $\approx$10 nm, it is not possible to resolve structural details within a few nanometers. (**b**) The shape of the stylus apex, the protrusions at the apex, the stylus apex radius (*r*), the stylus aspect ratio, and the distance (*d*) and height (*h*) between two protein structures are all crucial in determining the resolution and the quality of an image. For example, even if the apex radius is small, contouring a rod-shaped structure with a broad stylus will result in a poor topograph. If $r > d$, the stylus will not be able to contour the region in between two structures. Similarly if $h < r$, the stylus will be hindered and the structures will not be resolved (the stylus aspect ratio becomes a more critical parameter). (**c–h**) Examples of membrane proteins resolved by high-resolution AFM imaging: ATP synthase from (**c**) spinach chloroplast (30) and (**d**) *Ilyobacter tartaricus* (122), (**e**) photosynthetic complexes in high-light-adapted membranes of *Rhodospirillum photometricum* (123), (**f**) perfringolysin O pore complexes (124), (**g**) dimers of bovine rhodopsin in native disc membranes (125), and (**h**) extracellular surface of gap junction hemichannels (32). The contact mode topographs were imaged in buffer at an imaging force $\leq$100 pN.

Modulating the interaction forces—electrostatic double-layer repulsion and van der Waals attraction—between stylus–sample and/or stylus–support is a key factor in obtaining reliable surface dimensions in contact mode (and oscillating mode) imaging (47, 48). The electrostatic forces are balanced by adjusting the electrolyte concentration to minimize the lateral and vertical forces while imaging (49–52). Using low electrolyte concentration increases the repulsion between the stylus and the sample causing fuzzy images. For purple membranes, increasing the electrolyte concentration and the valency of the cations decreases the repulsion between the stylus and the sample. Also, at the same imaging force ($\leq$100 pN), the distance between the stylus and the membrane changes depending on the electrolyte concentration and the cation valency (47). This distance can be used to adjust the electrolyte concentration such that the van der Waals attraction is not greater than the electrostatic repulsion. If the van der Waals attraction is much higher, the cantilever will be pulled into the sample causing sample deformation. For stable imaging, the stylus position should be governed by an equilibrium of forces: the attractive force between the stylus and the sample should be compensated by the repulsive force between the two. The external force exerted by the cantilever spring should also be considered (Fig. 3a).

Due to the possibility of specimen damage, high resolution in the contact mode can be achieved only on sufficiently rigid samples (53). It has been estimated that forces as low as 50–100 pN are required for imaging membrane proteins and most biological samples (54, 55) (Fig. 3c–h). Comparison of AFM topographs with atomic structures obtained by X-ray and electron crystallography demonstrates that the imaging process minimally perturbs the protein structures if the contact force is kept around 100 pN.

1.6. Dynamic or AC Mode

A good way of avoiding deformation of soft biological samples is to use the long-range attractive forces to monitor the tip–sample interaction. These attractive forces are weaker than the repulsive force detected in the contact mode, and consequently different techniques are required to utilize them. In the dynamic modes, the cantilever is oscillated at its resonant frequency to minimize contact with the sample surface. A change in the resonance frequency, oscillation amplitude, or the frequency phase shift of the cantilever may be used to generate a topograph. The two main types of AC modes are the TappingMode™ (also known as the intermittent or the oscillating mode) and the true noncontact mode.

1.6.1. TappingMode (Intermittent or Oscillating Mode)

In the TappingMode, the tip–sample separation is modulated while the sample is scanned by a cantilever oscillating at or close to its resonance frequency (100–400 kHz) with amplitudes ranging between 10 and 100 nm. The cantilever is oscillated by applying a small sinusoidal electrical signal causing the main piezoelectric tube to vibrate in the vertical *z* direction, in turn causing the cantilever to vibrate by viscous coupling. Alternatively, the oscillation is induced by a small piezoelectric transducer at the fixed end of the cantilever or by an oscillating magnetic field (the cantilever must be coated with a magnetic material). The feedback control ensures a constant tip–sample interaction during imaging by keeping the oscillation amplitude constant by maintaining a constant RMS of the oscillation signal.

As mentioned above, scanning in the contact mode is not suitable for imaging weakly immobilized structures since these are often swept away by the cantilever while raster scanning the surface (56). The TappingMode overcomes this disadvantage. The oscillating cantilever touches the sample only once at each minimum at the end of its downward movement. This reduces the contact time, the friction, and the lateral forces ($\approx$1 pN) considerably compared to the contact mode, making it ideal for studying soft or elastic samples. A variety of macromolecules have been observed using the TappingMode, which could not be imaged before (57–60). Unfortunately, images obtained in the TappingMode lack the spatial resolution of $\leq$1 nm exclusively acquired in the contact mode. Nevertheless, the use of drive frequencies close to the resonance frequency of the cantilever enables precise control of the cantilever oscillation and to obtain high-resolution topographs of native proteins.

1.6.2. Noncontact Mode

As the name implies, in this mode the oscillating cantilever never actually touches the sample surface but hovers a few nanometers above it. The main difference between this mode and the TappingMode is that the cantilever is oscillated at a frequency which is slightly above the cantilever's resonance frequency with an amplitude of a few nanometers (<10 nm) to obtain an AC signal from the cantilever. A clear advantage of such a technique is that no force is exerted on the sample thus preventing sample damage. Recent progress has illustrated the use of NC mode in high-resolution imaging of membrane proteins in buffer solutions (45, 61).

The conventional AFM imaging modes offer low temporal resolution. It takes $\approx$100 s in the contact mode and $\approx$5–10 min in the noncontact mode to acquire a topograph of reasonable resolution (512 × 512 pixels). Therefore, for monitoring dynamic processes occurring within milliseconds (the time scale of most biological processes), fast imaging techniques are required. The recent development of high-speed AFM (HS-AFM) is a milestone in high-resolution imaging (17–19, 62). HS-AFM has been able to follow the assembly and dynamics of single bacteriorhodopsin molecules (63). Very recently, HS-AFM has provided live insight into the working of a rotorless ATP synthase (18, 64). This study convincingly exemplifies the power of AFM in understanding the working principles of biological machineries.

1.7. Single-Molecule Force Spectroscopy

Single-molecule 'push-pull' techniques have helped us to learn about the structure, function, dynamics, and interactions of a variety of biological systems in the context of their mechanical properties (5, 65–68). Single-molecule force spectroscopy (SMFS) with AFM has emerged as an indispensable technique to characterize the mechanical properties of proteins (28, 29, 69), ligand–receptor interactions (70), mechanoenzymatic catalysis (12, 20, 71, 72), and even single-bond mechanochemistry (73). The cantilever spring is used as a handle to manipulate single proteins and cells. In the most commonly used constant-velocity or velocity-clamp mode, the sample is brought into contact with the stylus (or stylus in contact with the sample) and retracted away at a certain velocity (approach–retract cycle). Probabilistically, during certain approach–retract cycles, a protein molecule is attached to the stylus and is pulled away from the surface and stretched. Pulling induces mechanical stress in the molecular system and eventually forces a structural transition such as the unfolding of a nucleic acid or a protein or the dissociation of a molecular complex. The sequence of transition events that follows when a protein is under an external force is, (1) elongation of the molecular linker and (2) unfolding that (3) causes a relaxation of the polypeptide chain and consequently a drop in the force.

The elongation, unfolding, and relaxation of the polypeptide chain result in a so-called force–distance (F–D) trace (Fig. 4a, b). The distance between distinct peaks in an F–D trace reflects a gain in distance after unfolding and stretching of a folded structure in the protein, and the amplitude is a measure of the force required to unfold it. This "fingerprint" of a protein reveals its different unfolding populations.

In contrast to water-soluble proteins, which unfold cooperatively under force (66, 74), membrane proteins encounter barriers along the entire structure and unfold in steps (28, 29). It is therefore possible to monitor the unfolding of a membrane protein via a stochastic "macroscopic" unfolding pathway constituted of structural segments. These structural segments are mechanically stable entities, and their stability is modulated by inter- and/or intra-protein interactions. On changing the experimental conditions, for example, temperature, salt concentration, pH, and mutations, which may have an effect on a protein's conformational state, new structural segments in an unfolding pathway are detected. Force spectroscopy thus provides a powerful means of comparing minute changes in the unfolding pathways of a protein in different conformations.

1.7.1. Theoretical Background of Mechanical Unfolding

The rupture force measured in unfolding experiments is not an absolute value that relates to the strength of a bond. The conceptual basis of forced unfolding lies in the lifetime of a bond. Weak bonds and structures have limited lifetimes, $\tau_{\text{off}}(=1/k_{\text{off}})$, and will dissociate under almost any level of force if pulled on for limited periods of time.

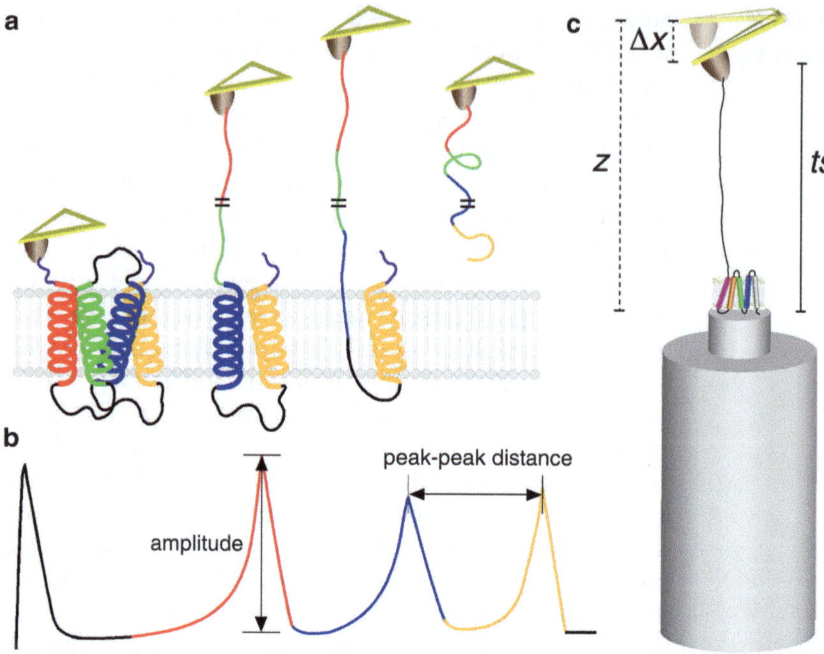

Fig. 4. Mechanical unfolding of a membrane protein. The AFM cantilever is a sensitive handle to dissect a membrane protein helix by helix. (**a**) The cantilever is approached close to the membrane surface to pick up a single protein from its terminal end. If attached, the terminal forms a linker between the stylus and the protein. Upon retracting from the surface, the cantilever exerts a mechanical force on the protein causing it to unfold. (**b**) The unfolding signal is akin to a fingerprint characterized by peaks and valleys and can be used to identify a protein. The peak amplitude denotes the force required to unfold a single helix or a pair of helices (including the connecting loops), whereas the peak-to-peak distance is a measure of the length of the unfolded polypeptide chain. (**c**) The actual length of the unfolded polypeptide is given by the tip–sample distance (tss) and not by the distance (z) traveled by the piezoelectric actuator. tss is obtained by subtracting the cantilever deflection (Δx) from z (tss $= z - \Delta x$).

A small force provides a push over the thermal barrier decreasing the lifetime of a bond. In force spectroscopy experiments, single bonds exist far from equilibrium and only have nonzero strength on time scales shorter than the time (τ_{off}) needed for spontaneous entropy-driven dissociation. If pulled apart faster than τ_{off}, the contribution of thermal energy in bond dissociation is less thereby causing an increase in the measured force. In essence, higher loading rates give rise to higher unfolding or detachment forces, and lower loading rates to lower unfolding forces (75, 76).

Like folding/unfolding reactions of many small water-soluble proteins under no force (77, 78), mechanical unfolding experiments too can be understood by invoking the classical two-state model. A two-state model assumes that a protein adopts only two states, a low-energy folded state and a high-energy unfolded state separated by a single transition barrier that should be overcome for the interconversion of the two states (79, 80) (Fig. 5a).

If the transition state has a free energy, G^*, then the unfolding rate constant, k_{u}, is

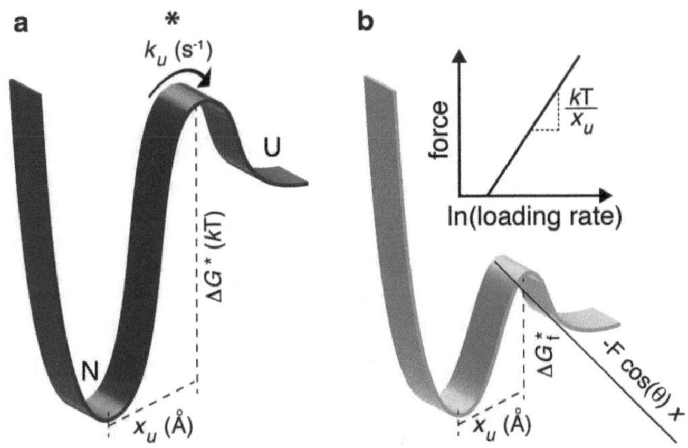

Fig. 5. Energy profile under force. (**a**) A two-state energy profile (*black*) of a protein depicting the native state (N) in a low-energy valley and the unfolded state (U) in a high-energy state. Under no external force, i.e., at equilibrium, N will have to cross a transition barrier (*) of height ΔG^* to reach state U. The distance between the native state and the transition state is denoted by x_u, and the rate of crossing over the barrier by k_u. (**b**) A force vector, $-F\cos(\theta)\bullet x$ (where θ is the angle at which the force is applied relative to the microscopic reaction coordinate x), tilts the energy profile (*light gray*) lowering the height of the transition barrier, ΔG_f^* ($\Delta G^* > \Delta G_f^*$). Although a mechanical potential increases the k_u, it is assumed that the x_u does not change. When a protein is unfolded dynamically i.e., at increasing pulling speeds, an exponential increase in force with loading rate is expected (*inset*, linear on a semilogarithmic scale). The slope of such a force regime determines the x_u, and the *y*-axis intercept denotes the k_u. Thus, the dynamic force spectrum of a protein is used to determine two very important coordinates defining its energy profile.

$$k_u = \left(\frac{1}{\tau_D}\right)\exp\left(-\frac{\Delta G^*}{k_B T}\right) \quad \Delta G^* = G^* - G_n \tag{1}$$

τ_D is the relaxation time, G_n the free energy of the native state, k_B is the Boltzmann constant, and T is the temperature. This model serves as an excellent starting point for understanding a reaction under force. A mechanical potential applied by an external force, F, along a microscopic reaction coordinate, x, tilts the energy landscape along this coordinate thus lowering the energy barrier at the transition state (Fig. 5b). In other words, an external force acts as a "mechanical catalyst" and lowers the barrier of the underlying energy landscape, expressed as,

$$\Delta G_f^* = \Delta G^* - Fx_u \tag{2}$$

ΔG_f^* and ΔG^* represent the free energy difference under an applied force and zero force, respectively. x_u is the distance of the transition state from the folded native or intermediate state of the protein.

Also, the rate of escape (or unfolding), k_{off} or k_u, of a molecule from its potential well increases exponentially with force (81),

$$k_{off} \approx \left(\frac{1}{\tau_{off}}\right) \exp\left(\frac{F}{F_u}\right) \tag{3}$$

where $F_u = \frac{k_B T}{x_u}$. F_u is small because thermal energy $k_B T \approx 4.1$ pN nm, at room temperature, and $x_u \approx 0.1$–1 nm. A relation similar to that of Bell's (Eq. 3) is obtained on combining Eqs. 1 and 2,

$$k_{u(F)} \cong \left(\frac{1}{\tau_D}\right) \exp\left(-\frac{\Delta G^* - Fx_u}{k_B T}\right) = k_u \exp\left(\frac{Fx_u}{k_B T}\right) \tag{4}$$

$k_{u(F)}$ is the unfolding rate (or the dissociation rate) of a protein (or a bond) under force.

A useful extension of the constant-velocity mode is dynamic force spectroscopy (DFS). As the name suggests, DFS involves dynamic unfolding of a protein, i.e., at different pulling velocities (nm s^{-1}) or loading rates (pN s^{-1}) (82). DFS is used to map the energy landscape of a protein (74, 83) or a receptor–ligand pair (5) (in general any molecular interaction). Two important parameters, the distance of the transition state to the folded state or the intermediate state, x_u, and the unfolding rate of these states, k_u, are obtained from DFS measurements. These parameters provide insight into the molecular mechanistic details of a biological process.

A plot of the most probable unfolding force (F_p) versus $\log_e$ (loading rate) maps the most prominent energy barriers traversed in the energy landscape along the force-driven pathway (76, 80). The most probable unfolding force is expressed as,

$$F_p = \frac{k_B T}{x_u} \ln\left(\frac{x_u r_f}{k_B T \cdot k_u}\right) \tag{5}$$

r_f is the loading rate and the other symbols have the same meaning as before. The dynamic force spectrum is fit to Eq. 5 to obtain the x_u (from the slope) and k_u (y-intercept). These parameters are used to expose the energy differences between the transition barriers of proteins (39, 83–85). Importantly, these key coordinates of the energy profile can be used to estimate the stiffness (κ) of the individual structural segments (α-helices, β-sheets, loops) countered during the unfolding process. Recent studies point to an apparent correlation between κ and a protein's function (86–89).

2. Materials

2.1. Reagents

1. Nanopure™ or Millipore™ water (≈18.2 mΩ/cm).
2. Purple membranes from *H. salinarum* (Munich Innovative Biomaterials GmbH, Sigma-Aldrich).

3. Adsorption buffer for purple membranes: 300 mM KCl, 10 mM Tris–HCl, pH 7.8.
4. Imaging buffer for purple membranes: 150 mM KCl, 10 mM Tris–HCl, pH 7.8.
5. All salts and electrolytes for making buffers should be analytical (p.a.) grade (e.g., from Sigma-Aldrich, Merck).
6. Ethanol/propanol.
7. Sodium dodecyl sulfate (SDS) or Hellmanex®.

2.2. Equipment

1. AFM (Bruker, JPK Instruments, Asylum Research).
2. Fluid cell (depending on the AFM, from Bruker, JPK instruments).
3. Oxide-sharpened V-shaped or rectangular Si_3N_4 cantilevers (Bruker, Olympus, MikroMasch).
4. Vibration isolation table.
5. Punch and die set (for punching Teflon and mica discs; Precision Brand Products Inc., Middlesex, UK).
6. Teflon foil.
7. Muscovite mica (Shree GR Exports, Kolkata, India; Mica Corporation, New York, USA).
8. Steel discs (Agar Scientific, Stansted, UK).
9. Loctite 770 (Koenig, Dietikon, Switzerland).
10. Araldite (water-insoluble, two-component epoxy glue; Conrad Electronic, Hirschau, Germany).

3. Methods

3.1. AFM Setup

1. A commercial (or in-house built) AFM should be optimized for high-resolution imaging and SMFS measurements. The AFM should be placed on an active or passive vibration isolation table. It may be necessary to cover the AFM with a glass bell jar or enclose it in a chamber to avoid acoustic disturbances. Electrical interference increases the noise in the recorded signals, and care must be taken that electrical equipment (e.g., transformers) is not placed too close to the AFM (see Notes 1–5).
2. Calibrate the piezoelectric actuator of the AFM in the x, y, and z dimensions using a calibration grid. An AFM equipped with a 50 μm $x - y$ piezoelectric scanner with a closed-loop 20 μm z-axis (or less) can be used for SMFS measurements. Different piezoelectric scanners are used for high-resolution imaging, for example, E-scanner (10 μm × 10 μm in $x - y$ and 2 μm in z) and J-scanner (125 μm × 125 μm in $x - y$ and 5 μm in z).

Special scanners are required for high-speed imaging, which is not discussed in this review.

3. Mount the AFM head with the fluid cell on the piezoelectric scanner (Bruker instruments). Secure the fluid cell with the screws (not too tightly or else the fluid cell may crack) so that it does not move inside the head (see Note 6).
4. Align the laser on the upper tip of the cantilever to get a high photodiode signal. Laser alignment on a cantilever is a relatively straightforward procedure when using a camera. If the instrument is not equipped with a camera, a paper strip can be used to position the laser beam at the cantilever tip (detailed method may be found in the manual of the AFM supplier) (see Notes 7, 8).
5. Allow the instrument to thermally equilibrate until a stable signal is reached (30 min–1 h, but could take up to 6 h).

3.2. Choosing a Cantilever

The heart of the AFM is the cantilever with a stylus mounted at its end. This nano-tool does the "feeling" and is inarguably the most important part in both imaging and force spectroscopy applications. Therefore, a cantilever should be chosen for a specific application after carefully considering all the experimental parameters. AFM styli and cantilevers used for biological applications are usually made from either silicon or silicon nitride using many of the microfabrication techniques that have been developed for integrated circuit manufacture, such as lithographic photo-masking, etching, and vapor deposition. Using the micro machining approach, sharp styli can be integrated onto cantilevers, a prerequisite for high-resolution imaging of biological samples. The structure and geometry of a stylus, respectively, determine the nature and the area of interaction with a surface (Fig. 3a, b). Cantilevers are often coated with a thin layer of gold or aluminum to provide high reflectivity ensuring efficient deflection of the laser beam from the cantilever.

To suit specific applications, cantilevers are made in different shapes and sizes. The most commonly used ones are triangular or V-shaped and rectangular or beam shaped with a length of 100–200 μm and a thickness of 0.1–5 μm. The stylus is usually tetrahedral or pyramidal in shape of radius as small as <7 nm and height between 3 and 15 μm, with conducting or nonconducting properties (excellent sources of information on cantilevers and styli are http://www.veeco.com and http://www.atomicforce.de/Olympus-Cantilevers.php). A simple reason for using a V-shaped cantilever is to minimize torsional or twisting motion of the cantilever while scanning a sample, making it the lever of choice for purely topographical imaging. A simple beam or rectangular geometry cantilever can be used for frictional measurements due to a greater degree of rotational freedom making it sensitive to lateral forces.

Cantilevers with wide-ranging κ values (0.006–42 N/m) are available (e.g., Olympus). The force resolution, the thermal noise, and the spring constant of a cantilever are all correlated and are critical parameters to consider while choosing a cantilever. Low values of cantilever stiffness represent high sensitivity to force for each nanometer deflection of the transducer. The subtle drawbacks of low stiffness are that the probe is susceptible to thermal fluctuations and the response time of the cantilever can be slow in an overdamped viscous-water environment.

$$\partial x^2 \sim \frac{k_B T}{\kappa} \qquad (6)$$

A sluggish response time is a major concern in optimizing the frame rate in high-resolution imaging (although the frame rate is also dependent on the response time of the piezoelectric scanner and the feedback loop (90, 91)). An increase in the thermal fluctuations will add unwanted noise to the system making the deciphering of fast events extremely difficult. Unfortunately, increasing the cantilever stiffness results in large thermal fluctuations of the applied force (75),

$$\partial f^2 \sim k_B T \cdot \kappa \qquad (7)$$

This is especially undesirable in force spectroscopy as it compromises the sensitivity of the measurements.

It is also crucial to select a cantilever of the right geometry and physical dimensions (length, width, and thickness). For example, κ for normal bending of a rectangular cantilever is given by (92),

$$\kappa = \frac{E \cdot w \cdot t^3}{4l^3} \qquad (8)$$

w is the width, l is the length, t is the thickness of the cantilever, and E denotes the Young's modulus of the material.

The quality factor, Q, is an important parameter when choosing a cantilever for imaging and force spectroscopy. In simple words, Q is a measure of the energy losses of the cantilever through dissipation processes of hydrodynamic and air damping (93). Cantilevers with high Q values need less energy input to be excited at resonance and exhibit a sharp, high resonance peak. Whereas cantilevers with low Q values exhibit a broad frequency spectrum and require a higher energy input for matching the oscillation amplitude of high Q cantilevers. The Q factor is not the same in vacuum, air, and aqueous environments owing to the increase in damping experienced in air and aqueous media.

The operator should carefully choose a cantilever depending on the specific information required. For example, for imaging applications, the choice of a cantilever would depend on the type of sample, the imaging mode (contact, oscillating, or noncontact), and the biological process being monitored (fast or slow time scale), to mention a few. A few common determining parameters would be the spring constant (κ), the resonance frequency (f_r), and the quality factor (Q) of the cantilever (8). Soft cantilevers with high resonance frequencies are best suited for contact mode imaging. Low κ values (0.05–0.5 N/m) prevent deformation of the sample surface, and high resonance frequencies are imperative in tracking structural features while matching the scanning speed. In the noncontact mode and the TappingMode, a cantilever experiences considerable damping owing to the up–down cycles (one contact/oscillation cycle). Therefore, stiff cantilevers ($\kappa > 10$ N/m) are suitable for imaging in these modes. Using cantilevers with high κ values prevents attraction by van der Waals interactions on close approach to the sample, and provides enough potential energy to retract once the cantilever has come in contact with the sample (94). Because the cantilever does not come in continuous contact with the sample surface during imaging in the noncontact mode, sample deformation by stiff cantilevers is not a concern. (However, using cantilevers of too high stiffness would compromise the deflection sensitivity and hence the structural details.) However, small cantilevers (10–50 μm), which suffer less viscous damping and have high resonance frequencies and low spring constants, can be used for high-resolution imaging (17, 95, 96).

The limits of the AFM system should also be considered while choosing a cantilever. For example, a high resonance frequency cantilever chosen for high scanning speeds may not always give optimum results if the scanner speed and the time resolution of the feedback loop are not matched. The stylus radius should be <10 nm to obtain the best resolution. Although manufacturers provide values of stylus radius, such values denote the radius at the base of the stylus and are not always accurate of the tip radius, which determines the resolution of an image. However, a sharp stylus may not always be needed especially when obtaining fine structural details is not the aim. For many applications, for example, cell measurements (97), tipless (without stylus) or round styli (silica or latex microspheres stuck at the cantilever tip) are preferred. In such applications a sharp stylus may rupture the soft biological samples.

For force spectroscopy measurements, the common criteria for choosing a cantilever should be low Brownian noise, high resonance frequency, and high sensitivity (small κ). Cantilevers with high resonance frequencies permit the use of high sampling rates enabling the detection of fast events which may not be detectable otherwise. Cantilevers with low deflection sensitivity may not be able to measure small forces in the piconewton range (<100–300 pN). Too long cantilevers should not be used; short cantilevers experience low viscous damping and therefore are suitable for pulling at high speeds (95). The minimum force, $F_{\min}$, that a cantilever can measure in a liquid is limited by its thermal noise and can be expressed as (93),

$$F_{\min} = \sqrt{4k_B T \cdot B \cdot R} \tag{9}$$

B is the measurement bandwidth, and

$$R = \frac{\kappa}{2\pi \cdot f_r \cdot Q} \tag{10}$$

In force spectroscopy, the stylus radius is not a critical parameter, and cantilevers with round or blunt styli may be used. Biolevers (A-lever −60 μm long, $\kappa = 0.03$ N/m, and B-lever −100 μm long, 0.006 N/m) are routinely used for force spectroscopy experiments.

3.2.1. Force Detection and Cantilever Calibration

Measuring accurate forces with the AFM requires knowledge of both the deflection sensitivity and the spring constant (κ) of a cantilever. Therefore, a cantilever needs to be calibrated for both these parameters. Signals are obtained by monitoring the cantilever deflection (in volts, V) as a function of the vertical displacement of the piezoelectric actuator (z in nm). This "raw" voltage-displacement signal is converted into an F–D trace by using two simple conversions. Firstly, the sensitivity of the AFM detector, measured by the slope of the retraction curve in the region where the stylus and the sample are in contact (V nm^{-1}), is used to convert the deflection voltage into cantilever deflection (Δx in nm). Secondly, Δx is converted into force (F) using Hooke's law ($F = -\kappa \cdot \Delta x$) (see below). Another correction is made to obtain the true length of the unfolded protein. The force measured is displayed as a function of z, the distance traveled by the piezoelectric actuator. However, the end-to-end distance of a stretched molecule is smaller than z due to Δx, which can be several nanometers for soft cantilevers ($\kappa < 0.1$ N/m) used in force spectroscopy experiments. To correct for the cantilever deflection, the force is presented versus the tip–sample separation (tss), i.e., the distance between the stylus and the sample surface (tss $= z - \Delta x$) (Fig. 4c). (The user can change the units in which the signal is shown. Most software incorporate one or more channels to visualize signals in different units simultaneously.)

To translate Δx (nm) to units of force (pN), it is necessary to determine κ of the cantilever. As outlined above, κ depends critically on the physical dimensions of a cantilever. Therefore, to account for the manufacturing variability of the cantilevers, it is very important to accurately determine κ using empirical methods for quantitative force measurements.

A number of methods have been developed for determining κ (92, 98–101). Routinely, analysis of the thermal fluctuations of the vibrating lever is used to determine the stiffness of the cantilever (102, 103). This analysis is based on the equipartition theorem (104). The cantilever is treated as a simple harmonic oscillator whose power spectrum of thermal fluctuations is used to derive the spring constant. In brief, the cantilever is raised several microns (≥20 μm) from the surface and its natural frequency of vibration (resonance frequency) is monitored for 2–3 s. The measured variance of the deflection $\langle x^2 \rangle$ can be used to calculate the spring constant,

$$\frac{1}{2}k_B T = \frac{1}{2}\kappa\langle x^2 \rangle \quad (11)$$

Using this approach, the spring constant of a cantilever is calibrated in either air or an aqueous medium (the cantilever should be calibrated in the same medium in which the experiment would be performed). Most new AFM software include functions to determine the deflection sensitivity and κ (e.g., Bruker, JPK instruments).

3.3. Preparation of Supports

AFM measurements do not require any special staining, labeling, or fixation of the sample. A prerequisite for obtaining reliable information from imaging and SMFS experiments is the preparation of special supports for sample immobilization. A good support should be chemically inert, usually hydrophilic, and have an atomically flat surface such that the membrane patch lies flat on the surface. This is important for obtaining subnanometer resolution in high-resolution imaging. A surface with high features may cause the membrane patch to adsorb unevenly leading to poor resolution. Materials that are commonly used are mica, highly ordered pyrolytic graphite (HOPG), gold, and glass. Mica has been the support of choice for membrane proteins (105). It has an effective net negative charge (-0.0025 C/m^2 at neutral pH) and is hydrophilic (106). The material used for making supports would however depend on the application, for example, glass slides are ideal for simultaneous force spectroscopy and optical measurements.

For imaging and force spectroscopy measurements of bacteriorhodopsin, mica is the recommended supporting surface (107). A freshly cleaved mica surface provides a ready-to-use, atomically flat surface. The preparation of mica supports for adsorption of membrane patches is described in detail below (Fig. 6).

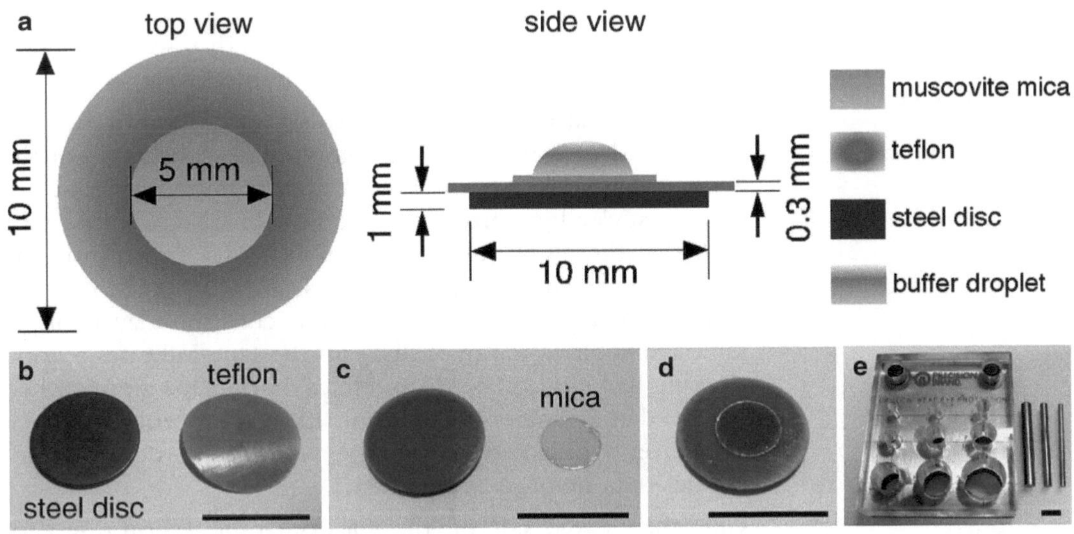

Fig. 6. Sample support for single-molecule imaging and force spectroscopy. (**a**) *Top view* and *side view* showing the dimensions and arrangement of mica and Teflon discs on a magnetic steel disc. (**b**) A magnetic steel disc (10 mm diameter) is used as the base, and a circular piece of Teflon (12 mm diameter) stuck on the steel disc (**c**). The Teflon disc prevents buffer spillage into the piezoelectric actuator. (**d**) A circular mica (5 mm diameter) is cut and stuck in the center of the Teflon disc. The membrane patch in an aqueous buffer is adsorbed on the mica as shown in (**a**). (**e**) A punch–die set used to cut Teflon and mica discs of the required diameters. Scale bars, (**b–d**) 10 mm, (**e**) 1 mm.

1. First, treat the magnetic steel discs (10 mm diameter) with 37% hydrochloric acid for ≤5 min in a glass beaker or a glass Petri dish.
2. Rinse the discs intensively with Millipore water (rinse when the disc edges start showing a green tinge).
3. Punch Teflon discs (≤0.2 mm thick, 12 mm diameter) using a punch–die set. Glue the Teflon onto the cleaned steel discs with a Teflon compatible instant glue (e.g., Loctite 770). Leave the stuck Teflon overnight for the glue to dry. Care should be taken to avoid any air bubbles between the Teflon and the steel disc. Air bubbles cause vibrations, which interferes with imaging at the nanometer scale.
4. Punch a 0.2–0.3 mm thick sheet of muscovite mica into discs of 5 mm diameter or more, but less than the diameter of the Teflon disc. Care should be taken that very thick sheets of mica are not used for punching discs. If the mica used is too thick, punching will cause cracks on the surface (easily detectable owing to the reflective nature of mica). Mica discs with cracks should not be used.
5. It is important that the mica surface to be glued onto the Teflon disc is atomically flat. Using a flat surface prevents

vibrations and thus noise in the measurements while scanning. Minimizing noise is more important in high-resolution imaging where subnanometer ($\leq$1 nm) resolution is desired.

6. To obtain an atomically flat surface, the upper layer of the mica disc should be cleaved using a colored tape (label tape) or the Scotch Magic Transparent Tape. Place the mica disc on a clean surface, stick a long piece of tape on the mica, remove the tape such that the disc is just visible, hold the disc down from the exposed edge with the tip of tweezers, and remove the tape in a sudden waxing motion. A one-piece circular mica layer (of the same diameter as the mica disc) on the tape indicates a good cleaving technique. The cleaved mica on the tape gleams when observed at an angle. If a crescent shaped or a broken piece of mica layer is obtained on the tape, repeat the procedure until the desired one-piece is obtained.
7. Glue the cleaved side of the mica onto the Teflon disc with Araldite (or any other epoxy glue). Care should be taken that no air bubbles are entrapped between the mica and Teflon discs while gluing the two surfaces. The disc should be left overnight for the glue to harden.
8. Importantly, prior to adsorbing a membrane sample, the mica surface of the support must be cleaved using a tape to obtain an atomically flat surface (see Notes 9, 10). The cleaving procedure is the same as explained above for cleaving the mica disc (step 6).

3.4. Contact Mode Imaging

3.4.1. Low-Resolution Imaging

1. Adsorb purple membrane (30 μl of ≈15 μg/ml in the adsorption buffer) onto a mica support, and incubate for 10–30 min. Remove the adsorption buffer carefully using a pipette, and rinse the mica surface at least 6 times with 30–40 μl of the imaging buffer. The wash steps are important to remove any unadsorbed membrane and protein. Insufficient washing after sample adsorption is a major reason of frequent cantilever contamination. The adsorption step may need to be optimized for different membrane proteins. The adsorption times will vary for different proteins, ranging from a few minutes to 1–2 h, depending on the buffer (electrolyte, concentration, pH) and/or the protein concentration.
2. Once a stable signal is reached after equilibration (Subheading 3.1), "engage" the cantilever stylus with the surface. Set the scan size to 0 μm. A scan size of 0 μm while engaging the cantilever prevents sample damage and cantilever contamination.
3. Retract (not disengage) the cantilever by 20–25 μm from the surface, switch to the imaging mode of the software, and approach the cantilever again to the surface keeping the scan

size 0 μm. The cantilever should be approached slowly by increasing the set point in small steps until the stylus just touches the mica surface.

4. Decrease the set point again to lose contact between the stylus and the surface, and increase slowly again to contact the surface. This procedure ensures that minimal force ($\leq$100 pN) is being applied to the surface and that the membrane patches are not destroyed while scanning. Increasing the force would cause the protein and membrane structures to deform reversibly (54); upon increasing the force too much, the damage is irreversible.
5. Scan the surface (20–50 μm × 20–50 μm) to ensure (a) the flatness of the mica, (b) that enough membrane patches are adsorbed on the surface, and (c) the stylus is not contaminated. A good indication of a contaminated cantilever is fuzzy or streaky images.
6. To further ensure the quality of the stylus and that the optimum electrolyte concentration and pH are used, record F–D traces on an empty mica area (where no membrane patches are adsorbed) in the force mode of the instrument (47) (Subheading 3.5). With a clean cantilever no sawtooth-like signals would be observed. If F–D traces are recorded on a membrane patch, sawtooth-like signals denoting the unfolding of the membrane protein may be obtained. For steps to be taken if the cantilever is contaminated, see Notes 11–13.
7. For imaging, switch to the imaging mode, increase the scan size, set the scan rate to $\approx$4 Hz, and adjust the force to a minimum as explained before (steps 3 and 4). Once a minimal force is reached, optimize the integral and proportional gains. A good way of doing this is to increase the gains in small steps until the cantilever starts oscillating. An oscillating cantilever is detected by fringes in an image. Decrease the gains slightly until the cantilever ceases to oscillate; use these as the optimum gain values. Once the force and gains are optimized, zoom on a membrane patch and image ($\approx$4 Hz).
8. Maintain the force at a minimum by continuously adjusting the set point as the sample is scanned.
9. For troubleshooting in case of no membrane patches or too many membrane patches, see Notes 14 and 15.

3.4.2. High-Resolution Imaging

1. Once a membrane patch has been selected by low-resolution imaging (Subheading 3.4.1), magnify the patch by zooming into it. Record F–D traces on the membrane patch (Subheading 3.5) to check if the electrolyte concentration and pH provide an electrostatic repulsion of $\approx$50–100 pN between the stylus and the membrane. This is important for stable imaging

of high-resolution topographs but should also be considered for low-resolution imaging. Once forces between the the stylus and the sample are balanced, scan the membrane sample at a slightly higher contact force of ≈75–125 pN (8, 47).

2. Molecular details can be observed by zooming into a membrane patch. The zooming should be done in discrete steps (e.g., 20 μm, 5 μm, 1 μm, 100 nm). This ensures that the required membrane area high in protein density is selected by reducing the chance of missing it due to spatial errors caused by piezoelectric scanner drifts (a problem encountered especially when the instrument is not stable).
3. The imaging force, gains, and the scan rate should be adjusted each time after changing the frame size (the procedure is the same as explained above in 3.4.1).
4. To increase the vertical (z) resolution, decrease the z-limit of the scanner at smaller frame sizes. Ideally, the cantilever should not be in contact with the surface while decreasing the z-limit to prevent damaging the sample and the stylus.
5. The number of scan lines per unit area should also be increased with magnification to observe structural details. As the image is magnified, increase the scan rate without compromising the resolution. Before increasing the scan frequency, set the contact force and the gains to a minimum.
6. If an increase in structural details is not observed, it means that the maximum resolution has been reached with those set of parameters (cantilever, scanning frequency, force, and gains). The resolution is also influenced by the nature of the membrane sample. The sample should be free of aggregates, and the adsorption protocol should result in flat patches (controlled by salt concentration, pH).
7. Upon zooming, if the image is streaky and no structural features are observed, immediately reduce the force to lose contact of the stylus with the surface. Increase the force slowly in steps until the stylus just contacts the surface and see if features appear. If no structures are observed, zoom out of the area and image again to ensure that the contact force was not too high and did not destroy the sample. If the membrane looks scratched, choose an unperturbed area of the sample. If no membrane patches are observed on zooming out, the salt concentration/pH may be adjusted to balance the repulsive and attractive forces.
8. A common source of loss of resolution in AFM imaging is high scan speeds. Too high scan speeds do not allow enough time for the cantilever to contour the sample.

9. The stylus can also introduce artifacts (108). A common artifact is the observation of a double image. This is due to small outgrowths or depositions on the stylus, which scan the sample along with the stylus but with an offset, thus giving rise to two identical images.
10. If the recorded images show noise interference, check that the AFM is well isolated against mechanical, electrical, and acoustic noise (see Note 5). Avoid using audio systems, iPods, etc. Any temperature fluctuations in the room may also cause mechanical drifts in the instrument, making it difficult to maintain a constant set point. If vibrational noise is observed while scanning at low contact force, optimize the gains and the scan speed.
11. During high-resolution imaging, software parameters and cantilevers have to be changed continuously if the desired results are not obtained. However, sometimes a stylus gets better after a few scans and the image resolution improves (8). It is therefore important to strike a balance between perseverance and common sense for getting results in a reasonable time (see Notes 16, 17).

3.4.3. Oscillating Mode Imaging

1. The main rules of operation in the oscillating mode are similar to those in the contact mode. However, an important difference is that the cantilever should be driven to oscillate at its resonance frequency in the imaging buffer (46). Optimum oscillation frequency is important to obtain the correct response owing to force differences along the sample topography (109).
2. Engage the cantilever with the surface keeping the scan size 0 μm. Adjust the set point to minimal force, and optimize the gains and the scan speed before scanning the membrane surface (Subheading 3.4.1). It is possible to achieve high-resolution topographs with drive amplitudes of 1 nm.
3. As in contact mode imaging, any mechanical or thermal drift should be compensated by adjusting the set point (109).
4. The reasons for poor image resolution in oscillating mode are mainly the same as in contact mode. A specific difference is related to imaging at a frequency far from the cantilever's resonance frequency causing imaging artifacts. If poor resolution is obtained, check the sample in contact mode, calibrate the cantilever resonance, and drive the oscillation at the correct frequency and amplitude (1–2 nm) (8).

3.5. Single-Molecule Force Spectroscopy

SMFS can be performed in three modes: constant velocity (66, 68), force ramp, and force clamp (110, 111). For carrying out measurements in any of the three modes, the protein is adsorbed onto a functionalized or nonfunctionalized surface. As for high-resolution imaging, for SMFS too, mica, HOPG, gold, and glass offer good options in minimizing protein denaturation, and for creating surfaces with new chemical and physical properties (105). Membrane patches for SMFS are routinely adsorbed onto mica surfaces. Gold is a good alternative for attaching proteins (usually multimodular water-soluble proteins) by forming a gold–thiol (Au–S) linkage between the engineered cysteines in a protein and the gold surface (112, 113).

1. Survey a membrane topograph, and zoom into an area ($\approx$1 μm) to check if the membrane contains proteins or not (Subheading 3.4.1). If the membrane protein sample is well prepared, and the imaging parameters are optimum (as explained above), densely packed proteins or a lattice assembly would be observed. Zoom in further on the membrane ($\approx$500 nm) where the density of proteins is high, and switch to the force spectroscopy mode. It is not always necessary to image the proteins at a high-resolution before force spectroscopy.
2. To attach a single protein to the stylus, bring the cantilever close to the surface and apply a force of 100–500 pN (max 1 nN) for $\leq$1 s. This facilitates a nonspecific attachment of the protein to the cantilever stylus (see Note 18). Retract the actuator at a certain velocity (constant-velocity mode) to unfold the protein. Not every approach–retract cycle will give an F–D signal indicative of protein unfolding. In the case of bacteriorhodopsin, a success rate of 1–10% could be expected. Therefore, thousands of approach–retract cycles are needed to acquire data for reliable statistical analysis.
3. A force >1 nN should not be applied on a membrane patch as this would destroy the protein and the membrane irreversibly giving dubious F–D traces. A protein such as bacteriorhodopsin of radius $\approx$2 nm (area $\approx$12.6 nm^2), under a cantilever exerting a force of 1 nN, will experience a pressure $\approx$0.08 GPa. This $\approx$800$\times$ the atmospheric pressure! If a very high force is repeatedly applied on a membrane patch, a hole through the membrane can be observed upon imaging. In such a case, a new area on the membrane patch should be chosen and force spectroscopy performed with minimal force ($\leq$500 pN).
4. Cantilevers can also be contaminated during SMFS measurements. A simple indication of a contaminated cantilever is F–D traces of irreproducible lengths and patterns. To ascertain if the stylus is contaminated, choose an empty area on the mica and record F–D traces. If sawtooth signals similar to those obtained on unfolding proteins are still observed, the cantilever is contaminated and should be cleaned (see Notes 11–13).
5. Because cantilevers from different wafers usually exhibit different surface properties, they will stick differently to a

protein, i.e., predominantly to a loop or to a terminus of the protein. The extended lengths of the polypeptides would therefore be different in each case (Subheading 3.6). Force spectroscopy should therefore be performed with different cantilevers from different wafers, and the extended polypeptide lengths and unfolding forces compared from different data sets. Using different cantilevers is important for imparting high confidence to the data.

6. Once a maximum diode signal is obtained, and the deflection sensitivity and the spring constant measured, the laser beam on the cantilever should not be realigned. Because the cantilever deflection is measured by the reflection of the laser beam on the photodiode, changing the laser position, for example, shifting it from the cantilever tip toward the base, would give erroneous deflection sensitivity. Using this deflection sensitivity value will in turn give incorrect unfolding force values.

3.6. Analysis of F–D Traces

Owing to the high spatial and temporal resolution and sensitivity of SMFS, it is possible to decipher the unfolding pathways of a membrane protein and determine the mechanical and kinetic characteristics of the on-pathway intermediates (28). The distance between distinct peaks in an F–D trace reflects the gain in distance after unfolding and/or stretching a (folded) structure (α-helix, β-sheet, or β-strand) in the protein; the peak amplitude is a measure of the force required to unfold that structure (Fig. 4a, b). An unfolding event occurring at a given extension is thus specified by its distance from the preceding peak and the amplitude of the force peak at that extension. These "fingerprints" of a protein reveal the different populations of its unfolding events.

3.6.1. Determining the Pulling Terminus and Membrane Sidedness

As a first step in determining the unfolding pathway of a membrane protein, it is important to determine the terminus from which the protein is unfolded. Depending on the number of helices in the membrane protein, the N- and C-termini of the protein will be on the same (even number helices) or opposite (odd number helices) sides of the membrane bilayer. Recent examples illustrate that the membrane sidedness can be determined by high-resolution imaging (Fig. 7) (68, 114). Although relatively straightforward in the case of bacteriorhodopsin, it may not always be easy to conclusively assert the sidedness of the protein (36, 87, 115).

Antibodies targeted to specific loops or termini may be used to ascertain the membrane side (116). Enzymatic cleavage of specific loops and performing SMFS is a powerful way to determine the preferred orientation of adsorption of the membrane patch on a surface (117). In the case where both the termini of a membrane protein are on the same side of the membrane, it becomes much more important to determine the membrane sidedness (118, 119).

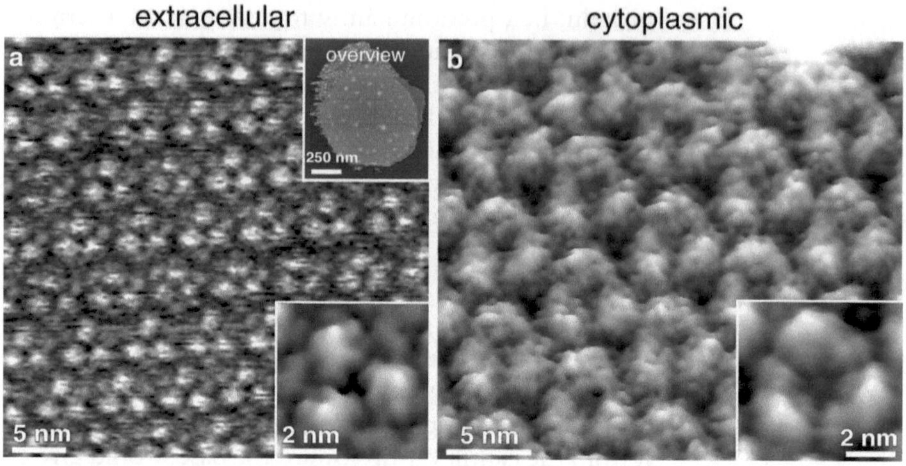

Fig. 7. Low-resolution and high-resolution topographs of bacteriorhodopsin. (**a**) Extracellular and (**b**) cytoplasmic sides of the bacteriorhodopsin lattice observed by high-resolution imaging. A purple membrane patch of diameter ≈600 nm (overview) lying flat on a mica surface (*top right inset*). In both extracellular and cytoplasmic sides, the trimer assembly of bacteriorhodopsin is clearly visible, and the resolution affords demarcation of individual monomers in the trimer (*bottom right insets*). These contact mode images were obtained in buffer at an imaging force ≤100 pN.

This is to ensure that the membrane patches are not predominantly adsorbed in a way that both the termini are facing the mica. When both the termini are on the same side, similar experiments to those described above, i.e., imaging many patches to get contour maps of the two sides, using antibodies targeted to the termini, and/or enzymatically cleaving a loop, should be employed to get unambiguous results.

The first example of unfolding a β-barrel membrane protein, OmpG, stresses the use of protein engineering in determining the pulling terminal end (119). Instead of cleaving a loop, the length of an extra-membranous loop was reduced by deleting 9 amino acids resulting in shorter F–D traces of the mutant OmpG compared to the wild-type protein. Providing detailed experimental and analytical procedures on how to determine the pulling end of a membrane protein using each of the above methods is beyond the scope of this chapter. The reader is referred to the following references (68, 85, 117–119).

3.6.2. F–D Trace Analysis

As explained previously, in SMFS experiments the cantilever stylus can attach nonspecifically to any of the solvent-exposed parts of the folded membrane protein resulting in F–D traces of various lengths (Fig. 8a). This necessitates an extremely careful analysis procedure of the F–D traces. The stochastic nature of the picking process inevitably leads to an experimental situation where most of the F–D signals have to be discarded due to various artifacts. (The stylus can potentially attach to misfolded or unfolded molecules on/in the membrane giving spurious F–D traces.) Besides

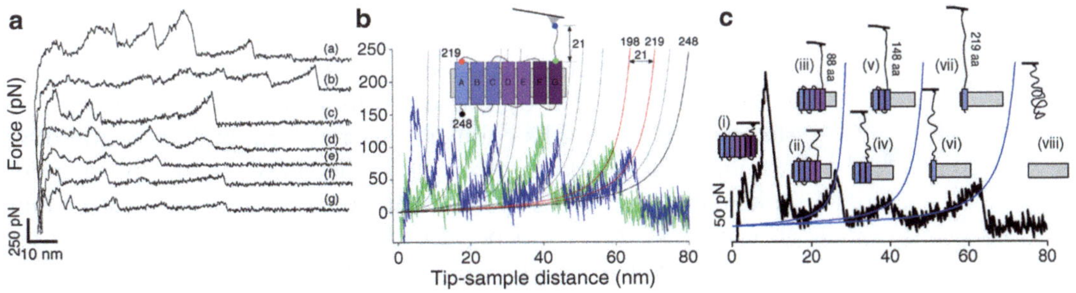

Fig. 8. Analysis and selection of F–D traces. (**a**) A variety of F–D traces are obtained during the mechanical unfolding of bacteriorhodopsin in an SMFS experiment. Traces due to attachment of multiple molecules or misfolded molecules to the cantilever stylus show peaks which do not fit to the WLC function and/or have forces higher than the average (traces **a**–**d**). (trace **e**) Many a time, F–D traces with a few well-defined peaks but shorter lengths are recorded. It may be assumed that in such cases the stylus detaches from the protein terminus during unfolding. All the traces falling in the above-mentioned and other ambiguous categories are not included in the analysis. (traces **f**, **g**) Representative F–D traces of bacteriorhodopsin unfolding from the C-terminal end. (**b**) The criteria used to select traces with well-defined peaks and the correct length are represented. To determine the correct length of an F–D trace, it is important to consider the length of the force-bearing terminus and the last structural segment to be unfolded. The cantilever stylus can attach at any position along the length of the terminus. In the two extreme cases—attachment of the stylus at the two ends of the terminus (*blue* and *green dots*)—the lengths of the F–D traces will differ by 21 amino acids (21×0.36 nm ≈ 7.6 nm), i.e., the length of the bacteriorhodopsin C-terminus. The last structural segment detected during the unfolding of a membrane protein is usually a complete α-helix or a part of it. In the case of bacteriorhodopsin, the last structural segment is a transmembrane part of the α-helix A (when pulled from the C-terminus). However, in the example shown, the whole α-helix A is considered as the last structural segment to unfold, and the force peak for the same is detected at amino acid position 219 (*red dot*). The F–D traces shown in blue and green correspond to the unfolding of bacteriorhodopsin from either end of the C-terminus till the beginning of α-helix A (*red dot*). The WLC fits (*red*) show that the observed difference in the lengths ($219 - 198 = 21$) is equal to the terminal length. The numbers at the end of the WLC fits denote the lengths (amino acid count from the C-terminal) of the unfolded polypeptides. The black WLC trace (248) corresponds to the complete unfolded length of the bacteriorhodopsin (between the *blue dot* at the C-terminus and the black dot at the N-terminus). (**c**) WLC fits to the force peaks give an estimate of the lengths of the unfolded structural segments. These structural segments determine the intermediates in the unfolding pathways of a membrane protein. The major unfolding pathway of bacteriorhodopsin, i.e., pairwise unfolding of helices, is elucidated from an F–D trace (120).

attaching at various regions of the molecule, the stylus can pick up none, one, or a few molecules of interest, thus contributing very different F–D traces encoded with varied information (Fig. 8a). Simultaneous pickup of multiple proteins creates parallel and, therefore, additive springs between the cantilever and the surface leading to higher forces and steep slopes to each peak. F–D traces with these characteristic peaks should be discarded from analysis to ensure that data originating only from single molecules are analyzed. Thus, the first step in analyzing F–D traces is to sort out the "good" traces from the "bad" ones.

A suitable and unambiguous criterion that is used to select F–D traces of bacteriorhodopsin (and membrane proteins in general) is the overall length of the signal, which reflects the tip–sample distance at which the last force peak occurs (68, 120) (Fig. 8b). The contour length (L) of a fully stretched protein upon unfolding can be estimated from its sequence length (l_s),

$$L = l_s \times l_p \tag{12}$$

l_p is the polypeptide persistence length ≈0.3–0.4 nm (66, 68). Pulling on a protein by one of its loops results in F–D traces of shorter overall lengths compared to a protein pulled by one of its termini. Also, if the molecular contact between the polypeptide linker and the stylus is broken while the protein is being stretched, shorter F–D traces will be recorded. Figure 8b demonstrates the method applied to determine the length of bacteriorhodopsin molecules upon forced unfolding from its C-terminal end. Depending on the amino acid residue at the C-terminal end where the cantilever stylus attaches, the length of the resulting unfolded polypeptide would differ. It has been shown for bacteriorhodopsin that F–D traces exhibiting lengths between 60 and 70 nm result from unfolding by pulling at their C- or N-termini (68, 114, 120).

All force signals exhibiting overall lengths of 60–70 nm and similar unfolding patterns should be selected and aligned (superimposed) manually (or using an algorithm) (e.g., in IGOR Pro, WaveMetrics, Inc., Oregon, USA). The alignment of F–D traces is a non-trivial task owing to the large number of peaks in an F–D trace (3–9 in bacteriorhodopsin). Alignment of the F–D traces therefore requires an objective procedure making use of an internal reference. Taking the adhesion peak at a stylus–membrane separation of ≤25 nm as the reference, the F–D traces should be superimposed in the first instance. This method of alignment also compensates for attachment at multiple and/or different sites of the C-terminus. However, the adhesion peak may not always be a good criterion for an internal standard. A first analysis of the F–D traces would help to determine if there is a repetitive unfolding pattern (observed in all the membrane proteins unfolded till date) (28, 86, 87, 119). One or more force peaks would occur in all the F–D traces (e.g., 3 force peaks in bacteriorhodopsin have an occurrence probability of 100%). These peaks should be used as reference points (along with the adhesion peak) to ascertain the reproducibility of the F–D traces and to realign the signals. In many published papers, the F–D traces have been aligned manually using the criteria described above. Algorithms have been developed for F–D signal alignment, peak fitting, and force determination. However, in spite of the time-consuming and cumbersome nature of manual alignment, it offers a reliable method to confirm data reproducibility.

The high spatial resolution (≈2 nm) of SMFS is crucial in identifying the unfolded structural segments, i.e., whether the force peaks correspond to one or more helices, to loops, or to a combination of helices and loops. The force peaks in an F–D trace can be modeled using the wormlike chain (WLC) function (121),

$$F(x) = \left(\frac{k_B T}{l_p}\right) \cdot \left[0.25\left(1 - \frac{x}{L}\right)^{-2} - 0.25 + \frac{x}{L}\right] \quad (13)$$

A fit of the WLC function to the force peaks gives an estimate of the contour lengths (L in nm) of the unfolded structural segments. The number of amino acids from the force-bearing terminus is counted to determine the size of the structural segment corresponding to the respective force peak (Fig. 8c). A correction factor compensating for the bilayer thickness (≈4 nm) should be considered and 11 amino acids added to L (from WLC) while assigning the structural segments which originate on the opposite side of the bilayer (86, 117, 120). Once the structural segments are determined, they are mapped onto the secondary structure of the membrane protein and the complete unfolding pathway elucidated (28). For a comprehensive overview of the unfolding pathway, the mechanical stability of each segment is determined from the force peak amplitude. The statistical significance of certain combinations of force peaks over others indicates the dominance of energetically favorable pathways (39, 119). Crucially, to elucidate the unfolding pathways of a protein, it is important to have knowledge of its secondary structure.

4. Notes

1. An AFM is an expensive and extremely sensitive instrument, and therefore utmost care must be exercised to maintain its sensitivity and life. The easiest parts to break are the head and the piezoelectric actuator. The head is fitted with a laser, a reflecting mirror, and a quadrant photodiode, all of which are fragile and sensitive components. Dropping the head or the actuator may result in erroneous measurements. If dropped, the instrument must be recalibrated using the standard calibration grids provided by the company. It is also advisable to check the calibration using a model protein such as bacteriorhodopsin or titin.
2. Single-molecule experiments are sensitive to dust and extraneous particles. The AFM and the laboratory should be kept clean and dust free.
3. Single-molecule AFM measurements are prone to molecular contaminations and small surface irregularities. Therefore, the mica support and the piezoelectric scanner surface should be cleaned with propanol using Kimwipes® before mounting

the support onto the piezoelectric actuator. Water should never be used! If spilled, water may cause a short circuit in the actuator's electronics.

4. The mirror in the head should also be cleaned periodically to prevent noise in the measurements. Clean the mirror gently using a cotton swab (Q-tip) dabbed in propanol.
5. The AFM should be isolated to minimize electrical, mechanical, and acoustic noise, and thermal drift. The instrument should always be placed on a vibration-damping table and should not be surrounded by electrical equipment. If interfering electrical noise is observed, each electrical device should be switched off to determine the noise source. Covering the AFM with a glass bell jar is a convenient way to cancel acoustic noise. If a closed cabinet is used, lining the walls of the cabinet with thick sponge pieces offers an additional effective way to damp acoustic noise.
6. The fluid cell should be cleaned before and after every experiment. Avoid touching the fluid cell with bare hands to prevent transferring any fats from the hands onto the fluid cell. Powder-free gloves should be used to avoid particulate contamination. Clean the fluid cell carefully with a mild dishwashing detergent (e.g., Pril, Henkel GmbH, Germany), gently dabbing the upper and the lower (where the cantilever is fixed) surfaces with fingertips. Wash the cell with copious amounts of distilled water making sure that no detergent residue is left on the fluid cell and inside the buffer reservoirs (residual detergent on the fluid cell may dissolve in the sample buffer disrupting the lipid bilayers). Rinse the fluid cell with ethanol followed by Millipore water alternatively at least 3 times. Dry the fluid cell under a stream of filtered nitrogen using a nozzle (a pipette tip fitted at the end of a tube connected to a nitrogen outlet).
7. Cantilevers from a single wafer usually give similar diode signals after proper laser alignment. If the diode sum signal obtained is less than expected, check if the cantilever is fixed properly to the fluid cell. If the laser beam is being reflected from the supporting mica surface, the signal obtained will also be less. An indication of the laser being reflected from the supporting surface is the observation of a diffuse laser spot on a paper strip (this is easily seen when the paper method is used for aligning the laser). A scratched coating on the cantilever surface may also cause a low diode signal.
8. It is important to handle the cantilevers with utmost care while breaking them off the wafer. General tips to follow are, always wear gloves, and the tools (e.g., tweezers) to handle the cantilevers and the surface (e.g., glass slides) used for breaking the cantilevers from the wafer should be cleaned before starting the

procedure. Always place the cantilevers on a Gel-Pak® box with the stylus bearing side facing up.

9. It is a good practice to scan the freshly cleaved mica surface before adsorbing a membrane sample to check for the flatness of the surface, and to ensure if the stylus has the desirable properties. Check the integrity of the freshly cleaved surface and the cantilever stylus by imaging a large area (50–100 μm × 50–100 μm) in buffer in the contact mode (Subheading 3.4.1).

10. Before adsorbing the sample, also record F–D traces on the mica surface (Subheading 3.5). If the cantilever and the surface are clean, a sharp transition at the stylus-surface contact will be observed. A contaminated stylus would give irreproducible F–D traces. Only those supports that show a height variation of ≤5 nm, and do not produce any F–D traces should be used to adsorb membrane patches.

11. If a cantilever is contaminated while imaging or force spectroscopy, the following steps can be taken:

    (a) Select a spot on the mica without any membranes. Scan the surface (20 Hz) with increasing force than used for imaging (starting from 100 to 1,000 pN). Scanning at contact forces >1 nN should be avoided as this may result in blunting of the stylus.

    (b) Using long ramps (1–5 μm) in the force spectroscopy mode is also useful in removing contaminants by pulling them off the stylus. A clean stylus is indicated by sharp transitions in the F–D traces (without any force peaks) at the contact region between the stylus and the surface. If a cantilever is contaminated, short and long F–D traces (up to 5 μm or more) will be observed. Continue the approach–retract cycle of the cantilever until no F–D traces are observed.

    (c) If the above steps do not get rid of the contaminants, clean the mica surface or change to a new support. Ascertain that the buffers and other solutions are free of extraneous matter; filter through a 0.2 μm filter and store in clean glassware rinsed with Millipore water. The sample should be free of any degraded membrane and protein fragments which can be removed by centrifugation.

    (d) Cantilevers from certain wafers are "stickier" than from others. Although a useful feature for force spectroscopy to facilitate easy pickup of molecules by nonspecific attachment, sticky cantilevers are also prone to faster contamination rendering imaging difficult. Therefore, using a cantilever from a new wafer may overcome frequent contamination.

12. Cantilevers may be reused after cleaning in Hellmanex® (2% v/v) or in SDS (1–2% v/v). Soak the cantilevers in

Hellmanex (50 °C) or SDS (room temperature) overnight (or longer) in a Petri plate (glass is preferred when using Hellmanex®). Pick the cantilevers gently using tweezers, and rinse in Millipore water kept in 6–8 different Petri dishes or a 96-well plate. The cantilevers may also be rinsed in ethanol followed by water. It should however be noted that this procedure does not always get rid of the contaminants. The cleaning efficiency can be increased by placing the cantilevers (in detergent) in a sonication bath for 1–2 h with intermittent sonication. It is also crucial to thoroughly rinse the cantilevers to avoid any detergent from dissolving into the sample buffer on the support.

13. If a cantilever is being reused, its spring constant should be recalibrated and not assumed from a previous experiment. The resonance frequency, f_r, of a simple harmonic oscillator, for example, a rectangular cantilever, can be described using the equation

$$f_r = \frac{1}{2\pi}\sqrt{\frac{\kappa}{m}} \tag{14}$$

m is the mass on the end of the cantilever spring. Thus, if m becomes $m + m_a$ (m_a being the extra mass due to contamination), f_r will change accordingly giving a wrong value for κ. It is commonly observed that a cantilever is contaminated but the contaminant does not interfere while recording F–D traces (this may happen if denatured proteins are stuck at the base of the stylus). Such a contaminated cantilever may be detected upon recalibration. On the other hand, a contaminant that may have been present on a 'clean' cantilever to start with may be removed upon cleaning, thus changing the spring constant.

14. If no membrane patches are observed, the following should be considered:
    (a) A common cause of not observing membrane patches is high force. Applying a high contact force ruptures the membrane patches, and no height variations are discernible. Therefore, optimize the force by decreasing the set point until the stylus is no longer in contact with the surface and increase in small steps again.
    (b) Alternatively, increasing the sample concentration, adsorption time, and/or optimizing the salt concentration, pH of the buffer may help.

15. On the other hand, if too many membrane patches or stacks of those are observed, decreasing the sample concentration and/or the adsorption time is most likely to solve the problem. Mild sonication of the sample may help, but care needs to be exercised since membrane patches may rupture to very small fragments rendering them useless for any imaging or force spectroscopy applications.

16. High-resolution imaging is a test of perfect methodology. The importance of proper calibration of the piezoelectric actuator, optimum gains and contact force, clean cantilevers and setup, and perfectly made flat supports cannot be overstressed.
17. Some common sources of imaging artifacts are: misalignment of the laser on the cantilever tending to give a low diode signal, improper installation of the fluid cell, i.e., the fluid cell is not secured tightly in the AFM head with the screws, and/or a is wobbly cantilever owing to a defective spring in the fluid cell. Teflon and mica discs, which are not stuck properly owing to incomplete drying of the glue, air bubbles between the discs, and nonuniform spreading of the glue, may all cause vibrations. If the height profile of an image is not in concurrence with the reported or expected value, check the calibration of the piezoelectric scanner.
18. In the case of membrane protein unfolding, using nonspecific attachment between the stylus and the protein is recommended instead of forming a specific Au–S linkage between a gold-coated stylus and an engineered cysteine in one of the protein termini. Nonspecific attachment has been shown to provide equivalent results and allows a much higher throughput as compared to specific attachment (120). Nonspecific protein–stylus attachment causes less frequent contamination of the cantilever than specific attachment. This is mainly owing to the easy removal of the nonspecifically attached protein from the stylus during the approach–retract cycles.

Acknowledgments

KTS was supported by an independent Marie Curie Intra-European Fellowship. KTS expresses gratitude to Prof. Daniel Müller and Dr. Stefania Mari for their invaluable input.

References

1. Binnig G, Rohrer H, Gerber C, Weibel E (1982) Surface studies by scanning tunneling microscopy. Phys Rev Lett 49:57–61
2. Binnig G, Quate CF, Gerber C (1986) Atomic force microscope. Phys Rev Lett 56:930–933
3. Bongrand P, Capo C, Mege J-L, Benoliel A-M (1988) Use of hydrodynamic flows to study cell adhesion. In: Bongrand P (ed) Physical basis of cell adhesion. CRC Press, Boca Raton, FL, pp 125–156
4. Leckband DE, Israelachvili JN, Schmitt FJ, Knoll W (1992) Long-range attraction and molecular rearrangements in receptor-ligand interactions. Science 255:1419–1421
5. Merkel R, Nassoy P, Leung A, Ritchie K, Evans E (1999) Energy landscapes of receptor-ligand bonds explored with dynamic force spectroscopy. Nature 397:50–53

6. Ashkin A (1997) Optical trapping and manipulation of neutral particles using lasers. Proc Natl Acad Sci U S A 94:4853–4860
7. Junker JP, Ziegler F, Rief M (2009) Ligand-dependent equilibrium fluctuations of single calmodulin molecules. Science 323:633–637
8. Müller DJ, Engel A (2007) Atomic force microscopy and spectroscopy of native membrane proteins. Nat Protoc 2:2191–2197
9. Engel A, Müller DJ (2000) Observing single biomolecules at work with the atomic force microscope. Nat Struct Biol 7:715–718
10. Muller DJ, Dufrene YF (2011) Force nanoscopy of living cells. Curr Biol 21:R212–R216
11. Muller DJ, Krieg M, Alsteens D, Dufrene YF (2009) New frontiers in atomic force microscopy: analyzing interactions from single-molecules to cells. Curr Opin Biotechnol 20:4–13
12. Alegre-Cebollada J, Perez-Jimenez R, Kosuri P, Fernandez JM (2010) Single-molecule force spectroscopy approach to enzyme catalysis. J Biol Chem 285:18961–18966
13. Zhuang X, Rief M (2003) Single-molecule folding. Curr Opin Struct Biol 13:88–97
14. Brujic J, Hermans R, Walther KA, Fernandez JM (2006) Single molecule force spectroscopy reveals signatures of glassy dynamics in the energy landscape of ubiquitin. Nat Phys 2:282–286
15. Cyr DM, Venkataraman B, Flynn GW (1996) STM investigations of organic molecules physisorbed at the liquid–solid interface. Chem Mater 8:1600–1615
16. Gross L, Mohn F, Moll N, Liljeroth P, Meyer G (2009) The chemical structure of a molecule resolved by atomic force microscopy. Science 325:1110–1114
17. Ando T, Kodera N, Takai E, Maruyama D, Saito K, Toda A (2001) A high-speed atomic force microscope for studying biological macromolecules. Proc Natl Acad Sci U S A 98:12468–12472
18. Uchihashi T, Iino R, Ando T, Noji H (2011) High-speed atomic force microscopy reveals rotary catalysis of rotorless F1-ATPase. Science 333:755–758
19. Shibata M, Yamashita H, Uchihashi T, Kandori H, Ando T (2010) High-speed atomic force microscopy shows dynamic molecular processes in photoactivated bacteriorhodopsin. Nat Nanotechnnol 5:208–212
20. Puchner EM, Gaub HE (2009) Force and function: probing proteins with AFM-based force spectroscopy. Curr Opin Struct Biol 19:605–614
21. Alberts B, Johnson A, Lewis J, Raff M, Roberts K, Walter P (2002) Molecular biology of the cell, vol 4. Garland Science, New York
22. Bowie JU (2005) Solving the membrane protein folding problem. Nature 438:581–589
23. White SH (2009) Biophysical dissection of membrane proteins. Nature 459:344–346
24. Booth PJ (2003) The trials and tribulations of membrane protein folding in vitro. Biochim Biophys Acta 1610:51–56
25. Sachs JN, Engelman DM (2006) Introduction to the membrane protein reviews: the interplay of structure, dynamics, and environment in membrane protein function. Ann Rev Biochem 75:707–712
26. Engelman DM (2005) Membranes are more mosaic than fluid. Nature 438:578–580
27. Muller DJ (2008) AFM: a nanotool in membrane biology. Biochemistry 47:7986–7998
28. Kedrov A, Janovjak H, Sapra KT, Muller DJ (2007) Deciphering molecular interactions of native membrane proteins by single-molecule force spectroscopy. Annu Rev Biophys Biomol Struct 36:233–260
29. Engel A, Gaub HE (2008) Structure and mechanics of membrane proteins. Annu Rev Biochem 77:127–148
30. Seelert H, Poetsch A, Dencher NA, Engel A, Stahlberg H, Müller DJ (2000) Structural biology: proton powered turbine of a plant motor. Nature 405:418–419
31. Müller DJ, Engel A (1999) Voltage and pH-induced channel closure of porin OmpF visualized by atomic force microscopy. J Mol Biol 285:1347–1351
32. Yu J, Bippes CA, Hand GM, Muller DJ, Sosinsky GE (2007) Aminosulfonate modulated pH-induced conformational changes in connexin26 hemichannels. J Biol Chem 282:8895–8904
33. Kedrov A, Krieg M, Ziegler C, Kuhlbrandt W, Müller DJ (2005) Locating ligand binding and activation of a single antiporter. EMBO Rep 6:668–674
34. Park PS, Sapra KT, Kolinski M, Filipek S, Palczewski K, Muller DJ (2007) Stabilizing effect of Zn^{2+} in native bovine rhodopsin. J Biol Chem 282:11377–11385
35. Janovjak H, Kessler M, Oesterhelt D, Gaub HE, Müller DJ (2003) Unfolding pathways of native bacteriorhodopsin depend on temperature. EMBO J 22:5220–5229
36. Cisneros DA, Oberbarnscheidt L, Pannier A, Klare JP, Helenius J, Engelhard M, Oesterhelt F, Muller DJ (2008) Transducer binding

establishes localized interactions to tune sensory rhodopsin II. Structure 16:1206–1213
37. Sapra KT, Besir H, Oesterhelt D, Muller DJ (2006) Characterizing molecular interactions in different bacteriorhodopsin assemblies by single-molecule force spectroscopy. J Mol Biol 355:640–650
38. Sapra KT, Doehner J, Renugopalakrishnan V, Padros E, Muller DJ (2008) Role of extracellular glutamic acids in the stability and energy landscape of bacteriorhodopsin. Biophys J 95:3407–3418
39. Sapra KT, Balasubramanian GP, Labudde D, Bowie JU, Muller DJ (2008) Point mutations in membrane proteins reshape energy landscape and populate different unfolding pathways. J Mol Biol 376:1076–1090
40. Kedrov A, Janovjak H, Ziegler C, Kühlbrandt W, Müller DJ (2006) Observing folding kinetics and pathways of single antiporters. J Mol Biol 355:2–8
41. Stewart MP, Toyoda Y, Hyman AA, Muller DJ (2011) Force probing cell shape changes to molecular resolution. Trends Biochem Sci 36:444–450
42. Krieg M, Arboleda-Estudillo Y, Puech PH, Kafer J, Graner F, Muller DJ, Heisenberg CP (2008) Tensile forces govern germ-layer organization in zebrafish. Nat Cell Biol 10:429–436
43. Müller DJ, Dufrêne YF (2011) Atomic force microscopy: a nanoscopic window on the cell surface. Trends Cell Biol 21:461–469
44. Müller DJ, Hand GM, Engel A, Sosinsky G (2002) Conformational changes in surface structures of isolated connexin26 gap junctions. EMBO J 21:3598–3607
45. Hoogenboom BW, Hug HJ, Pellmont Y, Martin S, Frederix PLTM, Fotiadis D, Engel A (2006) Quantitative dynamic-mode scanning force microscopy in liquid. Appl Phys Lett 88:193109
46. Hansma PK, Cleveland JP, Radmacher M, Walters DA, Hillner PE, Bezanilla M, Fritz M, Vie D, Hansma HG, Prater CB, Massie J, Fukunaga L, Gurley J, Elings V (1994) Tapping mode atomic force microscopy in liquids. Appl Phys Lett 64:1738–1740
47. Müller DJ, Fotiadis D, Scheuring S, Müller SA, Engel A (1999) Electrostatically balanced subnanometer imaging of biological specimens by atomic force microscope. Biophys J 76:1101–1111
48. Butt H-J, Jaschke M, Ducker W (1995) Measuring surface forces in aqueous solution with the atomic force microscope. Bioelect Bioenerg 38:191–201
49. Butt H-J (1992) Measuring local surface charge densities in electrolyte solutions with a scanning force microscope. Biophys J 63:578–582
50. Butt H-J (1992) Electrostatic interaction in scanning probe microscopy when imaging in electrolyte solutions. Nanotechnology 3:60–68
51. Müller DJ, Engel A (1997) The height of biomolecules measured with the atomic force microscope depends on electrostatic interactions. Biophys J 73:1633–1644
52. Czajkowsky D, Allen M, Elings V, Shao Z (1998) Direct visualization of surface charge in aqueous solution. Ultramicroscopy 74:1–5
53. Baker AA, Helbert W, Sugiyama J, Miles MJ (2000) New insight into cellulose structure by atomic force microscopy shows the i(alpha) crystal phase at near-atomic resolution. Biophys J 79:1139–1145
54. Müller DJ, Büldt G, Engel A (1995) Force-induced conformational change of bacteriorhodopsin. J Mol Biol 249:239–243
55. Müller DJ, Engel A (2002) Conformations, flexibility, and interactions observed on individual membrane proteins by atomic force microscopy. Methods Cell Biol 68:257–299
56. Karrasch S, Dolder M, Hoh J, Schabert F, Ramsden J, Engel A (1993) Covalent binding of biological samples to solid supports for scanning probe microscopy in buffer solution. Biophys J 65:2437–2446
57. Bezanilla M, Drake B, Nudler E, Kashlev M, Hansma PK, Hansma HG (1994) Motion and enzymatic degradation of DNA in the atomic force microscope. Biophys J 67:2454–2459
58. Fritz M, Radmacher M, Allersma MW, Cleveland JP, Stewart RJ, Hansma PK, Schmidt CF (1995) Imaging microtubles in buffer solution using tapping mode atomic force microscopy. SPIE 2384:150–157
59. Guthold M, Zhu X, Rivetti C, Yang G, Thomson NH, Kasas S, Hansma HG, Smith B, Hansma PK, Bustamante C (1999) Direct observation of one-dimensional diffusion and transcription by *Escherichia coli* RNA polymerase. Biophys J 77:2284–2294
60. Elie-Caille C, Severin F, Helenius J, Howard J, Muller DJ, Hyman AA (2007) Straight GDP-tubulin protofilaments form in the presence of taxol. Curr Biol 17:1765–1770
61. Hoogenboom BW, Suda K, Engel A, Fotiadis D (2007) The supramolecular assemblies of - voltage-dependent anion channels in the native membrane. J Mol Biol 370:246–255

62. Viani MB, Pietrasanta LI, Thompson JB, Chand A, Gebeshuber IC, Kindt JH, Richter M, Hansma HG, Hansma PK (2000) Probing protein–protein interactions in real time. Nat Struct Biol 7:644–647
63. Yamashita H, Voïtchovsky K, Uchihashi T, Contera SA, Ryan JF, Ando T (2009) Dynamics of bacteriorhodopsin 2D crystal observed by high-speed atomic force microscopy. J Struct Biol 167:153–158
64. Junge W, Muller DJ (2011) Seeing a molecular motor at work. Science 333:704–705
65. Florin E-L, Moy VT, Gaub HE (1994) Adhesion forces between individual ligand-receptor pairs. Science 264:415–417
66. Rief M, Gautel M, Oesterhelt F, Fernandez JM, Gaub HE (1997) Reversible unfolding of individual titin immunoglobulin domains by AFM. Science 276:1109–1112
67. Cui Y, Bustamante C (2000) Pulling a single chromatin fiber reveals the forces that maintain its higher-order structure. Proc Natl Acad Sci U S A 97:127–132
68. Oesterhelt F, Oesterhelt D, Pfeiffer M, Engel A, Gaub HE, Müller DJ (2000) Unfolding pathways of individual bacteriorhodopsins. Science 288:143–146
69. Borgia A, Williams PM, Clarke J (2008) Single-molecule studies of protein folding. Annu Rev Biochem 77:101–125
70. Hinterdorfer P, Dufrene YF (2006) Detection and localization of single molecular recognition events using atomic force microscopy. Nat Methods 3:347–355
71. Wiita AP, Perez-Jimenez R, Walther KA, Grater F, Berne BJ, Holmgren A, Sanchez-Ruiz JM, Fernandez JM (2007) Probing the chemistry of thioredoxin catalysis with force. Nature 450:124–127
72. Zhang X, Halvorsen K, Zhang CZ, Wong WP, Springer TA (2009) Mechanoenzymatic cleavage of the ultralarge vascular protein von Willebrand factor. Science 324:1330–1334
73. Alegre-Cebollada J, Kosuri P, Rivas-Pardo JA, Fernández JM (2011) Direct observation of disulfide isomerization in a single protein. Nat Chem 3:882–887
74. Dietz H, Rief M (2004) Exploring the energy landscape of GFP by single-molecule mechanical experiments. Proc Natl Acad Sci U S A 101:16192–16197
75. Evans E (2001) Probing the relation between force-lifetime-and chemistry in single molecular bonds. Annu Rev Biophys Biomol Struct 30:105–128
76. Evans E (1998) Energy landscapes of biomolecular adhesion and receptor anchoring at interfaces explored with dynamic force spectroscopy. Faraday Discuss 111:1–16
77. Baldwin RL, Rose GD (1999) Is protein folding hierarchic? II. Folding intermediates and transition states. Trends Biochem Sci 24:77–83
78. Baldwin RL, Rose GD (1999) Is protein folding hierarchic? I. Local structure and peptide folding. Trends Biochem Sci 24:26–33
79. Rief M, Fernandez JM, Gaub HE (1998) Elastically coupled two-level-systems as a model for biopolymer extensibility. Phys Rev Lett 81:4764–4767
80. Evans E, Ritchie K (1997) Dynamic strength of molecular adhesion bonds. Biophys J 72:1541–1555
81. Bell GI (1978) Models for the specific adhesion of cells to cells. Science 200:618–627
82. Evans E (1999) Looking inside molecular bonds at biological interfaces with dynamic force spectroscopy. Biophys Chem 82:83–97
83. Janovjak H, Struckmeier J, Hubain M, Kedrov A, Kessler M, Müller DJ (2004) Probing the energy landscape of the membrane protein bacteriorhodopsin. Structure 12:871–879
84. Nevo R, Brumfeld V, Kapon R, Hinterdorfer P, Reich Z (2005) Direct measurement of protein energy landscape roughness. EMBO Rep 6:482–486
85. Kedrov A, Hellawell AM, Klosin A, Broadhurst RB, Kunji ER, Müller DJ (2010) Probing the interactions of carboxy-atractyloside and atractyloside with the yeast mitochondrial ADP/ATP carrier. Structure 18:39–46
86. Ge L, Perez C, Waclawska I, Ziegler C, Muller DJ (2011) Locating an extracellular K+-dependent interaction site that modulates betaine-binding of the Na+-coupled betaine symporter BetP. Proc Natl Acad Sci U S A 108:E890–898
87. Bippes CA, Zeltina A, Casagrande F, Ratera M, Palacin M, Muller DJ, Fotiadis D (2009) Substrate binding tunes conformational flexibility and kinetic stability of an amino acid antiporter. J Biol Chem 284:18651–18663
88. Sapra KT, Park PS, Palczewski K, Muller DJ (2008) Mechanical properties of bovine rhodopsin and bacteriorhodopsin: possible roles in folding and function. Langmuir 24:1330–1337
89. Damaghi M, Sapra KT, Koster S, Yildiz O, Kuhlbrandt W, Muller DJ (2010) Dual energy landscape: the functional state of the beta-barrel outer membrane protein G molds its

unfolding energy landscape. Proteomics 10:4151–4162

90. Schitter G, Stark RW, Stemmer A (2004) Fast contact-mode atomic force microscopy on biological specimen by model-based control. Ultramicroscopy 100:253–257
91. Mahmood IA, Moheimani SOR (2009) Making a commercial atomic force microscope more accurate and faster using positive position feedback control. Rev Sci Instrum 80:063705
92. Meyer E, Hug HJ, Bennewitz R (2003) Scanning probe microscopy—the lab on a tip. Springer, Berlin
93. Yasumura KY, Stowe TD, Chow EM, Pfafman T, Kenny TW, Stipe BC, Rugar D (2000) Quality factors in micron- and submicron-thick cantilevers. J Microelectromech Syst 9:117–125
94. Zong Q, Innis D, Kjoller K, Elings VB (1993) Fractured polymer/silica fiber surface studied by tapping mode atomic force microscopy. Surf Sci Lett 290:L688–L692
95. Viani MB, Schäffer TE, Paloczi GT, Pietrasanta LI, Smith BL, Thompson JB, Richter M, Rief M, Gaub HE, Plaxco KW, Cleland AN, Hansma HG, Hansma PK (1999) Fast imaging and fast force spectroscopy of single biopolymers with a new atomic force microscope designed for small cantilevers. Rev Sci Ins 70:4300–4303
96. Viani MB, Schäfer TE, Chand A, Rief M, Gaub H, Hansma PK (1999) Small cantilevers for force spectroscopy of single molecules. J Appl Phys 86:2258–2262
97. Friedrichs J, Helenius J, Muller DJ (2010) Quantifying cellular adhesion to extracellular matrix components by single-cell force spectroscopy. Nat Protoc 5:1353–1361
98. Cleveland JP, Manne S, Bocek D, Hansma PK (1993) A nondestructive method for determining the spring constant of cantilevers for scanning force microscopy. Rev Sci Instrum 64:403–405
99. Sader JE, Larson I, Mulvaney P, White LR (1995) Method for calibration of atomic force cantilevers. Rev Sci Instrum 60:3131–3134
100. Gibson CT, Watson GS, Myhra S (1996) Determination of the spring constants of probes for force microscopy/spectroscopy. Nanotechnology 7:259–262
101. Torii A, Sasaki M, Hane K, Okuma S (1996) A method for determining the spring constant of cantilevers for atomic force microscopy. Meas Sci Technol 7:179–184
102. Butt H-J, Jaschke M (1995) Calculation of thermal noise in atomic force microscopy. Nanotechnology 6:1–7
103. Florin E-L, Rief M, Lehmann H, Ludwig M, Dornmair C, Moy VT, Gaub HE (1995) Sensing specific molecular interactions with the atomic force microscope. Biosens Bioelectron 10:895–901
104. Howard J (2001) Mechanics of motor proteins and the cytoskeleton. Sinauer Associates Inc., Sunderland, MA
105. Muller DJ, Engel A (2008) Strategies to prepare and characterize native membrane proteins and protein membranes by AFM. Curr Opin Colloid Int Sci 13:338–350
106. Pashley RM (1981) Hydration forces between mica surfaces in Li^+, Na^+, Na^+ and Cs^+ electrolyte solutions: a correlation of double layer and hydration forces with surface cation exchange properties. J Colloid Interface Sci 83:531–546
107. Müller DJ, Amrein M, Engel A (1997) Adsorption of biological molecules to a solid support for scanning probe microscopy. J Struct Biol 119:172–188
108. Schwarz UD, Haefke H, Reimann P, Guntherodt HJ (1994) Tip artefacts in scanning force microscopy. J Microsc 173:183–197
109. Möller C, Allen M, Elings V, Engel A, Müller DJ (1999) Tapping-mode atomic force microscopy produces faithful high-resolution images of protein surfaces. Biophys J 77:1150–1158
110. Oberhauser AF, Hansma PK, Carrion-Vazquez M, Fernandez JM (2001) Stepwise unfolding of titin under force-clamp atomic force microscopy. Proc Natl Acad Sci U S A 98:468–472
111. Fernandez JM, Li H (2004) Force-clamp spectroscopy monitors the folding trajectory of a single protein. Science 303:1674–1678
112. Dietz H, Bertz M, Schlierf M, Berkemeier F, Bornschlogl T, Junker JP, Rief M (2006) Cysteine engineering of polyproteins for single-molecule force spectroscopy. Nat Protoc 1:80–84
113. Cisneros DA, Muller DJ, Daud SM, Lakey JH (2006) An approach to prepare membrane proteins for single-molecule imaging. Angew Chem Int Ed Engl 45:3252–3256
114. Kessler M, Gaub H (2006) Unfolding barriers in bacteriorhodopsin probed from the cytoplasmic and the extracellular side by AFM. Structure 14:521–527
115. Cisneros DA, Oesterhelt D, Müller DJ (2005) Probing origins of molecular

interactions stabilizing the membrane proteins halorhodopsin and bacteriorhodopsin. Structure 13:235–242

116. Müller DJ, Schoenenberger CA, Büldt G, Engel A (1996) Immuno-atomic force microscopy of purple membrane. Biophys J 70:1796–1802

117. Sapra KT, Park PS, Filipek S, Engel A, Müller DJ, Palczewski K (2006) Detecting molecular interactions that stabilize native bovine rhodopsin. J Mol Biol 358:255–269

118. Kedrov A, Ziegler C, Janovjak H, Kuhlbrandt W, Müller DJ (2004) Controlled unfolding and refolding of a single sodium-proton antiporter using atomic force microscopy. J Mol Biol 340:1143–1152

119. Sapra KT, Damaghi M, Koster S, Yildiz O, Kuhlbrandt W, Muller DJ (2009) One beta hairpin after the other: exploring mechanical unfolding pathways of the transmembrane beta-barrel protein OmpG. Angew Chem Int Ed Engl 48:8306–8308

120. Müller DJ, Kessler M, Oesterhelt F, Möller C, Oesterhelt D, Gaub H (2002) Stability of bacteriorhodopsin alpha-helices and loops analyzed by single-molecule force spectroscopy. Biophys J 83:3578–3588

121. Bustamante C, Marko JF, Siggia ED, Smith S (1994) Entropic elasticity of lambda-phage DNA. Science 265:1599–1600

122. Stahlberg H, Muller DJ, Suda K, Fotiadis D, Engel A, Meier T, Matthey U, Dimroth P (2001) Bacterial Na(+)-ATP synthase has an undecameric rotor. EMBO Rep 2:229–233

123. Scheuring S, Sturgis JN (2005) Chromatic adaptation of photosynthetic membranes. Science 309:484–487

124. Czajkowsky DM, Hotze EM, Shao Z, Tweten RK (2004) Vertical collapse of a cytolysin prepore moves its transmembrane beta-hairpins to the membrane. EMBO J 23:3206–3215

125. Fotiadis D, Liang Y, Filipek S, Saperstein DA, Engel A, Palczewski K (2003) Atomic-force microscopy: rhodopsin dimers in native disc membranes. Nature 421:127–128

Chapter 6

Optimized Negative-Staining Protocol for Examining Lipid-Protein Interactions by Electron Microscopy

Mark Garewal, Lei Zhang, and Gang Ren

Abstract

A large number of proteins are capable of inserting themselves into lipids, and interacting with membranes, such as transmembrane proteins and apolipoproteins. Protein–lipid interactions have been identified as one of the keys in understanding biological processes, while the structure of proteins at the lipid-binding stage can provide evidence to help identify their roles and critical functions. However, structure determination of proteins at the lipid-binding stage is rather difficult, because conformational and compositional heterogeneities of the protein–lipid complexes are major barriers to unravel their structures using traditional methods, such as X-ray crystallography. Electron microscopy (EM) is an alternative approach to determine protein structure and has demonstrated a capability in visualizing lipid–protein interactions directly. Among various EM techniques, negative-staining (NS) is an easy, rapid, qualitative approach that is a well-established technique, frequently used in research laboratories. Conventional NS protocols, unfortunately, often generate artifacts with lipid-related proteins, such as the rouleau formation of lipoproteins. To overcome this artifact formation, Ren and his colleagues recently developed an optimized NS protocol that was validated by comparing images of lipoproteins from cryo-electron microscopy (cryo-EM). The optimized NS protocol could produce "near native-state" particle images and high contrast images of the protein in its lipid-binding state that is favorable for three-dimensional (3D) reconstruction by single-particle analysis and individual-particle electron tomography (IPET), suggesting this optimized protocol can be used widely to examine the structure of proteins at lipid-binding stage.

Key words: Lipoprotein structure, Lipoprotein morphology, Electron microscopy, Optimized negative-staining protocol, Negative-staining electron microscopy

1. Introduction

Protein–lipid interactions can be found in membrane proteins and apolipoproteins with lipids, which can function as pumps, transporters, cell-to-cell communication messengers (1), or lipid transfer vehicles for lipid metabolism (2). To understand the mechanisms of these biological complexes, studying the structure of a protein at the lipid-bound stage is crucial. To determine their structures, however, difficulties arise due to their dynamic heterogeneity in size, shape, and component. For example, the dynamic nature of

Jörg H. Kleinschmidt (ed.), *Lipid-Protein Interactions: Methods and Protocols*, Methods in Molecular Biology, vol. 974, DOI 10.1007/978-1-62703-275-9_6, © Springer Science+Business Media New York 2013

lipoproteins plays a vital, functional role for cholesteryl transport in cardiovascular disease (CVD).

Structure determination of proteins at the lipid-binding stage, however, is rather difficult by X-ray crystallography because of conformational and compositional heterogeneity. Electron microscopy (EM), as an alternative approach, has been used more frequently to determine protein structure, and among various EM techniques, cryo-crystallography has been successful in determining a "native-state" structure of proteins in the lipid-bound form under frozen hydrated cryogenic conditions (3, 4). For example, Walz et al. used cryo-crystallography uncovered details of the interaction between a lipid and a membrane protein, aquaporin (AQP0), at a resolution of 1.9 Å (3). This technique is quite difficult, requires the most advanced level of equipment and considerable amount of expertise, and, most importantly, requires two-dimensional crystallization, which is still considered an art.

Single-particle cryo-EM is another alternative approach that has become a most popular method to study protein structure. This approach, used as the only technology capable of directly visualizing proteins at the native state, involves imaging hundreds to thousands of particles embedded in vitreous ice. By this method, the images of the particles, which have the same space orientation, are grouped and averaged to reduce the noise level and increase image contrast prior to reconstruction into a 3D density map (4, 5). Usually, such averaging and classification can improve contrast (6, 7); however, it has limited accuracy with heterogeneous particle populations (8), such as high-density lipoprotein (HDL). Another restriction of the cryo-EM approach is radiation damage of the sample. Hence, imaging is performed under a low-illumination dose and under low-temperature conditions. As a result, cryo-EM images contain a low signal-to-noise ratio with high amounts of noise (9).

Negative-staining (NS) is an easy, rapid, qualitative approach and a well-established technique, frequently used in many research laboratories. NS-EM is an approach that directly visualizes individual particles, such as proteins, viruses, and even very thin cell slice specimens. It also permits the study of morphology and structure of these particles, along with lipid-bound forms of apolipoprotein (8, 10); hence, NS-EM has been frequently used (11–13). NS has improved resistance to radiation when compared to cryo-EM (10), and due to the protein being fixed and coated with a layer of heavy metal ions in an emulsion of high ionic strength, it provides exceedingly high contrast (11, 12). However, in many experiments artifacts from NS have been noticed (14–17). For example, a rouleau forms in lipid-bound forms of apoE4 with the conventional NS protocol, and it is believed to be an artifact, because rouleaux are not observed in serum or solutions of phosphate-buffered or Tris-buffered saline (18–20). Phosphotungstic acid (PTA) is used with conventional NS at high salt concentrations in buffer, but NS

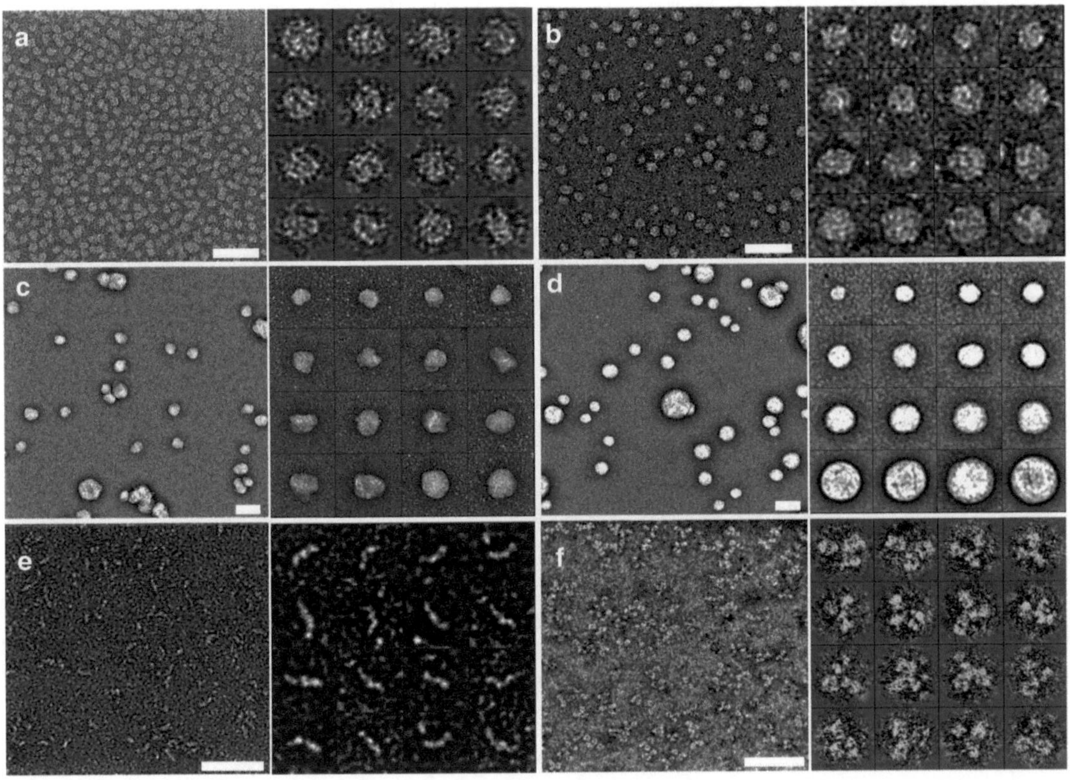

Fig. 1. Morphologies of proteins by optimized negative-staining EM. Micrographs of apoE4-POPC (**a**) (8), α-HDL from plasma (**b**) (10), IDL (**c**) (10), VLDL (**d**) (10), cholesteryl ester transfer protein (CETP, 53 kDa) (**e**) (22), and the IgG antibody (**f**) (25), obtained by EM using the optimized negative-staining protocol. Micrographs by EM for all specimens (**a-f**) mentioned above are shown individually (*left*), and 16 selected individual particles are shown from a larger EM micrograph for all specimens (**a-f**) mentioned above (*right*). Bar = 50 nm.

experiments with the apoE4•palmitoyl-oleoylphosphatidylcholine (POPC) phospholipid particle and liposome vesicles showed that the particles were stacked to each other by PTA, connected by lipid surfaces of neighboring particles (8). As a result, Ren and his colleagues developed an optimized NS protocol that minimizes rouleau formation usually seen in conventional NS-EM studies, and this method was utilized to report structure and morphology of apoE4•POPC reconstituted HDL (rHDL) (8); apoA-I 7.8, 8.4, and 9.6 nm discoidal rHDL (10); 9.3 nm spherical rHDL (10); human plasma HDL (10); low-density lipoprotein (LDL) (10); intermediate-density lipoprotein (IDL) (10); very-low-density lipoprotein (VLDL) (10); a hydrophobic glycoprotein such as cholesteryl ester transfer protein (CETP) (21, 22); and the antibody IgG (23–25) (Fig. 1). Benefits of the optimized NS protocol allow for the study of morphology and structure of individual particles due to minimization of rouleaux, and most of all images from EM with optimized NS provide increased structural resolution that is highly cooperative for 3D reconstruction models.

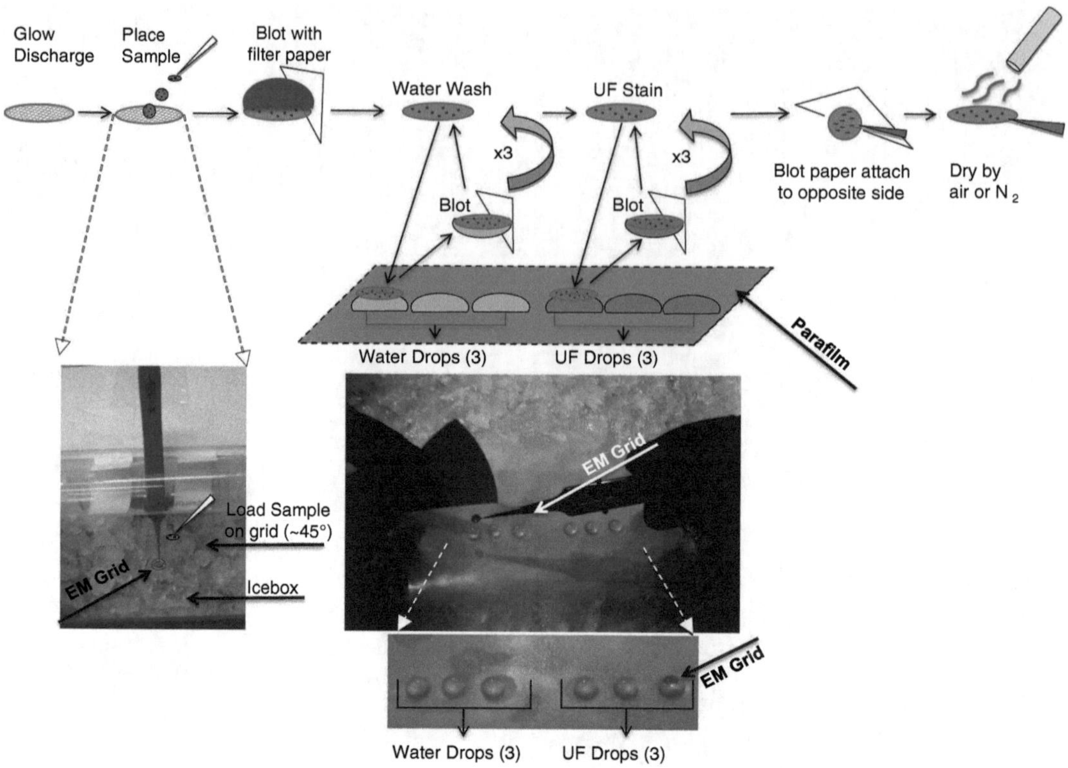

Fig. 2. Diagram of the optimized negative-staining protocol procedure. EM grid manipulations with stain, filter paper, and water contacts in chronological order (*top cartoon*), grid in icebox incubation at 4°C (*left bottom*), rapid contact with stain or water with filter paper from Parafilm in flat icebox chamber (*middle top*), 3 drops of water followed by 3 drops of stain on Parafilm in flat icebox chamber with EM grid lying on top of last UF stain drop (*middle bottom*).

The optimized NS protocol involves placing a drop of the lipoprotein solution on a glow-discharged carbon-coated copper grid and then removing excess solution by blotting with filter paper (Fig. 2). Immediately after, three washes with deionized water on the EM carbon-coated grid are performed (Fig. 2), and then it is stained with uranyl formate (UF) (Fig. 2). Following staining, the grid is blotted again with filter paper and then dried in air. Finally, it is stored at room temperature before being used for EM (8, 10) (Fig. 2).

2. Materials

1. Uranyl formate: $UO_2(CHO_2)_2 \bullet H_2O$.
2. 1 NORM-JECT 1 ml tuberculin syringe, Luer.
3. Sterile syringe filter: pore size of 0.02 μm (Anotop 10).
4. Protein sample, 2.5 μl (~0.005 mg/ml, protein).
5. Dulbecco's Phosphate-Buffered Saline.

6. Parafilm: 4 × 4″.
7. Ice.
8. Cu-300 CN: Thin carbon-coated 300 mesh copper grids (Pacific Grid-Tech, San Francisco, CA).
9. Dumont style#5 medical tweezers with clamping ring.
10. Ultrapure water: Obtained from Millipore Synthesis unit.
11. Filter papers: Qualitative circles, 90 mm (Whatman).
12. Aluminum foil.
13. Petri dishes.
14. EMS 100: Glow discharge unit.
15. Icebox: 4″ × 5″, with lid, insulated.
16. Flat ice chamber: leveled, insulated, uniformly flat, large enough to hold the 4″ × 4″ Parafilm, and contains a lid.

3. Methods

1. Prepare 100 ml of a 1% (w/v) solution of UF powder in deionized water and stir it overnight in a dark room at room temperature. Cover the bottle used to prepare the solution with aluminum foil. Filter 5 ml of the 1% solution with the NORM-JECT syringe and the Anotop filter of 0.02 μm, and aliquot it into 2 ml vials, wrapped in aluminum foil to keep the solution in the dark. Immediately after aliquoting the 1% UF solution, place the vials into liquid nitrogen by using a long-handled forceps (see Notes 1, 2).
2. Store the 2 ml vials of the 1% solution in an −80°C freezer until use.
3. Before use, thaw a vial in a 4°C water bath, and make sure it remains wrapped (cover it) in aluminum foil to keep the vial in the dark.
4. Once the UF is thawed and in liquid form, filter the UF again, using a 1 ml NORM-JECT syringe and the Anotop filter of 0.02 μm pore size, cover it with aluminum foil, and store it on ice or at 4°C (see Note 2).
5. Place ice in a uniformly leveled manner into the flat ice chamber, and cover it.
6. Designate 3 rows of 6 small circular regions in Parafilm. Place the Parafilm in the flat ice chamber and then place ~35 μl drops of deionized water in the first three circle regions. Subsequently place ~35 μl drops of the filtered UF in the next three small circle regions in each row (see Note 3).
7. Fill the icebox with ice, cover it, and let it stand for ~10 min.

8. Obtain carbon-coated grids with Dumont #5 medical tweezers with clamping ring, perform glow discharge with an EMS 100, and place the grids on a clean filter paper in a petri dish and cover it (see Note 4).
9. Open the icebox and hold the grid with tweezers at a 45° angle; place ~2.5 μl of the lipoprotein sample on the grid and incubate for 1 min (see Note 5).
10. After ~1 min, remove excess solution by gently touching the edge of the grid with filter paper. Wash the grid by briefly placing the surface of the grid with a drop (~35 μl) of deionized water on Parafilm and then blot with filter paper to remove the excess solution. The touching and blotting steps are to be performed three times, each with a clean drop of deionized water. Perform the same touching and blotting steps with three successive drops (~35 μl) of 1% UF solution applied on Parafilm, and remove the excess solution by blotting similarly with water. Contact the grid with the last UF drop with the sample side down for 1–3 min in the dark (close the lid of the flat ice chamber) before removing excess stain by blotting again with the entire backside parallel to the grid (non-carbon side) with filter paper. Subsequently, air-dry the sample by a low flow of nitrogen gas at room temperature (see Fig. 2, Note 6).
11. Store the grid on filter paper in a petri dish, and partially cover it for ~30 min.
12. Send the grid to EM or store it in a grid storage box (see Note 7).

4. Notes

1. UF is sensitive to low light. Hence, this procedure should be performed in the dark. UF is also radioactive and should be handled accordingly. The waste of the UF should be placed in an appropriate waste container in compliance with the appropriate waste management system guidelines of the lab.
2. During filtration, filter very slowly, and ensure the UF is covered by aluminum foil. Be careful to use the correct side of the filter when attaching it to the syringe. Discard the filter and the syringe as radioactive waste components.
3. Ensure that the Parafilm is leveled when placing it into the flat ice chamber. Use paperweights as necessary to hold the Parafilm down onto the ice. Make circles/wells with care and do not place them too close to each other. Rows of circles should be made according to the number of samples, but to have a few extra is always best practice in case of errors. Divide circles into two sections of 3, one for water and the other for UF. Again, UF is light sensitive, so close the lid whenever possible.

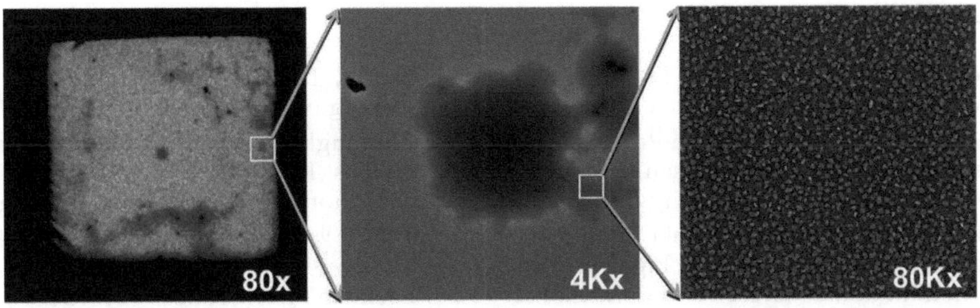

Fig. 3. The best EM imaging area. The best imaging area of the protein was generally from the area that contained the thicker stain. Micrographs showing cloudy areas to locate and obtain images by EM for lipoproteins. Cloud (*highlighted by box*) of lipoprotein to designate lipoprotein location at 80× magnification (*left*), same designated area of cloud (*highlighted by box*) of lipoprotein further magnified at 4Kx (*middle*), and same designated cloud area further magnified at 80Kx (*right*).

4. Before handling carbon-coated grids, clean the tweezers by lightly wiping them on unused filter paper. Handle grids with tweezers so that the tweezers grip the metal edge of the grid, do not bend grids, and ensure carbon-coated grids are always faced up. Refer to your instrument manual to perform the glow discharge appropriately with the EMS 100. Alternate glow dischargers or plasma cleaners maybe used, but ensure that glow discharge occurs by validating your process. Be careful not to clamp down too hard on grids, as they bend easily.
5. Ensure the sample (protein portion) concentration is ~0.01-0.005 mg/ml, and use Dulbecco's Phosphate-Buffered Saline to dilute if necessary. When placing a sample on the grid, be careful to approach the grid at an angle and close lid when possible in an icebox to slow down any reactivity of the grid with air (hence the low-temperature placement of the grid at ~4°C or less). Multiple grids can be held at a time in an icebox if desired, but contact with ice must be avoided at all times.
6. Prepare filter papers beforehand, cut them into small sections to avoid confusion, and designate one paper for water and one for UF. Properly discard UF-contaminated paper according to lab guidelines. Be extremely careful to blot by touching the side of grid with filter paper, and ensure the carbon side of the grid is contacting deionized water and UF stain. It is important to note that the thickness of the stain of the carbon-coated grid is not even. In some areas the stain is thicker than in other areas. However, the best images of proteins have generally been obtained from areas of thicker staining (Fig. 3). An area of thicker stain looks like a "cloud" on the grid when using low magnification (<400×).
7. For storage of the grids, excess heat, dryness, and low humidity conditions must be avoided.

References

1. Reichow SL, Gonen T (2009) Lipid–protein interactions probed by electron crystallography. Curr Opin Struct Biol 19:560–565
2. Gennis RB, Jonas A (1977) Protein-lipid interactions. Annu Rev Biophys Bioeng 6:195–238
3. Gonen T, Cheng YF, Sliz P, Hiroaki Y, Fujiyoshi Y, Harrison SC, Walz T (2005) Lipid–protein interactions in double-layered two-dimensional AQPO crystals. Nature 438:633–638
4. Ren G, Rudenko G, Ludtke SJ, Deisenhofer J, Chiu W, Pownall HJ (2010) Model of human low-density lipoprotein and bound receptor based on CryoEM. Proc Natl Acad Sci U S A 107:1059–1064
5. Kumar V, Butcher SJ, Oorni K, Engelhardt P, Heikkonen J, Kaski K, Ala-Korpela M, Kovanen PT (2011) Three-dimensional cryoEM Reconstruction of native LDL particles to 16 Å resolution at physiological body temperature. PloS One 6:e18841
6. Ludtke SJ, Baldwin PR, Chiu W (1999) EMAN: semiautomated software for high-resolution single-particle reconstructions. J Struct Biol 128:82–97
7. Frank J, Radermacher M, Penczek P, Zhu J, Li Y, Ladjadj M, Leith A (1996) SPIDER and WEB: processing and visualization of images in 3D electron microscopy and related fields. J Struct Biol 116:190–199
8. Zhang L, Song J, Newhouse Y, Zhang S, Weisgraber KH, Ren G (2010) An optimized negative-staining protocol of electron microscopy for apoE4•POPC lipoprotein. J Lipid Res 51:1228–1236
9. Orlova EV, Sherman MB, Chiu W, Mowri H, Smith LC, Gotto AM (1999) Three-dimensional structure of low density lipoproteins by electron cryomicroscopy. Proc Natl Acad Sci U S A 96:8420–8425
10. Zhang L, Song J, Cavigiolio G, Ishida BY, Zhang S, Kane JP, Weisgraber KH, Oda MN, Rye K-A, Pownall HJ, Ren G (2011) Morphology and structure of lipoproteins revealed by an optimized negative-staining protocol of electron microscopy. J Lipid Res 52:175–184
11. Ohi M, Li Y, Cheng Y, Walz T (2004) Negative-staining and image classification—powerful tools in modern electron microscopy. Biol Proced Online 6:23–34
12. Oliver RM (1973) Negative stain electron microscopy of protein macromolecules. Methods Enzymol 27:616–672
13. Woeste S, Demchick P (1991) New version of the negative stain. Appl Environ Microbiol 57:1858–1859
14. Bradley DE (1962) A study of the negative-staining process. J Gen Microbiol 29:503–516
15. Cunningham WP, Staehelin LA, Rubin RW, Wilkins R, Bonneville M (1974) Effects of phosphotungstate negative-staining on the morphology of the isolated Golgi apparatus. J Cell Biol 62:491–504
16. Egelman EH, Amos LA (2009) Electron microscopy of helical filaments: rediscovering buried treasures in negative stain. Bioessays 31:909–911
17. Melchior V, Hollingshead CJ, Cahoon ME (1980) Stacking in lipid vesicle-tubulin mixtures is an artifact of negative-staining. J Cell Biol 86:881–884
18. Innerarity TL, Pitas RE, Mahley RW (1979) Binding of arginine-rich (E) apoprotein after recombination with phospholipid vesicles to the low density lipoprotein receptors of fibroblasts. J Biol Chem 254:4186–4190
19. Peng D, Song C, Reardon CA, Liao S, Getz GS (2003) Lipoproteins produced by ApoE−/− astrocytes infected with adenovirus expressing human ApoE. J Neurochem 86:1391–1402
20. Schneeweis LA, Koppaka V, Lund-Katz S, Phillips MC, Axelsen PH (2005) Structural analysis of lipoprotein E particles. Biochemistry 44:12525–12534
21. Ren G, Zhang S, Cavigiolio G, Lei D, Krauss RM, Oda M, Weisgraber KH, Rye KA, Pownall HJ, Qiu X (2010) Cholesteryl ester transfer protein penetrates lipoproteins for cholesteryl ester transfer. Biophys J 98:36a
22. Zhang L, Yan F, Zhang S, Lei D, Charles MA, Cavigiolio G, Oda M, Krauss RM, Weisgraber KH, Rye K-A, Pownall HJ, Qiu X, Ren G (2012) Structural basis of transfer between lipoproteins by cholesteryl ester transfer protein. Nat Chem Biol 8:342–349.
23. Zhang L, Ren G (2010) Determining the dynamic protein structure by individual-particle electron tomography: an individual antibody structure at a nanometer resolution. Biophys J 98:441a–441a
24. Zhang L, Kaspar A, Woodnutt G, Ren G (2010) Monitoring the structural changes of conjugated antibodies by high-resolution electron microscopy and individual-particle electron tomography. Biophys J 98:440a–441a
25. Zhang L, Ren G (2012) IPET and FETR: Experimental Approach for Studying Molecular Structure Dynamics by Cryo-Electron Tomography of a Single-Molecule Structure. PLoS One 7:e30249.

Chapter 7

Examining Protein–Lipid Complexes Using Neutron Scattering

Luke A. Clifton, Cameron Neylon, and Jeremy H. Lakey

Abstract

Studying the structure of protein–lipid complexes, be they in vesicles, planar bilayers, monolayers, or nanodiscs, poses two particular challenges. Firstly such complexes are often dynamic. Secondly we need to resolve the lipid and protein structures within the complex. Neutron scattering is well placed to help in both respects since it deals with molecules in large, complex, dynamic structures and can easily differentiate between different molecular species. This comes from the great penetrating power of neutrons and their sensitivity to the difference between hydrogen (H) and deuterium (D). Both membrane proteins and lipids can be produced with varying degrees of deuteration, thus allowing us to dissect complexes with great accuracy. Two main scattering techniques are immediately applicable to the study of protein–lipid interactions. Neutron reflection exploits the constructive interference, which occurs when neutrons are reflected from different points in a layer. An everyday example is the rainbow of colors reflected from an oil film on water, which result from varying film thickness and the angle of reflection. Neutrons because of their short wavelengths (4–15 Å) and H/D sensitivity can, in reflectometry mode, provide accurate cross sections of lipid monolayers and bilayers. Small-angle neutron scattering (SANS) can resolve the structures of protein–lipid complexes if they are present as homogeneous dispersions. This is easiest with detergent micelles, but increasingly methods are being developed whereby vesicles, nanodiscs, etc., can be resolved. Again the ability to deuterate proteins and lipids enables SANS to resolve the inner structure of big, dynamic, lipid–protein complexes. The recent introduction of advanced neutron beam lines means that the technique is now within the grasp of a broad cross section of researchers.

Key words: Neutron scattering, Reflectometry, Small-angle scattering, Protein–lipid complexes, Deuteration, Lipid monolayers

1. Introduction

Membrane protein structures are difficult to study, and many methods have been developed to make them more accessible. The removal of transmembrane domains to make X-ray crystallography easier is one approach, while significant developments in crystallization of proteins dissolved in detergents have begun to make the determination of entire structures of integral membrane proteins a

Jörg H. Kleinschmidt (ed.), *Lipid-Protein Interactions: Methods and Protocols*, Methods in Molecular Biology, vol. 974, DOI 10.1007/978-1-62703-275-9_7, © Springer Science+Business Media New York 2013

regular occurrence. Cryo-EM of two-dimensional crystals comes closest of the direct structural methods to recreate the native lipid bilayer of the cell membrane. There is increasing realization of the role of the lipids in membrane protein function, and methods, which resolve the structure and dynamics of both components, are required. These include confocal fluorescence microscopy at one extreme and solid-state NMR at the other. In between these, with the ability to resolve structures and dynamics at the molecular level, sits neutron scattering. This technique requires either a nuclear reactor or a "spallation source" to produce the neutron beam. Therefore, it is often found at large national facilities serving a mixed community of visiting biologists, soft matter chemists, solid-state physicists, and engineers. Access to this technique is usually obtained via a panel or peer-reviewed application process (see Note 1).

For structural biology, the benefits of neutron scattering are its ability to penetrate samples causing no damage and its differential sensitivity to the two isotopes of hydrogen protium and deuterium (1). For protein–lipid interactions, two types of samples are used. Planar samples of mixed proteins and lipids spread over several cm^2, which are investigated by reflectometry methods, and protein–lipid complexes, which are investigated by small-angle scattering in solution.

2. Materials

In addition to normal laboratory reagents, neutron scattering often requires the use of deuterated materials. Deuterated water is available from many sources while speciality chemicals can be obtained from Cambridge Isotope Laboratories (www.isotope.com), Merck Chemicals (www.merck-chemicals.com), and Sigma-Aldrich (www.sigmaaldrich.com). Deuterated detergents are available from Anatrace which is now part of Affymetrix (www.affymetrix.com), while Avanti (avantilipids.com) supplies the widest range of deuterated lipids (see Notes 2 and 3). Deuterated proteins can be made in any well-equipped protein facility, but certain skills are required. Deuteration facilities exist to create bespoke chemicals and proteins which are usually associated with specific allocations of beam time at neutron sources, e.g., Bio-Deuteration Laboratory, Oak Ridge National Laboratory, Tennessee, USA (weisskl@ornl.gov); Australian National Deuteration Facility (www.ansto.gov.au); the D-Lab at ILL, Grenoble, France (www.ill.eu/sites/deuteration/); and the Oxford Isotope Facility at ISIS, UK (www.isis.stfc.ac.uk/apply-for-beamtime). Quartz cuvettes for SANS studies were from Hellma.

3. Methods

3.1. Examining Protein–Lipid Complexes at Interfaces Using Neutron Reflectometry

Due to the high penetrating power of neutrons, neutron reflectometry (NR) has been used in a wide range of interfacial studies, from examining the structures of data storage materials (2) to improving the effectiveness of industrial surfactants (3). NR is also playing an increasing role in determining the complex structure of protein-bound or embedded model lipid membranes, where isotopic labeling (in particular deuterium labeling of samples) is used to highlight both the contribution of the protein and lipid components to the overall interfacial structure.

Perhaps the easiest and therefore most assessable biological membrane models to prepare and examine are monolayers spread at the air/liquid interface. Neutron reflection studies have been used previously to examine antimicrobial peptide (4), protein (5), DNA (6), and viral protein (7) interactions with lipid monolayers and bilayers, with the structures derived from these studies providing information on protein–lipid and head-group/tail binding selectivity and molecular orientation of the bound species. Collecting multiple reflectometry data sets of an interfacial structure with differential isotopic substitution of the sample components can yield information on the relative contributions and distribution of the protein and the lipid components to across interface (4, 8, 9) (see Note 4).

3.2. Technical Overview of Reflectometry

Specular neutron reflectometry (as the name suggests) measures the reflection of a collimated beam of neutrons from an interface of interest (Fig. 1)

1. *The Beam Line.* In general NR instruments consist of a series of beam choppers (which define the wavelength band with the correct neutron wavelength range required), slits which collimate the incoming beam, monitors to assess the neutron wavelength spectrum prior to the sample, a sample mount, and a detector.

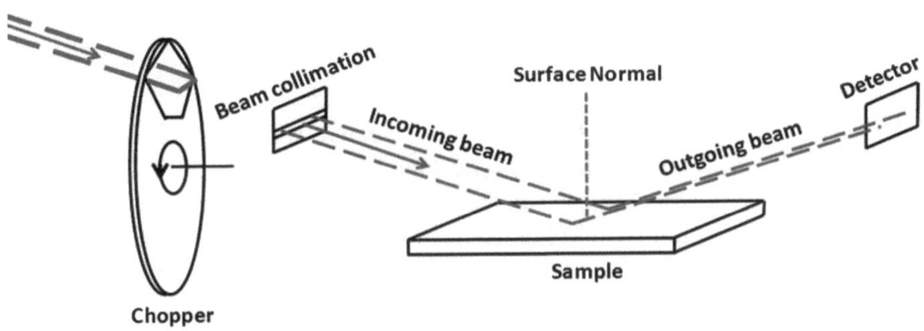

Fig. 1. A simplified diagram of a neutron reflectivity instrument showing a pulsed collimated ribbon beam of neutrons as it undergoes specular reflection from a sample surface.

Table 1
A list of some neutron reflectometry instruments and their parent facilities

Reflectometer	Parent neutron source
Liquids reflectometer	Spallation Neutron Source, Oak Ridge National Laboratory, Tennessee, USA
Amor	Paul Scherrer Institute, Switzerland
CRISP, SURF, INTER, OffSpec, PolREF	ISIS Spallation Neutron Source, Rutherford Appleton Laboratory, UK
D17, superADAM and Figaro	Institut Laue–Langevin, Grenoble, France
D3 reflectometer	Chalk River Laboratories, Canada
Platypus	Opal Facility, ANSTO, Sydney, Australia
NG1, NG7, AND/R	NIST Center for Neutron Research, Maryland, USA
SPEAR	Lujan Institute, Los Alamos National Laboratory, New Mexico, USA
EROS and PRISM	CEA Laboratoire Léon Brillouin, Saclay, France
BL16	Materials and Life Science Experimental Facility, J-PARC, Japan Proton Accelerator Research Complex, Japan
REFSANS	FRM II, Garching, Germany

The instrument design will vary between reactor and a spallation sources and the range of functions of the instrument. No two neutron reflectometers are identical in design. Table 1 gives a list of neutron sources where time on neutron reflectometry instrumentation can be applied for.

2. *Data Analysis Theory.* The information gained from neutron reflection measurements describes the scattering length density (SLD or "neutron refractive index") profile across the interface (i.e., along the z axis). The reflectivity profile obtained from a surface is a one-dimensional data set of reflected neutron intensity versus wavevector transfer which is described by

$$Q_z = \frac{4\pi}{\lambda} \sin \theta, \tag{1}$$

where Q_z is the wave vector transfer, λ is the neutron wavelength, and θ is the incidence angle of the beam with respect to the sample surface. Figure 2 shows an annotated example of neutron reflectivity data from an air-/deuterium oxide (D_2O, heavy water)-only interface (Fig. 2a) and an interface containing a 60 Å thick homogeneous film (Fig. 2b).

By comparing a and b in Fig. 2, it is noted that the presence of the interfacial film significantly alters the neutron reflectometry profile compared to a D_2O surface only, with fringes in the

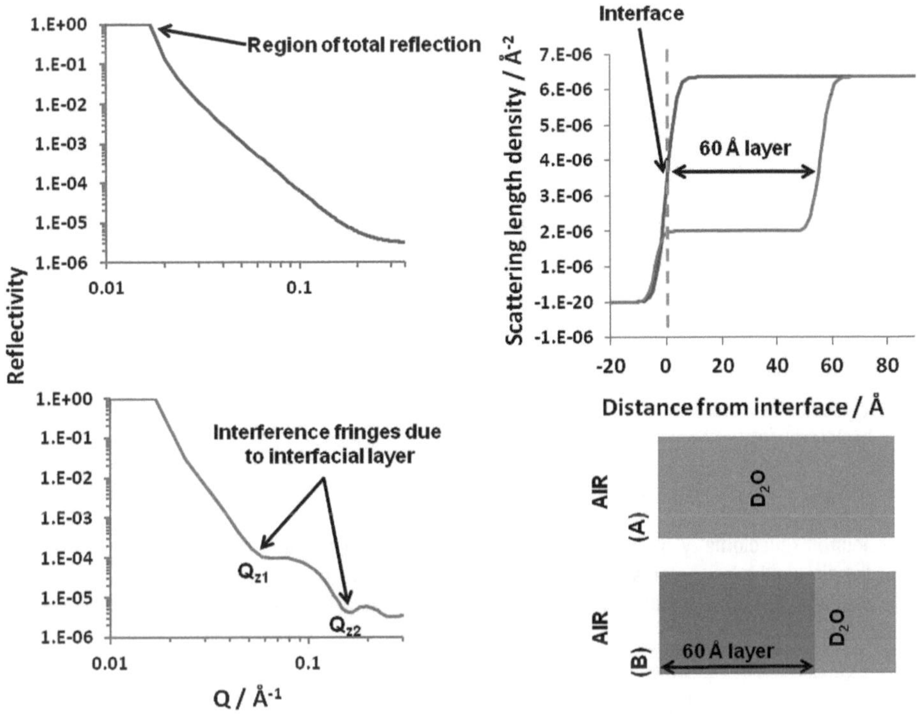

Fig. 2. Simulated neutron reflectometry data from an air/liquid interface of D_2O only (**a**) and an interface containing a 60 Å thickness interfacial film (**b**). The real space distribution of this data as scattering length density versus distance across the interface and a cartoon representation of the interfacial structures that these data describe are shown (**c**).

profile appearing due to interference between neutrons reflected from the top and bottom of the interfacial layer. The distance in Q between these interference fringes is directly related to the thickness of the interfacial layer by

$$Q_{z_2} - Q_{z_1} = \frac{2\pi}{d}, \tag{2}$$

where $Q_{z_2} - Q_{z_1}$ is the separation between the interference fringes and d is the thickness of the interfacial film.

As well as interference fringes which are used to determine the thickness of interfacial films, the intensity of the neutron reflection can be used to determine the amount of material present at a given interface. Indeed, in general, interference fringes will not be present in neutron reflectometry data for interfacial films below ~35 Å in thickness. However, reflectivity will still be observed, and the intensity versus Q_z profile of the reflectivity data will be determined by the difference in SLD between the two bulk phases, the SLD of the interfacial film, and its thickness. Figure 3 shows three simulated neutron reflectivity profiles for a fully deuterated lipid monolayer at the air/liquid interface on a solution subphase of null reflective

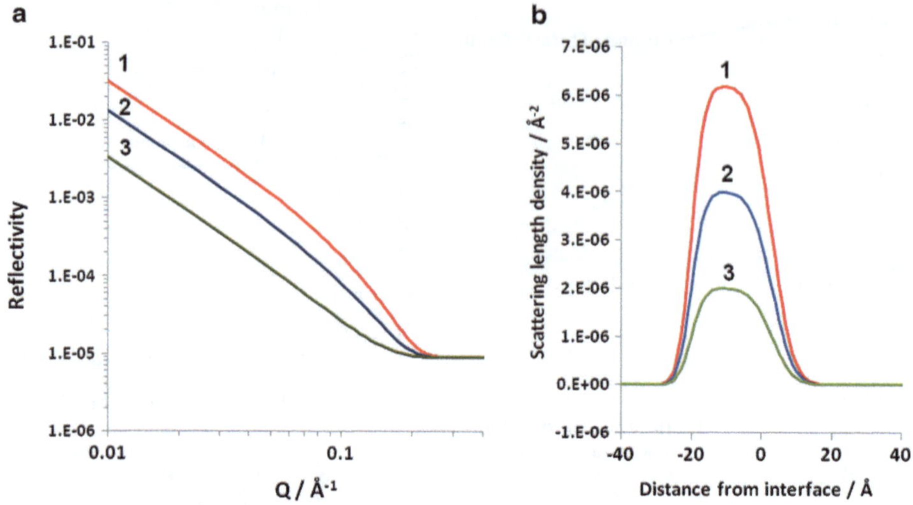

Fig. 3. Model neutron reflectometry profiles (**a**) of tail deuterated 1,2-dipalmitoyl-*sn*-glycero-3-phosphocholine (DPPC) monolayers at the air/liquid interface on a null reflective water subphase with surface coverages of 32 % (3, *green line*), 64 % (2, *blue line*), and 100 % (1, *red line*), and the scattering length density versus distance profiles the reflectivity data describe (**b**).

water (NRW, also known as air contrast matched water, 8.2 % D_2O, 91.8 % H_2O); this solution, like air, has a scattering length density of 0 and therefore does not contribute to the neutron reflectivity (see Note 3).

At a theoretical coverage of 100 % deuterated lipid (Fig. 3, reflectivity profile 1), we observe the highest intensity neutron reflectivity profile (Fig. 3, reflectivity profile 1), whereas with lower film coverage, we observe lower intensity in the neutron reflectivity from the interfacial monolayers (Fig. 3, reflectivity profiles 2 and 3).

4. Methods to Control Neutron Scattering Contrast in Biological Studies

The key benefit of neutron scattering techniques in the biological sciences is the application of *contrast*. Neutrons scatter constructively from interfaces, and differences in SLD between the sample and its environment can highlight the complete interfacial film or a component of it. Deuterium (^2H) labeling of the samples can be highly useful in analyzing complex structures because, while hydrogen, which is 99.98 % ^1H (protium), possesses a negative neutron scattering length (-3.74×10^{-5} Å), deuterium has a large positive value (6.64×10^{-5} Å) (see Table 2). This large difference means that chemically similar samples possessing differing levels of hydrogen and deuterium labeling will show large differences in their neutron reflectivity or scattering profile. Similarly the relatively

Table 2
Neutron scattering lengths of hydrogen deuterium and the most abundant elements found in proteins and lipids

Element	Scattering length ($\sum b$)/10^{-5} Å
Hydrogen	−3.74
Deuterium	6.671
Carbon	6.646
Nitrogen	9.36
Oxygen	5.803
Sulfur	2.847
Phosphorous	5.13

high SLD for nitrogen means that biological materials such as protein and DNA differ strongly in their SLD from hydrogenous materials such as lipids providing a natural contrast between protein and DNA, and protein and lipid, which can be exploited experimentally. Absolutely key is the fact that the major biological solvent, and thus the solution environment for experiments, is water. The SLD of H_2O is -0.56×10^{-6} Å^{-2}, whereas that of D_2O is 6.35×10^{-6} Å^{-2}. By manipulating the D_2O content of the solvent being used, it is possible to create solution conditions that "match" the SLD of specific components of complex biological systems (see Note 2). By removing any SLD difference, the component no longer scatters differently to the background and is rendered "neutron invisible."

In both small-angle neutron scattering and neutron reflection, the signal depends on differences in scattering length density that the beam encounters. In SANS the important difference in SLD is between (and within) the sample and the buffer. In NR the important differences are between the bulk phases and the layers that the beam encounters at the point of reflection. The use of contrast in experimental design is discussed in detail in the context of each technique. Here we discuss the differences in scattering length density between different biological materials

Figure 4 shows a graph of scattering length density versus % deuteration for lipid acyl chains, protein, and water. The increase in deuterium compared to hydrogen in a molecule increases its scattering length density and therefore its ability to scatter/reflect neutrons. The use of both hydrogenated and deuterated molecules in a complex interfacial film/solution structure can be used to highlight individual components. Hydrogenated materials are highlighted when examined in deuterated solution (such as

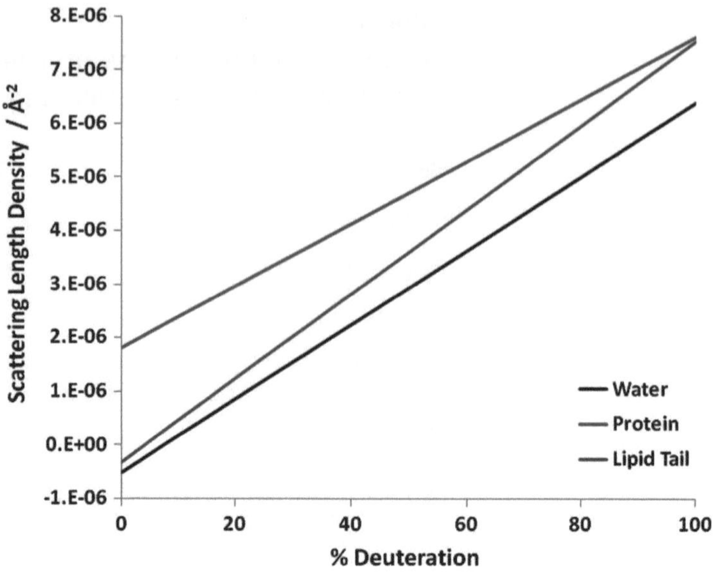

Fig. 4. Neutron scattering length density versus % deuteration for water, protein*, and lipid acyl chains. *Values for protein in H_2O solution.

D_2O), and deuterated material is highlighted against hydrogenated solutions (such as H_2O) due to the difference in contrast between the species under study and the solution in which it is examined.

The SLD of molecules with exchangeable hydrogen groups (such as those which possess amine, amide, and hydroxide groups) is solution D_2O concentration dependent. In general biological molecules such as proteins, ribonucleic acids, and carbohydrates contain a high degree of exchangeable hydrogen groups and therefore possess D_2O concentration-dependent neutron scattering length densities (see Fig. 5 for a graph of neutron SLD vs. solution D_2O %).

In NR, collecting multiple SLD profiles of an interfacial film, in which successive samples have altered isotopic substitution of one or more components, one can build up a contrast series which allows you to not only produce unique "model to data fits" of the interfacial structure but also determine the individual contributions of components within the complex structure.

Figure 6 shows the simulated neutron reflectivity profiles for a protein-only monolayer on three differing isotopic contrasts of solution subphases. The difference between the reflectivity profiles of the same film on D_2O and NRW is clear and is due to the difference in the neutron reflectivity of the solution subphase below the interfacial film. NRW has an SLD of 0 so does not contribute to the neutron reflectivity, so any observed reflection comes directly from the interfacial monolayer (Fig. 6a, 3). Conversely D_2O has a comparatively high SLD and is the main contributor to reflectivity of contrasts on this subphase; the

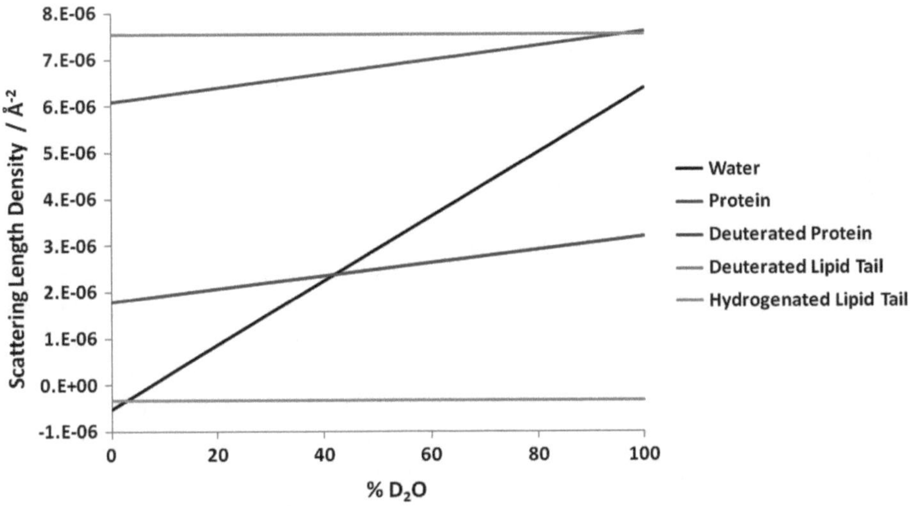

Fig. 5. Neutron scattering length density versus % D_2O for water, hydrogenated and per deuterated protein, and hydrogenated and deuterated lipid tail.

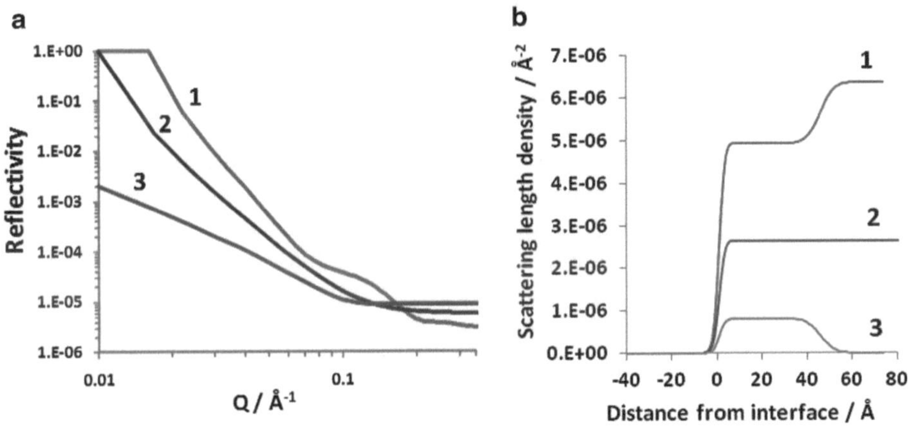

Fig. 6. Simulated neutron reflectometry (**a**) and the scattering length density versus distance profiles (**b**) of a 50 Å thick protein layer at the air/liquid interface. Reflectivity and SLD profiles are shown for this monolayer on 100% D20 (1), 42 % D20 (2), and NRW 8.2% D20 (3) subphases.

interfacial monolayer is identified in this contrast through the shape of the intensity versus *Q* profile and fringes within the data. By having both the D_2O and NRW contrasts in this simple example, both the quantity of protein at the interface and the film thickness are accurately defined. Conversely, at 42 % D_2O, the SLD of the solution matches that of the protein film so although reflectivity is observed (Fig. 6a, 2), due to the difference in SLD between the solution and the air, the profile contains no information relating to the monolayer structure as the monolayer and the solution have matched SLD.

4.1. Methodology for Examining a Protein-Adsorbed Lipid Monolayer Structure with Neutron Reflectometry

Due to the wide range of interfaces which it can probe, neutron reflectometry has been used to examine the interaction of proteins with both lipid monolayers at the air/liquid interface and bilayer structures at the solid/liquid interface. Both systems have their relative advantages and disadvantages for protein interaction studies.

4.1.1. Overview

Lipid monolayer films at the air–liquid interface have been used in a wide range of applications, most notably as model biological membranes for the understanding of the physical properties of membranes and to model biologically or biotechnologically relevant membrane interactions (10, 11) (see Note 4).

The advantages of the various bilayer membrane models (12–14) used in neutron reflectometry studies are in general the similarity these systems have to real biological membranes. While lipid monolayers at the air/liquid interface can only represent one leaflet of a cell membrane, they have nonetheless been important in membrane interaction studies. This is due, in part, to the ease of creating and controlling this class of interfacial film and the wide range of techniques available to study monolayers at the air–liquid interface (15).

4.2. Preparation for Examining Protein-Adsorbed Lipid Monolayers by NR

1. Neutron Reflectometry Instrument Time: The time available for the experiment will be set either in days or hours. To make efficient use of the facility, data is collected over a 24-h period. For analysis of protein–lipid monolayers at the air/liquid interface, individual NR profiles are likely to need collection times between 2 and 8 h. Researchers wishing to apply for neutron reflectivity instrument time are recommended to contact a reflectometry instrument scientist at the facility where they wish to apply, who will advise on technical aspects of a proposed experiment as well as assisting with data collection and analysis (see Note 1).
2. H_2O and buffers can be from any good source; at most concentrations, the buffer contributes minimally to the SLD so its choice is entirely to suit the experiment.
3. D_2O 99.9% purity is readily available from a variety of suppliers.
4. Mix the buffers according to volume to obtain v/v percentages of D_2O (see Note 3).
5. Milligram quantities of the protein and lipids under study should be obtained. If possible, these should include hydrogenated and deuterated forms of the lipid and/or protein components. This will allow a suitable isotopic contrast series of reflectometry profiles to be obtained.

Table 3
A list of potential isotopic contrasts used in protein/lipid monolayer studies with neutron reflection with associated contrast cartoons of protein-adsorbed lipid monolayers showing which interfacial film components contribute to the neutron reflectivity for a given contrast

#	Isotopic contrast	Main information content of reflectivity data	Contrast cartoon
1	Deuterated-lipid: Hydrogenated-protein on null reflective water	Lipid component	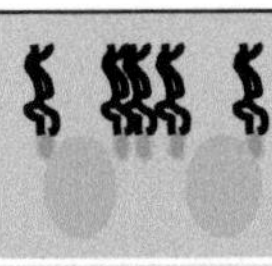
2	Hydrogenated-lipid: Hydrogenated-protein on null reflective water	Protein component	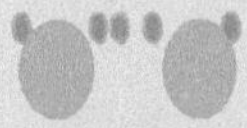
3	Hydrogenated-lipid: Hydrogenated-protein on D_2O	Solution and hydrogenous material—both lipid and protein components	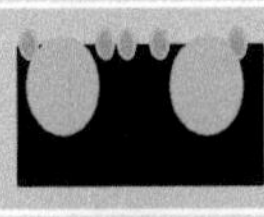
4	Deuterated-lipid: Hydrogenated-protein on D_2O	Solution and protein components	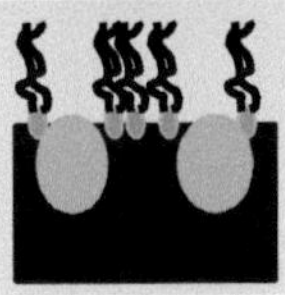
5	Hydrogenated-lipid: Deuterated- protein on null reflective water	Protein components	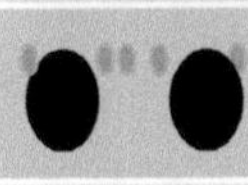
6	Deuterated-lipid: Deuterated-protein on null reflective water	Deuterated material—both lipid and protein components	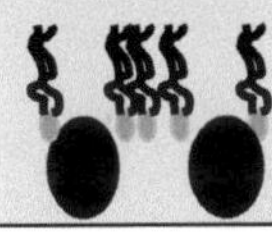

6. Deciding which isotopic contrasts to analyze for a given system is a large consideration when planning a neutron reflection experiment and is of particular importance to protein-bound lipid layers. Each contrast should either highlight a different component of the interfacial structure or highlight structural length scales via fringes in that reflectivity profile. The contrast series should not only produce unique solution to the interfacial structure during data analysis (see later) but also allow for the contribution of each component to the interfacial complex to be derived. Table 3 gives a list of potential reflectometry contrasts, with the major component information obtained from each contrast listed.
7. The availability of deuterated materials often decides the contrast series for an individual experiment (see Note 5). In Table 3, you may notice that the two solution subphases suggested for monolayer structural examination using neutron reflection are

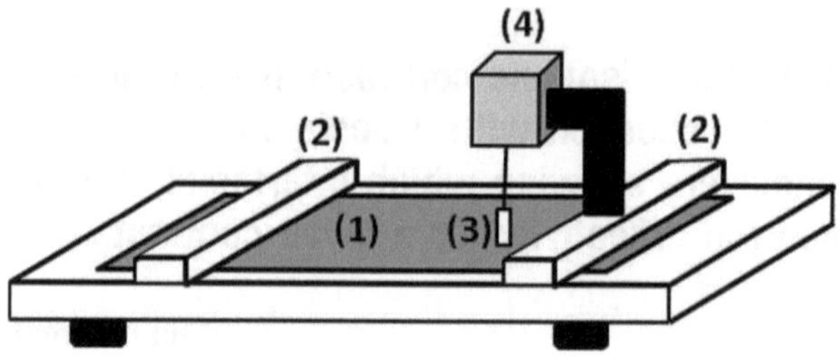

Fig. 7. A cartoon of a Langmuir trough showing solution surface (1), movable barrier for compression of the surface film (2), and a Wilhelmy plate (3) attached to a surface pressure sensor (4) used to monitor the compression and stability of the interfacial film and change in surface pressure during protein–lipid interactions.

NRW (SLD of 0) and D_2O (SLD of 6.35×10^{-6} Å^{-2}). NRW is helpful as the solution subphase on which the air/liquid interfacial monolayer floats since it makes no contribution to the neutron reflectivity, and therefore, any reflectivity comes directly from the interfacial film only. As a result, contrasts on NRW are useful in quantitative examination of the protein/lipid components at the interface (see 4.2.11–4.2.13 for more details). A D_2O subphase will highlight hydrogenous material. In air–liquid interface measurements, contrasts on D_2O can help accurately determine the interfacial thickness. However, this is an oversimplification as reflectivity profiles on both NRW and D_2O can be used to determine both interfacial layer thicknesses and amounts at the interface.

8. Analysis of the protein–lipid monolayer by other methods prior to neutron reflectometry can be highly useful to provide complementary data and help optimize the experimental conditions. A review of techniques used in the study of air/liquid interfacial monolayers has been written by Dynarowitztka et al. (15). The most basic of these measurements will be surface pressure measurements, which can both be run concurrently and, in the case of surface pressure versus time or area isotherms, immediately prior to NR measurements (see Note 4).

9. *Sample Environment*: The correct sample environment is vital. A Langmuir trough with interfacial monolayer compression monitored by surface pressure measurements is the best sample environment to examine protein–lipid monolayers (see Fig. 7 for a diagram of this apparatus). Prior to collecting neutron reflectivity measurements, this apparatus will need to be placed into the sample position on the neutron reflectometer and calibrated. During all measurements, care should be taken to ensure that the surface area is large enough to both allow the neutron beam to pass over the sample surface without hitting the trough barriers or the Wilhelmy plate, allowing concurrent collection of surface pressure and neutron reflection data (see Note 6).

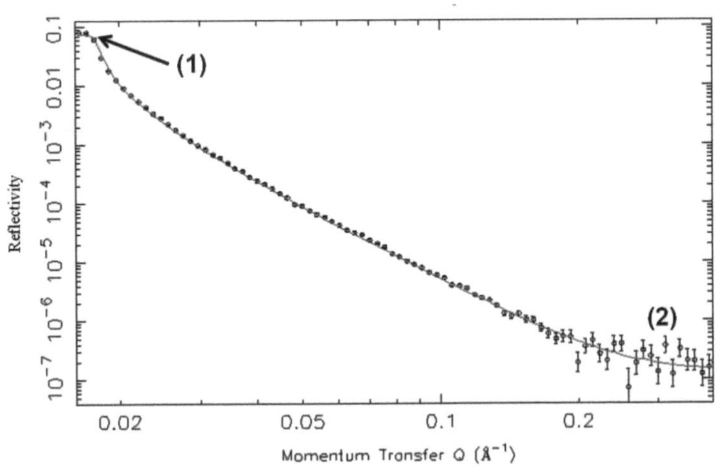

Fig. 8. Neutron reflectivity profile from a clean D2O surface collected on the SURF reflectometer (ISIS). Experimental reflectivity data (circle) and the model fit to this data (line) are shown. From this data the scale factor for data normalization can be determined from the critical edge (region of total reflectivity, (1)), as well as the interfacial roughness (in this case 4 Å) and background (2).

10. *Data Collection*: Initially the NR profile of a clean D_2O surface is collected using the collimation settings and incident angle(s) which are to be used during the experimental runs (Fig. 8). This data is used to determine the scale factor needed to normalize the experimental data, as well as allowing examination of the reflectivity profile to check that instrument and sample environment are optimum for the experiment prior to sample data collection (see Note 7).

    To enhance the Q range and maximize the data coverage, the incident angle of the neutron beam to the sample surface is changed (due to the relationship between Q_z and the incident angle, θ, see Eq. 1). Profiles collected should be partially overlapping in Q range to allow the data to be easily "stitched" together to provide the final reflectivity profile. Between each angle slit, settings are changed to keep the data resolution constant. For solid samples, the sample angle can be changed, but since liquids need to be kept flat, a neutron reflective mirror is employed to deflect the incoming neutron beam to the desired incident angle (see Fig. 9).
11. *Data Analysis*: Neutron reflectometry contrast series are fitted using procedures which convert the information contained within the neutron reflectivity profiles into real space descriptions of scattering length density across the interface.
12. *Kinematic or Born Approximation*: In this data analysis approach, it is assumed that the scattered wave from the sample has a negligible effect on the incident beam, and multiple

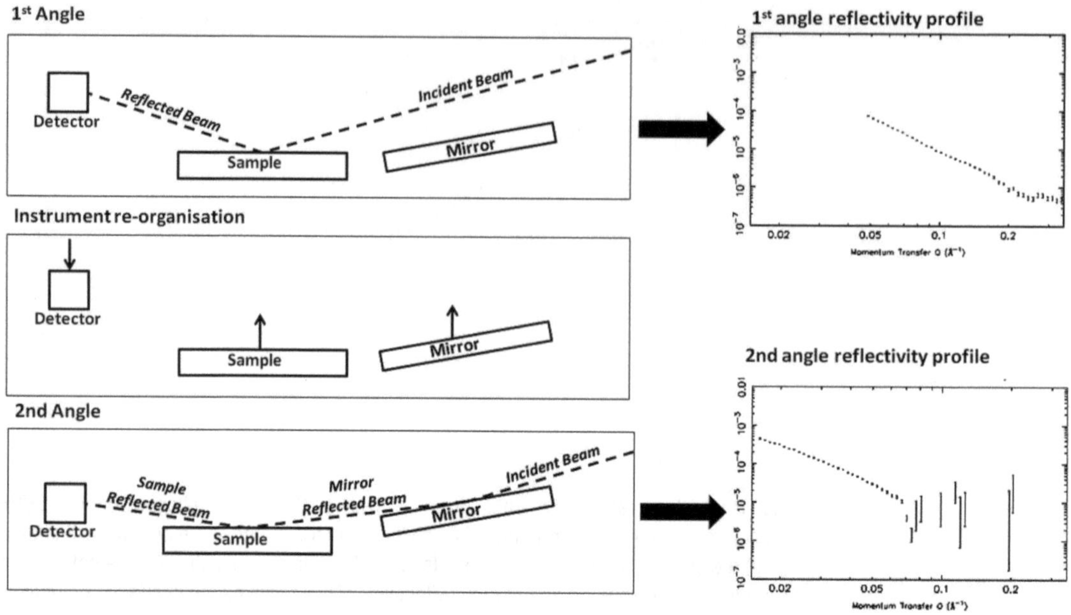

Fig. 9. A simplified diagram showing the instrument organization and change between and the reflectometry data obtained at two different incident angles (in this case 1.5° and 0.5°). A super mirror is used to change the incident angle of the neutrons and therefore the *Q* range probed.

scattering is ignored. This is the equivalent of a weak scattering assumption. In this limit, the scattering cross section or reflectivity is given by the Fourier transform of the scattering length density function. Therefore, the kinematic or Born approximation provides a basis for NR data analysis routines (16, 17). A detailed description of its use to fit neutron reflectometry data can be found in a review article by Lu and Thomas (18).

13. *Layer Structures Fitted Using Abeles Matrix Formalism*: In this approach, the structure across the interface is modeled as a series of slabs, with each individual layer characterized by its scattering length density, thickness, and roughness. Table 4 lists some of the software packages used to analyze neutron reflectometry data. When fitting a contrast series, the individual reflectivity profiles are constrained to fit the same layer structure, but the scattering length density is adjusted between individual contrasts to obtain suitable model data fits to the experimental reflectivity profiles. During the analysis of a multiple contrast reflectometry data set, the experimental data is initially fitted to the simplest model possible (i.e., least number of layers) of the interfacial film. Below is a list of reflectometry data analysis packages with web addresses where information relating to these can be found. As a single layer is simplest possible structure for an interfacial film, this is the first model to which data fitting should be attempted. The model structure

Table 4
Reflectometry packages and associated web addresses

Analysis package	Web address
Motofit	http://sourceforge.net/projects/motofit/
RasCAL	http://sourceforge.net/projects/rscl/
Reflfit	http://www.ncnr.nist.gov/programs/reflect/data_reduction/software/reflfit.html
Surface	http://www.isis.stfc.ac.uk/instruments/surf/data-analysis/surf-data-analysis2565.html
StochFit	http://stochfit.sourceforge.net/

is then increased in complexity until satisfactory model data fits to the experimental reflectivity profiles are obtained. The model structure can then be interpreted in terms of the real interfacial structure based on the scattering length density profile and the relationship between the isotopic contrasts. Often when examining the structure of protein-adsorbed lipid monolayers, two or more layer structures are required to adequately fit the reflectivity data from a contrast series (4), because lipid headgroup-adsorbed or partially embedded proteins can often form a protein-only layer below the lipid film (8).

4.3. Using the Scattering Length Density Profile to Determine the Complex Interfacial Structure

1. *Interpreting the Scattering Length Density Profile*:
The scattering length density profile obtained from layer model fitting of a reflectometry contrast series can be interpreted in terms of molecular structure. Figure 10 shows model reflectivity profiles of a protein-adsorbed lipid monolayer using three different isotopic contrasts of sample and solution subphase to highlight different components of the interfacial film. The relationship between the individual contrasts and the SLD versus distance profile indicates the relative positions of protein, lipid, and solvent components with respect to the interface. In the example shown in Fig. 10, the SLD versus distance profile obtained from d-lipid on NRW shows an SLD peak located between −20 and +3 Å from the interface. Since the reflectivity of this sample is dominated by the deuterated lipid (see Table 3), this region can be assumed to represent the lipid component of the interface. However, h-lipid on NRW contrast shows a uniform distribution from −20 to 49 Å; the protein component dominates the reflectivity of this contrast; therefore, the SLD profile can be interpreted to show that the protein has partially embedded the lipid region from −20 to

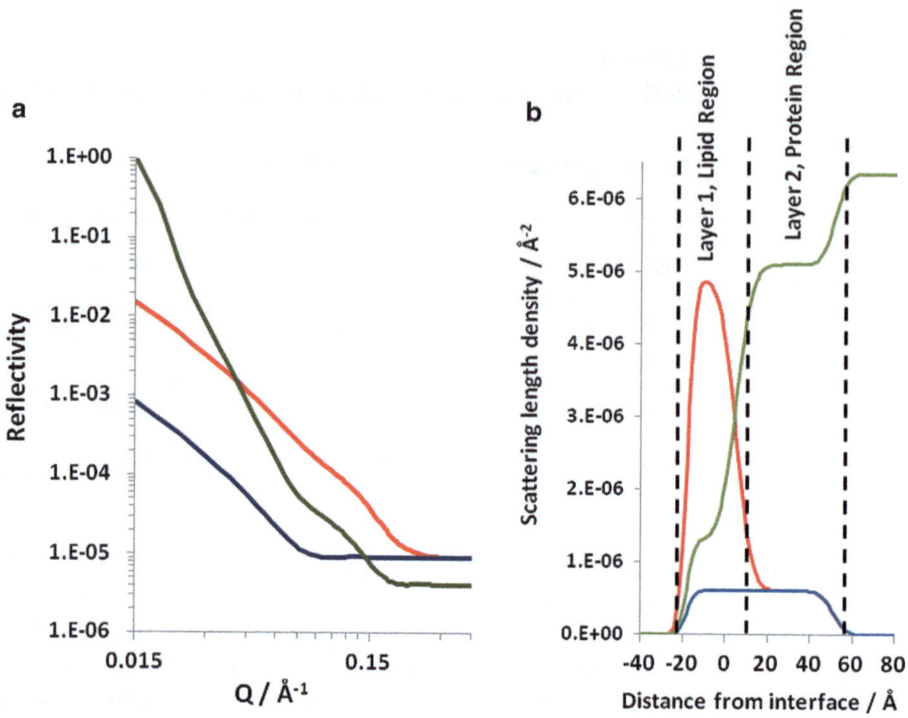

Fig. 10. Model neutron reflectivity profiles of a protein-adsorbed 1,2-dipalmitoyl-*sn*-glycero-3-phosphatidylglycerol (DPPG) monolayers (**a**) and the scattering length density versus distance profiles this data describes (**b**). The three isotopic contrasts shown are deuterated lipid (d-lipid), hydrogenated protein on NRW (*red*); hydrogenated lipid (h-lipid), hydrogenated protein on NRW (*blue*); and hydrogenated lipid, hydrogenated protein on D_2O (*green*).

3 Å (combining data obtained from the SLD vs. distance profiles of d-lipid on NRW and h-lipid on NRW) with the remainder of the protein in solution below the lipid film (where d- and h-lipid on NRW contrasts show the same SLD). The h-lipid on D_2O contrasts in agreement with this showing a low SLD from −20 to 3 Å (where both protein and lipid are found) with a higher scattering length density region from 3 to 49 Å, with the SLD reaching solution values in all contrasts below that. From this example, it is clear that NR allows for determination of the complex structure of protein–-lipid monolayers, with the use of multiple contrasts allowing for the identification of the contribution of the relative position of the protein and lipid components in this film (see Note 5).

2. *Calculating the Contribution of Protein–Lipid and Solvent to the Interfacial Structure*:

   Comparing the SLDs obtained for each contrast in each layer of the structure of protein–lipid monolayers at the air/liquid interface allows for the relative contributions of the protein, and lipid to each layer of an interfacial structure can be derived, as the SLD for each contrast is related to the volume fraction of each component of the interfacial structure by

Table 5
Summary of known scattering length densities of the lipid and protein components in the model data sets shown in Fig. 10

Lipid/protein/solvent	Scattering length density/10^{-6} Å^{-2}
h-DPPG in NRW	0.3
h-DPPG in D_2O	0.5
d-DPPG in NRW	6.07
Protein in NRW	1.86
Protein in D_2O	3.21
NRW	0
D_2O	6.35

$$\rho_{(layer)} = \rho_{(lipid)}\varphi_{(lipid)} + \rho_{(protein)}\varphi_{(protein)} + \rho_{(solvent)}\varphi_{(solvent)}, \quad (3)$$

where φ is the volume fraction and ρ is the scattering length density. When starting calculations relating to the relative contributions of individual components to an interfacial film, the theoretical SLDs of components should be calculated (see Table 5 for an example of the model values for the protein–lipid monolayer shown in Fig. 10).

The SLDs of the different components in the interfacial structure are shown in Table 5. An example of using different solute contrasts is shown by examining the SLDs of the three contrasts used in the model protein-adsorbed data set (Fig. 10). For example, the h-lipid on NRW and d-lipid on NRW contrasts can be used to determine the contribution of the lipid component to an individual layer, as the SLD's protein and solution components of these two contrasts will be identical; therefore, any difference in SLD between these two contrasts in a given layer of the structure will be solely due to the lipid components whose contribution (volume fraction, φ) to a given layer can be calculated from these two contrasts by

$$\varphi_{lipid} = \frac{(\rho_{d\text{-lipid contrast}} - \rho_{h\text{-lipid contrast}})}{(\rho_{d\text{-lipid}} - \rho_{h\text{-lipid}})}, \quad (4)$$

where φ_{lipid} is the volume fraction of lipid in the layer and $\rho_{\text{d-lipid contrast}}$ and $\rho_{\text{h-lipid contrast}}$ describe the scattering length densities obtained from a given layer of the model data fits of the d-lipid on NRW and h-lipid on NRW contrasts, respectively. $\rho_{\text{d-lipid}}$ and $\rho_{\text{h-lipid}}$ are the calculated SLDs of the d- and h-lipid, respectively. Once the lipid volume fraction of the layer is determined, the protein content of that layer can be determined from the same two contrasts by

$$\varphi_{protein} = \frac{(\rho_{h\text{-lipid}} - (\varphi_{lipid} \times \rho_{h\text{-lipid}}))}{\rho_{protein}}, \tag{5}$$

where φ_{lipid} is the volume fraction of lipid in the layer, $\rho_{\text{d-lipid contrast}}$ and $\rho_{\text{h-lipid contrast}}$ describe the SLD obtained from a given layer of the model data fit for the d-lipid on NRW and h-lipid on NRW contrasts, respectively, and $\rho_{\text{d-lipid}}$ and $\rho_{\text{h-lipid}}$ are the calculated SLDs of the d- and h-lipid, respectively. The solvent content of a given layer can be calculated in the case of the data sets shown in Fig. 10; however, in this case all three components will change SLD between the two contrasts due to labile hydrogen exchange on the protein and lipid head group, so the known lipid and protein volume fraction calculated in Eqs. 4 and 5 is employed in the calculation by

$$\varphi_{solvent} = \frac{\rho_{h\text{-lipid on } D_2O}((\varphi_{lipid} \times \rho_{lipid\ on\ D_2O}) + (\varphi_{protein} \times \rho_{protein\ on\ D_2O}))}{\rho_{D_2O}}, \tag{6}$$

where φ_{solvent} is the volume fraction of solvent in the layer and $\rho_{\text{h-lipid on D2O}}$ describes the SLD of the layer in the h-lipid on D_2O contrast, $\rho_{\text{lipid on D2O}}$ and $\rho_{\text{h-protein on D2O}}$ are the calculated SLDs of the h-lipid and h-protein components, respectively, and φ_{lipid} and φ_{protein} are the volume fractions of the protein and lipid components of the structure calculated from Eqs. 4 and 5, respectively. By employing Eqs. 3, 4, 5, and 6, the contributions of the protein, lipid, and solvent to the model reflectometry contrast series are determined and shown in Table 6.

Equations 4, 5, and 6 are useful in calculating the contribution of protein, lipid, and solvent from the three contrasts shown for the model protein-adsorbed lipid monolayer given in Fig. 10. Using different contrasts will require the experimenter to adapt the equations used to suit the experiment conducted and contrast series obtained. The equations used to derive the relative contributions of the components to the interfacial structure will be experiment specific, and Eqs. 3, 4, 5, and 6 show the methodology employed to calculate the contribution of protein, lipid, and solvent to the protein–lipid monolayer.

4.4. NR of Lipid–Protein Complexes at the Solid–Water Interface

As mentioned above, it is possible to assemble lipid bilayers on flat solid surfaces under buffer. Examples include tethered lipid bilayers on gold (19), formation of bilayers on silicon from vesicle deposition (20), and floating lipid bilayers.

As well as the improved biological accuracy of these model membranes compared to monolayers, an advantage of examining surface-adsorbed bilayers at the solid/liquid interface is the ease with which solution contrast can be changed for the same sample, as the sample is loaded into or assembled in flow cells. Changing the H_2O/D_2O ratio within the cell allows multiple contrasts to be

Table 6
Parameters obtained from the best two-layer model structure of a protein-adsorbed DPPG monolayer

	Contrast	Thickness/Å	$\rho/10^{-6}$ Å^{-2}	φ_{lipid}	$\varphi_{protein}$	$\varphi_{solvent}$
Layer 1	d-DPPG on NRW	23	4.9	0.78	0.19	0.05
	h-DPPG on NRW	23	0.6	–	–	–
	h-DPPG on D_2O	23	1.3	–	–	–
Layer 2	h-DPPG on NRW, Layer 2	46	0.6	–	0.32	0.68
	d-DPPG on NRW, Layer 2	46	0.6	–	–	–
	h-DPPG on D_2O, Layer 2	46	5.1	–	–	–

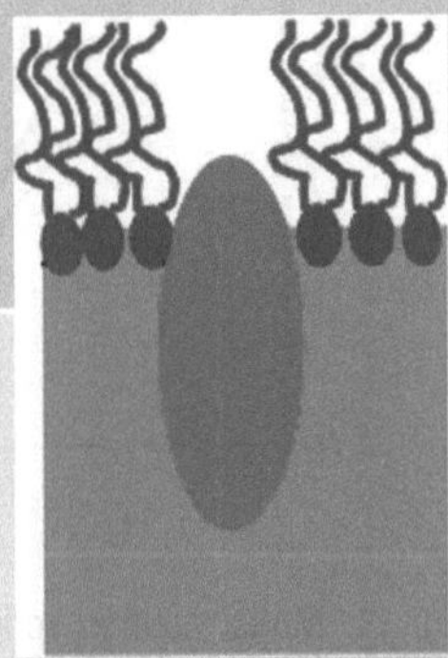

collected. Figure 11 shows model NR profiles for three solution contrasts of a hydrogenated 1,2-dipalmitoyl-sn-glycerol-3-phosphocholine (DPPC) bilayer deposited on silicon and the SLD profiles this data describes. Note how the neutron reflectometry profile of the bilayer in 100 % D_2O is highly sensitive to the presence of the total hydrogenated bilayer, whereas the 100 % H_2O contrast is sensitive to the lipid head-group regions but less so toward the hydrophobic core of the bilayer due to the scattering length density of the lipid acyl chains and H_2O close in value.

Floating lipid bilayers offer the chance to observe membrane bilayers in a close to free state. A lower lipid monolayer is chemically grafted onto silicon with the lipid head groups uppermost. Then a bilayer is deposited on this surface such that a significant water layer remains between the two structures (14). This enables the bilayer to remain fully fluid and largely unconstrained permitting membrane proteins to be inserted. Such systems also increase the complexity of the SLD profile so modeling becomes more demanding with increased numbers of layers to fit. One method which simplifies fitting is magnetic contrast neutron reflectometry which makes use of a magnetic (e.g., iron) layer below the solid surface which exploits another feature of neutrons, their sensitivity to the magnetic dipole. This approach doubles the number of contrasts which can be gained from one experiment without requiring solvent changes or selective deuteration (21). Otherwise, data treatment and layer modeling are similar to those described for lipid monolayers.

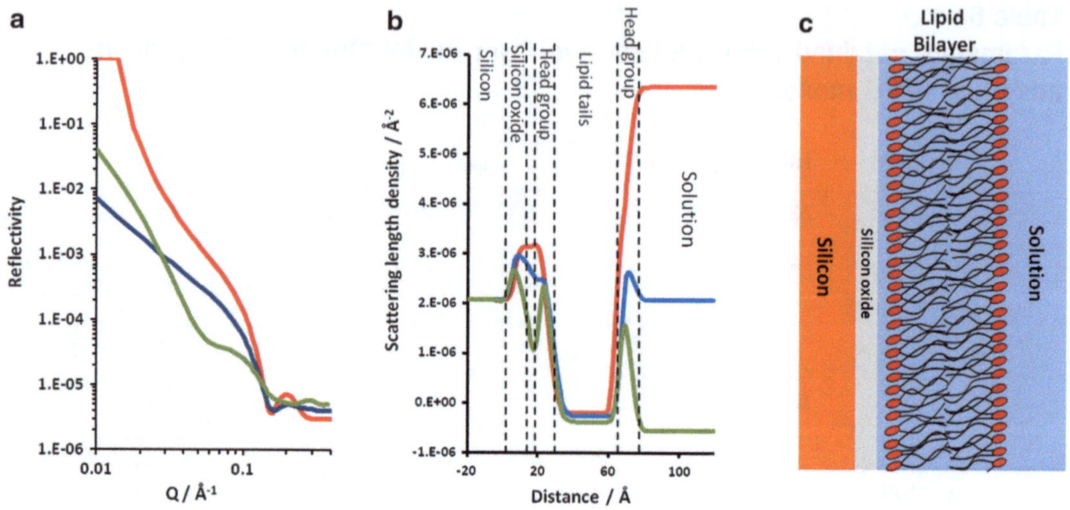

Fig. 11. Model neutron reflectivity profiles of a silicon surface-adsorbed hydrogenated DPPC bilayer in three solution contrasts H_2O/D_2O contrasts (**a**), the scattering length density profiles these profiles describe (**b**). The three solution contrast shown are for the h-DPPC bilayer in 100 % D_2O (*red*), silicon matched water (38 % D_2O, 62 % H_2O, *blue*), and 100 H_2O (*green*). A cartoon of the interfacial structure this data describes is shown (**c**).

4.4.1. In Conclusion

Neutron reflectometry is an emerging technique for the structural analysis of protein–lipid interactions. The differing neutron scattering properties allow for determination of the structural contribution of individual components to a complex structure.

4.5. Examining the Membrane Protein Complexes Using Small-Angle Neutron Scattering

4.5.1. Introduction: Small-Angle Scattering and SANS

The term small-angle scattering covers a variety of experimental types that utilize X-rays, photons, or neutrons to examine structures larger in scale than atomic distances. In particular these techniques are valuable for determining large-scale structures in solution. While light scattering is a widespread laboratory technique and small-angle X-ray scattering (SAXS) is available in specialized laboratories, the requirements for large-scale infrastructure to generate neutrons and high intensity X-rays mean that the majority of both SAXS and SANS experiments involve the use of large facilities. Access is via reviewed applications as with NR. Scattering at both small and wide angles is a more general case of diffraction, but for those coming from the biological sciences, it can be useful to describe the technique starting from the spots on diffraction patterns familiar from protein crystallography. These spots arise as a result of constructive interference of incident radiation caused by regularities in the structure of the sample. In the case of an ideal single crystal, the spots arise because the arrangement of molecules within the crystal is regular *in all three dimensions.* That is, for any given atom within the molecular structure, the equivalent atoms are found at regular positions and distances in all three dimensions. Diffraction spots arise because the distances between atoms within the molecules are repeated in a regular fashion across the crystal lattice. The larger the lattice, the sharper the points.

In solution, while the interatomic distances are still repeated, the lattice is gone. The sharp diffraction points are now smeared out. The familiar protein diffraction pattern also provides an example of this in the form of the water ring, the result of regular water–water distances within the *amorphous* water trapped in the crystal. What would be the sharp diffraction peaks arising from the unit cell dimensions in a single water crystal are spread out to become a ring. Because the scattering is isotropic (i.e., it forms symmetric rings on the detector), it is usual to radially average the image data that is often collected and plot solution scattering data as a one-dimensional plot of intensity versus angle (usually expressed as momentum transfer Q as different instruments operate at different – or multiple – wavelengths).

In principle the complete atomic structure of a protein is captured by the scattering pattern at wide angle. However the overlap of the peaks arising from many similar interatomic distances and the weak signal mean that wide-angle scattering is not a widely used technique for structural determination. However the intermolecular distances that describe the overall shape of the molecule and its size give rise to useful experimental information, particularly when combined with other data, that can be used to resolve larger scale structures. In addition the use of contrast enables SANS in particular to be very useful in defining how molecules within a complex are arranged with respect to each other (see Note 5).

4.5.2. The SANS Instrument

As there are two sources of neutrons (reactor and spallation sources), there are broadly two classes of SANS instrument with similar overall geometry. For both classes a collimated beam of neutrons is passed through a "pinhole" (in reality around one square centimeter) into a sample where it is scattered. The scattered beam is then passed, through a window, into a vacuum tank containing an image detector where the scattered neutrons are collected. The majority of neutrons do not interact with the sample at all so the detector is protected from this direct beam by a beam stop. The difference in instruments lies in the type of beam brought to the sample.

A classical single-wavelength pinhole SANS instrument is very similar to a synchrotron SAXS instrument. A continuous white beam of neutrons, generally from a cold source on a nuclear reactor, is transported from the source via neutron guides to the instrument. The white beam passes through a monochromator to select the desired wavelength of neutrons and then moves into a section of guides and collimation that leads to the pinhole and sample. The detector will collect data continuously noting the position where each scattered neutron strikes and building up an image over the collection period. With the known geometry of the detector and sample and the known wavelength of incident neutrons, the image can be directly converted to intensity versus *Q*.

On a spallation neutron source, the time-averaged white beam (i.e., containing all the wavelengths) intensity is much lower, although the peak intensity is higher, than a reactor source. So selecting a small wavelength range would involve throwing most of the neutrons away. However since the beam is pulsed it is possible to utilize the whole of the white beam by using a time-of-flight approach. The neutron wavelength is inversely related to its velocity. Since the time of the start of the neutron pulse and distance to the detector is known, it is possible to determine the wavelength of any incident neutron based on its arrival time at the detector. This has two advantages. Firstly it means the whole of the beam is being used. Secondly the wavelength range means that a wider range of *Q* can be recorded for the same size of detector as *Q* is dependent on both angle and wavelength.

Which type of instrument is better depends on your experimental requirements. Most biological science work has traditionally been done on single-wavelength reactor instruments as spallation source instruments have not traditionally had the signal to noise required for relatively dilute and weakly scattering biological samples. A new generation of spallation source instruments is now challenging this with optimized instruments running with sources designed to have optimal characteristics for probing these large-scale structures. In general measurements that do not require a large *Q* range or which require time resolution will benefit from single-wavelength instruments and experiments where the widest simultaneous *Q* range is advantageous (e.g., with samples that exhibit structure over wide length ranges but may not be very stable), then white beam time-of-flight instruments have advantages.

4.5.3. SANS Experiments on Protein–Lipid Complexes

1. The SANS experiment is broadly similar for all instruments. Data will be collected on some combination of calibration standards, background samples, and the actual sample of interest. At most facilities the calibration samples (which vary from facility to facility) will generally be run by the instrument scientist supporting your experiment, and the results from these runs will be used to set up the process for "reducing" your data. This should result in an instrument independent data set with all background signals removed. The image (or images and time) data that is collected will be converted to scattering intensity versus *Q* where intensity is on an absolute scale. This reduction process requires a scattering pattern from the sample (and a background if desired), the neutron transmission of the sample (and background if appropriate), and the transmission of the empty instrument. Reduction tools differ greatly from instrument to instrument (and from instrument scientist to instrument scientist), but in most cases these will be

set up for you so that you need only add the identity of new SANS runs and transmissions as they are created to do the reduction. Most SANS instruments have large sample changers taking from 10 to 40 samples at a time. With each individual measurement taking from a few minutes up to several hours, it is generally possible to set up a series of samples to run for hours between sample changes. Experiments are generally scheduled for a day to a few days, although some facilities are using shorter allocation times. The complexity of experiments with membrane–protein systems means that these experiments generally take several days (see Note 8).

2. The purpose of doing a SANS experiment on a protein–membrane system in most cases will be to determine the shape of the protein, while using contrast to mask the contribution from the lipid or detergent that is stabilizing the protein in solution. The experimental design therefore aims to make it possible to match the lipid or detergent component out making it possible to focus specifically on the protein. Early experiments involved using deuterated lipids which could be matched out in buffers containing close to 100 % D_2O. However these experiments are challenging as the use of lipid vesicles means that the sample is largely empty buffer with the protein confined to the small volume of lipid membrane. The high average concentrations of protein required to generate a reasonable signal make it difficult to balance signal quality with protein stability in the membrane, and we have observed apparent protein aggregation in the membrane in the systems we have examined. Vesicle–peptide interactions have been successfully studied via SANS (22), although even here the focus is more on the effect of the peptide on the membrane than the peptide configuration itself.

3. To obtain higher concentrations of protein, it is necessary to move to isolated membrane mimics, and most work has focused on the use of detergents. Hydrogenated detergents with their high hydrogen content have fairly low scattering length densities and can be matched out with buffer D_2O contents of around 10–15 %. However this match point is relatively close to that of most proteins (~40 % D_2O) meaning that the signal from the protein is suboptimal. The ideal approach to this experimental design would be to use either deuterated protein (difficult to achieve for many membrane proteins) or deuterated detergents to enhance the contrast difference between protein and stabilizing surfactant. Unfortunately few of the most commonly used detergents for membrane protein stabilization are available in deuterated form.

4. Clifton et al. (23), using SDS to stabilize the colicin N–OmpF complex, describe an elegant experiment in which carefully selected mixtures of deuterated and hydrogenated SDS are

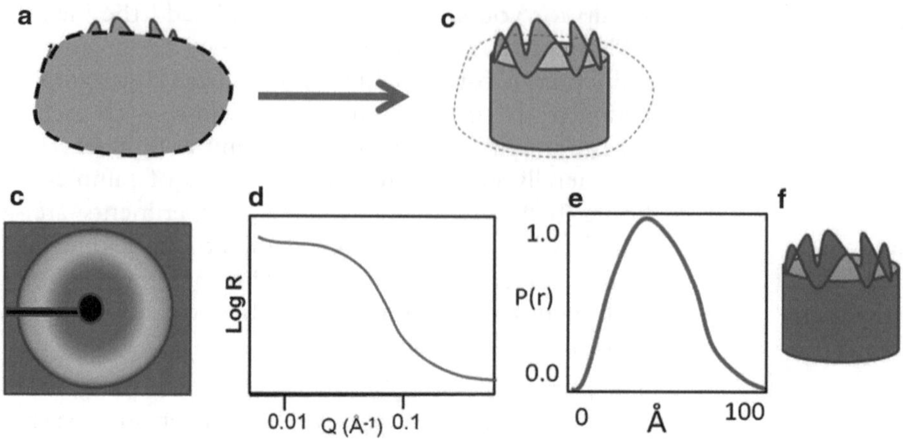

Fig. 12. Stages in SANS data analysis of a membrane–lipid complex. (**a**) A membrane protein is prepared in H_2O and normal detergent or lipid is largely amorphous to neutron scattering. (**b**) The detergent/lipid is matched by solution contrast variation and rendered invisible such that the scattering signal arises solely from the protein. (**c**) The scattering data is collected as a radially symmetrical data set on a 2D detector. Note the beam stop shadow. (**d**) The next stage is to average this data to provide a 1D data set of scattering versus *Q*. The data can then be analyzed by Guinier methods (see text) and used (**e**) to provide a *P*(*r*) plot which represents the distribution of interatomic distances in the particle. Finally (**f**) a 3D model is derived using rigid body modeling or ab initio methods.

used to match the solvent contrast at various points, which are to distinguish the contribution from hydrogenated and deuterated protein. This form of experiment where the aim is to exactly match out one component of a system to make it "invisible" is challenging even in two-component systems. The first challenge is that both surfactant and lipid molecules often differ in the scattering length density of their head and tail regions. Careful mixing of fully deuterated and tail deuterated lipids can provide a mixture with consistent match points corresponding to around 90–100 % D_2O (Fig. 12a, b), but this is difficult. Among the best examples of contrast matching a lipid vesicle is that shown by Garg et al. (see Fig. 12a) (24) in a study of cholesterol transport; however this is matched at 48.6 % D_2O, very close to the match point of a protein.

5. *Nanodiscs as Isolated Membrane Mimics*: A system with significant promise in tackling these problems is lipid nanodiscs. Small-angle scattering experiments on protein–lipid nanodiscs have shown that discs themselves provide excellent scattering data and a detailed model of their structure is being developed (25) that has the potential to support structural studies of proteins embedded in the nanodiscs or association with them. Similar developmental work on bicelles, formed via a mixture of normal and short-chain lipids (26), and SMALPs (27) where a lipid disc is stabilized by a polymer also shows promise. Amphipols, amphiphilic polymers which keep membrane proteins

soluble in the absence of detergents or lipids, have also been successfully used in neutron scattering studies (28, 29).

6. The lipid complexes have a number of advantages. Firstly in addressing the problems faced in using vesicles, they provide a lipid platform that can be prepared at high concentration but also with a high relative protein proportion. Lateral aggregation of proteins in the membrane is prevented by the isolation of the discs from each other. Protein–protein aggregation via solvent exposed portions may of course still be an issue. The second advantage is that the isolated discs are available in size ranges that are convenient for gel filtration chromatography, making it practical to prepare highly purified samples with high homogeneity. Generally protein containing discs can be purified away from empty discs. Finally the use of lipids and (in the case of the SMALPs (27)) polymers that can be part or completely deuterated opens up a range of contrast variation strategies that are not accessible with detergents.

4.5.4. Data Analysis of SANS Experiments

1. Solution SANS data (Fig. 12c) is generally presented as a one-dimensional plot of *Q* versus intensity (Fig. 12d). The scattering intensity is generally at its highest at low values of *Q* and falls away before reaching a background level which is determined by the signal to noise of the instrument. The key information is in the overall intensity of scattering, and the shape of the falling intensity at *Q* increases. This can provide information on the size, concentration, contrast, shape, and internal organization of the species in solution. In practice there are essentially two strategies for analyzing SANS data on protein–lipid/detergent complexes, and these mirror the approaches generally used for the analysis of biological scattering data (see Note 9).
2. The first approach is to use existing models to predict scattering patterns and to compare these to experimental data. This may take the form of distinguishing between two different models or might involve the generation of large numbers of model structures to identify classes of candidates. In most cases the models will be atomic resolution structures built from known structures, but it is also possible to use simpler shape models where structural detail is not available.
3. However in cases where insufficient structural detail is available to build an atomic resolution model, the approach of choice is to use ab initio methods to attempt to reconstruct a low-resolution structure of the protein directly from the experimental data. The ATSAS suite of tools developed at EMBL is the most popular method for this form of analysis; however the use of these tools needs care, and it is recommended that an experienced user is consulted when designing the experiment

and throughout the analysis. A variety of training courses on the use of this software are also available.

4. *Extracting the Scattering of the Protein Component*: Whichever analytical strategy is chosen, the first step in data analysis will generally be to isolate the contribution to the scattering from the different components in the system. This stage of analysis needs to be considered carefully in the design of the experiment. The simplest approach, that of contrast matching the lipid or surfactant component, requires that the experimental matching be near perfect, and this is challenging to achieve in practice. If the match point is slightly off extracting, the contribution of the protein becomes difficult or impossible to achieve with sufficient precision.

5. An alternative approach is to use a contrast variation technique. Here, instead of explicitly matching the components, a range of different D_2O contents are used from 100 % H_2O to 100 % D_2O (see Note 3). The overall scattering at any given D_2O content is given by the theoretical scattering from each component, the experimental contrast for that component, and a cross term that reflects the spatial relationship between the components. Here $P(Q)$ represents the theoretical scattering from protein, lipid, and cross term, respectively, and $\Delta\rho$ is the difference in scattering length density between solution and each component.

$$I(Q) = \Delta\rho_p^2 P(Q_p) + \Delta\rho_l^2 P(Q)_l = \Delta\rho_c \Delta\rho_a P(Q)_x. \quad (7)$$

In the contrast match case, $\Delta\rho_l$ is zero so only the protein component is visible. By using a range of D_2O contents (i.e., different $\Delta\rho$), it is possible to solve this equation to determine the various $P(Q)$, which are the theoretical scattering patterns for each component. This minimizes the effect of small errors in calculated contrast and D_2O content. In addition the cross term can provide useful information on the spatial relationship between the protein and lipid that may be useful in model building (30). Web services for analyzing contrast variation series by this approach have been described (31).

6. *Model building*: Once the scattering of interest has been analytically isolated (Fig. 11d), a common analysis approach is to compare this to the scattering predicted from models. The most common method for converting a structure to predicted scattering pattern is to use CRYSON from the ATSAS software suite. It is important that the model be complete; missing loops or domains will have a significant effect on the predicted scattering pattern and render any comparison meaningless. Model building itself is beyond the scope of this chapter. Where one

component of the system has a known structure, then checking the experimental scattering from this component (obtained by matching the other parts or via contrast variation) to the predicted scattering is an excellent check of the quality of the data and that analysis has proceeded correctly. For instance, Clifton et al. (23) show that the scattering from OmpF–ColN complex where the colicin is completely contrast matched is essentially identical to that predicted from the crystal structure of OmpF. This provides confidence that the sophisticated contrast match strategy is working in this experiment. Conversely where the scattering does not match that predicted this is evidence of a change in conformation. However, in this case it is important to ensure appropriate controls are in place to demonstrate that the difference is not an artifact of the contrast strategy or of sample heterogeneity.

Similarly SANS provides an effective way of distinguishing between a small set of possible models provided there is sufficient structural differences. Differences in the mid-Q range, particularly the shape and position of the main fall off in scattering intensity and position and the presence of specific features at slightly higher Q, are most diagnostic. Differences at high Q are not diagnostic, both because these can easily be the result of conformational variability and because the signal is generally reaching the noise level of the measurement.

It is in principle possible to generate very large numbers of models from structural components and to use this to identify conformations that are consistent with the experimental data. This approach has been implemented successfully with proteins in solution, but we are not aware of published studies on protein–membrane or protein–detergent complexes. This approach requires very high-quality data to provide sufficient confidence in model refinement, and this is challenging for membrane protein systems.

4.5.5. Ab Initio Model Construction

As the scattering is a function of the shape of the species in solution, it is in principle possible to reconstruct possible shapes of the species directly from the data. As one-dimensional data is being converted to a three-dimensional model, there will be multiple models that can generate the same scattering. However by applying reasonable constraints on compactness and connectivity of proteins, it is often possible to identify a single plausible model that is consistent with the data (Fig. 12e, f). Again tools from the ATSAS suite are the most commonly used for this form of analysis. These are relatively easy to use but interpretation of the output data requires care, and it is recommended that an experienced user is consulted in the planning of experiments as well as their analysis. Clifton et al. (23) used MONSA (32) which uses multiple contrast data sets on the same system to reconstruct the structure of two-component systems, to determine an overall structure of the OmpF–ColN complex. Note the careful comparison with available crystal structures and other information that is used to support the conclusions derived from this analysis.

4.5.6. SANS Summary

Small-angle neutron scattering is a natural technique to use to determine the structure of membrane protein systems due to the benefits of contrast variation. SANS can provide information on the shape and size of the specific components in complex mixtures. However as a technique it requires significant quantities of sample and gives relatively low signal to noise. This means that careful consideration of how to get sufficient material of interest into the sample in a stable form (see Note 3) is critical to experimental success.

Critical to the success of a SANS experiment is the careful design of the contrast variation strategy and its effect on analysis approach. For two-component (protein–lipid, protein–detergent) systems, a contrast variation strategy is recommended, particularly if one of the components can be obtained in deuterated form. For more complex systems such as that studied by Clifton et al. (23), careful experimental optimization is required to ensure that contribution to scattering from "matched" components is removed (see Note 5).

The simplest form of experiment, and the one with the most chance of success, is one in which the aim is to distinguish between two possible structures. This is the most straightforward form of analysis, where scattering predicted from one model matches better than the other. It is possible to determine low-resolution structures without prior information and in principle to distinguish between large numbers of classes of high resolution model, but this is more challenging.

SANS remains a challenging technique to apply successfully to biological system, and in combination with the experimental challenges of working with membrane systems, this means these experiments are certainly not straightforward (see Note 6 and 7). However the benefits that SANS brings in being able to obtain the contribution from different components of the system and using this to distinguish between possible models are significant and unique. There is significant potential to expand the use of SANS in the study of these systems.

5. Notes

1. It is often advisable for research groups new to neutron reflectometry or SANS to collaborate with one of the scientists responsible for the SANS/NR instrument where the studies will take place. The instrument scientist will have a great depth of knowledge about the workings of the beam line, data collection, experimental strategy, and data analysis and probably will have been assigned to contact your experiment due their knowledge of the research area the study falls into.
2. When working with protein solutions in various solution contrasts, check that the protein is soluble in both D_2O and H_2O buffer solutions prior to your experimental beam time.
3. There are many techniques complementary to neutron reflectivity such as X-ray reflectivity, Brewster angle microscopy, imaging ellipsometry, and quartz crystal microbalance methods. Always include data obtained on complementary techniques in your application for instrument time. This data will greatly enhance your case for instrument time and provide complementary data to allow you to better understand your system under study. It is very important to make the case why this can only be solved by neutrons.
4. Techniques complementary to SANS are dynamic light scattering, size exclusion chromatography, and small-angle X-ray scattering. One or more of these techniques should be used to examine your protein–lipid complex samples prior to submitting your application for SANS instrument time.
5. Using deuterated and hydrogenated protein contrasts in your series is advantageous in both NR and SANS studies of protein–lipid complexes. In advance of your experimental beam time, see if the facility you are going has a support team who can assist you in the deuteration of your protein of interest.
6. Make sure you order your deuterated lipids far in advance of instrument time. They are often produced to order so they may take time to arrive.
7. Instrument time is allocated in days, and research groups will be expected to monitor the experiments on a 24-h basis; this means that experimenters should be aware they may be expected to work unsociable hours. Bring enough people to man the experiment at all times.
8. Check that the facility where you are going has the equipment you need in its sample preparation and support laboratories. Enquire about automatic sample changers which might allow you some more sleep.

9. *SANS Sanity Checks*: A number of simple graph transforms can be carried out on SANS data to provide first-stage information on how an experiment is going and ensure that samples are behaving. They can also provide the first steps in guiding model building. The most common form of this is a Guinier analysis, which is used to determine the overall size of the scattering object in solution. The Guinier equation provides an analytical approximation for a scattering curve at low values of Q that depends only on a parameter of overall size, the radius of gyration. This readily provides important helpful checks: firstly that the sample has not obviously aggregated, secondly that it is approximately the expected size, and finally to provide evidence that there are no particle–particle interactions, which is an assumption made by most downstream analysis processes.

The Guinier approximation provides the equation:

$$I \approx I_o e^{(-Q^2 R_g^2/3)}. \tag{8}$$

Rearrangement gives

$$\ln I = \ln(I_o) - \left(\frac{R_g^2}{3}\right) Q^2. \tag{9}$$

meaning that a graph of the natural log of intensity versus Q^2 should give a straight line where the slope is negative and $R_g = \sqrt{-3\,gradient}$. If this graph curves upward as Q tends to zero, this is generally a sign of aggregation. If the R_g is significantly different from the expected radius of the system, then this is either a sign of some form of interaction (if larger than expected) or possibly of repulsive interactions between particles (if smaller).

For most analysis techniques, it crucial that such repulsive interactions be ruled out. This can be achieved by a range of techniques, including AUC and light scattering. We use a Malvern Zetasizer dynamic light scattering DLS device which has proven very reliable in revealing solutions which go on to show aggregation in SANS. DLS calculation of molecular size is dependent upon a knowledge of viscosity, and here it should be noted that the viscosity of D_2O at 20°C is 1.251 mPa s and H_2O is 1.002 mPa s. Thus we assume a linear relationship between D_2O percentage and viscosity.

One can also perform preliminary small-angle scattering measurements at a range of concentrations and show that the experimental R_g does not change. Interactions of this sort will reduce the experimental R_g at low concentrations. This kind of experiment can be done during beam time but is best carried out beforehand. Often these measurements can be performed on a laboratory small-angle X-ray scattering instrument which then provides a helpful check of sample integrity. The optimal sample concentration for the main experiment is the maximum concentration at which the experimental R_g is the same as that measured at high dilution.

Acknowledgments

We thank John Webster for helping in providing helpful advice on instrumentation and data modeling procedures.

References

1. Lakey JH (2009) Neutrons for biologists: a beginner's guide, or why you should consider using neutrons. J R Soc Interface 6(Suppl 5): S567–S573
2. Zabel H, Theis Brohl K, Wolff M, Toperverg BP (2008) Polarized neutron reflectometry for the analysis of nanomagnetic systems magnetics. IEEE Trans Magnetics 44:1928–1934
3. Penfold J, Staples EJ, Tucker I, Thompson LJ (1999) The adsorption of mixed surfactants at the hydrophilic silica surface from aqueous solution: studied by specular neutron reflection. Int J Thermophys 20:19–34
4. Lad MD, Birembaut F, Clifton LA, Frazier RA, Webster JRP, Green RJ (2007) Antimicrobial peptide-lipid binding interactions and binding selectivity. Biophys J 92:3575–3586
5. Clifton LA, Green RJ, Hughes AV, Frazier RA (2008) Interfacial structure of wild-type and mutant forms of puroindoline-b bound to DPPG monolayers. J Phys Chem B 112:15907–15913
6. Callow P, Fragneto G, Cubitt R, Barlow DJ, Lawrence MJ (2009) Interaction of cationic lipid/DNA complexes with model membranes as determined by neutron reflectivity. Langmuir 25:4181–4189
7. Kent MS, Murton JK, Sasaki DY, Satija S, Akgun B, Nanda H, Curtis JE, Majewski J, Morgan CR, Engen JR (2010) Neutron reflectometry study of the conformation of HIV Nef bound to lipid membranes. Biophys J 99:1940–1948
8. Clifton LA, Sanders MR, Hughes AV, Neylon C, Frazier RA, Green RJ (2011) Lipid binding interactions of antimicrobial plant seed defence proteins: puroindoline-a and β-purothionin. Phys Chem Chem Phys 13:17153–17162
9. Miano F, Zhao X, Lu JR, Penfold J (2007) Co-adsorption of human milk lactoferrin into the dipalmitoylglycerolphosphatidylcholine phospholipid monolayer spread at the air/water interface. Biophys J 92:1254–1262
10. Le Blanc RM (2006) Molecular recognition at Langmuir monolayers. Curr Opin Chem Biol 10:529–536
11. Mowald H (1993) Surfactant layers at water surfaces. Rep Prog Phys 56:653–685
12. Wacklin HP, Tiberg F, Fragneto G, Thomas RK (2007) Distribution of reaction products in phospholipase A2 hydrolysis. Biochim Biophys Acta 1768:1036–1049
13. Wacklin HP, Tiberg F, Fragneto G, Thomas RK (2005) Phospholipase A2 hydrolysis of supported phospholipid bilayers: a neutron reflectivity and ellipsometry study. Biochemistry 44:2811–2821
14. Hughes AV, Howse JR, Dabkowska A, Jones RAL, Lawrence MJ, Roser SJ (2008) Floating lipid bilayers deposited on chemically grafted phosphatidylcholine surfaces. Langmuir 24:1989–1999
15. Dynarowicztka PA, Dhanabalanb A, Oliveira ON Jr (2001) Modern physicochemical research on Langmuir monolayers. Adv Colloid Interface Sci 91:221–293
16. Bowers J, Zarbakhsh A, Webster JRP, Facility I, Hutchings LR, Richards RW (2001) Structure of a spread film of a polybutadiene-poly (ethylene oxide) linear diblock copolymer at the air-water interface as determined by neutron reflectometry. Langmuir 17:131–139
17. Shen H-H, Thomas RK, Chen C-Y, Darton RC, Baker SC, Penfold J (2009) Aggregation of the naturally occurring lipopeptide, surfactin, at interfaces and in solution: an unusual type of surfactant? Langmuir 25:4211–4218

18. Lu JR, Thomas RK (1998) Neutron reflection from wet interfaces. J Chem Soc Faraday Trans 94:995–1018
19. Nanda H, Datta SAK, Heinrich F, Losche M, Rein A, Krueger S, Curtis JE (2010) Electrostatic interactions and binding orientation of HIV-1 matrix studied by neutron reflectivity. Biophys J 99:2516–2524
20. Chenal A, Prongidi-Fix L, Perier A, Aisenbrey C, Vernier G, Lambotte S, Haertlein M, Dauvergne MT, Fragneto G, Bechinger B, Gillet D, Forge V, Ferrand M (2009) Deciphering membrane insertion of the diphtheria toxin T domain by specular neutron reflectometry and solid-state NMR spectroscopy. J Mol Biol 391(5):872–883
21. Holt SA, Le Brun AP, Majkrzak CF, McGillivray DJ, Heinrich F, Lösche M, Lakey JH (2009) An ion channel containing model membrane: structural determination by magnetic contrast neutron reflectometry. Soft Matter 5:2576–2586
22. Qian S, Heller WT (2011) Peptide-induced asymmetric distribution of charged lipids in a vesicle bilayer revealed by small-angle neutron scattering. J Phys Chem B 115:9831–9837
23. Clifton LA, Johnson CL, Solovyova AS, Callow P, Weiss KL, Ridley H, Le Brun AP, Kinane CJ, Webster JRP, Holt SA, Lakey JH (2012) Low resolution structure and dynamics of a colicin-receptor complex determined by neutron scattering. J Biol Chem 287(1):337–346
24. Garg S, Porcar L, Woodka AC, Butler PD, Perez-Salas U (2011) Non-invasive neutron scattering measurements reveal slower cholesterol transport in model lipid membranes. Biophys J 101:370–377
25. Wadsäter M, Simonsen JB, Lauridsen T, Tveten EG, Naur P, Bjørnholm T, Wacklin H, Mortensen K, Arleth L, Feidenhans'l R, Cárdenas M (2011) Aligning nanodiscs at the air-water interface, a neutron reflectivity study. Langmuir 27:15065–15073
26. Pabst G, Kucerka N, Nieh MP, Rheinstadter MC, Katsaras J (2010) Applications of neutron and x-ray scattering to the study of biologically relevant model membranes. Chem Phys Lipids 163:460–479
27. Knowles TJ, Finka R, Smith C, Lin Y-P, Dafforn T, Overduin M (2009) Membrane proteins solubilized intact in lipid containing nanoparticles bounded by styrene maleic acid copolymer. J Am Chem Soc 131:7484–7485
28. Gohon Y, Dahmane T, Ruigrok RWH, Schuck P, Charvolin D, Rappaport F, Timmins P, Engelman DM, Tribet C, Popot JL, Ebel C (2008) Bacteriorhodopsin/amphipol complexes: structural and functional properties. Biophys J 94:3523–3537
29. Popot JL, Berry EA, Charvolin D, Creuzenet C, Ebel C, Engelman DM, Flotenmeyer M, Giusti F, Gohon Y, Herve P, Hong Q, Lakey JH, Leonard K, Shuman HA, Timmins P, Warschawski DE, Zito F, Zoonens M, Pucci B, Tribet C (2003) Amphipols: polymeric surfactants for membrane biology research. Cell Mol Life Sci 60:1559–1574
30. Whitten AE, Trewhella J (2009) Small-Angle Scattering and Neutron Contrast Variation for Studying Bio-Molecular Complexes. In: Lee JW, Foote RS (eds) Micro and nano technologies in bioanalysis: methods and protocols. Springer, New York, pp 307–323
31. Whitten AE, Cai S, Trewhella J (2008) MULCh: modules for the analysis of small-angle neutron contrast variation data from biomolecular assemblies. J Appl Crystallogr 41:222–226
32. Svergun DI (1999) Restoring low resolution structure of biological macromolecules from solution scattering using simulated annealing. Biophys J 76:2879–2886

Chapter 8

Circular-Dichroism and Synchrotron-Radiation Circular-Dichroism Spectroscopy as Tools to Monitor Protein Structure in a Lipid Environment

Koichi Matsuo and Kunihiko Gekko

Abstract

Circular-dichroism (CD) spectroscopy is a powerful tool for the secondary-structure analysis of proteins. The structural information obtained by CD does not have atomic-level resolution (unlike X-ray crystallography and NMR spectroscopy), but it has the great advantage of being applicable to both nonnative and native proteins in a wide range of solution conditions containing lipids and detergents. The development of synchrotron-radiation CD (SRCD) instruments has greatly expanded the utility of this method by extending the spectra to the vacuum-ultraviolet region below 190 nm and producing information that is unobtainable by conventional CD instruments. Combining SRCD data with bioinformatics provides new insight into the conformational changes of proteins in a membrane environment.

Key words: Circular-dichroism spectroscopy, Conformational change, Lipid environment, Membrane protein, Neural network, Secondary-structure analysis, Sequence-based prediction, Synchrotron radiation, Vacuum-ultraviolet circular dichroism

1. Introduction

Knowledge of the structures of membrane proteins is essential to understanding their biological functions such as in intracellular communication and stimulation responses of cells. However, there remain a huge number of membrane proteins whose tertiary structures are unknown due to (1) the technical difficulties of having to crystallize proteins in the membrane environment (or detergent micelle) for X-ray crystallography and (2) only small proteins being analyzable by NMR spectroscopy. Circular-dichroism (CD) spectroscopy, which is a well-established technique in structural biology, does not provide atomic resolution (unlike X-ray crystallography and NMR spectroscopy), but it has been widely used for analyzing the structure and conformational changes of membrane

Jörg H. Kleinschmidt (ed.), *Lipid-Protein Interactions: Methods and Protocols*, Methods in Molecular Biology, vol. 974,
DOI 10.1007/978-1-62703-275-9_8, © Springer Science+Business Media New York 2013

proteins (1–6) because CD spectra are very sensitive to local structures and are measurable for any proteins at low concentration under various solvent conditions or in a lipid environment. Proteins exhibit characteristic CD spectra in the visible to vacuum-ultraviolet (VUV) regions. Most folded proteins exhibit significant optical activity in the near-ultraviolet (UV) region (250–300 nm), owing primarily to the presence of aromatic residues and disulfide bonds that sensitively reflect the tertiary structure. In the far-UV region (<250 nm), these spectral characteristics are determined primarily by the secondary structures of protein. The principles, techniques, and applications of CD spectroscopy have been comprehensively reviewed in many books (7–9).

The secondary-structure analysis of proteins by CD spectroscopy has been markedly improved by (1) the development of programs (e.g., DSSP and Xtlsstr) for assigning the secondary structures from atomic coordinates (10, 11), (2) the advancements in software (e.g., CONTIN, SELCON3, and CDSSTR) for analyzing CD spectra (12), and (3) the extension of CD measurements to the VUV region (13). The short-wavelength limit of CD spectroscopy can be extended down to ~160 nm by using synchrotron radiation (SR) as a high-flux source of photons, which yields much more detailed information that cannot be obtained with a conventional CD spectrophotometer (14–16). VUV CD spectroscopy using SR (SRCD) is a powerful tool for estimating not only the contents but also the numbers of α-helix and β-strand segments with high accuracy (17, 18). Furthermore, combining SRCD data with bioinformatics (e.g., neural-network (NN) algorithms) can considerably improve the sequence-based prediction of secondary structures (19), which can also be used to test the predicted tertiary-structure model of a protein. This chapter addresses recent developments in CD and SRCD spectroscopy as tools to monitor protein secondary structures in a lipid environment, focusing on nonintegral membrane proteins.

1.1. Basic Principles

CD is defined as the difference (ΔA) between the absorbance of left- and right-handed circularly polarized light, each of which follows the Beer–Lambert law at a given wavelength $A_L = \varepsilon_L lC$ and $A_R = \varepsilon_R lC$

$$A = A_L - A_R = (\varepsilon_L - \varepsilon_R)lC = \varepsilon lC \qquad (1)$$

where ε_L and ε_R are the molar absorptions of the sample for left- and right-handed circularly polarized light, respectively, C is the molar concentration of the sample, and l is the path length of the optical cell (in cm). CD instruments usually represent the CD in $\Delta\varepsilon$ or a molar ellipticity, $[\theta]$, with the latter being widely used for proteins and polypeptides. $[\theta]$ is related to $\Delta\varepsilon$ by

$$\varepsilon = [\theta]/3,298 \qquad (2)$$

and is obtained by normalizing the raw ellipticity, θ, with the sample concentration and the path length of the cell:

$$[\theta] = 100\theta/lC \quad (3)$$

For proteins (or other polymers), the molar concentration unit is based on the mean residue weight (MRW), which is the molecular weight of the protein divided by the number of amino acid residues per molecule. Thus, $[\theta]$ can be obtained in the unit of degrees cm^2 $dmol^{-1}$ by

$$[\theta] = \theta \times (MRW)/10cl \quad (4)$$

where c is the concentration of the sample in mg/ml.

Consider the hen-egg lysozyme protein, which has a molecular weight of 14,314 and 129 residues—its MRW value is 111 (i.e., 14,314/129). When θ is 100° at 222 nm, c is 0.1 mg/ml, and l is 1 cm, the value of $[\theta]_{222\ \mathrm{nm}}$ is calculated to be 11,100 (i.e., 100 × 111/(10 × 0.1 × 1)) degrees cm^2 $dmol^{-1}$ from Eq. 4, which corresponds to $\Delta\varepsilon_{222\ \mathrm{nm}} = 3.37$ (i.e., 11,100/3,298) from Eq. 2.

2. Materials

2.1. Preparation of Protein Solution

The sample preparation methods are essentially the same in CD and SRCD measurements, although the protein concentration must be adjusted according to the path length of the cell. The most important factor is the protein purity because the presence of impurities in the protein solution will affect the CD spectrum, inevitably leading to incorrect data on the secondary structure. The purity should be at least 95% as judged by mass spectroscopy or sodium dodecyl sulfate– polyacrylamide gel electrophoresis. The protein solution should also be free from aggregates that cause light scattering and abnormal absorption. Aggregates can be removed by centrifugation (at 14,000 rpm (20,400 G) for 20 min) and/or by filtration (0.2 μm Millipore filter).

The use of a solvent with a high absorbance should be avoided, especially when CD spectra are measured down to the VUV region. Chloride and carboxylate ions exhibit strong absorption in the VUV region. Phosphate, acetate, and Tris–HCl buffers are available at high concentrations below 50 mM. HEPES buffer can be used only at concentrations below 5 mM. Chloride ions may be replaced by fluoride or sulfate ions. If the use of chloride ions is unavoidable for a reason such as protein stability or solubility, their concentration should be below 5 mM. Organic solvents such as acetonitrile, ethanol, methanol, and trifluoroethanol, which may be useful for increasing the solubility of membrane proteins, can be used at moderate concentrations in the VUV region. The absorption by buffer or solvent should be confirmed as being sufficiently low by checking the high-tension voltage of the CD instrument before preparing protein solutions.

2.2. Preparation of Liposomes

While the liposome preparation varies with the type of lipid, the following methods are typically used (6, 20–23):

1. Lipid is dissolved in chloroform (or a chloroform/methanol mixture with a volume ratio of 3:1) and the solvent is evaporated under a stream of nitrogen gas and in a vacuum for at least 3 h. The obtained lipid film is dispersed in buffer and further hydrated by vibrating on a vortex to produce multilamellar vesicles (MLVs).
2. Sonic-disintegration and extrusion methods are often used to prepare small unilamellar vesicles (SUVs) from MLVs. In the sonic-disintegration method, SUVs are prepared by sonicating an MLV solution with a probe sonicator to near optical transparency, and the residual MLVs and the titanium particles released from the probe are removed by centrifugation at 14,000 rpm (20,400 G) for 20 min. The average diameter of the obtained SUVs can be estimated by measuring light scattering (e.g., using the DynaPro device). In the extrusion method, SUVs are prepared by filtering an MLV solution through polycarbonate membrane. The liposome size can be adjusted by using membranes with pore sizes from 1.0 to 0.03 μm (Avanti Polar Lipids). The uniformity of the liposomes can be improved by performing three freeze–thaw cycles.
3. The SUV liposome is mixed with protein in the buffer at a given mole ratio of protein and lipid by vibrating on a vortex, and the mixture is then incubated at room temperature overnight before performing CD measurements.

2.3. Determination of Protein Concentration

Accurate knowledge of the protein concentration is necessary for obtaining reliable CD spectra (Eq. 4) and secondary-structure data of proteins. Amino acid analysis is the preferred method for determining the concentration of proteins with a known sequence. In this method, the protein is completely hydrolyzed into amino acids by heating at 110°C for 24 h in a 6-N HCl solution, and the protein concentration is determined from the amount of abundant amino acids. Unfortunately, this method requires special techniques and is expensive and generally cannot be applied to membrane protein samples containing detergents and phospholipids.

The Bradford (24), Lowry (25), and bicinchoninic acid (26) methods, which are based on the chromogenic reactions of ligands or dyes bound to proteins, are more convenient for determining protein concentrations. However, these methods might not be sufficiently accurate because the response is dependent on the proteins involved and is affected by the coexistence of substances such as phospholipids.

The most convenient and practical method is to measure the protein absorbance at 280 nm (A_{280}), which is attributed to absorptions by Trp, Tyr, and Cys residues (27, 28) (see Note 1). This method calculates a protein concentration based on its extinction coefficient (absorbance at $l = 1$ cm and $c = 1$ mg/ml), which is predetermined by standardizing the absorption of a protein solution whose concentration has been determined precisely by another methods such as a dry-weight method. When this value is unknown, the molar extinction coefficient of the protein ($\varepsilon_{\text{protein}}$) can be estimated from those of Trp, Tyr, and Cys as follows:

$$\varepsilon_{protein} = (\varepsilon_{Trp} \times N_{Trp}) + (\varepsilon_{Tyr} \times N_{Tyr}) + (\varepsilon_{Cys} \times N_{Cys}), \quad (5)$$

where ε_{Trp}, ε_{Tyr}, and ε_{Cys} are the molar extinction coefficients of Trp, Tyr, and Cys at 280 nm, respectively, and N_{Trp}, N_{Tyr}, and N_{Cys} are the numbers of corresponding amino acid residues per protein molecule. Since the values of ε_{Trp}, ε_{Tyr}, and ε_{Cys} are known (27), we can calculate $\varepsilon_{\text{protein}}$ and the protein concentration (c_{protein}) using the following equations:

$$\varepsilon_{protein} = 5,690 N_{Trp} + 1,280 N_{Tyr} + 120 N_{Cys} \quad (6)$$

and

$$c_{protein}(mg/ml) = (molecular \times weight) \times (A_{280}/\varepsilon_{protein}). \quad (7)$$

For example, hen-egg lysozyme contains six Trp, three Tyr, and eight Cys residues per molecule (the molecular weight is 14,314). Then, $\varepsilon_{lysozyme} = 5,690 \times 6 + 1,280 \times 3 + 120 \times 8 = 38,940$. If A_{280} is 0.500, $c_{lysozyme} = (0.500/38{,}940) \times 14,314 = 0.183\, mg/ml$. This absorption method cannot be applied to proteins that do not contain Trp and Tyr residues because the calculated extinction coefficients contain significant errors, as expected from Eq. 6. Any disturbance to the absorbance at 280 nm such as by cofactors and light scattering also introduces large errors into the calculated protein concentration. The contribution of light scattering to the protein absorbance may be estimated by plotting the logarithm of the absorbance against the logarithm of the wavelength above 310 nm and then extrapolating the line down to 280 nm (29).

3. Methods

3.1. CD Instruments

Two types of CD instruments are in use today: (1) conventional CD instruments use a laboratory-based light source and work in the far-UV region down to about 190 nm and (2) SRCD instruments use SR as an intense light source and work in the VUV region down to about 120 nm, and they are designed individually using commercially available components and are set up in the beam lines of SR facilities. CD instruments using H_2 discharge as the conventional source—which were constructed by some CD experts (30, 31)—are no longer in use.

3.1.1. Conventional CD Instruments

Conventional CD spectrophotometers equipped with xenon lamps are commercially available and are now ubiquitous. These instruments employ similar optical devices but somewhat different data-processing systems. Most optical devices consist of a monochromator, a polarizer (POL), a photoelastic modulator (PEM) for producing right- and left-handed circularly polarized light, and a photomultiplier tube (PMT) for detecting the signal. The CD signal is proportional to the ratio of the intensities of the AC and DC signals on the PMT. Most instruments keep the DC signal constant by controlling the voltage to the PMT. The intensities and signs of the AC signal are detected by a phase-sensitive lock-in amplifier (LIA) and displayed as the CD value or θ in the unit of millidegrees. The sample chamber is purged by nitrogen gas and equipped with a Peltier temperature-control unit. The short-wavelength limit is typically around 180 nm, except in water or in a lipid environment when it is around 190 nm or longer.

Optical cells constructed from fused quartz are often used for conventional CD measurements. Cells with various shapes (square and round) and path lengths (0.1–10 mm) are commercially available. For path lengths of 1 mm or less, demountable cells are recommended because they are relatively easy to fill with the sample solution and clean on the inside.

3.1.2. SRCD Instruments

The SRCD spectrophotometer has been developed worldwide since the first one was constructed in 1980 (32, 33), and proof-of-principle and validation studies have been performed for applications in structural biology. In 2010 there were 14 operational SRCD beam lines, including at the NSLS, ISA, SRS, HiSOR, BSRF, NSRRC, Soleil, TERAS, BESSYII, ANKA, and Diamond synchrotrons (9). These beam lines employ various insertion devices and photon energies (which are three to six orders of magnitude more intense than xenon lamps in the VUV wavelength region around 180 nm). The recent development of SRCD is summarized in reviews (34, 35) and in a book by Wallace and Janes (9). Here we present the optical system of the SRCD spectrophotometer and the optical cell that are operating at Hiroshima Synchrotron Radiation Center (HiSOR). The storage-ring energy of this facility is 0.7 GeV, which is suitable for VUV and soft-X-ray experiments.

Figure 1 shows a block diagram of the SRCD spectrophotometer at HiSOR. All optical devices are set up in two separate chambers: a polarization modulation chamber and a sample chamber. Both chambers are evacuated to 10^{-4} Pa by a turbo molecular pump in order to avoid the absorption of light by air and water

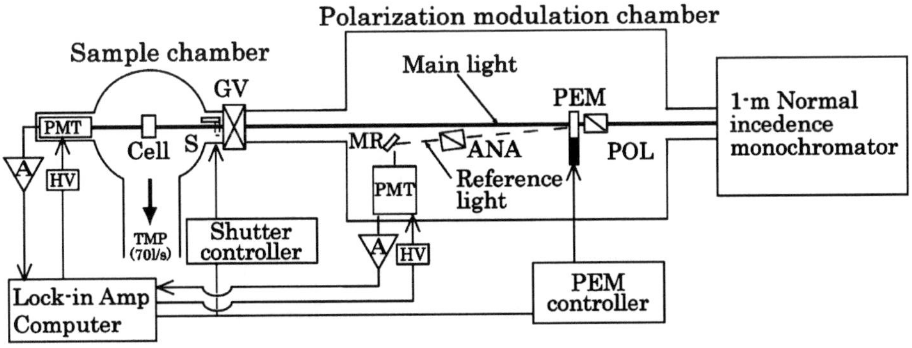

Fig. 1. Block diagram of an SRCD spectrophotometer. *A* preamplifier, *Amp* amplifier, *ANA* analyzer, *GV* gate valve, *HV* high-voltage supply, *MR* mirror, *PEM* photoelastic modulator, *PMT* photomultiplier tube, *POL* polarizer, *S* shutter, *TMP* turbo molecular pump (Adapted from Ref. 16).

vapor; this contrasts with most other SRCD instruments being purged with nitrogen gas. The sample chamber is separated from the polarization modulation chamber by a gate valve, which allows the sample to be exchanged while maintaining a high vacuum in the modulation chamber. The incident light from a monochromator is separated into two orthogonal linearly polarized light beams by an MgF_2 polarizer. Both linearly polarized light beams are modulated into circularly polarized light at 50 kHz by an LiF PEM. The main light beam is directed to the sample cell, and the CD signal is detected using the main PMT, which is covered by an MgF_2 window. The CD signal is rectified and amplified by the LIA and finally recorded on a personal computer. Another light beam, which is detected by a reference PMT, is used as a reference signal to accurately control the PEM and stabilize the LIA. Although it is generally difficult to obtain optimal driving voltages for the PEM due to refractive-index dispersion, this double-beam system—a so-called servo-control system (16, 36, 37)—achieves achromatic modulation and compensation for the thermal drift of the PEM in the VUV region. The offset signal due to the strain birefringence of the MgF_2 window on the main PMT can be minimized by rotating the PMT using a differentially pumping rotary feed-through apparatus. Damage to the sample by the SR irradiation can be avoided, when necessary, using a shutter that opens automatically only when signals are being accumulated.

Figure 2 shows a schematic of the optical cell for SRCD measurements, which consists of a stainless-steel container with a cylindrical screw and two CaF_2 (or MgF_2) windows that are 20 mm in diameter and 1 mm thick (see Note 2). The optical path length can be adjusted over the range from 1.3 μm (without a spacer) to 100 μm (with donut-shaped Teflon spacers) (see Note 3). The sample solution in the cell is sealed with three fluoride-rubber O-rings (38) (see Note 4). The optical cell is assembled by first setting up two O-rings and a CaF_2 disc on a stainless-steel bottom block.

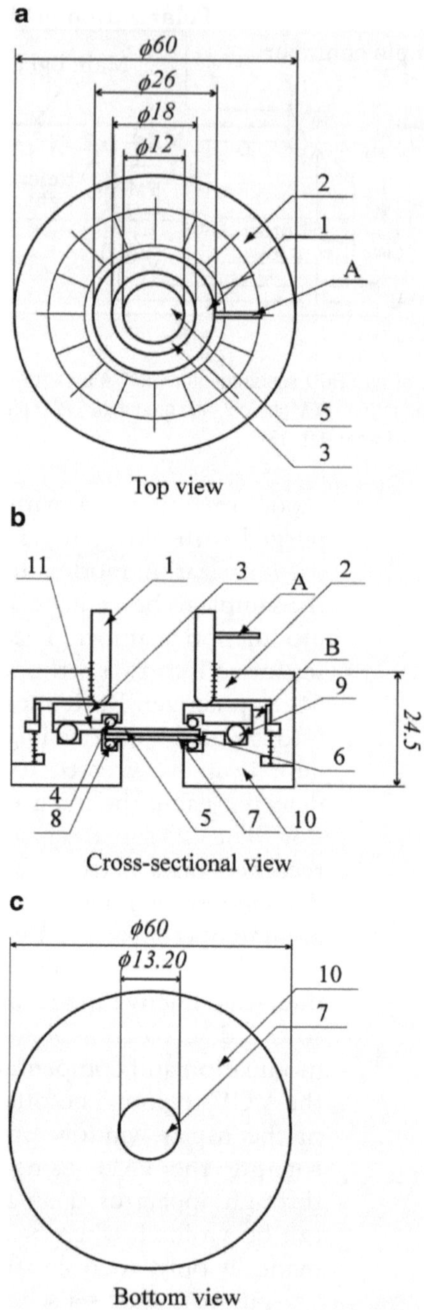

Fig. 2. Block diagram of the optical cell: top (**a**), cross-sectional (**b**), and bottom (**c**) views. *1*, cylindrical screw; *2*, stainless-steel cap; *3*, stainless-steel cover; *4*, *8*, and *9*, fluoride-rubber O-rings; *5* and *7*, CaF_2 discs; *6*, Teflon spacer; *10*, stainless-steel bottom block; *11*, space for overflow; *A*, needle; *B*, disc stopper. Linear dimensions in italics are expressed in millimeters (Adapted from Ref. 38).

A spacer to produce the required path length is then placed on the CaF_2 disc, and the sample solution is mounted at the center of this disc. The sample solution is then covered by another CaF_2 disc and sealed by a stainless-steel cover with an O-ring (see Note 5). Any of the sample that overflows through the spacer is trapped in a separate space. Finally, these assembled components are tightly fixed by a stainless-steel cap and the cell windows are pressed uniformly by turning the needle of a cylindrical screw to a position indicated on the steel cap. The CaF_2 discs and the spacer are fixed by a stopper on the bottom block. These accessories are useful for reassembling the cell with good reproducibility (see Note 6). Before assembling the cell, the CaF_2 discs and the spacer are cleaned with distilled water and ethanol (38). This sample cell is designed so that it can also be used in a conventional CD instrument to enable easy calibration of the path length and the sample concentration. The temperature of the sample is regulated between −20 and 70°C by the Peltier thermoelectric element (see Note 7).

3.1.3. Calibration of CD Instruments

The wavelength and intensity of the spectra must be calibrated in conventional CD and SRCD spectrophotometers using standard materials (see Note 8). For accuracy, calibration should be performed at more than one wavelength in order to confirm the linear wavelength dependence of the intensity over the full wavelength range of a measured protein spectrum. D-(+)-10-camphorsulfonic acid and ammonium camphorsulfonate are typically used as the standards because they have two characteristic CD peaks at 290.5 nm ($\Delta\varepsilon = +2.36$) and 192.5 nm ($\Delta\varepsilon = -4.72$) (39).

3.2. CD Measurements

CD measurements can be made using a conventional instrument following the attached manual describing the adjustable parameters, such as the scan speed, time constant, bandwidth, and accumulation time. The scan speed and time constant are important because the resolution of the CD spectrum is dominantly affected by the combination of these two parameters. A sample with strong absorption often produces a large amount of noise in the CD spectrum. The signal-to-noise (S/N) ratio can be improved by increasing the time constant and accumulation time within a limited range. When the sample solution scatters light, its contribution to the CD spectrum may be reduced by using a short-path-length cell and by positioning the sample cell near the detector of the CD instrument. The CD spectrum of the solvent needs to be measured before performing the sample measurements since even water (which has no chirality) generally exhibits small baseline signals over a wide wavelength range due to the inevitable optical constraints of CD instruments. The true CD spectrum of the sample is then obtained by subtracting the solvent spectrum from the observed sample spectrum.

SRCD measurements are basically the same as conventional CD measurements. Since a short-path-length cell (<50 μm) is used in SRCD measurements to reduce the absorption of the solvent in the VUV region (see Note 9), SRCD measurements require a higher protein concentration than conventional CD measurements. The effects of increased light scattering in the VUV region might be reduced by ensuring that the sample cell and detector (i.e., the PMT) are as close as possible. The S/N ratio can also be improved by performing repeated scans (e.g., 16 accumulations), although long-time data acquisition should be avoided in order to protect the sample from damage induced by SR irradiation when using high-flux beam lines (greater than 2 GeV) (see Note 10).

Accurate knowledge of both the path length of the cell and the protein concentration is needed to obtain reliable CD spectra. Although the path lengths of commercially available cells are indicated by their manufacturers, it is desirable to check their path lengths using the standard sample of known concentration and absorption (or ellipticity) at a given wavelength. When using short-path-length cells (<50 μm) for SRCD measurements, the interference-fringe method using a UV spectrophotometer is useful, although it is only applicable to an empty cell. An alternative method employs confocal laser microscopy to determine the gap between the windows with an accuracy of 0.1 μm for the solution-mounted cell, but this is expensive and requires a high degree of technical skill. A more convenient method is to normalize the CD spectra measured using a short-path-length cell (<50 μm) with those measured using a longer-path-length cell (e.g., 1 mm) of known accuracy in the wavelength region where their spectra overlap. This method is available for very-short-path-length cells, including those without spacers.

3.3. Secondary-Structure Analysis of Proteins

CD spectra of a protein in the far-UV and VUV regions sensitively reflect secondary structures such as α-helices, β-sheets, turns, and unordered structures. Therefore, we can estimate the contents and the numbers of segments of these secondary structures of an unknown protein from its CD spectra using some analytical programs. These secondary-structure data can be further combined with bioinformatics to improve the sequence and tertiary-structure predictions. These advancements in CD spectroscopy are due to developments in analytical programs and the extension of CD measurements to the VUV region. As demonstrated in Fig. 3, the SRCD spectra of proteins and the secondary-structure components (α-helix, β-sheet, turn, and unordered structures) exhibit characteristic peaks below 195 nm, thereby providing much more information on the secondary structures compared with conventional far-UV CD spectra. The secondary-structure analysis by CD spectroscopy mainly consists of the following steps (as depicted by the protocol in Fig. 3):

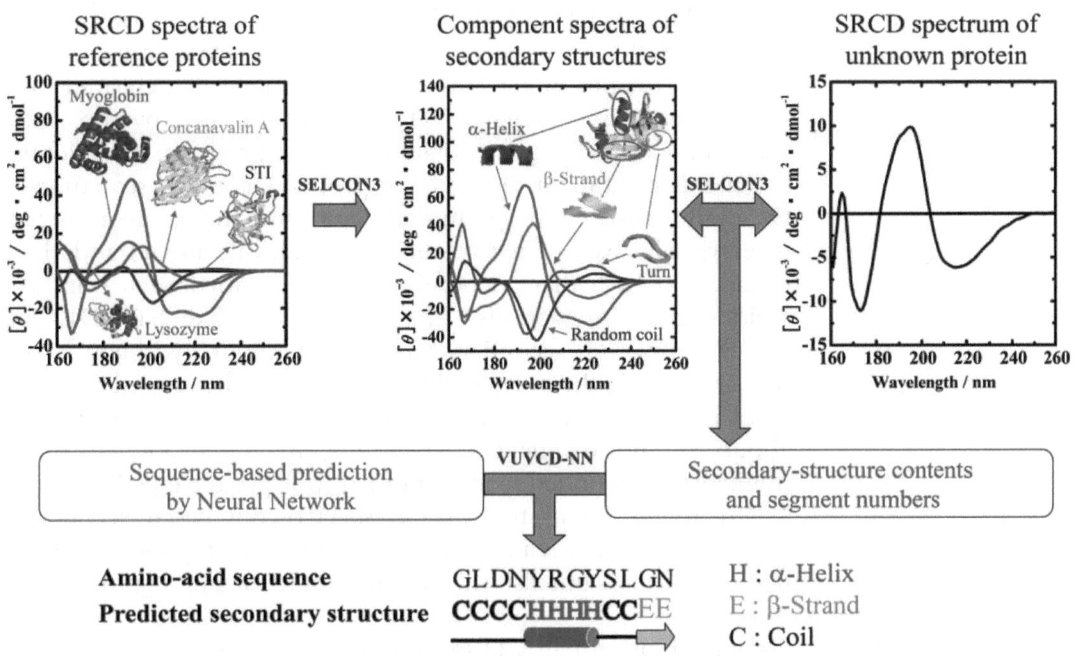

Fig. 3. Scheme for the secondary-structure analysis of proteins by CD spectroscopy.

1. Assign the secondary structures from atomic coordinates using various programs (e.g., DSSP, STRIDE, and Xtlsstr) (10, 11, 40).
2. Estimate the secondary-structure contents using analytical software (e.g., CONTIN, SELCON3, and CDSSTR) (12).
3. Estimate the number of secondary-structure segments (41).
4. Predict the secondary-structure sequences based on bioinformatics (e.g., using an NN) (19).
5. Test the tertiary-structure models predicted by homology modeling or ab initio calculations (22).

Each analytical step is described in more detail in Subheadings 3.3.1–3.3.6.

3.3.1. Secondary-Structure Assignment from Atomic Coordinates of Proteins

Estimating the secondary structures of an unknown protein requires their component spectra to be extracted from a set of CD spectra of reference proteins with known secondary structures. The secondary structures are assigned from the atomic coordinates of proteins determined by X-ray crystallography or NMR spectroscopy, which are registered in the Protein Data Bank (PDB). DSSP, which is the most widely used assignment program (10), defines the secondary structures based on the pattern of hydrogen-bonded geometrical features and the solvent exposure and thus can assign the α-, 3_{10}-, and π-helices, β-strands, turns, bends, and β-bridges at the residue level (see Note 11). Another general-purpose program is STRIDE, which assigns the protein secondary structures by combining the hydrogen-bond energy and statistically derived backbone torsion angle (40). This program can determine the same types of secondary structures as determined by the DSSP program, except for bend structures. Both programs are adopted as general assignment methods for protein secondary structures in the PDB. Xtlsstr is a program that determines the secondary structures visually based on two dihedral angles (ζ and τ) and three distances (two hydrogen bonds and one nonhydrogen bond) in amide–amide interactions. This program (unlike STRIDE and DSSP) can assign secondary structures without hydrogen bonds that are outside the normal dihedral angles of the secondary structures in a Ramachandran plot (11), thus providing information about α- and 3_{10}-helices, β-strands, hydrogen-bonded turns, nonhydrogen-bonded turns, and PPII (polyproline type II) helices. Xtlsstr might be useful for CD-spectroscopy-based secondary-structure analysis because CD recognizes not only the pattern of hydrogen bonds but also the dihedral angles of amide groups (11).

3.3.2. Database of Protein CD Spectra

A database of protein CD spectra needs to be available in order to analyze the secondary-structure contents and numbers of segments. Many databases containing a wide range of secondary structures are available on the Internet (e.g., CDPro (http://lamar.colostate.edu/~sreeram/CDPro/) and DichroWeb (http://dichroweb.cryst.bbk.ac.uk/html/home.shtml)) (12, 42), but most of them comprise the CD spectra of soluble proteins, and their utility in analyzing the CD spectra of membrane proteins has been questioned. Park et al. suggested that membrane proteins are immersed in a medium that has a much lower dielectric constant than are soluble proteins, which affects the π–π^* and n–π^* transitions (43). Furthermore, the average chain length of α-helices integrated in plasma membrane (i.e., 22–28 residues) is about twice that of α-helices in soluble proteins. Wallace et al. revealed that soluble and membrane proteins exhibit somewhat different spectral characteristics by comparing the CD spectra of two soluble proteins (myoglobin and concanavalin A) and two membrane proteins (a mechanosensitive channel and a ferric enterobactin receptor) (44). On the other hand, Sreerama and Woody (who developed the CDPro software including seven protein CD databases) found no evidence for wavelength shifts between the CD spectra of soluble and membrane proteins in a comparison of the CD spectra of 30 membrane proteins reported by Park et al. (43) with the SP43 soluble protein reference set in CDPro (45). Thus, while CD spectra might not differ significantly between membrane and soluble proteins, it is known that combining their data sets improves the predictive accuracy of secondary-structure contents and numbers of segments in both protein types (42, 45).

3.3.3. Calculation of Secondary-Structure Contents

Many programs for analyzing secondary-structure contents use a reference CD data set comprising many proteins with known atomic coordinates (46). Typical programs are CONTIN (47), SELCON3 (48), CDSSTR (49), CCA (50), and K2D (51), which employ analytical techniques such as ridge regression (47), variable selection (49), singular value decomposition (48), convex constraint analysis (50), and NN methods (51). These programs generally use the following matrix equation:

$$F = XC \tag{8}$$

where **F** and **C** are matrices consisting of the secondary-structure contents and the CD spectra of n reference proteins, respectively. Transformation matrix **X** is solved using a least-squares method, and then the secondary-structure contents of an unknown protein can be estimated from its CD spectrum. CONTIN, SELCON3, and CDSSTR are freeware programs (http://lamar.colostate.edu/~sreeram/CDPro/) that can provide the secondary-structure fractions under some general restrictions: the sum of the fractions is between 95 and 105%, each fraction is greater than –2.5%, and the root-mean-square deviation between the reconstructed and experimental CD is less than 0.25 $\Delta\varepsilon$. The accuracy of this method can be improved by (1) increasing the number of reference proteins, (2) selecting an appropriate reference set from the large available set of reference proteins, (3) stabilizing the least-squares analysis of **X** for the large set of reference proteins, and (4) extending the short-wavelength limit of the CD spectra. It is the last one of these factors that an SRCD instrument addresses.

The improvement of the secondary-structure estimation was evaluated using the root-mean-square deviation (RMSD) (δ) and Pearson correlation coefficient (r) between X-ray and CD estimates of α-helix, β-strand, turn, and unordered-structure contents. As the short-wavelength limit of CD spectra decreases from 200 to 160 nm, the δ values for α-helices, β-strands, and overall performance decrease and the r values increase up to 0.937 and 0.826 for α-helices and β-strands, respectively, indicating an improved estimation of secondary-structure contents (17, 18). The good performance of SRCD spectroscopy increases the number of secondary-structure components to eight (52), which are useful for estimating the numbers of α-helix and β-strand segments.

3.3.4. Estimation of Numbers of Secondary-Structure Segments

The numbers of α-helix and β-strand segments have been estimated from the CD spectra using two methods. Pancoska et al. used a matrix descriptor of secondary-structure segments for the NN-based analysis of proteins (53). Sreerama et al. estimated these numbers from the distorted residues in α-helices and β-strands, assuming that on average there are four and two distorted residues per α-helix and β-strand, respectively (41). In this method, the secondary structures are classified into six types: regular α-helix (α_R), distorted α-helix (α_D), regular β-strand (β_R), distorted β-strand (β_D), turn, and unordered structure; in this case matrix **F** in Eq. 8 involves six secondary-structure vectors of n reference proteins. The numbers of α-helix and β-strand segments calculated from the contents of α_D and β_D of 31 reference proteins agree well with those obtained from X-ray structures: the r values between the SRCD and X-ray estimates for the numbers of α-helix and β-strand segments are 0.954 and 0.849, respectively, which correspond to root-mean-square differences of 2.6 and 4.0. Thus, SRCD spectroscopy provides superior performance compared with conventional CD spectroscopy in estimating not only the secondary-structure contents but also the numbers of secondary-structure segments of proteins (17, 18).

3.3.5. Prediction of Secondary-Structure Sequences

CD spectroscopy in principle yields no information on the sequences of the secondary structures, but sequence-based prediction of α-helices, β-strands, and other structures is possible by exploiting the correlations between the X-ray structures and aminoacid sequences of many proteins without having to perform other experiments. Various types of algorithm (Chou–Fasman, GOR, and Lim methods) have been developed for this purpose (54–56), and their predictive accuracy has been greatly improved by employing various computational techniques, including NNs (57, 58), Profile Network from HeiDelberg (59), discrimination of secondary-structure class (60), Predator (61), and Jpred (62). However, the accuracy of these computational predictions of the secondary-structure contents and numbers of segments is lower than that of SRCD spectroscopy—the accuracy of these sequence-based algorithms should be improved by combining them with the SRCD data. Based on this idea, we developed a method for incorporating SRCD data into the NN algorithm developed by Jones (63) to improve sequence-based prediction (VUVCD–NN method) (19).

The computational protocol for the VUVCD–NN method is shown in Fig. 4. The α-helix and β-strand weights (W_α and W_β) of 20 amino acids are first calculated by the NN method (step 1 in the figure) (see Note 12), and the positions of α-helices and β-strands are then assigned based on the amino acid sequence in descending order of their weights until the determined numbers of α-helix and β-strand residues converge to those estimated in SRCD analysis. Those residues that are assigned as neither α-helices nor β-strands are considered to be other structures (step 2). Overlapping of α-helix and β-strand segments is avoided using the following selection rules: Overlapping regions (residues) are assigned to α-helices for target proteins that have more than twice as many α-helix residues as β-strand residues, while they are assigned to β-strands for target proteins that have more than twice as many β-strand residues as

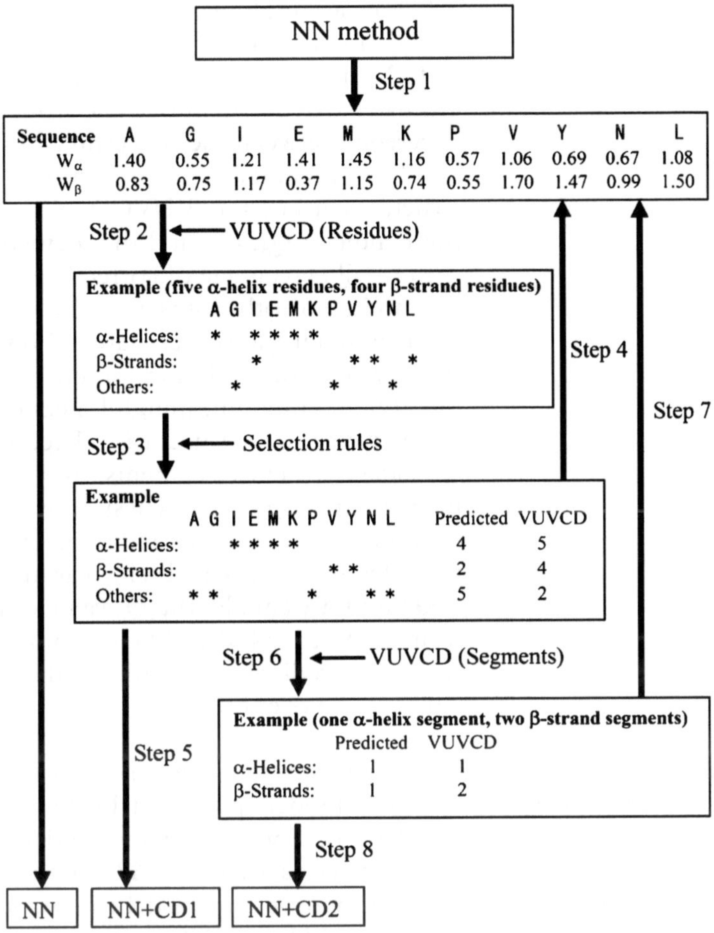

Fig. 4. Computational protocol for combining SRCD data with the NN algorithm (VUVCD–NN method). Details are provided in the text (Adapted from Ref. 19).

α-helix residues. For other target proteins, overlapping regions are assigned to either α-helices or β-strands by directly comparing the weights of the overlapping residues. Furthermore, short stretches comprising less than four α-helix residues and less than two β-strand residues are classified as other structures since these short-peptide-length limits are used as the cutoff parameters of secondary structures in the DSSP program (10) (step 3). The procedures in steps 2 and 3 are repeated until the predicted numbers of α-helix and β-strand residues (contents) agree with those estimated in SRCD analysis (step 4). The assignment is complete if the two estimates are consistent or if the sequence alignment minimizes the difference between the two estimates (step 5). Next, the numbers of α-helix and β-strand segments estimated in SRCD analysis are used in the NN calculation (step 6). If the predicted numbers of α-helix and β-strand segments do not agree with those estimated in SRCD analysis, the weights (W_α) of 20 amino acids for α-helices are multiplied by factors from 0.1 to 2.0 until the predicted numbers of segments converge to those obtained in the SRCD estimates (step 7). The assignment is complete if the two estimates are consistent or if the sequence alignment minimizes the difference between the two estimates (step 8).

The accuracy of predicting sequence-based secondary structures using the VUVCD–NN method can be tested with three indices: Q_3, C_{sr}, and the segment overlap value (SOV). Q_3 is the overall percentage of correctly predicted residues for α-helices, β-strands, and other structures (10); C_{sr} is the correlation coefficient between the success rates of different secondary structures, thereby reflecting a higher-ordered predictive accuracy (64); and the SOV is a more meaningful measure of the secondary-structure predictive accuracy since it is related to the types and positions of secondary-structure segments, in contrast to a per-residue assignment such as Q_3 (65). These three indices evidently increase by introducing the contents and numbers of segments and by selecting the reference data set obtained from the SRCD spectra down to 160 nm; for example, the Q_3 value of the NN algorithm increased from 70.9 to 74.9%. When combined with the currently best sequence-prediction algorithms, the predictive accuracy could be enhanced to over 80%.

3.3.6. Testing Tertiary-Structure Models

Secondary-structure information can be used to test the tertiary-structure models predicted by homology modeling or ab initio calculations. There are many model-prediction methods, such as the Modeler package (66–68), but it is generally difficult to select the most appropriate model to use. Secondary-structure information obtained by SRCD spectroscopy cannot prove which model is correct, but it can provide a novel criterion for eliminating certain models. Thus, SRCD spectroscopy and VUVCD–NN methods are indispensable tools for analyzing the structure of proteins that are difficult targets in X-ray crystallography and NMR spectroscopy.

3.4. Application to Nonintegral Membrane Proteins

3.4.1. α-Amyloid Peptide βAP

α-Amyloid peptide βAP (residues 1–40) is a major component of Alzheimer's amyloid deposits and presumably exerts its neurotoxic action via interactions with neuronal membranes. Terzi et al. measured the CD spectra of βAP in the absence and presence of 1-palmitoyl-2-oleoyl-*sn*-glycero-3-phosphoglycerol (POPG) at βAP/POPG molar ratios of 1:11, 1:22, 1:55, and 1:110 and observed distinct changes in the spectra, which suggest the presence of an equilibrium conformation between coil and β-structured aggregate at low βAP/POPG ratios and the transition from a β-structure to an α-helix at βAP/POPG ratios higher than 1:55 (1). Bokvist et al. measured the CD spectra of βAP interacting with vesicles for different molar ratios of L-α-dimyristoylphosphatidylcholine (DMPC) to DMPG (L-α-dimyristoylphosphatidylglycerol) (4). Based on the contents of α-helices, β-strands, and unordered structures, they suggested that there are two different membrane-binding mechanisms of βAP to hinder and accelerate aggregations. Yagi-Utsumi et al. measured the SRCD spectra of βAP down to 175 nm in the absence and presence of monosialotetrahexosylganglioside (GM1) at βAP/GM1 ratios of 1:15 and 1:30 and found that the contents of α-helix, β-strand, turn, and unordered structures depend on this ratio (23). Combining these data with NMR spectroscopy, they suggested that GM1 clusters promote specific βAP–βAP interactions that vary with the sizes and curvatures of the clusters.

3.4.2. α-Synuclein

α-Synuclein (α-syn) plays an important role in lipid transport and synaptic membrane biogenesis. Jo et al. measured the CD spectra of α-syn in various membrane environments: 1-palmitoyl 2-oleoyl phosphatidylcholine (POPC), 1-palmitoyl 2-oleoyl phosphatidylserine (POPS), phosphatidylcholine (PC), phosphatidylethanolamine (PE), phosphatidic acid (PA), phosphatidylinositol, and phosphatidylserine (PS) (2). From the dependence of α-helix content on the type of membranes, they showed that α-syn binds to acidic phospholipid vesicles (PS and PA), with this being significantly augmented by the presence of PE (neutral phospholipid). This result suggests that the interaction of α-syn with cellular membranes in vivo sensitively depends on the lipid compositions of the membranes. Perrin et al. measured the CD spectra of α-syn in various vesicles (PS/PA, PS/PC, POPC/1-palmitoyl 2-oleoyl phosphatidic acid, and POPC/POPS) and found a general correlation between the lipid-induced α-helix contents and the degree of binding to vesicles, which was applied to searching the lipid-binding capacity of the mutants associated with Parkinson's disease (3).

3.4.3. α_1-Acid Glycoprotein

Human α_1-acid glycoprotein (AGP) exhibits characteristic binding abilities to numerous drugs as well as to steroid hormones. The drug-binding capacity of AGP is known to decrease upon interaction with biomembranes, indicating a large conformational change of AGP (20). AGP is a glycoprotein containing five glycan chains comprising about 40% of its total mass (36 kDa), and this has made it a difficult target in X-ray crystallography and NMR spectroscopy. SRCD spectroscopy is highly advantageous for such glycoproteins because it can detect the CD spectra of sugar moieties in the VUV region below 200 nm.

Matsuo et al. measured the SRCD spectra of AGP down to 160 nm in the absence and presence of liposomes of L-α-phosphatidyl-DL-glycerol (PG) (22). As shown in Fig. 5, a liposome environment induces a large change in the CD spectra corresponding to the conformational transition of AGP from a β-strand-rich to α-helix-rich structure, although these spectra are affected by the presence of the glycan moiety. To eliminate this affect, Matsuo et al. measured the SRCD spectra of the constituent sugars—L-fucose, D-mannose,

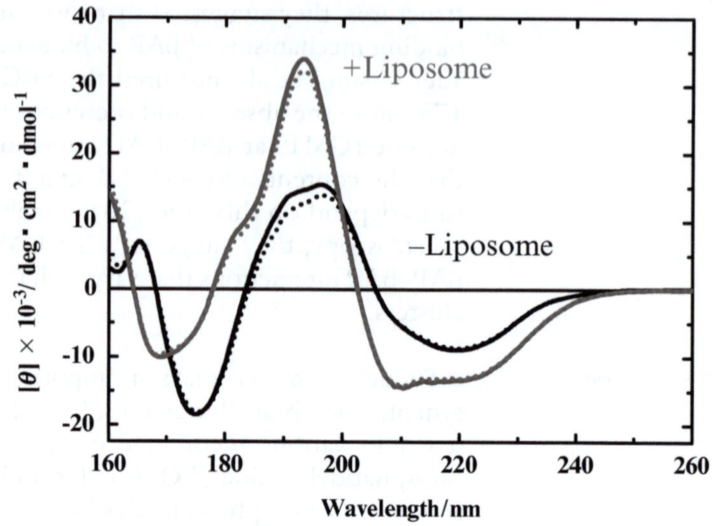

Fig. 5. SRCD spectra of AGP in the absence and presence of liposomes at 25°C. *Dotted lines* show spectra of the protein moiety obtained by subtracting the spectra of the glycan moiety (inset of Fig. 8.6) from the observed AGP spectra (Adapted from Ref. 22).

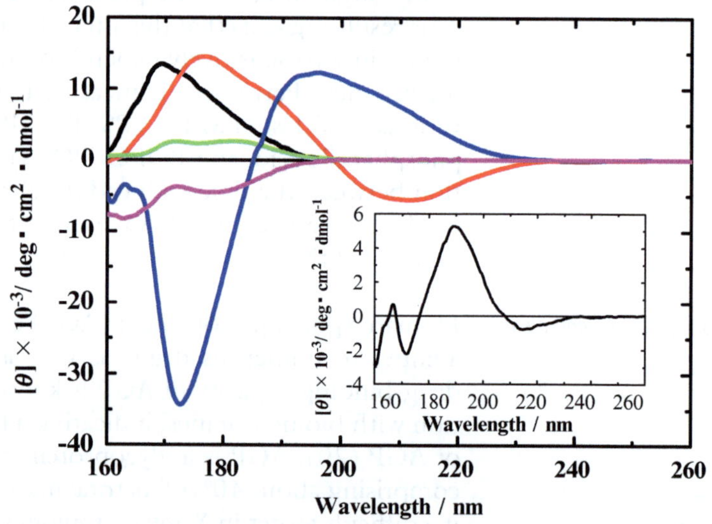

Fig. 6. SRCD spectra of the constituent sugars of AGP in water at 25°C: mannose (*black line*), fucose (*green*), galactose (*pink*), *N*-acetylglucosamine (*red*), and *N*-acetylneuraminic acid (*blue*). The inset shows the SRCD spectrum of the glycan moiety calculated from those of the constituent sugars (Adapted from Ref. 22).

D-galactose, *N*-acetylglucosamine, and *N*-acetylneuraminic acid—and constructed the SRCD spectrum of the glycan moiety by combining these constituent spectra while considering the sugar compositions in the glycan chains (Fig. 6). The SRCD spectra of the protein moiety of AGP, which were obtained by subtracting the glycan

Table 1
Secondary-structure contents and segment numbers of AGP without glycan chains in native (N) and membrane-binding (I) states (22)

State	α-Helices Content (%)	α-Helices Number	β-Strands Content (%)	β-Strands Number	Turns (%)	Unordered structures (%)
N –Glycan	14.4	3	37.7	10	19.3	28.8
Modeler	11.5	3	36.6	10	23.5	28.4
I –Glycan	49.9	8	5.6	2	19.3	25.4

spectrum from the observed AGP spectra, are shown by dotted lines in Fig. 5, which indicates that the contributions of the glycan chains are small but not negligible in secondary-structure analysis.

The secondary-structure contents and numbers of segments of AGP were estimated using the SELCON3 program from the SRCD spectra of the protein moiety down to 160 nm. As listed in Table 1, AGP consists of 37.7% β-strands (ten segments) and 14.4% α-helices (three segments) in the absence of liposomes (pH 7.4) and of 49.9% α-helices (eight segments) and 5.6% β-strands (two segments) in the presence of liposomes (pH 4.5), confirming the conformational change from a β-strand-rich to an α-helix-rich structure without significant changes in the contents of turns and unordered structures. These secondary-structure data are in good agreement with those calculated from the tertiary-structure model of native AGP predicted by the Modeler package (Fig. 7), confirming the validity of the predicted model.

The positions of secondary structures on the amino acid sequence were predicted using the VUVCD–NN method (see Subheading 3.5). As shown in Fig. 8a, native AGP includes 11 β-strands and 3 α-helices, while liposome-bound AGP consists of 3 α-helices and 2 β-strands. The numbers of segments thus predicted at the sequence level do not necessarily agree with those predicted from the SRCD data, which is due to sequence-alignment minimization being performed in the VUVCD–NN method. Interestingly, most of progesterone binding sites changed to helices or unordered structures in the presence of liposomes, suggesting complete collapses of the stereo-structures of the binding sites. Considering the net charge and hydrophobicity of each helix and the evidence that W25 and W160 interact with liposomes (20), Matsuo et al. proposed the following model for the membrane-binding mechanism of AGP (Fig. 8.8b): (1) positively charged AGP with a ligand approaches the anionic membrane surface via an electrostatic

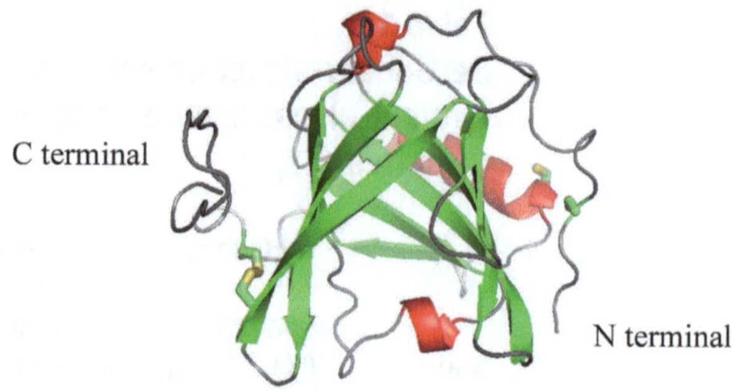

Fig. 7. Ribbon model of the native structure of AGP constructed by the Modeler package: α-helices (*red*), β-strands (*green*), and other structures (*gray*). AGP has a β-barrel-like conformation comprising ten antiparallel β-strands and three α-helices (Adapted from Ref. 22).

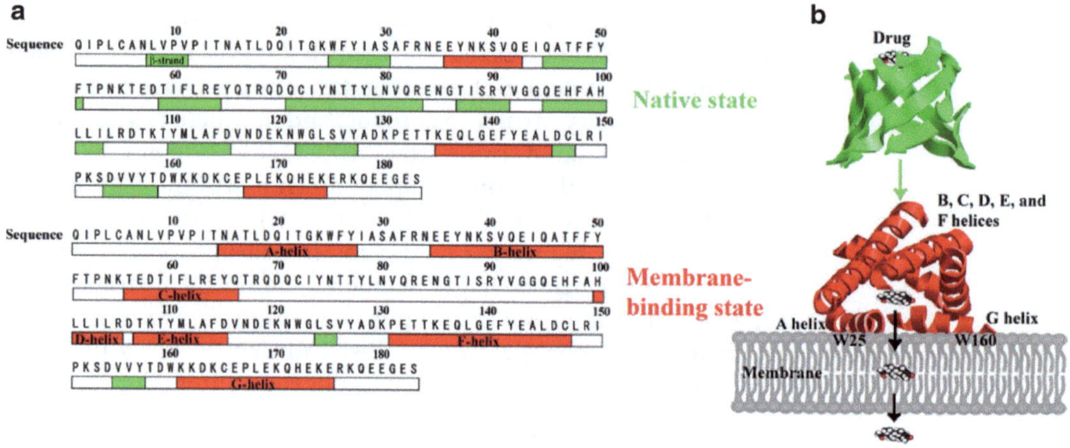

Fig. 8. Conformational change of AGP by interaction with liposomes. (**a**) Sequence-based secondary structures of AGP in the native and membrane-binding states predicted by the VUVCD–NN method. The α-helices, β-strands, and other structures are shown in *red*, *green*, and *white*, respectively. (**b**) Speculative model of the membrane-binding mechanism and structure–function relationship of AGP (Adapted from Ref. 22).

interaction and (2) W25 and W160 are inserted into liposomes via hydrophobic interactions, forming A- and G-helices in contact with the membrane surface and subsequently other helices, accompanied by the release of ligand into the membrane. This model provides the sequence-level information on the membrane-binding mechanism and the structure–function relationship of AGP, which are more generally related to the protein-mediated uptake and membrane-transport mechanisms of hormones and drugs.

3.4.4. Others

Vernier et al. measured the CD spectra of apo-myoglobin in the presence of vesicles (PA and PC) at various pH values to investigate whether the interaction of this protein with the vesicles facilitates the uptake of heme from the outer mitochondrial membrane (69).

On the other hand, Zhang and Keiderling (70) and Zhang et al. (71) measured the CD spectra of β-lactoglobulin at various PG concentrations to reveal a general mechanism of protein–membrane interactions. Arnulphi et al. examined the α-helix contents of human apolipoprotein A-I bound to POPC in the presence of sphingomyelin (SM) and cholesterol (Chol) (72) and found that raft-like lipid domains composed of POPC, SM, and Chol reduce folding of this protein compared with Chol-free model membranes.

3.5. Application to Integral Membrane Proteins

3.5.1. Photoreceptor Cyclic-Nucleotide-Gated Channels

Photoreceptor cyclic-nucleotide-gated (CNG) channels are nonselective cation channels located in the plasma membrane of the outer segments, where certain transmembrane proteins function as important components of the visual transduction system. Matveev et al. applied CD spectroscopy to examine the influence of two mutations (R377W and F488L) of the CNGA3 subunit on the structure and function of the channels (5). These mutations decreased the α-helicity of the protein, while the allosteric transitions by 8-pCPT-cGMP, cGMP, and cAMP were diminished in the two mutants. From these results they suggested that the observed alterations in the local secondary structure exert adverse effects on the channel's activity and cellular processing.

3.5.2. Protein-Dependent Manganese Transporter

MntH (*Escherichia coli* protein-dependent manganese transporter) is an integral transmembrane protein comprising 11 transmembrane segments. Nunuková et al. synthesized three peptides corresponding to the sixth segment among them and measured the CD spectra of these peptides in the presence of $MnCl_2$ and lipid membrane composed of 1,2-diphytanoyl-*sn*-glycero-3- phosphatidylcholine and 1,2-diphytanoyl-*sn*-glycero-3-[phospho-rac-(1-glycerol)] at a molar ratio of 3:1 (6). They found that these peptides interact with the lipid membrane to induce α-helices and that their helicities are further increased by adding $MnCl_2$. They also measured the CD spectra of the peptides in trifluoroethanol solution and assessed their abilities to form manganese-conducting channels.

3.5.3. Equinatoxin II

Equinatoxin II (EqtII) is a member of the actinoporin family of sea-anemone toxins that function by forming pores in cell membranes via a multistep mechanism. Miles et al. measured the SRCD spectra of EqtII down to 180 nm in the absence and presence of SUVs with different lipid compositions (dioleoylphosphatidylcholine (DOPC), SM, and Chol) (21). All of the SRCD spectra were similar in the far-UV region, but they exhibited large differences in the peak positions and intensities in the VUV region, depending on the lipid composition. The secondary-structure contents of EqtII estimated by SRCD revealed that EqtII is in the "bound" state of high lytic activity in the presence of DOPC/SM vesicles and in the "free" state in the presence of DOPC or DOPC/Chol vesicles, indicating the lipid specificity for structural rearrangements of EqtII associated with toxicity and lysis functions.

3.5.4. Others

SRCD spectroscopy has often been used to monitor structural changes of other integral membrane proteins (73–75). Miles et al. measured the SRCD spectra of pig Na, K-ATPase down to 175 nm in the membrane-bound state at various temperatures and showed that the loss of enzymatic activity is correlated with changes in the protein secondary structures (73). Cronin et al. investigated the effects of deglycosylation of sodium channels on the secondary structures using an SRCD spectrophotometer (74). Furthermore, McKibbin et al. used SRCD spectroscopy to monitor structural changes of rhodopsin and opsin aged for 1–6 h in DMPC/DHPC (L-α-1,2-dihexanoyl-*sn*-glycero-3-phosphocholine) vesicles at room temperature (75).

3.6. Future Prospects

We have demonstrated that CD spectroscopy is a powerful technique for monitoring the conformational changes of nonintegral and integral membrane proteins. Developments in SRCD instruments and in programs for analyzing CD spectra have yielded detailed secondary-structure information—in terms of the contents and numbers of segments, and the amino acid sequences—that cannot be obtained by conventional CD instruments, thereby providing new insight into the structure–function relationships of proteins in a membrane environment. The structures of a huge number of membrane proteins have barely been analyzed by X-ray crystallography or NMR spectroscopy, and hence CD and SRCD spectroscopy will play important roles in the structural biology of membrane proteins.

4. Notes

1. Since a high protein concentration is used in CD measurements, the protein solution must be diluted with solvent to the concentration at which $A_{280} = 0.5 \sim 1.0$. The protein concentration is calculated from the A_{280} value multiplied by the dilution factor determined using the volume or weight of the protein solution and the solvent.
2. Optical crystal discs of CaF_2 and MgF_2 are commercially available. However, the *c*-axis-cut disc should be customized for MgF_2 to eliminate the birefringence effects of the cell, which affect the baseline and CD signal in the VUV region.
3. A homemade die can be used to cut the donut-shaped spacer from a Teflon sheet of the desired thickness (5–100 μm).
4. The O-rings, 4, 8, and 9 in Fig. 2 might be coated with a high-vacuum grease to completely seal the sample solution.

5. The formation of air bubbles in the sample cell can be avoided by using tweezers to carefully and slowly place the cover disc on the surface of the sample solution.
6. Reproducible CD data can be obtained in separate measurements by always fixing the discs at the same position marked on their edge when assembling the optical cell.
7. The temperature of sample solution can be determined with an accuracy of ±1°C from the calibrated temperature difference between the heat conductor and the cell windows. Water within a sufficiently narrow space does not freeze at subzero temperatures, and hence the sample solution can be cooled down to −17 and −10°C without freezing by using cells with path lengths of 25 and 50 μm, respectively. No frost is formed on the cell windows because the cell is kept under vacuum.
8. It is desirable to routinely calibrate the instrument.
9. A 50-μm path length is available for measurements down to 175 nm, but measurements below 175 nm must be carried out without a spacer in order to reduce the effect of light absorption by water.
10. Damage to the sample can be detected by monitoring spectral changes during the measurements (1–3 h).
11. The structures denoted as "H" and "E" in DSSP are designated as α-helices and β-strands, respectively, and 3_{10}-helices ("G"), π-helices ("I"), turns ("T"), bends ("S"), β-bridges ("B"), and blanks ("") are grouped into "others" in the VUVCD–NN method.
12. The training data set is prepared from the protein structures in the PDB using a resolution cutoff of 2.0 Å for X-ray structures and by selecting the subsets of residues with an RMSD for the C_α–C_α distance of less than 1.0 Å for NMR structures. Small proteins comprising less than 30 amino acid residues and proteins containing more than 5% unidentified sequences are not used. Proteins with sequence similarities greater than 25% are eliminated using Hobohm algorithm #1 (76). Finally, a data set of 607 chains is obtained as a training set for the NN algorithm by manually eliminating any chain breaks. The secondary structures of these training proteins and their amino acid sequence information are determined by the PSSM (position-specific scoring matrices) program (63) and used to calculate the weights and biases of 20 amino acids for α-helices and β-strands with 15 window positions and no hidden layer using the public free simulation package Stuttgart Neural Network Simulator (version 4.2) (77) with a momentum term of 0.5 and a learning rate of 0.0001. The threshold criterion for assigning α-helices and β-strands is 0.37.

References

1. Terzi E, Hölzemann G, Seelig J (1997) Interaction of Alzheimer β-amyloid peptide (1–40) with lipid membranes. Biochemistry 36:14845–14852
2. Jo E, McLaurin J, Yip CM, St G-HP, Fraser PE (2000) alpha-Synuclein membrane interactions and lipid specificity. J Biol Chem 275:34328–34334
3. Perrin RJ, Woods WS, Clayton DF, George JM (2000) Interaction of human alpha-synuclein and Parkinson's disease variants with phospholipids. Structural analysis using site-directed mutagenesis. J Biol Chem 275:34393–34398
4. Bokvist M, Lindström F, Watts A, Gröbner G (2004) Two types of Alzheimer's beta-amyloid (1–40) peptide membrane interactions: aggregation preventing transmembrane anchoring versus accelerated surface fibril formation. J Mol Biol 335:1039–1049
5. Matveev AV, Fitzgerald JB, Xu J, Malykhina AP, Rodgers KK, Ding XQ (2010) The disease-causing mutations in the carboxyl terminus of cone cyclic nucleotide-gated channel CNGA3 subunit alter the local secondary structure and interfere with the channel active conformational change. Biochemistry 49:1628–1639
6. Nunuková V, Urbánková E, Jelokhani-Niaraki M, Chaloupka R (2010) Ion channel activity of transmembrane segment 6 of Escherichia coli proton-dependent manganese transporter. Biopolymers 93:718–726
7. Fasman GR (1996) Circular dichroism and the conformational analysis of biomolecules. Plenum, New York
8. Berova N, Nakanishi K, Woody RW (2000) Circular dichroism: principles and applications, 2nd edn. Wiley-VCH, New York
9. Wallace BA, Janes RW (2009) Modern techniques for circular dichroism and synchrotron radiation circular dichroism spectroscopy. IOS, Amsterdam
10. Kabsch W, Sander C (1983) Dictionary of protein secondary structure: pattern recognition of hydrogen-bonded and geometric features. Biopolymers 22:2577–2637
11. King SM, Johnson WC (1999) Assigning secondary structure from protein coordinate data. Proteins 35:313–320
12. Sreerama N, Woody RW (2000) Estimation of protein secondary structure from circular dichroism spectra: comparison of CONTIN, SELCON, and CDSSTR methods with an expanded reference set. Anal Biochem 287:252–260
13. Toumadje A, Alcorn SW, Johnson WC Jr (1992) Extending CD spectra of proteins to 168 nm improves the analysis for secondary structures. Anal Biochem 200:321–331
14. Sutherland JC, Emrick A, France LL, Monteleone DC, Trunk J (1992) Circular dichroism user facility at the National Synchrotron Light Source: estimation of protein secondary structure. Biotechniques 13:588–590
15. Wallace BA (2000) Conformational changes by synchrotron radiation circular dichroism spectroscopy. Nat Struct Biol 7:708–709
16. Ojima N, Sakai K, Fukazawa T, Gekko K (2000) Vacuum-ultraviolet circular dichroism spectrophotometer using synchrotron radiation: optical system and off-line performance. Chem Lett 29:832–833
17. Matsuo K, Yonehara R, Gekko K (2004) Secondary-structure analysis of proteins by vacuum-ultraviolet circular dichroism spectroscopy. J Biochem 135:405–411
18. Matsuo K, Yonehara R, Gekko K (2005) Improved estimation of the secondary structures of proteins by vacuum-ultraviolet circular dichroism spectroscopy. J Biochem 138:79–88
19. Matsuo K, Watanabe H, Gekko K (2008) Improved sequence-based prediction of protein secondary structures by combining vacuum-ultraviolet circular dichroism spectroscopy with neural network. Proteins 73:104–112
20. Nishi K, Maruyama T, Halsall HB, Handa T, Otagiri M (2004) Binding of alpha1-acid glycoprotein to membrane results in a unique structural change and ligand release. Biochemistry 43:10513–10519
21. Miles AJ, Drechsler A, Kristan K, Anderluh G, Norton RS, Wallace BA, Separovic F (2008) The effects of lipids on the structure of the eukaryotic cytolysin equinatoxin II: a synchrotron radiation circular dichroism spectroscopic study. Biochim Biophys Acta 1778:2091–2096
22. Matsuo K, Namatame H, Taniguchi M, Gekko K (2009) Membrane-induced conformational change of alpha1-acid glycoprotein characterized by vacuum-ultraviolet circular dichroism spectroscopy. Biochemistry 48:9103–9111
23. Yahi-Utsumi M, Matsuo K, Yanagisawa K, Gekko K, Kato K (2011) Spectroscopic characterization of intermolecular interaction of amyloid beta promoted on GM1 micelles. Int J Alzheimers Dis. doi:10.4061/2011/925073
24. Bradford MM (1976) A rapid and sensitive method for the quantitation of microgram quantities of protein utilizing the principle of

protein-dye binding. Anal Biochem 72:248–254
25. Lowry OH, Rosebrough NJ, Farr AL, Randall RJ (1951) Protein measurement with the folin phenol reagent. J Biol Chem 193:265–275
26. Smith PK, Krohn RI, Hermanson GT, Mallia AK, Gartner FH, Provenzano MD, Fujimoto EK, Goeke NM, Olson BJ, Klenk DC (1985) Measurement of protein using bicinchoninic acid. Anal Biochem 150:76–85
27. Edelhoch H (1967) Spectroscopic determination of tryptophan and tyrosine in proteins. Biochemistry 6:1948–1954
28. Gill SC, von Hippel PH (1989) Calculation of protein extinction coefficients from amino acid sequence data. Anal Biochem 182:319–326
29. Wallace BA, Mao D (1984) Circular dichroism analyses of membrane proteins: an examination of differential light scattering and absorption flattening effects in large membrane vesicles and membrane sheets. Anal Biochem 142:317–328
30. Johnson CW Jr (1978) Circular dichroism spectroscopy and the vacuum ultraviolet region. Annu Rev Phys Chem 29:93–114
31. Pysh ES (1976) Optical activity in the vacuum ultraviolet. Annu Rev Biophys Bioeng 5:63–75
32. Snyder PA, Rowe EM (1980) The first use of synchrotron radiation for vacuum ultraviolet circular dichroism measurements. Nucl Instrum Methods 172:345–349
33. Sutherland JC, Keck PC, Griffin KP, Takacs PZ (1982) Simultaneous measurement of absorption and circular dichroism in a synchrotron spectrometer. Nucl Instrum Methods 195:375–379
34. Wallace BA (2009) Protein characterisation by synchrotron radiation circular dichroism spectroscopy. Q Rev Biophys 42:317–370
35. Wallace BA, Gekko K, Hoffmann SV, Lin YH, Sutherland JC, Tao J, Wien F, Janes RW (2010) Synchrotron radiation circular dichroism (SRCD) spectroscopy: an emerging method in structural biology for examining protein conformations and protein interactions. Nucl Instr Methods Phys Res A 649:177–178
36. Matsuo K, Matsushima Y, Fukuyama T, Senba S, Gekko K (2002) Vacuum-ultraviolet circular dichroism of amino acids as revealed by synchrotron radiation spectrophotometer. Chem Lett 31:826–827
37. Matsuo K, Fukuyama T, Yonehara R, Namatame H, Taniguchi M, Gekko K (2005) Vacuum-ultraviolet circular dichroism spectrophotometer using synchrotron radiation. J Electron Spectrosc Relat Phenom 144–147:1023–1025
38. Matsuo K, Sakai K, Matsushima Y, Fukuyama T, Gekko K (2003) Optical cell with a temperature-control unit for a vacuum-ultraviolet circular dichroism spectrophotometer. Anal Sci 19:129–132
39. Takakuwa T, Konno T, Meguro H (1985) A new standard substance for calibration of circular dichroism: ammonium *d*-10-camphorsulfonate. Anal Sci 1:215–218
40. Frishman D, Argos P (1995) Knowledge-based protein secondary structure assignment. Proteins 23:566–579
41. Sreerama N, Venyaminov SY, Woody RW (1999) Estimation of the number of alpha-helical and beta-strand segments in proteins using circular dichroism spectroscopy. Protein Sci 8:370–380
42. Abdul-Gader A, Miles AJ, Wallace BA (2011) A reference dataset for the analyses of membrane protein secondary structures and transmembrane residues using circular dichroism spectroscopy. Bioinformatics 27:1630–1636
43. Park K, Perczel A, Fasman GD (1992) Differentiation between transmembrane helices and peripheral helices by the deconvolution of circular dichroism spectra of membrane proteins. Protein Sci 1:1032–1049
44. Wallace BA, Lees JG, Orry AJ, Lobley A, Janes RW (2003) Analyses of circular dichroism spectra of membrane proteins. Protein Sci 12:875–884
45. Sreerama N, Woody RW (2004) On the analysis of membrane protein circular dichroism spectra. Protein Sci 13:100–112
46. Woody RW (1995) Circular dichroism. Methods Enzymol 246:34–71
47. Provencher SW, Glöckner J (1981) Estimation of globular protein secondary structure from circular dichroism. Biochemistry 20:33–37
48. Sreerama N, Woody RW (1993) A self-consistent method for the analysis of protein secondary structure from circular dichroism. Anal Biochem 209:32–44
49. Johnson WC (1999) Analyzing protein circular dichroism spectra for accurate secondary structures. Proteins 35:307–312
50. Perczel A, Hollósi M, Tusnády G, Fasman GD (1991) Convex constraint analysis: a natural deconvolution of circular dichroism curves of proteins. Protein Eng 4:669–679
51. Böhm G, Muhr R, Jaenicke R (1992) Quantitative analysis of protein far UV circular dichroism spectra by neural networks. Protein Eng 5:191–195

52. Klose DP, Wallace BA, Janes RW (2010) 2Struc: the secondary structure server. Bioinformatics 26:2624–2625
53. Pancoska P, Janota V, Keiderling TA (1999) Novel matrix descriptor for secondary structure segments in proteins: demonstration of predictability from circular dichroism spectra. Anal Biochem 267:72–83
54. Chou PY, Fasman GD (1974) Conformational parameters for amino acids in helical, beta-sheet, and random coil regions calculated from proteins. Biochemistry 13:211–222
55. Garnier J, Osguthorpe DJ, Robson B (1978) Analysis of the accuracy and implications of simple methods for predicting the secondary structure of globular proteins. J Mol Biol 120:97–120
56. Lim VI (1974) Structural principles of the globular organization of protein chains. A stereochemical theory of globular protein secondary structure. J Mol Biol 88:857–872
57. Qian N, Sejnowski TJ (1988) Predicting the secondary structure of globular proteins using neural network models. J Mol Biol 202:865–884
58. Holley LH, Karplus M (1989) Protein secondary structure prediction with a neural network. Proc Natl Acad Sci U S A 86:152–156
59. Rost B, Sander C (1993) Prediction of protein secondary structure at better than 70% accuracy. J Mol Biol 232:584–599
60. King RD, Sternberg MJ (1996) Identification and application of the concepts important for accurate and reliable protein secondary structure prediction. Protein Sci 5:2298–2310
61. Frishman D, Argos P (1996) Incorporation of non-local interactions in protein secondary structure prediction from the amino acid sequence. Protein Eng 9:133–142
62. Cuff JA, Barton GJ (1999) Evaluation and improvement of multiple sequence methods for protein secondary structure prediction. Proteins 34:508–519
63. Jones DT (1999) Protein secondary structure prediction based on position-specific scoring matrices. J Mol Biol 292:195–202
64. Matthews BW (1975) Comparison of the predicted and observed secondary structure of T4 phage lysozyme. Biochim Biophys Acta 405:442–451
65. Zemla A, Venclovas C, Fidelis K, Rost B (1999) A modified definition of Sov, a segment-based measure for protein secondary structure prediction assessment. Proteins 34:220–223
66. Martí-Renom MA, Stuart AC, Fiser A, Sánchez R, Melo F, Sali A (2000) Comparative protein structure modeling of genes and genomes. Annu Rev Biophys Biomol Struct 29:291–325
67. Simons KT, Bonneau R, Ruczinski I, Baker D (1999) Ab initio protein structure prediction of CASP III targets using ROSETTA. Proteins 37:171–176
68. Takada S (2001) Protein folding simulation with solvent-induced force field: folding pathway ensemble of three-helix-bundle proteins. Proteins 42:85–98
69. Vernier G, Chenal A, Vitrac H, Barumandzadhe R, Montagner C, Forge V (2007) Interactions of apomyoglobin with membranes: Mechanisms and effects on heme uptake. Protein Sci 16:391–400
70. Zhang X, Keiderling TA (2006) Lipid-induced conformational transitions of beta-lactoglobulin. Biochemistry 45:8444–8452
71. Zhang X, Ge N, Keiderling TA (2007) Electrostatic and hydrophobic interactions governing the interaction and binding of beta-lactoglobulin to membranes. Biochemistry 46:5252–5260
72. Arnulphi C, Sánchez SA, Tricerri MA, Gratton E, Jonas A (2005) Interaction of human apolipoprotein A-I with model membranes exhibiting lipid domains. Biophys J 89:285–295
73. Miles AJ, Wallace BA, Esmann M (2011) Correlation of structural and functional thermal stability of the integral membrane protein Na, K-ATPase. Biochim Biophys Acta 1808:2573–2580
74. Cronin NB, O'Reilly A, Duclohier H, Wallace BA (2005) Effects of deglycosylation of sodium channels on their structure and function. Biochemistry 44:441–449
75. McKibbin C, Farmer NA, Jeans C, Reeves PJ, Khorana GH, Wallace BA, Edwards PC, Villa C, Booth PJ (2007) Opsin stability and folding: modulation by phospholipid bicelles. J Mol Biol 374:1319–1332
76. Hobohm U, Scharf M, Schneider R, Sander C (1992) Selection of representative protein data sets. Protein Sci 1:409–417
77. Zell A, Mamier G, Vogt M, Mache N, Hubner R, Doring S, Herrmann KW, Soyez T, Schmalzl M, Sommer T, Hatzigeorgiou A, Posselt D, Schreiner T, Kett B, Clemente G, Wieland J (1995) Stuttgart Neural Network Simulator, Version 4.2. University of Stuttgart, Stuttgart, Germany. http://www.ra.cs.uni-tuebingen.de/downloads/SNNS/

Chapter 9

Structural Characterization of Membrane Proteins and Peptides by FTIR and ATR-FTIR Spectroscopy

Suren A. Tatulian

Abstract

Fourier transform infrared (FTIR) spectroscopy is widely used in structural characterization of proteins or peptides. While the method does not have the capability of providing the precise, atomic-resolution molecular structure, it is exquisitely sensitive to conformational changes occurring in proteins upon functional transitions or upon intermolecular interactions. Sensitivity of vibrational frequencies to atomic masses has led to development of "isotope-edited" FTIR spectroscopy, where structural effects in two proteins, one unlabeled and the other labeled with a heavier stable isotope, such as ^{13}C, are resolved simultaneously based on spectral downshift (separation) of the amide I band of the labeled protein. The same isotope effect is used to identify site-specific conformational changes in proteins by site-directed or segmental isotope labeling. Negligible light scattering in the infrared region provides an opportunity to study intermolecular interactions between large protein complexes, interactions of proteins and peptides with lipid vesicles, or protein–nucleic acid interactions without light scattering problems often encountered in ultraviolet spectroscopy. Attenuated total reflection FTIR (ATR-FTIR) is a surface-sensitive version of infrared spectroscopy that has proved useful in studying membrane proteins and lipids, protein-membrane interactions, mechanisms of interfacial enzymes, and molecular architecture of membrane pore or channel forming proteins and peptides. The purpose of this article was to provide a practical guide to analyze protein structure and protein-membrane interactions by FTIR and ATR-FTIR techniques, including procedures of sample preparation, measurements, and data analysis. Basic background information on FTIR spectroscopy, as well as some relatively new developments in structural and functional characterization of proteins and peptides in lipid membranes, are also presented.

Key words: Infrared spectroscopy, Protein structure, Lipid structure, Membrane, Protein-membrane interactions, Transmembrane orientation, Protein isotope labeling

1. Introduction

FTIR spectroscopy has become one of the most popular techniques employed in structural characterization of proteins and peptides. The utility of this method is based, first of all, on the wide spectral range of a single spectrum, which covers the vibrational frequencies of various chemical groups occurring not only in a polypeptide

Jörg H. Kleinschmidt (ed.), *Lipid-Protein Interactions: Methods and Protocols*, Methods in Molecular Biology, vol. 974, DOI 10.1007/978-1-62703-275-9_9, © Springer Science+Business Media New York 2013

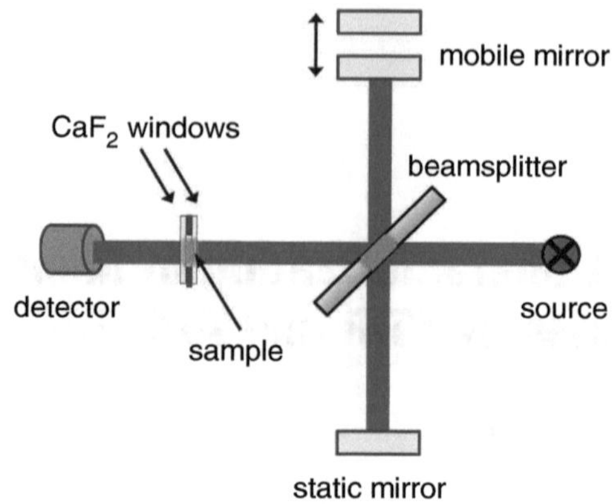

Fig. 1. Schematic presentation of the infrared experimental setup. Description is provided in the main text.

chain but practically in all organic and inorganic compounds. This allows for analysis of interactions of proteins with lipids, nucleic acids, drugs, and other molecules; the spectra of complex systems provide structural information on all components that generate absorbance bands at specific spectral regions. Moreover, since various structural or functional groups of biomolecules have distinct chemical compositions and hence absorb at specific frequencies, information on the individual functional groups of molecules is obtained from FTIR spectra without the need of labeling with molecular probes. Other advantages of FTIR spectroscopy include the absence of light scattering problems, often encountered in UV spectroscopy such as circular dichroism or fluorescence. This is especially important when working with membrane proteins or peptides in the presence of lipid vesicles, which strongly scatter at lower wavelengths in the UV region. The kinetics of amide hydrogen/deuterium exchange (HX) of a protein upon exposure to D_2O is readily measured by FTIR and provides information on the structural flexibility and the degree of solvent accessibility, that is, the structural dynamics of proteins. Polarized ATR-FTIR offers additional advantages such as increased sensitivity and an opportunity to determine the orientation of membrane-bound proteins in addition to their secondary structure and dynamic properties. Sensitivity of infrared vibrations to atomic masses allows one to gain site-specific structural information on proteins by stable isotope labeling, without gross perturbations like those often induced by labeling with relatively bulky chemical groups. Lastly, the ease of FTIR experiments and the affordable cost of instrumentation make it a facile yet powerful technique for protein structural studies.

FTIR instrumentation is relatively simple (Fig. 1). The infrared light source, such as a heated tungsten filament, radiates polychromatic light in the infrared region. The beam is directed to a Michelson interferometer that contains a beam splitter, for example, germanium, which passes 50 % of the beam and reflects the other 50 %. The split beams hit two mirrors, one static and the other oscillating back and forth with certain frequency and amplitude. The two beams, reflected from the static and mobile mirrors, recombine at the beam splitter, undergo interference because of the optical path-length difference (Δl), and travel to the sample. In direct transmission FTIR, the sample is normally placed between two infrared-transparent windows, such as CaF_2, separated by a thin (typically 5–50 μm) spacer. The beam passes through the sample and travels to the detector, such as a high-sensitivity, liquid nitrogen-cooled Hg/Cd/Te (MCT) detector, which simply records the light intensity. For each individual wavelength within the beam, the beams returning from the static and mobile mirrors undergo constructive interference at zero Δl as well as constructive or destructive interference at $\Delta l = m\lambda$ or $\Delta l = (m + 1/2)\lambda$, respectively, where m is any integral number. The result is that the light that reaches the detector is a sum of cosine functions, known as the interferogram. Fourier transformation of the interferogram results in a transmission spectrum, that is, the intensity of transmitted light as a function of frequency (or the wavenumber). (The wavenumber is $W = 1/\lambda = v/c$, where λ is the wavelength, v is the temporal frequency of the vibration, and c is the speed of light.) The resolution of the spectrum is directly proportional to the total displacement of the mobile mirror, so at a constant speed of the mobile mirror, collection of higher resolution spectra (such as ~1 cm^{-1}) will take longer time. Also, the quality (i.e., the signal-to-noise ratio) is proportional to $\sqrt{N}$, where N is the number of scans per spectrum, so again recording of a "good" spectrum will take longer time. (Normally, 1,000 scans can be collected in ~15 min) When two transmission spectra are recorded, one for the (protein) sample and the other for the blank buffer or any other "reference," the absorbance spectrum is calculated either automatically or manually as $A = -\lg(T_s/T_r)$, where T_s and T_r are the transmissions of the sample and the reference, respectively.

A polypeptide chain generates several infrared-active amide vibrational modes (Table 1), among which the amide I and amide II modes have been analyzed most extensively because of their exceptional sensitivity to the secondary structure and the kinetics of HX, respectively. Both are delocalized vibrational modes and include contributions from various chemical bonds within the amide group. The amide I mode occurs between 1,700 and 1,600 cm^{-1} and is generated primarily by the C=O stretching vibration, plus CN and CCN out-of-plane bending vibrations and minimally by the NH in-plane bending mode (1). The resulting amide I oscillators are coupled to each other via covalent bonds, H-bonding, and through space, all of which are distinct for different secondary structures (1, 2). Consequently, different secondary structures of polypeptides produce amide I bands at distinct

Table 1
Description of various amide vibrational modes and the wavenumbers (*W*) in H_2O and D_2O[a]

Amide mode	W_{H2O} (cm^{-1})	W_{D2O} (cm^{-1})	Approximate contributions from individual chemical bonds (%)[b]
Amide I	1,700–1,600	1,700–1,600[c]	CO s (76), CN s (11), CCN d (8), NH ib (5)
Amide II	1,570–1,540	1,480–1,460	NH ib (45), CN s (30), CO ib (10) CC s (8), NC s (7)
Amide III	~1,250	~960	NH ib (50), CC s (20), CN s (18), CO ib (12)
Amide IV	640–620	640–620[c]	CO ib (45), CC s (35), CNC d (20)
Amide V	735–715	520–500	CN t (62), NH ob (38)
Amide VI	650–600	650–600[c]	CO ob (85), CN t (15)
Amide VII	230–200	165–150	NH ob (68), CN t (20), CO ob (12)
Amide A	~3,300	2,500–2,400	Fermi resonance between NH s and amide II
Amide B	3,100–3,050	~2,400	Fermi resonance between NH s and amide II

[a]Adapted from refs. (1, 102)
[b]s, stretching; d, deformation; ib, in-plane bending; ob, out-of-plane bending, t, torsion
[c]These modes involve minimal contribution from the NH group and can undergo a slight (~5–15 cm^{-1}) downshift upon amide NH deuteration

frequencies, as shown in Table 2. The amide I mode is by far the most useful absorbance band in structural characterization of proteins and peptides. A protein that contains various secondary structures generates an asymmetric, composite amide I band. The fractions of distinct secondary structures in the protein are identified by curve-fitting procedures on the amide I band, as described below. Briefly, the fraction of each secondary structure in the protein (or peptide) is determined as the fraction of respective amide I spectral component. Corrections should be made with respect to the extinction coefficients of various secondary structures, as summarized in Table 3.

The amide II mode occurs around 1,570–1,540 cm^{-1} and includes a significant contribution from the NH in-plane bending, along with CN stretching and other vibrations of the amide group, and hence is highly sensitive to the kinetics of amide HX (i.e., amide NH to ND conversion). The amide III mode occurs in the 1,300–1,200 cm^{-1} region as a result of NH and C=O in-plane bending vibrations and CC stretching vibration, and hence is sensitive to amide HX (1). Although amide III band exhibits sensitivity to the polypeptide chain secondary structure and has been used in protein and peptide structural studies (3), this mode is affected by side chains and hence is likely to vary depending on the protein's amino acid composition (1, 4). Amide IV to amide VII modes

Table 2
Characteristic amide I wavenumbers (*W*) of various secondary structures in H_2O and D_2O

Secondary structure	W_{H20} (cm^{-1})	W_{D20} (cm^{-1})
α-Helix	1,658–1,647	1,655–1,638
α_{II}-Helix[a]	1,666–1,658	1,658–1,652
α-Helical coiled coil[b]	1,640–1,630	1,640–1,630
3_{10}-Helix	1,665–1,655[c]	1,665–1,655[c]
3_1-Helix[d]	1,639	1,632
β-Helix[e]	1,640–1,630	~1,630
Parallel β-sheet	1,638–1,632	1,636–1,630
Antiparallel β-sheet	1,638–1,632 (strong) and 1,695–1,675 (weak)	1,636–1,630 (strong) and 1,680–1,670 (weak)
Intermolecular (aggregated) β-sheet	1,627–1,615[f]	1,625–1,613[f]
α-Pleated sheet[g]	1,653–1,650	1,653–1,650
β-Turns	1,685–1,655	1,675–1,640
γ-Turns	1,690–1,650	1,690–1,650
Irregular ("unordered") structure	1,660–1,652	1,648–1,640

[a]α_{II}-Helix is similar to the regular α-helix (α_I-helix) but has tilted amide plane, which weakens the intramolecular H-bonding, resulting in a stronger C=O bond and higher amide I frequency (98, 103)
[b]From refs. (104, 105)
[c]Observed and calculated amide I wavenumbers for 3_{10} helix were in 1,665–1,655 cm^{-1} region (1, 106), but several researchers reported lower wavenumbers in the 1,640–1,630 cm^{-1} region (107, 108)
[d]The 3_1-helix is an extended helical structure, found, for example, in polyglycine, forming hexagonal lattice stabilized by intermolecular H-bonding (1)
[e]The β-helix is a helical structure of variable geometry found in polypeptides with alternating L-, D-chirality of amino acids such as gramicidin A or poly(γ-benzyl-LD-glutamate) (109, 110)
[f]From refs. (111–114)
[g]The α-pleated sheet is an extended, pleated structure with C=O and NH groups at opposite sides of the chain, predicted by Pauling and Corey (115) and found in amyloid peptides (116, 117) as well as certain proteins such as the selectivity filter of K^+-channel. (The amide I frequency for the non-deuterated form was calculated by Wu et al. (118))

occur at lower frequencies (750–200 cm^{-1}) and are used in protein structural studies less frequently because of their low intensities and moderate sensitivity to protein structure (1). Some of them involve large contributions from the NH vibration and can be used to study the amide HX kinetics. For example, the amide V mode results from NH out-of-plane bending and CN torsion vibrations and

Table 3
The peak absorbance extinction coefficients (ε) and the integrated absorbance extinction coefficients (*B*) for amide I and amide II modes of proteins in H_2O at 25°C[a]

Amide mode (second. structure)	ε ($M^{-1}cm^{-1}$)	*B* ($M^{-1}cm^{-2}$)
Amide I (α-helix)	700 ± 100	$(7.6 \pm 1.1) \times 10^4$
Amide II (α-helix)	300 ± 50	$(2.9 \pm 0.6) \times 10^4$
Amide I (β-sheet)	975 ± 10[b] 180 ± 20[c]	$(6.9 \pm 0.8) \times 10^4$ [b] $(0.5 \pm 0.05) \times 10^4$ [c]
Amide II (β-sheet)	340 ± 10	$(3.3 \pm 0.3) \times 10^4$
Amide I (irregular)	330 ± 50	$(4.7 \pm 0.5) \times 10^4$
Amide II (irregular)	220 ± 50	$(3.7 \pm 0.8) \times 10^4$

[a]Adapted from ref. (99). Disproportional changes in ε and *B* indicate change in the band lineshape (the band can become higher and slimmer, or shorter and broader, without changes in the area). For α-helix in D_2O, $\varepsilon_{\text{Amide I}}$ decreases to 465 ± 20 M^{-1} cm^{-1}, while for β-sheet and irregular structures, $\varepsilon_{H2O} \approx \varepsilon_{D2O}$
[b]Major, low-frequency β-sheet component (1,638–1,632 cm^{-1})
[c]High-frequency β-sheet component (1,695–1,675 cm^{-1})

undergoes a strong downshift upon amide HX. Amide A and amide B modes are generated by Fermi resonances between amide II and NH stretching vibrations and are located around 3,300 and 3,100 cm^{-1}, respectively. They are exceptionally sensitive to protein amide HX kinetics and in a few cases have been used for protein secondary structure characterization (5, 6). Certain amino acid side chains absorb appreciably in amide I and amide II regions and should be taken into consideration during data analysis. The absorbance wavenumbers generated by certain functional groups of amino acid side chains and respective extinction coefficients are summarized in Table 4. More details on amide modes and other infrared spectroscopic features of proteins and peptides can be found elsewhere (1, 7).

FTIR of proteins and peptides is normally conducted on samples in buffers made using D_2O rather than H_2O because the H–O–H bending vibration of H_2O strongly absorbs around 1,645 cm^{-1} and obscures the most important conformation-sensitive amide I band of the protein or the peptide (8–10). In addition, the broad and intense H–O–H stretching bands between 3,500 and 3,200 cm^{-1} mask the amide A mode (Table 5). D_2O is transparent in these regions; it weakly absorbs in the amide II region (1,555 cm^{-1}), and HOD absorbs in the deuterated amide II region (1,450 cm^{-1}), which should be taken into consideration in amide II band analysis. While the molar extinction coefficients of H_2O or D_2O vibrational modes are approximately an order of

Table 4
FTIR wavenumbers [W (cm^{-1})] of various vibrational modes of certain amino acid side chains and terminal amino and carboxyl groups, the peak absorbance extinction coefficients [ε (M^{-1} cm^{-1})], and the integrated absorbance extinction coefficients [B (M^{-1} cm^{-2})] in H_2O and D_2O[a]

Amino acid	Mode[b]	W_{H2O}	ε_{H2O}	B_{H2O}	W_{D2O}	ε_{D2O}	B_{D2O}
Arginine	$CN_3H_5^+$ as	1,673	420	4.3×10^4	1,608	460	3.1×10^4
	$CN_3H_5^+$ ss	1,633	300	3.6×10^4	1,686	500	3.4×10^4
Asparagine	C=O s	1,678	310	2.7×10^4	1,648	570	5.2×10^4
	NH_2 d	1,622	160	2.5×10^4			
Aspartic acid	COO^- as	1,574	380	5.5×10^4	1,584	820	8.8×10^4
	COOH s	1,716	280	4.1×10^4	1,713	290	3.5×10^4
Glutamine	C=O s	1,670	360	3.1×10^4	1,635	550	5.8×10^4
	NH_2 d	1,610	220	3.5×10^4			
Glutamic acid	COO^- as	1,560	470	7.1×10^4	1,567	830	8.9×10^4
	COOH s	1,712	220	3.6×10^4	1,706	280	3.4×10^4
Histidine	Ring	1,596[c]	70	3.0×10^3	1,620[d]	6	2.5×10^2
					1,623[e]	16	8.0×10^2
					1,600[e]	35	1.5×10^3
Lysine	NH_3^+ ad	1,629	130	1.8×10^4			
	NH_3^+ sd	1,526	100	1.3×10^4			
Phenylalanine	Ring	1,494	80	2.0×10^3	1,607	10	5.0×10^2
					1,596	10	5.0×10^2
Tryptophan	Ring				1,545	10	5.0×10^2
Tyrosine	Ring	1,518[f]	430	1.0×10^4	1,615[f]	160	5.0×10^3
		1,602[g]	160	7.0×10^3	1,515[f]	500	1.1×10^4
		1,498[g]	700	2.5×10^4	1,603[g]	350	1.8×10^4
					1,500[g]	650	2.9×10^4
α-Amino group	NH_2 d	1,560	450	7.5×10^4			
	NH_3^+ ad	1,630	210	3.8×10^4			
	NH_3^+ sd	1,515	200	4.3×10^4			
α-Carboxyl group	COOH s	1,735	170	2.1×10^4	1,720	230	2.7×10^4
	COO^- as	1,598	240	3.5×10^4	1,594	830	8.4×10^4

[a]Adapted from refs. (119, 120)
[b]as = asymmetric stretching; ss = symmetric stretching, ad = asymmetric deformation, sd = symmetric deformation, d = deformation, s = stretching
[c]Imidazole absorbance parameters in H_2O did not change upon protonation (119)
[d]Non-protonated imidazole ring in D_2O produces one weak band
[e]Protonated imidazole ring in D_2O produces two bands
[f]The (−OH) form of the ring produces one narrow band in H_2O and two bands in D_2O
[g]Deprotonated ($-O^-$) ring produces two bands in both H_2O and D_2O

Table 5
FTIR vibrational wavenumbers (*W*) of H_2O, D_2O, and HOD; the peak absorbance extinction coefficients (ε); and the integrated absorbance extinction coefficients (*B*) at 25°C[a]

Vibrational mode[b]	W (cm^{-1})	ε (M^{-1} cm^{-1})	B (M^{-1} cm^{-2})
HOH as	3,490 ± 20	122 ± 20[c]	(28 ± 4) × 10^3 [c]
HOH ss	3,280 ± 20	106 ± 16 [c]	(25 ± 3) × 10^3 [c]
HOH a	2,125 ± 10	3.4 ± 0.4	(3 ± 0.3) × 10^2
HOH b	1,645 ± 4	20 ± 2	(1.6 ± 0.3) × 10^3
DOD as	2,540 ± 20	120 ± 20 [c]	(19 ± 2) × 10^3 [c]
DOD ss	3,450 ± 20	110 ± 12 [c]	(17 ± 2) × 10^3 [c]
DOD a	1,555 ± 20	1.9 ± 0.05	(1.2 ± 0.2) × 10^2
DOD b	1,215 ± 20	16 ± 1	(1 ± 0.1) × 10^3
HOD ss	3,380 ± 20	120 ± 20 [c]	(8 ± 1) × 10^3 [c]
HOD as	2,500 ± 20	108 ± 15[c]	(7 ± 1) × 10^3 [c]
HOD b	1,450 ± 20	18 ± 2	(1.6 ± 0.4) × 10^3

[a]Adapted from refs. (8–10)
[b]as = asymmetric stretching, ss = symmetric stretching, a = association, b = bending
[c]Asymmetric and symmetric stretching extinction coefficients apply to the whole molecule rather than a single OH or OD vibration

magnitude weaker than those of amide I or amide II bands (Tables 3 and 5), the large molar excess of the solvent (~55 M of solvent vs. $\leq 10^{-4}$ M of protein or peptide) makes the solvent contribution too large and hence strongly reduces the reliability of solvent subtraction procedures. Exposure of the peptide to D_2O initiates HX of the amide NH group. The vibrational frequency of a diatomic molecule can be approximately expressed as

$$\nu = \frac{1}{2\pi}\sqrt{k\left(\frac{1}{m_1}+\frac{1}{m_2}\right)}, \tag{1}$$

where k is the spring constant, proportional to the chemical bond strength, and m_1 and m_2 are the atomic masses. Deuteration of the amide NH group results in a downshift of the amide A mode from ~3,300 cm^{-1} to 2,500–2,400 cm^{-1}; a shift of amide V and amide III modes from 725 cm^{-1} down to 510 cm^{-1}, and from ~1,250 to 960 cm^{-1}, respectively; a smaller but significant shifts of the amide II mode (~90 cm^{-1}); and a ~10 cm^{-1} downshift of the amide I mode. A set of FTIR spectra measured within several hours following exposure of the sample to D_2O allows determination of the secondary structure of the protein/peptide as well as the amide HD kinetics. The latter reflects both the rigidity of protein's secondary structure and the degree of solvent accessibility. Deuteration of amide NH group occurs only when (a) it is exposed to D_2O and (b) is not H-bonded to C=O or other groups. Thus, a backbone NH group involved in a secondary structure such as α-helix of β-sheet can only undergo deuteration upon a transient breakage of the H-bond ("local breathing") provided it is exposed to the D_2O solvent. Since the breakage occurs less frequently in rigid secondary structures, a faster amide HX may reflect a more flexible structure, a higher degree of solvent accessibility, or a combination of the two. The information gained from amide HX measurements often is considered to reflect the protein's "dynamic structure" or the "structural dynamics," that is, the degree of molecular flexibility plus the tightness of the three-dimensional packing.

Analysis of the FTIR spectra of proteins or peptides usually provides global structural information, that is, the fractions of various secondary structures, without identification of their locations within the protein molecule. There are at least two procedures that can be used to locate the experimentally determined secondary structures along the polypeptide chain. First, application of a secondary structure prediction algorithm to the amino acid sequence identifies stretches that are most likely in α-helical, β-strand, turn, or irregular conformations. Internet-based secondary structure prediction programs are available, such as Psipred, provided by University College London Department of Computer Science (http://bioinf.cs.ucl.ac.uk/psipred/), or Jufo, provided by Professor Jens Meiler's lab of Vanderbilt University (http://www.meilerlab.org/index.php/servers/show). If the predicted and experimental secondary structure contents are similar, then the FTIR-derived secondary structures can be assigned to the predicted regions. In the absence of consensus, site-specific structural information can be obtained by isotope labeling of desired regions of the protein or the peptide. This can be done by using synthetic, semisynthetic, or metabolic methods, described in detail below. Labeling of single or multiple amino acids with isotopes such as ^{13}C results in a downshift of the vibrational frequency of the amino acid (s) containing the heavier atom, according to Eq. 1, and thus allows spectral resolution and site-specific structural characterization of the protein. Isotopic enrichment is also used to study protein–protein interactions. When one of the interacting proteins is ^{13}C-labeled and the other is not labeled, then the sample containing both proteins generates a split amide I band, the higher and lower frequency components corresponding to unlabeled and ^{13}C-labeled proteins. Curve-fitting of both components and comparison of the data with those obtained on each protein separately allows evaluation of the structural changes occurring in the two proteins upon interaction.

Membrane proteins have been studied by FTIR spectroscopy using both conventional, "direct transmission" technique and the ATR-FTIR technique (7, 11, 12). In the former case, the protein is either solubilized in a membrane-mimetic environment such as a detergent or reconstituted in lipid vesicles, followed by placing the sample between two IR windows and measurement of the spectra. For integral proteins or peptides, which are highly hydrophobic and therefore reside exclusively within the detergent or the membrane, this method is completely justified and provides valuable structural information. For peripheral proteins, which are water soluble yet bind to membranes when they are present, the direct transmission method may not be an optimal choice because a fraction of the protein may reside in the aqueous phase. The spectra will reflect the structures of both free and membrane-bound species, preventing an unambiguous determination of membrane-induced conformational effects.

ATR-FTIR spectroscopy provides several advantages in structural characterization of membrane proteins. High-quality spectra can be obtained on a single lipid bilayer containing the reconstituted protein, requiring only ~100 μg of lipid and several times lower amounts of protein. Stacked bilayers can be easily prepared on the internal reflection element (IRE), which significantly increases the signal intensity and, correspondingly, the signal-to-noise ratio, and prevents interactions of the membrane-incorporated protein with the solid substrate. A pair of ATR-FTIR spectra, measured at *p* and *s* polarizations of the incident light, allows evaluation of the secondary structure of the membrane-bound protein by curve-fitting of the amide I band, the orientation of the protein from the linear dichroic ratios, the amide HX kinetics from the time-dependent changes in the amide II band, the phase state of the lipid from the acyl chain methylene stretching frequencies, the lipid order parameter from the acyl chain dichroism, the hydration state of the membrane from the lipid carbonyl stretching frequency and lineshape, membrane binding affinity and stoichiometry of the protein, and more (7, 11–16). This method is particularly useful to study the mechanisms of interfacial enzymes, that is, the enzymes that bind to membranes, undergo activation, and acquire their substrate, such as lipid or fatty acid molecules, from the membrane (13, 16–18). Using segmentally ^{13}C-labeled PLA_2, the angular orientation of membrane-bound enzyme has been determined, which, in combination with fluorescence and modeling results, produced the mode of membrane binding of the protein at an atomic resolution (19).

Below, a more detailed description of FTIR and ATR-FTIR spectroscopy of proteins and peptides in lipid membranes is provided, including experimental and data analysis protocols. A number of examples are presented to demonstrate what these techniques can provide in terms of structural, and sometimes functional, characterization of membrane proteins or peptides. As such, this article is a practical guide of FTIR and ATR-FTIR spectroscopy of membrane proteins and peptides rather than a review of literature.

2. Materials

2.1. Buffers

Buffers for FTIR experiments can be prepared using H_2O or D_2O. Since H_2O strongly absorbs in the 1,700–1,600 cm^{-1} region and thereby obscures the conformation-sensitive amide I band (Table 5), D_2O is the preferred solvent in FTIR spectroscopy of proteins or peptides. In certain cases, for example, when the amide II band is to be measured in an unperturbed state (i.e., without amide deuteration), measurements can be conducted in H_2O-based buffers, which are prepared by conventional methods. D_2O-based buffers can be prepared using a regular pH meter, provided certain precautions are exercised. A typical, simple buffer for protein or peptide FTIR studies is described below:

1. 100 mM NaCl, 20 mM Hepes, pH* 6.8.

Here, pH* 6.8 is the pH meter reading and corresponds to real deuterium ion concentration equivalent to pD = 7.2. The isotope effect results in 0.4 pH unit difference between pD and pH*: pD = pH* + 0.4 (20, 21). The procedures of buffer preparations are as follows:

1. Weigh the amounts of NaCl and Hepes required for a certain volume (e.g., 200 mL) of buffer and dissolve in ~150 mL of D_2O.
2. Calibrate the pH meter with standard buffers.
3. Rinse the electrode with D_2O to remove traces of H_2O, and immerse the electrode into the solution while stirring with a magnetic stir bar.
4. Titrate the solution with a NaOH solution of relatively high concentration (knowledge of the exact NaOH concentration is not necessary) until the desired pH meter reading (i.e., pH*) is reached. This should be 0.4 U below the pH of the medium to be imitated (see above). Other buffers, such as Tris or phosphate, can be used (Tris will be titrated with HCl or other acids). Ideally, NaOD and DCl should be used to minimize the presence of protons, and hence formation of H_2O, in the solution, but using NaOH or HCl has not led to significant problems.
5. Once the target pH* is reached, add D_2O to achieve the final volume of the buffer and store at 4°C.

2.2. Lipids

Lipids for preparation of artificial membranes, such as vesicles or supported bilayers, are available from various vendors. Using synthetic lipids may be preferred over the lipids isolated from biological tissues, such as bovine brain or liver, because these may contain trace contaminants of biological origin. When modeling a certain type of cellular membrane, such as endoplasmic reticulum, mitochondrion, or plasma membrane, it is advisable to use a mixture of lipids that mimic the respective membrane lipid composition, including both the headgroups and the degree of hydrocarbon chain unsaturation.

2.3. Proteins

Proteins are usually obtained by overexpression in prokaryotic or eukaryotic systems and purification using chromatographic methods. For structural studies, the purity of the sample is critical; hence care should be taken to reach a single spot in SDS gels upon staining with silver (22). Coomassie staining is less sensitive and may not detect relatively significant contaminations. Proteins uniformly labeled with certain isotopes, such as ^{13}C or ^{15}N or both, can be obtained by growing the plasmid-transfected cells in a minimal medium (e.g., M9) that contains 2.0–4.0 g/L of U-$^{13}C_6$-D-glucose and/or 5.0 g/L $^{15}NH_4Cl$ as single metabolic sources of carbon or nitrogen, respectively (23, 24). This will work with bacteria that can synthesize all 20 amino acids. When mammalian cells are used as an expression system, the medium should contain not only U-$^{13}C_6$-D-glucose or $^{15}NH_4Cl$ but also the isotope-labeled essential amino acids, which cannot be synthesized by mammalian cells. For example, the Bioexpress-6000 (Mammalian) medium from Cambridge Isotope Laboratories contains isotope-labeled Gln, His, Ile, Leu, Lys, Met, Thr, Val, Phe, and Trp.

In some cases, a certain amino acid in the protein can be labeled with a desired isotope by using an auxotroph, such as an *Escherichia coli* strain that is unable to make the amino acid (25, 26). The labeled amino acid should be present in the cell culture medium so the cells will incorporate it in all proteins during biosynthesis. Multiple *E. coli* strains that are auxotroph for any one of the 20 amino acids have been engineered by gene knockout methods (see http://cgsc.biology.yale.edu/Auxotrophs.php.) If the amino acid is present at multiple positions of the target protein, correspondingly multiple sites will be labeled, which may not be an optimal situation for gaining site-specific structural information. Insertion of an isotopically or otherwise modified amino acid into a desired site of a protein is more difficult. One of the widely used methods is the nonsense suppression protein engineering method (27–31). At the target site in the sequence of plasmid DNA, the codon is replaced with one of the three stop (nonsense) codons, such as TAG. A tRNA with a respectively modified anticodon loop (e.g., $tRNA_{CUA}$) that is complementary to the nonsense codon (in this case, UAG in mRNA) is created by chemical or recombinant methods. The modified tRNA is then charged with an unnatural or labeled amino acid either chemically or enzymatically, using a mutated aminoacyl tRNA synthetase. Addition of the $tRNA_{CUA}$ loaded with the modified amino acid to the growth medium of cells transfected with the plasmid containing TAG at the target site will produce the protein with the modified amino acid at the desired site. TAG is not used by bacteria frequently as a stop codon, so $tRNA_{CUA}$ will interact only with the UAG codon in the mRNA generated by the expression plasmid. On the other hand, the premature termination of translation can be excluded by thermal inactivation or removal of the release factors specific to the utilized stop codon, using cell-free expression systems. Incorporation of more than one modified amino acid in a protein has been achieved by using codons containing three, four, or five nucleotides (32, 33).

Labeling just one residue in a protein with ^{13}C, or ^{13}C and ^{15}N, may not generate a strong enough signal to be distinguished from the amide I band generated by the unlabeled part of the protein. In addition, distinct amide I frequencies generated by each secondary structure are determined by vibrational coupling between neighboring amide units, implying that the FTIR signal from a lonesome ^{13}C-labeled (uniformly or carbonyl ^{13}C=O-labeled) amino acid cannot provide reliable information on the local secondary structure, even when it is spectrally resolved. These drawbacks can be remedied by labeling a whole segment of the protein rather than just one residue. Segmental labeling of proteins with stable isotopes for FTIR studies has been achieved by protein semisynthesis methods (19, 34). Isotope-edited FTIR is described in detail in Subheading 5 below.

2.4. Peptides

Peptides can be obtained by recombinant expression as proteins. However, numerous companies offer peptide synthesis services at an affordable price, which makes the custom order of synthetic peptides the preferred option of obtaining pure peptides of any sequence, containing 50 and more amino acid residues. The peptide sample often contains trifluoroacetic acid (TFA) that is being used for cleavage of the peptide from the solid phase support after synthesis and/or for purification by HPLC. TFA strongly absorbs in the amide I region, around 1,674 cm^{-1}; therefore it should be removed before analysis of the peptide by FTIR. This can be done easily by two or three cycles of dissolving the peptide in HCl (2–100 mM HCl has been used) or 0.1 % phosphoric acid solution followed by lyophilization, which results in replacement of TFA with chloride or phosphate anions as a counterion of cationic amino acid side chains (2, 35, 36). It is advisable to use pure H_2O for one or two additional cycles to remove the traces of the acid.

The synthetic method offers additional advantages. For example, isotopically labeled or unnatural amino acids can be easily

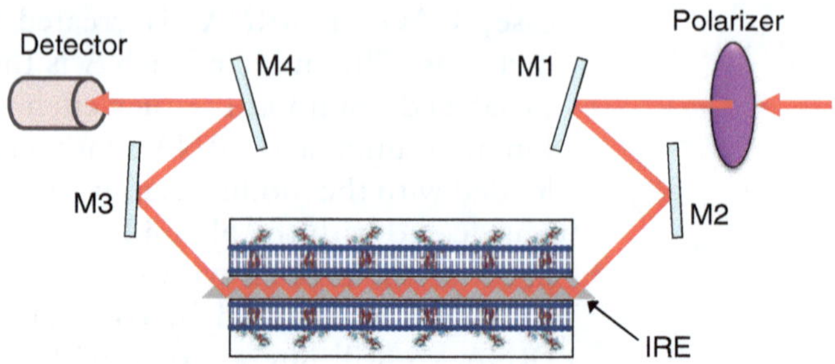

Fig. 2. A version of ATR-FTIR experimental setup for measurements of the spectra of a supported membrane with a reconstituted protein. The *gray trapezoid* is an internal reflection element (IRE), such as a germanium plate, which is covered with lipid bilayers at both surfaces (*blue*), containing the reconstituted integral protein shown in ribbon format. Mirrors M1 and M2 direct the incident light into the IRE, and mirrors M3 and M4 direct the light into the detector. Other details are described in the main text.

incorporated at any site in the peptide sequence during the synthesis (35–37). Also, any chemical groups, such as a C-terminal thioester for native chemical ligation experiments, can be added. As in the case of proteins, the amide I signal from one labeled residue may be lost in the large amide I band from the unlabeled part of the peptide and may not reflect the local secondary structure; hence labeling of a number of consecutive residues has been employed to gain site-specific structural information on peptides by FTIR (15, 35, 36, 38).

2.5. Direct Transmission FTIR Accessories

In addition to an FTIR spectrometer, certain accessories are used in a simple, direct transmission FTIR experiment. These include (1) a liquid sample holder, which usually slides vertically into the stand on the baseplate in the sample compartment of the spectrometer, (2) two infrared-transparent windows, such as CaF_2 or BaF_2 windows, (3) a spacer, that is, a thin ring that fits the circumference of the windows, made of Teflon or another inert and durable material (the reasonable thickness of the spacer is in the range 5–50 μm). If the spectrometer is using an MTC detector, approximately half a gallon of liquid nitrogen will be needed to cool it down.

2.6. ATR-FTIR Accessories

ATR-FTIR measurements require (1) an ATR system composed of four vertically mounted mirrors directing the incident light into the sample and then to the detector (Fig. 2); (2) a vertical stage that holds the ATR system and allows adjustment of the height of the mirrors to maximize the signal; (3) an internal reflection element (IRE) that is transparent to mid-infrared light and has a greater refractive index than the lipid–protein sample, such as germanium, zinc selenide, or thallium bromoiodide (a.k.a. KRS-5) (diamond meets all IRE requirements but is prohibitively expensive and hence nonpractical); (4) a demountable, perfusable sample cell composed of two halves designed to sandwich the IRE between them; (5) a polarizer, such as aluminum wire grid deposited on KRS-5 substrate; and (6) an argon plasma cleaner to treat the IRE before use. The size of the IRE may be $5.0 \times 2.0 \times 0.1$ cm^3, cut at 45° aperture angles at the 2.0 cm sides. If the substrate-supported membrane is to be prepared using a Langmuir–Blodgett method, a monolayer trough will be required.

3. Methods

3.1. Direct Transmission FTIR Measurements

3.1.1. Acquisition of Protein Spectra

As explained above, FTIR of proteins or peptides is conducted using D_2O-based buffers. The main reason for this is that D_2O is transparent in the amide I region, while H_2O strongly absorbs around 1,645 cm^{-1} (Table 5). Using D_2O has additional advantages. The α-helical and irregular structures absorb in overlapping regions in both H_2O and D_2O (Table 2). However, the unordered structure undergoes amide HX much faster (seconds) than the α-helix (minutes to hours) (14). Thus, a protein sample that is exposed to D_2O for about 0.5–1.0 h produces an amide I band where the irregular component is mostly exchanged while the α-helical component is still unexchanged, thus allowing for spectral resolution and estimation of the relative contents of both structures in the protein. Prolonged exposure to D_2O will result in exchange of all secondary structures, again preventing distinction between α-helix and irregular structure. (Note that absorbance bands of deuterated proteins or peptides are denoted amide I', amide II', etc.)

The best strategy is to measure the spectrum in an H_2O-based buffer and then prepare a sample in a D_2O-based buffer and record spectra over time, for 1–2 h or more. The spectrum measured in H_2O will provide the amide II band intensity, and those measured in D_2O will show a gradual decrease in the amide II band intensity in the 1,570–1,540 cm^{-1} region, reflecting the amide HX kinetics. At each time point, the extent of amide HX will be given by the relative amide II intensity, that is, the ratio of amide II band area measured at a given time divided by the area measured in H_2O, provided same protein concentrations and optical path-length are used in both cases. If this is not the case, the spectrum measured in H_2O should be normalized relative to an absorption band that is not exchangeable and hence should have the same intensity in all spectra, measured in H_2O and D_2O. This spectral feature should not overlap with solvent bands. An extrinsic compound with relatively high extinction coefficient can be used for such purpose. Finally, the amide I band of a protein in D_2O also undergoes spectral downshift due to amide HX and hence can be used to determine amide HX kinetics as well (12, 14).

For direct transmission FTIR experiments, it is convenient to have the protein or the peptide sample in lyophilized form:

1. Dissolve a certain amount of the lyophilized sample, typically 30–100 μg, in ~50 μL of a D_2O-based buffer and place on a CaF_2 or other window in a demountable liquid sample cell. The spacer should be placed on the window in advance.
2. Place the second window on top of the sample, which forms a thin layer between the two windows separated by the spacer.
3. Assemble the cell and mount in the sample cell holder in the spectrometer. These procedures should be done quickly to decrease the time lapse between the start of HX and first measurement.
4. It is customary to visually inspect the sample against light to see if voids ("air bubbles") are present. If large voids are present, which may result from greasy windows, for example, the window should be cleaned and a new sample should be prepared.
5. Start collection of spectra immediately after placing the sample in the spectrometer.
6. The lid of the sample compartment should be closed and the instrument continuously purged with dry air. Additional passage of the air through an air drying column, such as silica-gel column, results in very efficient removal of humidity and hence allows one to obtain "clean" spectra free of water vibrational modes in the amide I and amide A regions.
7. Record consecutive transmission spectra by collecting any number of scans per spectrum. A large number of scans, such as 500–1,000, will generate a better quality spectrum but will take a longer time, reducing the time resolution of amide HX kinetics. Around 10–50 scans may yield spectra of acceptable quality, especially when a high-sensitivity MCT detector is used. About 20 spectra collected within the first 2 h of amide deuteration (i.e., exposure of the sample to D_2O) will provide enough data for characterization of the structure and amide HX kinetics of the protein.
8. Following the collection of a set of sample transmission spectra, wash the cell and load with the blank buffer.
9. Measure spectra of the buffer, that is, the reference (or background) at same parameters as the sample.
10. In addition, two spectra of the "open beam" can be measured, corresponding to different times of purging with dry air after closing the lid of the sample compartment. Using an empty cell with the two windows separated by the spacer is not recommended because these measurements are usually affected by interference fringes. The two transmission spectra of the open

beam can be used to calculate an absorbance spectrum, using the first as "sample" and the second as "reference," to obtain a spectrum of atmospheric water vapor.

11. Each protein sample transmission spectrum can be paired with a buffer transmission spectrum to get an absorbance spectrum that is minimally affected by humidity.
12. If upward or downward spikes of water vapor absorbance are still present, they can be removed by spectral subtraction or addition of the water vapor spectrum. These procedures usually result in high-quality spectra that do not need any further smoothing unless needed for second derivative or other procedures.

3.1.2. Reconstitution of Proteins and Peptides in Lipid Vesicles

Procedures involved in preparation of membrane-bound protein samples depend on the nature of the protein. For peripheral proteins, which are water soluble but bind to membranes when they are present, the lipid vesicles can be prepared, followed by mixing with the protein solution. The extrusion method is an efficient and easy way of preparation of unilamellar vesicles of defined size:

1. If the lipid is provided in powder form, weigh a certain amount of lipid powder and dissolve in chloroform to obtain a stock lipid solution; a 10-mM concentration is customary. If chloroform does not dissolve the lipid, try a 2:1 (v:v) mixture of chloroform with methanol.
2. Transfer a desired amount of lipid solution into a small (e.g., 2–4 mL) glass vial and remove the solvent under a stream of nitrogen, followed by desiccation for ~3 h (desiccation for up to 12 h will not hurt).
3. Add a desired aqueous buffer to the dry lipid, followed by vortexing. This yields multilamellar vesicles (liposomes), unless the lipid has an intrinsic property of forming other, nonlamellar structures. For example, cardiolipin tends to form an inverted hexagonal structure, especially in the presence of calcium ions (39).
4. Extrude the liposome suspension through polycarbonate or other membranes with defined pore size, such as 100 nm or larger, using an extruder (40). About ten cycles of extrusion yield unilamellar vesicles with a homogeneous size distribution.
5. Prepare the protein solution in the same or similar buffer and mix with the vesicles. Avoid osmotic shock, that is, make sure the buffers used for preparation of the vesicles and the protein solution have similar osmolarities. This will result in binding of the protein to the external surface of the vesicles (17).

Alternatively, the vesicles can be prepared in the presence of the protein, resulting in binding of the protein to both external and internal membrane surfaces. This latter option may not be an optimal choice, however, because extrusion may adversely affect the protein. If the spectra of a membrane-bound protein need to be measured, the lipid and protein concentrations should be chosen in a way that nearly all of the protein is bound to vesicles. This can be estimated using the protein-membrane binding constant and stoichiometry, measured separately (22, 40, 41). In the same time, the lipid/protein molar ratio should be kept within a reasonable range. In most cases, a considerable fraction of the protein may be free in solution rather than membrane bound, unless membrane binding is extremely strong. This implies that direct transmission FTIR is not the best choice for measuring the spectra of membrane-bound peripheral proteins; ATR-FTIR offers a number of advantages, as described below.

Reconstitution of integral proteins or peptides in vesicle membranes is relatively easy (14, 15):

1. Dissolve the peptide in an organic solvent, such as hexafluoroisopropanol, trifluoroethanol, or chloroform. Avoid using solvents that absorb in the amide I or amide II regions.
2. Mix the peptide solution with the solution of lipid in chloroform or chloroform/methanol (2:1, v/v) at a desired molar ratio.
3. Remove the solvent under a stream of nitrogen followed by desiccation.
4. Suspend the dry peptide/lipid sample in a buffer and extrude (see above).

This method works for most peptides. The hydrophobic nature of the peptide, which may correspond, for example, to the transmembrane stretch of an integral membrane protein, ensures its partitioning into the membrane. The organic solvents are likely to affect the peptide structure, but it may recover its membrane-bound structure following reconstitution.

Using organic solvents for membrane reconstitution of proteins may not always be advisable because they can denature the protein (42, 43). If the protein is overexpressed in a cell culture and purified, an appropriate detergent can be used as a component of the purification buffer, at a concentration two- to threefold above its critical micelle concentration (cmc). Detergents suitable for membrane protein reconstitution have been thoroughly studied (43, 44). The hydrophobic protein will partition into the detergent micelles, which can be isolated by size-exclusion or affinity chromatography or by density gradient centrifugation. The micelles then can be mixed with the preformed lipid vesicles, so the detergent concentration drops below its cmc, resulting in spontaneous partitioning of the protein into the vesicle membranes (14). Alternatively, the detergent-solubilized protein can be incubated with the vesicles and SM-2 Bio-beads (Bio-Rad) for several hours. The high affinity of the Bio-beads for the detergent results in detergent partitioning into the Bio-beads and transfer of the protein into the vesicle membranes (45, 46). The vesicles, containing the reconstituted protein, then can be separated from the Bio-beads and used for FTIR measurements. One drawback of these procedures is that if an H_2O-based buffer is used, amide I spectra may be masked by the H_2O bending mode, and if a D_2O-based buffer is used, the protein will be extensively deuterated before the first spectrum is recorded, preventing measurements of amide HX kinetics. Again, ATR-FTIR may be considered a better choice for analysis of the structure and dynamics of membrane proteins (see below).

3.2. Direct Transmission FTIR Data Analysis

3.2.1. Protein Secondary Structure

The secondary structure of a protein or a peptide is usually evaluated by curve-fitting of the amide I absorbance band. The number and locations of amide I band components can be obtained from the second derivative spectra, the downward peaks of which indicate the amide I components. Often the inverted second derivatives are used, where the upward peaks point to the locations of the components (Fig. 3). Once the number of the amide I components and their spectral locations (the wavenumbers) are determined, a curve-fitting program such as GRAMS can be used to reproduce the actual amide I components. Gaussian or Lorentzian lineshapes, or a linear combination of the two, are being used. The curve-fitting procedure is considered satisfactory when (a) the deviation between peak wavenumbers of the components and those predicted by the second derivative does not exceed the nominal spectral resolution, for example, 2 cm^{-1}; (b) the sum of all components (the "curvefit") fits the actual amide I band; and (c) the widths of the components are within reasonable limits, for example, between 15 and 40 cm^{-1}. The components corresponding to the side chains, that is, those within 1,615–1,600 cm^{-1} (1,570–1,555 cm^{-1} for a ^{13}C-labeled protein), are subtracted, and the sum of the areas of all remaining components is used as the total amide I area (A_T). The components that correspond to α-helix, β-sheet, and irregular structure are identified following standard assignment procedures (Table 2). The antiparallel β-sheet has two components, a major, low-frequency component at 1,638–1,630 cm^{-1} and a high-frequency component at 1,695–1,675 cm^{-1}, and the integrated extinction coefficient of the latter constitutes ~7 % of the former (Tables 2 and 3). Therefore, the total β-sheet fraction can be determined by multiplying the fraction of the low-frequency β-sheet component area ($A_{\beta,\text{low}}$) by 1.07. The sum of the components between 1,700 and 1,660 cm^{-1} (1,655–1,615 cm^{-1} for a ^{13}C-labeled protein), minus $0.07 \times A_{\beta,\text{low}}$, then can be assigned to various types of turn structures. According to the Beer–Lambert law,

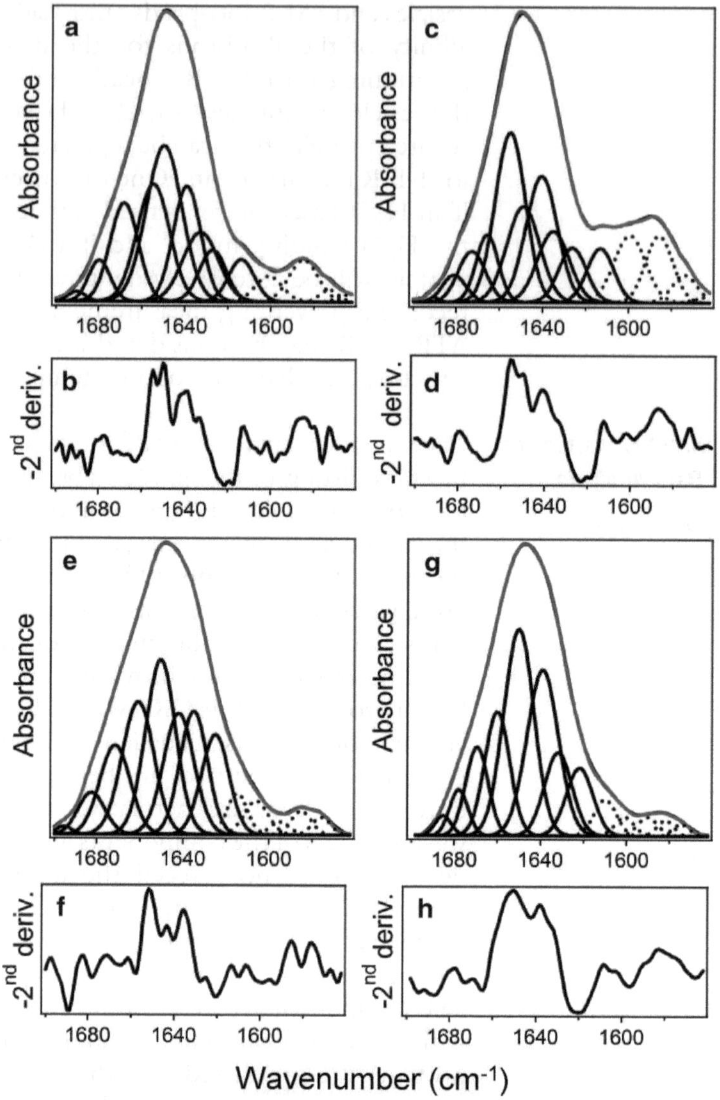

Fig. 3. Direct transmission (**a**, **c**) and ATR-FTIR (**e**, **g**) spectra of human group IIA PLA_2 (**a**, **e**) and its V3W mutant (**c**, **g**) either free in a buffer of 100 mM NaCl, 1 mM NaN_3, 1 mM EGTA, 50 mM Hepes in D_2O, pD 7.0 (**a**, **c**) or bound to a supported membrane composed of 70 % POPC and 30 % POPG, exposed to same buffer (**e**, **g**). ATR-FTIR spectra shown in panels **e**, **g** were obtained from the spectra measured at *p* and *s* polarizations of the infrared light as $A_p + 0.8A_s$. The components obtained by curve-fitting are shown for each spectrum; low-frequency components that are assigned to side chains are shown in *dotted lines*. The *gray lines* in (**a**), (**c**), (**e**), (**g**) are the sums of all components. Panels (**b**), (**d**), (**f**), (**h**) show the inverted second derivatives of spectra in panels (**a**), (**c**), (**e**), (**g**), respectively.

$$A_T = \sum l\varepsilon_i C_i, \quad (2)$$

where l is the optical path-length and ε_i and C_i are the extinction coefficient and the concentration of the *i*th structural component,

Table 6
FTIR vibrational wavenumbers (*W*), bandwidths at half height (*BW*), relative amide I extinction coefficients (ε_{rel}), peak absorbance extinction coefficients (ε), and the integrated absorbance extinction coefficients (*B*) of various secondary structures of proteins in H_2O solutions at 20–25°C[a]

Secondary structure	*W* (cm^{-1})	*BW* (cm^{-1})	ε_{rel}	ε (M^{-1} cm^{-1})	*B* (M^{-1} cm^{-2})
Ordered α-helix	1,652	29.6	1.00	700 ± 100	$(7.6 \pm 1.1) \times 10^4$
Unordered helix + random str.	1,676	27.8	0.64	448 ± 64	$(4.9 \pm 0.7) \times 10^4$
	1,646	47.3	1.51	1,060 ± 150	$(11.5 \pm 1.7) \times 10^4$
	1,620	9.2	2.85	2,000 ± 300	$(21.7 \pm 3.1) \times 10^4$
β-Sheet	1,691	15.3	0.13	90 ± 15	$(1.0 \pm 0.1) \times 10^4$
	1,633	27.9	0.82	600 ± 80	$(6.2 \pm 0.9) \times 10^4$
Turns	1,720	46.7	0.17	120 ± 20	$(1.3 \pm 0.2) \times 10^4$
	1,668	21.8	0.20	140 ± 20	$(1.5 \pm 0.2) \times 10^4$
	1,629	9.1	0.26	180 ± 30	$(2.0 \pm 0.3) \times 10^4$
	1,613	23.3	0.10	70 ± 10	$(0.8 \pm 0.1) \times 10^4$

[a]Data in first four columns are from ref. (3). The extinction coefficients in columns 5 and 6 are calculated using the α-helical extinction coefficient reported by Venyaminov and Kalnin (99) and the relative extinction coefficients reported by Vedantham et al. (3)

side chains excluded. Eq. 2 yields the following expression for the concentration of *i*th secondary structure:

$$C_i = \frac{A_T - \sum\limits_{j \neq i} l\varepsilon_j C_j}{l\varepsilon_i}. \quad (3)$$

The numerator in Eq. 3 is simply the area of the *i*th component, a_i; therefore, it can be rewritten as follows:

$$C_i = \frac{a_i}{l\varepsilon_i}. \quad (4)$$

The fraction of the *i*th secondary structure in the protein is

$$f_i = \frac{C_i}{C}, \quad (5)$$

where *C* is the total concentration of all secondary structural elements and is given as follows:

$$C = \sum \frac{a_i}{l\varepsilon_i}. \quad (6)$$

In Eq. 6, the summation is over all secondary structures. This approach requires the amide I areas and the extinction coefficients of all secondary structures (Tables 3 and 6). In some cases, the protein secondary structure can be presented as α-helix, β-sheet, irregular structure, and "other," primarily including turns and other structures, or α-helix, β-sheet, irregular structure, turns, and "other." The problem here is that the extinction coefficient for the "other" structure is not known, but an extinction coefficient, which is the average of those of known structures, may be an acceptable approximation. If the α-helix, β-sheet, irregular structure (ρ), and "other" approach is adopted, then Eq. 6 can be rewritten as follows:

$$C = \frac{a_\alpha}{l\varepsilon_\alpha} + \frac{a_\beta}{l\varepsilon_\beta} + \frac{a_\rho}{l\varepsilon_\rho} + \frac{a_{other}}{l\varepsilon_{other}} \quad (7)$$

If the turn structures are included along with its extinction coefficients, then an additional term for the turns should be added. Equations 4, 5, and 7 yield

$$f_i = \frac{a_i}{\varepsilon_i\left(\frac{a_\alpha}{\varepsilon_\alpha} + \frac{a_\beta}{\varepsilon_\beta} + \frac{a_\rho}{\varepsilon_\rho} + \frac{a_{other}}{\varepsilon_{other}}\right)} \quad (8)$$

In case of spectra of two proteins combined in one sample, one unlabeled (i.e., ^{12}C) and other ^{13}C-labeled, the amide I components belonging to each protein should be identified keeping in mind that ^{13}C-labeling causes a 40–50 cm^{-1} spectral downshift. Together with respective extinction coefficients, eight fractions are found: $f_{\alpha,12C}$, $f_{\beta,12C}$, $f_{\rho,12C}$, $f_{other,12C}$, $f_{\alpha,13C}$, $f_{\beta,13C}$, $f_{\rho,13C}$, and $f_{other,13C}$ using Eq. 8 and are corrected as follows. Each of the eight fractions is multiplied by the total number of amino acid residues in two proteins and then divided by the number of amino acid residues in the respective (i.e., ^{12}C or ^{13}C) protein to determine the protein secondary structure.

Figure 3 shows an example of amide I band curve-fitting for human group IIA PLA_2 and its V3W mutant free in solution and bound to supported bilayers, using the second derivatives. Amide I components at 1,655 and 1,649 cm^{-1}, which are present in the spectra of both proteins in solution, have been assigned to non-exchanged and deuterated α-helix. These components constitute 19 and 28 % for the wild-type PLA_2 and 26 and 16 % for the mutant, indicating a faster amide HX in the former case. Upon membrane binding, the 1,655 cm^{-1} component shifts to 1,661 cm^{-1} and reduces in intensity in both cases, indicating formation of a more flexible α-helical structure. Weaker helical H-bonding corresponds to a stronger main chain C=O bond and a higher amide I frequency (see Eq. 1). Thus, FTIR provides information on protein secondary structure and the quality of helices, such as their relative rigidity. Formation of more flexible helices in group IIA PLA_2 is confirmed by amide HX data described below.

3.2.2. Amide HX Kinetics

The kinetics of amide HX is most frequently analyzed based on the time-dependent decrease in the amide II band intensity upon exposure to D_2O. In general, every single amino acid in a protein can undergo HX at a distinct rate depending on H-bonding, local

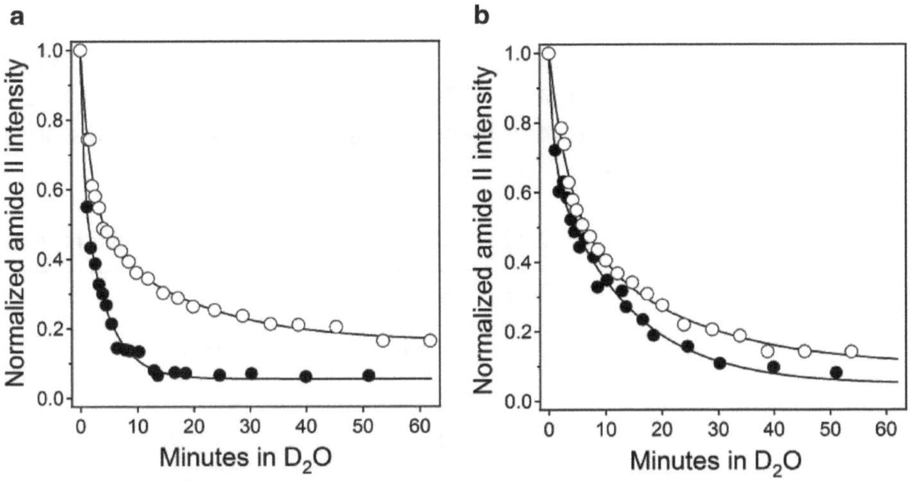

Fig. 4. Kinetics of amide HX for free (*open circles*) and membrane-bound (*closed circles*) human group IIA PLA_2 (**a**) and the V3W mutant (**b**), measured by polarized ATR-FTIR. The supported membrane was composed of 70 % POPC and 30 % POPG, and the buffer was 100 mM NaCl, 1 mM NaN_3, 1 mM EGTA, 50 mM HEPES (pH 7.0) in H_2O (at time 0) or D_2O (at time >0). The *solid lines* were calculated through Eq. 9, using the parameters summarized in Table 7.

mobility, structure, and steric protection from solvent. However, individual HX rate constants for each amino acid residue cannot be determined by conventional FTIR spectroscopy. Considering two to four populations, each characterized with a certain HX rate constant, usually allows a reasonable description of the overall HX kinetics of a protein. It is considered that each main chain amide group can be either in H form (non-exchanged) or D form (deuterated). The time dependence of the fraction of non-exchanged residues, [H/(H + D)], can be given as

$$\left[\frac{H}{H+D}\right]_t = \frac{A_{amideII,t}}{A_{amideII,0}} = a_0 + a_1 e^{-k_1 t} + a_2 e^{-k_2 t}, \quad (9)$$

where $A_{\mathrm{amideII},0}$ and $A_{\mathrm{amideII},t}$ are the amide II band areas of the protein in H_2O and at time t of exposure to D_2O, respectively; a_0, a_1, and a_2 are the fractions of exchange-resistant, slow exchanging, and fast exchanging residues, respectively; and k_1 and k_2 are the exchange rate constants of the latter two populations. Since the protein can undergo adsorption to or desorption from the membrane during the measurements, it is advisable to use the ratio of the amide II area divided by the amide I area instead of just A_{amideII}, especially when analyzing ATR-FTIR data. Amide HX kinetics of PLA_2 free in solution and bound to supported membranes has been characterized using this approach (Fig. 4, Table 7). The data of Fig. 4 and the quantitative parameters of HX presented in Table 7 clearly indicate a more efficient HX of the protein upon membrane binding. This cannot be interpreted in terms of solvent protection upon membrane binding since it would have an opposite effect.

Table 7
Characterization of the amide HX kinetics of human group IIA PLA_2 and its V3W mutant free in solution or bound to a supported membrane composed of 70 % POPC and 30 % POPG, at 20°C[a]

	Wild-type PLA_2		V3W mutant of PLA_2	
	Free	Membrane-bound	Free	Membrane-bound
a_0	0.16	0.06	0.10	0.05
a_1	0.35	0.59	0.46	0.64
a_2	0.49	0.35	0.44	0.31
k_1 (min^{-1})	0.057	0.24	0.051	0.083
k_2 (min^{-1})	0.51	2.03	0.29	1.20

[a]Data from ref. (22). The data are described in Fig. 4

Instead, membrane binding renders the protein structure more flexible, resulting in an increased rate of HX, which is consistent with the data of amide I band curve-fitting described above.

In cases when amide II band intensity decreases very rapidly, spectral shifts in the amide I region can be used to evaluate the amide HX kinetics (12, 14). The amide I spectrum of the membrane-bound protein measured in an H_2O-based buffer is subtracted from a series of spectra measured at various times of exposure to D_2O. If a spectrum in H_2O cannot be measured, a spectrum of a sample that is exposed to D_2O for a very short time, for example, ~1–2 min, can be used. Since upon deuteration the amide I band undergoes a spectral downshift, the difference spectra will contain negative and positive components. All amide I areas can be normalized before subtraction. If the sum of the absolute values of all components of the difference spectra at time t of exposure to D_2O is ΔA_t, then the kinetics of the HX can be described by

$$1 - \frac{\Delta A_t}{r} = \sum_{i=0}^{m} a_i e^{-k_i t}. \tag{10}$$

In Eq. 10, r is a proportionality coefficient relating spectral changes in the amide I and amide II regions:

$$r = \frac{\Delta A_t}{1 - \left(A_{amideII,t}/A_{amideII,0}\right)}.$$

and can be determined by measuring ΔA_t for the amide I band and $A_{\text{amideII},t}/A_{\text{amideII},0}$ at initial times of deuteration when the amide II band still can be measured reliably. Given the fact that FTIR does not allow determination of a large number of individual amino acid HX kinetics, using $m = 2$–3, that is, considering an exchange-resistant population and 2 or 3 populations that undergo HX with similar kinetics, is considered a reasonable approach. Amide HX kinetics of SKC1 K^+ channel reconstituted in lipid membranes has been evaluated using both methods, that is, changes in the amide II region (Eq. 9) and amide I region (Eq. 10) and yielded consistent results (14). Approximately 80 % of all amino acid residues of the channel were able to undergo amide HX within ~3 h, consistent with a model of the channel containing water-filled vestibules.

3.3. ATR-FTIR Measurements

3.3.1. ATR-FTIR Sample Preparation and Measurements

ATR-FTIR experiments of membrane proteins have been conducted using a single lipid monolayer, a bilayer, or a multilayer (7, 11, 12, 46–49). Axelsen and coworkers used a germanium IRE rendered hydrophobic by treating it with octadecyltrichlorosilane (49–51). Positioning of the IRE horizontally against a lipid monolayer at the air–water interface, with the protein adsorbed to the lipid from the aqueous phase, allowed recording of polarized ATR-FTIR spectra.

A lipid bilayer supported on the IRE surface can be prepared using at least two methods. In both cases, the IRE, such as a germanium plate, should be thoroughly cleaned with organic solvent(s) such as a chloroform/methanol (2:1, v/v) mixture and processed with an argon plasma cleaned for about 10 min. The "monolayer fusion" method involves the following steps:

1. Deposit a lipid monolayer at both surfaces of the IRE, using a Langmuir–Blodgett trough.
2. Assemble the IRE in the ATR flow-through cell and inject a lipid vesicle suspension in an H_2O-based buffer, with lipid concentration between 1 and 10 mM. The lipid polar headgroups adsorb to the IRE surface, and the hydrocarbon chains are facing air, so the monolayer is stable for many hours. Using phosphatidylcholine for monolayer deposition works very well; acidic lipids do not adsorb to germanium efficiently probably due to electrostatic effects. Upon injection of the vesicles, they spread on the hydrophobic monolayer surface, resulting in formation of a supported bilayer. This process is driven by hydrophobic contacts between the vesicles and the monolayer as well as the curvature strain in the vesicle membrane (lipids such as phosphatidylcholine tend to form lamellar rather than curved structures). Therefore, highly curved vesicles, for example, those obtained by sonication, undergo spreading more readily. However, extruded vesicles of 100 nm diameter will work as well.
3. Allow the system to equilibrate for ~1–2 h.

4. Flush the cell with the buffer to remove excess lipid. If the vesicles are prepared in the absence of any peptides or proteins, the resulting membrane is a pure lipid bilayer that can be used to examine the lipid order, membrane phase state, etc. (12, 13, 16).

In the second method, the sonicated vesicles are prepared using at least 20–30 mol % of an acidic lipid, in a buffer containing ~5 mM $CaCl_2$ and directly injected into the ATR cell containing the IRE. Supported bilayers are formed spontaneously, probably stabilized by Ca^{2+} ionic bridges between the acidic lipids and the plate. In experiments with peripheral proteins, the protein solution can be injected into the ATR cell, followed by its adsorption to the supported membrane. The kinetics of membrane binding of the protein can be determined from the time-dependent increase in the amide I band intensity (13, 18). If an H_2O-based buffer is used, the amide II band can be monitored instead of amide I. Once the protein binding to the membrane stabilizes, the ATR cell is flushed with a D_2O-based buffer and polarized ATR-FTIR spectra are collected to determine the protein secondary structure and orientation. For example, consecutive pairs of *p* and *s* polarized spectra can be collected for 1–2 h, which can be used to determine the protein's secondary structure, the kinetics of amide HX, and the orientation with respect to the membrane (see below).

In experiments with integral proteins, the protein is reconstituted in vesicles, as described above, and after spreading of the vesicles on the lipid monolayer or directly on the IRE, appears in the supported bilayer. Other procedures are similar to those used with peripheral proteins.

The samples for ATR-FTIR experiments can be prepared using lipid–protein or lipid–peptide multilayers rather than a single bilayer, as described (48, 52, 53). The vesicles with the reconstituted protein are prepared in an H_2O-based buffer, at a high lipid and protein concentration, evenly spread on one surface of the IRE and dried under nitrogen and under vacuum. These "dry" samples then can be used directly for collection of ATR-FTIR spectra or can be hydrated by exposing to a D_2O-saturated atmosphere or to a D_2O-based buffer. The final sample is expected to be a multilamellar structure, that is, stacked bilayers containing the incorporated protein, more than 1 μm thick. While both lipids and proteins keep a nominal amount of bound water that cannot be removed by desiccation, the hydrated samples are considered more biologically relevant. In case of peptides, the solutions of the lipid and the peptide can be combined at a desired molar ratio and spread at one surface of the IRE followed by removal of the solvent. The lipid–peptide film then can be humidified by a D_2O-saturated atmosphere or by a bulk D_2O-based buffer.

The first exposure of the sample to D_2O is the time of initiation of amide HX. Consecutive spectra measured at *p* and

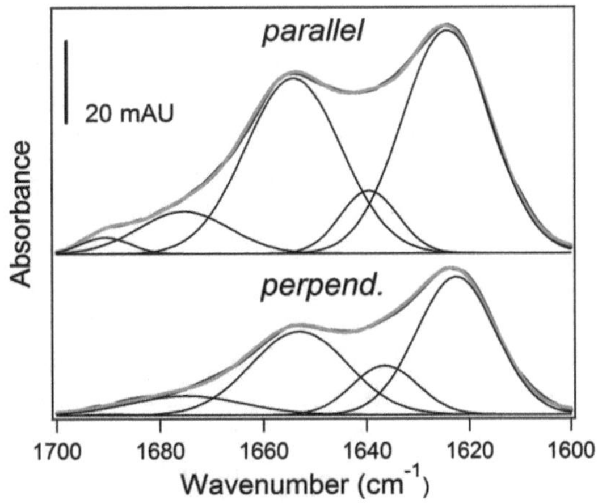

Fig. 5. ATR-FTIR spectra of a peptide derived from Bax protein in a supported POPC/POPG (7:3) multilayer at parallel and perpendicular polarizations, as indicated. The sample was exposed to a buffer of 150 mM NaCl, 10 mM Hepes in D_2O, pH* 6.8. The curvefit, that is, the sum of all amide I components shown under each spectrum, is in good fit with the measured spectrum. Major components at 1,655 cm^{-1} and 1,628 cm^{-1} indicate an α/β secondary structure.

s polarizations following exposure to D_2O are used to determine the protein's or peptide's secondary structure, orientation, and HX kinetics. Since most FTIR spectrometers are single beam instruments, reference spectra should be measured separately using a bare germanium plate or a sample prepared using plane lipid without any peptide or protein. Absorbance spectra at each polarization are obtained using the sample and reference transmission spectra at same polarization. Atmospheric water humidity spectra can be measured at both polarizations by collecting spectra at various times of purging with dry air, using a bare germanium plate. These water vapor spectra can be subtracted from sample absorbance spectra to clear the spectra from noise generated by residual humidity, when necessary. A pair of polarized ATR-FTIR spectra of a peptide in a 1.6 μm thick POPC/POPG multilayer is shown in Fig. 5, and the determination of the peptide's secondary structure and orientation based on the curve-fitted *p* and *s* polarized spectra is described in the next section.

3.3.2. ATR-FTIR Data Analysis

Polarized ATR-FTIR spectra can be used to evaluate the orientational order parameters of the lipid and the membrane-bound peptide or protein molecules, protein's and peptide's secondary structure, kinetics of amide HX, and more. The molecular order parameter can be determined from polarized ATR-FTIR experiments as (7, 47, 48, 54)

$$S = \frac{2B}{(3\langle\cos^2\alpha\rangle - 1)(B - 3E_z^2)}, \tag{11}$$

where $B = E_x^2 - R^{ATR} E_y^2 + E_z^2$, E_x, E_y, and E_z are the electric vector components of the evanescent wave at the surface of IRE normalized relative to the incident light amplitude; α is the angle between the transition dipole moment and the molecular axis; R^{ATR} is the ATR dichroic ratio: $R^{\text{ATR}} = A_p/A_s$, where A_p and A_s are the absorbance intensities at the peak of a band or integrated over the whole band for p and s polarizations of the infrared light, that is, parallel and perpendicular to the incidence plane, respectively; and the angular brackets indicate the average value. For the amide I (amide II) band, $\alpha = 38$–$40°$ (70–73°) for α-helix and ~90° (~0°) for β-strand (54–56); the latter only applies to the low-frequency β component at ~1,638–1,630 cm^{-1}. For methylene stretching vibrations of lipid hydrocarbon chains in all-trans conformation, $\alpha = 90°$. Two cases are considered, that is, a thin film sample, much thinner than the decay length of the evanescent wave, d_p, and a thick film sample, much thicker than d_p (in the amide I region, for a germanium-water system and a 45° incidence angle, $d_p \approx 320$ nm). For the thin film (57),

$$E_x = \frac{2\cos\gamma\sqrt{\sin^2\gamma - n_{31}^2}}{C}, \tag{12a}$$

$$E_y = \frac{2\cos\gamma}{\sqrt{1 - n_{31}^2}}, \tag{12b}$$

$$E_z = \frac{n_{32}^2 \sin 2\gamma}{C}, \tag{12c}$$

where γ is the incidence angle, $C = \sqrt{[(1 + n_{31}^2)\sin^2\gamma - n_{31}^2](1 - n_{31}^2)}$, and $n_{ij} \equiv n_i/n_j$, n_1, n_2, and n_3 are the refractive indices of IRE ($n_1 = 4$ for germanium), the thin layer of the sample ($n_2 \approx 1.43$ for a lipid layer), and the medium above the sample ($n_3 = 1$ for air and ~1.33 for water). If $\gamma = 45°$, then for a thin sample under air, $E_x = 1.411$, $E_y = 1.461$, and $E_z = 0.738$, and for a thin sample under water $E_x = 1.40$, $E_y = 1.51$, and $E_z = 1.375$. It is assumed that the sample is so thin that the evanescent field can be considered constant inside the sample. For a thick film, the evanescent wave does not exit the sample so only two phases are considered, the IRE and the sample. In this case, both n_2 and n_3 apply to the thick sample and $n_{32} = 1$, resulting in $E_z = 1.621$ (E_x and E_y stay the same). The average angle θ between the membrane normal and the molecular axis can be found from the relationship

$$S = \frac{1}{2}\left(3\langle\cos^2\theta\rangle - 1\right). \tag{13}$$

This theory can be applied to molecules that have a symmetry axis, such as lipid acyl chains in all-trans configuration or an α-helix.

In cases when a protein or a peptide contain a single α-helix, curve-fitting can be performed on amide I spectra measured at *p* and *s* polarizations, as shown in Fig. 5. The α-helical dichroic ratio, that is, the ratio of the areas of α-helical components at 1,655 cm^{-1} at *p* and *s* polarizations, then can be used to determine the helical orientation. Evaluation of the orientation of relatively large membrane-bound proteins, which may contain a variety of differently oriented helices and strands, is not easy. One way this can be achieved is isotope labeling of certain helices within the protein; combination of orientations of more than one helix allows positioning of the protein relative to the membrane (19). These procedures are described below.

Determination of the β-strand orientation is more complex. When the strands are arranged in a structure that is characterized by a central rotational axis, such as a β-barrel, the following holds (58):

$$\frac{1}{2}\left(3\langle\cos^2\delta\rangle - 1\right) = \frac{2B}{\left(3\langle\cos^2\chi\rangle - 1\right)\left(B - 3E_z^2\right)}, \tag{14}$$

where δ is the angle of the transition dipole moment of β-strands with respect to the barrel axis and χ describes the orientation of barrel axis with respect to the membrane normal. For β-strands, the transition dipole moments of amide I and amide II modes are oriented perpendicular and along the strand axis, respectively (58). Therefore,

$$\delta_I = \frac{\pi}{2} - \beta \tag{15a}$$

and

$$\delta_{II} = \beta \tag{15b}$$

where subscripts I and II indicate amide I and amide II bands, respectively, and β is the angle between strand axes relative to the central axis of the barrel. Substitution of (15a) and (15b) into Eq. 14 yields two equations that can be solved together, using the experimentally determined dichroic ratios for amide I and amide II bands, to evaluate the angles χ and β.

In polarized ATR-FTIR spectra of a membrane-bound protein, the absorbance intensity of each structural element is determined by its spatial orientation among other things. Therefore, protein's secondary structure cannot be determined using a spectrum obtained at a certain polarization. ATR-FTIR spectra measured at *p* and *s* polarizations are used to generate a "polarization-independent" spectrum: $A = A_p + GA_s$, where the scaling factor G is given as (59)

$$G = \frac{2E_z^2 - E_x^2}{E_y^2} \tag{16}$$

Using the values of E_x, E_y, and E_z shown above, we obtain $G = 0.80$ in case of a thin film between germanium plate and an aqueous medium and $G = 1.44$ for a thick film deposited on germanium. Once a corrected amide I spectrum is obtained, the curve-fitting can be performed and protein's secondary structure determined following the procedures described above. Curve-fitted p and s polarized amide I bands can also be used to obtain fractions of various secondary structures (59):

$$f_i = \frac{A_{p,i} + GA_{s,i}}{\sum\left(A_{p,i} + GA_{s,i}\right)}. \tag{17}$$

Application of both approaches to the polarized ATR-FTIR spectra of a membrane-reconstituted proteins or peptides yields similar results. For example, data shown in Fig. 5 indicate the following secondary structure for the membrane-bound peptide: $f_\alpha = 0.39, f_\beta = 0.40, f_t = 0.11$, and $f_\rho = 0.10$.

3.4. Isotope-Edited FTIR

3.4.1. Site-Directed Isotope Labeling

Since the conformation-sensitive amide I mode is generated primarily by the main chain C=O stretching vibration, site-directed labeling with ^{13}C or $^{13}C{=}^{18}O$ offers an opportunity to gain local structural information on peptides and proteins (2, 15, 35, 36, 60–64). The fact that through bond and through space vibrational couplings between interacting peptide group oscillators significantly contribute to the amide I frequencies should be given special consideration in interpretation of the signal generated by the labeled sites (2, 36, 63). Sensitivity of these couplings to the geometric arrangements of the oscillators renders the amide I frequencies dependent not only on distinct secondary structures but also on the distribution (e.g., consecutive vs. interrupted) of the isotope-labeled amide units (65). Introduction of one or two $^{13}C{=}O$ labels in a 14-residue β-sheet peptide resulted in a shift of the main amide I component from 1,628 cm^{-1} down to 1,611–1,606 cm^{-1} (2). The relative intensity of the downshifted component was significantly larger than the proportion of ^{13}C-labeled units, and this effect became more pronounced when the two ^{13}C labels were intervened by a single ^{12}C as compared to two adjacent ^{13}C labels. These anomalous intensity patterns were explained by a semiempirical treatment of vibrational couplings, including through covalent bonding, through H-bonding, and through space transition dipole couplings between ^{13}C and ^{12}C peptide units (2). Similar analysis on membrane-bound β-sheet peptides was performed by polarized ATR-FTIR, indicating that anomalously high intensities and frequencies of ^{13}C-generated amide I components can be explained by inter- and intra-strand couplings between ^{13}C and ^{12}C amide oscillators (63). The degree of spectral downshift and the relative intensity of the ^{13}C-labeled amide mode were correlated with the type of β-sheet (parallel vs. antiparallel) and the number of strands in the sheet (2, 36, 63).

In contrast to peptides in β-sheet structure, the ^{13}C-labeled amide units in α-helical peptides appear to behave like isolated oscillators (2). Indeed, published data indicate that the vibrational coupling between amide units (peptide groups) containing ^{13}C-labeled and unlabeled (^{12}C) carbonyls depends on the secondary structure; ^{13}C–^{12}C coupling occurs readily in a β-sheet structure but not in an α-helical structure (36, 62, 63). While a single ^{13}C-labeled residue can provide local structural characterization of a peptide in β-sheet conformation (36, 62, 63), identification of a local α-helical structure requires at least three consecutive ^{13}C-labeled amide units, which will provide a directly connected pair of coupled ^{13}C oscillators that will generate the diagnostic vibrational frequency and intensity (15, 35, 36, 65). FTIR studies on helical peptides containing three or four consecutively $^{13}C{=}O$-labeled residues revealed site-specific structural features, such as a higher thermal stability of the central region of an α-helix and relative conformational dynamics of C- vs. N-termini of peptides (35, 38).

Phospholamban transmembrane domain labeled with ^{13}C at one or two amino acids and reconstituted in supported lipid membranes was analyzed by polarized ATR-FTIR to gain insight into the local secondary structure and orientation (66). Such labeling resulted in ~45 cm^{-1} downshift in the amide I α-helical frequency from 1,658 to 1,614–1,612 cm^{-1}, as expected for decoupled oscillators, but the fact of decoupling raises questions about the meaning of the amide I mode of an isolated oscillator in terms of protein conformation or orientation. ATR-FTIR studies on glycophorin A transmembrane domain, ^{13}C-labeled at two or three consecutive sites, provided local structural and orientational information (67). The issue of helix or strand orientation determination by isotope-edited polarized ATR-FTIR was treated by Marsh (68). For a single site labeled peptide, linear dichroisms of more than one amide modes, such as amide I, II, and A, are required for helical orientation determination. For determination of the orientation of a β-sheet relative to the membrane plus the orientation of strands relative to the sheet, six dichroic ratios are required, which can be obtained from three different amide modes of two isotopically labeled residues.

Better amide I spectral resolution has been achieved by site-specific amino acid labeling with $^{13}C = {}^{18}O$ (69). $^{13}C = {}^{18}O$ labeled amino acids can be created by incubating the commercially available carbonyl ^{13}C-labeled amino acid in $H_2{}^{18}O$/dioxane (1:1 or 3:1, v/v) at 100°C at pH 1.0 for 1 h, followed by incorporation in the peptide sequence by chemical synthesis (69, 70). Care should be taken to avoid side chain modification of pH-sensitive amino acids. Labeling of the transmembrane domain of phospholamban with $^{13}C = {}^{18}O$ at either one of two centrally located adjacent residues resulted in a significant downshift of the amide I signal from 1,657 to 1,590 cm^{-1}, allowing determination of the local transmembrane orientation by polarized ATR-FTIR (69). Transmembrane orientations and helix packing details on ErbB-2 and other integral membrane proteins were deduced from ATR-FTIR experiments on $^{13}C = ^{18}O$ labeled peptides (53, 70, 71). Similar isotope editing strategies provided site-specific information on protein thermal stability and folding dynamics (72, 73), mechanisms of peptide aggregation (37, 74), and structural details of amyloid formation (75–77). Some of these studies used two-dimensional infrared spectroscopy (74–77), which is out of the scope of this text and has been reviewed elsewhere (78).

3.4.2. Structural Effects in Protein–Protein Interactions

Isotope-edited FTIR was used to gain insight into conformational changes in proteins upon their interactions. Uniform ^{13}C- or $^{13}C/^{15}N$-labeling of one of the two interacting proteins results in spectral resolution of their amide I bands and hence allows identification of conformational changes in both proteins (79–81). This strategy was utilized to detect structural changes in calmodulin and target peptides upon their interaction (80), as well as the structural effects involved in the chaperon-like activity of α-crystallin (81). Polarized ATR-FTIR studies on ^{13}C-labeled phospholamban and unlabeled sarcoplasmic reticulum Ca^{2+}-ATPase co-reconstituted in supported lipid bilayers identified that the inhibitory effect of phospholamban involved stabilization of α-helices of the Ca^{2+}-ATPase (46). More information on the analysis of protein–protein interactions by isotope-edited FTIR is provided elsewhere (82).

3.4.3. Segmental Isotope Labeling and Protein-Membrane Interactions

Labeling of a selected segment of proteins with ^{13}C can provide local conformational information. Moreover, for membrane proteins this strategy can provide the orientation of the membrane-bound protein. Using peptide ligation techniques (83–86), a semi-synthetic human pancreatic PLA_2 was engineered where the N-terminal helix was unlabeled, whereas the rest of the protein was uniformly ^{13}C-labeled (19). The procedure is schematically described in Fig. 6 and involved the following steps:

1. A peptide corresponding to the N-terminal α-helix of PLA_2 and containing a C-terminally added thioester group (Ala^1-Lys^{10}-$COSCH_2COOH$) was obtained by chemical synthesis.
2. A fragment of the protein starting with Cys^{11} was expressed in *E. coli* in a minimal medium containing 0.2 % uniformly ^{13}C-labeled D-glucose as a sole metabolic source of carbon.
3. The ^{13}C-labeled fragment Cys^{11}–Ser^{126} was reacted with a C-terminally thioesterified peptide in a buffer containing 6 M guanidinum-HCl, 100 mM Na-phosphate, 5 % β-mercaptoethanol, 1 mM EDTA, 4 % thiophenol, and 4 % benzyl mercaptan, pH 7.4. The concentrations of the C-terminal

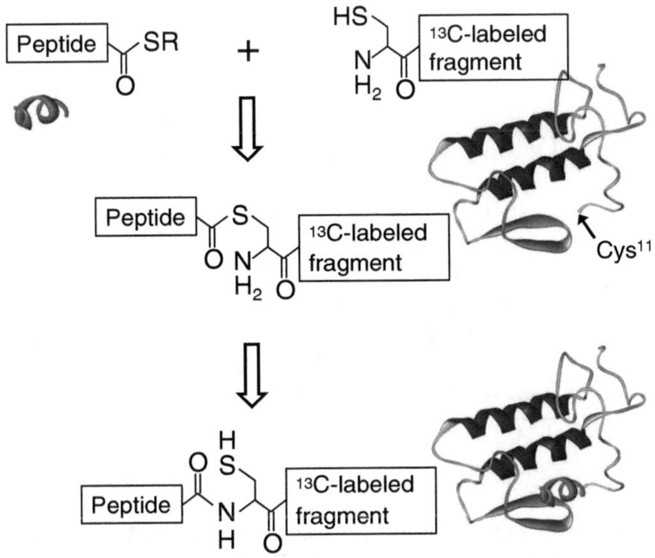

Fig. 6. Schematic description of production of a semisynthetic, segmentally ^{13}C-labeled protein using thioester-to-cysteine peptide ligation. Details are described in the main text.

fragment and the N-terminal peptide were 16 and 1.6 mg/ml, respectively, approximately corresponding to a 1:1 molar ratio.

4. The reaction was allowed to proceed for 6 h at 37°C, followed by refolding of the ligated protein by dialysis against 25 mM Tris–HCl, 5 mM $CaCl_2$, 5 mM L-cysteine, and 0.9 M guanidinum–HCl (pH 8.0), at 4°C. In thioester-to-cysteine-type peptide ligation, the thioester reacts with the thiol of cysteine side chain, followed by irreversible S–N acyl transfer and formation of a native peptide bond. Internal cysteines of the protein can also react with the thioesterified peptide, but these reactions are reversible under appropriately selected reducing conditions.
5. The ligated protein was purified by a Mono Q 5/50 column equilibrated with 2.5 mM KCl and 20 mM diethanolamine (pH 9.0).
6. This was followed by dialysis against 150 mM KCl, 25 mM Tris–HCl (pH 7.4), and additional purification using a size-exclusion HiLoad Superdex-75 column.
7. PLA_2 activity was measured using diheptanoyl-thiophosphatidylcholine as a substrate, as described elsewhere (40).

The semisynthetic, segmentally ^{13}C-labeled PLA_2 was dissolved in a D_2O-based buffer containing 100 mM NaCl, 1 mM NaN_3, 1 mM EGTA, 50 mM Hepes (pD 7.4) and injected into an ATR cell containing a supported membrane composed of 80 % 1-palmitoyl-2-oleoyl-*sn*-glycero-3-phosphocholine (POPC) and

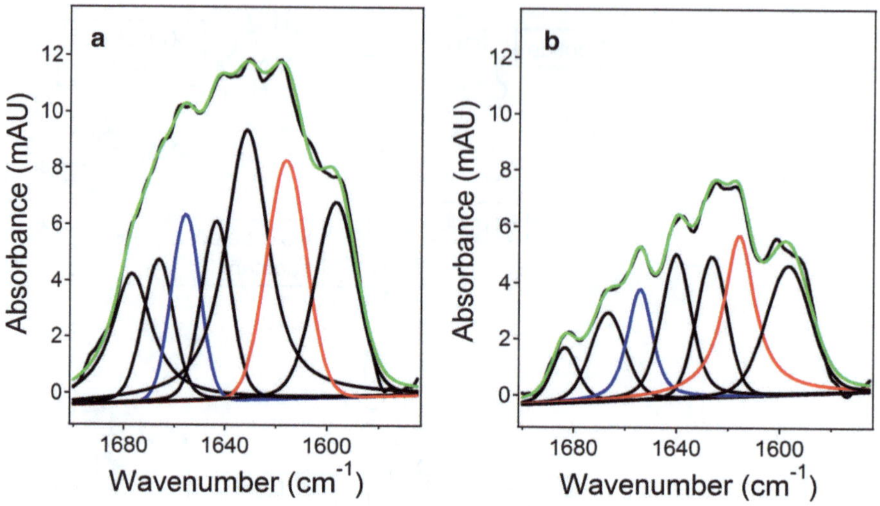

Fig. 7. ATR-FTIR spectra of a semisynthetic human pancreatic PLA_2 in which the N-terminal helix is unlabeled while the rest is uniformly ^{13}C-labeled, at *p* and *s* polarizations (**a** and **b**, respectively). The protein is bound to a POPC/POPG (4:1) bilayer supported on a germanium IRE, in a buffer of 100 mM NaCl, 1 mM NaN_3, 1 mM EGTA, 50 mM Hepes, pD 7.4. The *green line* is the curvefit, that is, the sum of all amide I components shown under the spectrum. The *blue* and *red* components are assigned to unlabeled and ^{13}C-labeled α-helices. Details are described in the main text.

20 % 1-palmitoyl-2-oleoyl-*sn*-glycero-3-phosphoglycerol (POPG). Polarized ATR-FTIR spectra of the membrane-bound protein, presented in Fig. 7, indicate relatively broad and complex amide I contours, resulting from the unlabeled and ^{13}C-labeled segments of the proteins, each of which is undergoing amide HX following distinct kinetics. Amide I components located around 1,656 cm^{-1} and 1,616 cm^{-1} were assigned to the unlabeled (^{12}C) and ^{13}C-labeled α-helices of PLA_2. Intensities of these components at *p* and *s* polarizations of the infrared light were used to determine the orientations of the unlabeled and ^{13}C-labeled α-helices. The interhelical angle between the two internal helices of pancreatic PLA_2 were shown to be ~6 °, using an analytic geometry algorithm (87), indicating that the two internal helices could be considered as one in terms of helical orientation. Determination of orientations of the N-terminal and the internal helices with respect to the membrane by polarized ATR-FTIR, combined with homology modeling and fluorescence quenching experiments, allowed identification of the orientation and the depth of membrane insertion of the membrane-bound protein molecule (PDB entry 1YSK). Similar protein engineering techniques, including protein semisynthesis and segmental ^{13}C-labeling, were used to generate a chimeric PLA_2 where the N-terminal unlabeled helix of human group IIA PLA_2 was ligated to a ^{13}C-labeled pancreatic (group IB) PLA_2 fragment (34). ATR-FTIR studies on the chimeric PLA_2 elucidated the regulatory role of the N-terminal helix in PLA_2 membrane binding mode and activity.

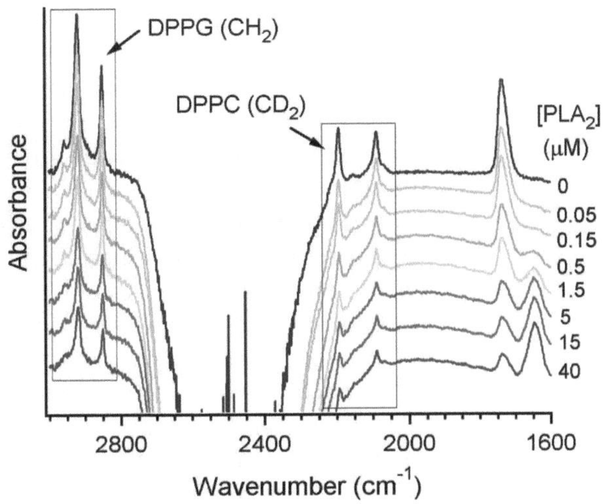

Fig. 8. ATR-FTIR spectra of a lipid bilayer composed of 50 % 1,2-(d_{62})dipalmitoyl-*sn*-glycero-3-phosphocholine (with deuterated acyl chains) and 50 % unlabeled 1,2-dipalmitoyl-*sn*-glycero-3-phosphoglycerol supported on a germanium IRE at various concentration of a snake venom PLA_2 in a buffer containing 2 mM $CaCl_2$. The CH_2 and CD_2 stretching vibrational bands of the unlabeled and deuterated lipids are shown in rectangular frames. Note the strong spectral downshift of methylene stretching vibrations upon deuteration, as well as the increase in protein amide I band (1,700–1,600 cm^{-1}) and the decrease in lipid methylene and carbonyl (~1,735 cm^{-1}) stretching bands with increasing PLA_2 concentration, indicating binding of PLA_2 to the supported membrane and lipid hydrolysis.

3.4.4. FTIR Studies on Isotopically Labeled Lipids

Isotope-edited FTIR spectroscopy has also been applied to study phase transitions, phase separation, hydration, and other properties of phospholipid membranes. Selective ^{13}C-labeling of lipid carbonyl groups at *sn*-1 or *sn*-2 positions identified their individual H-bonding and hydration properties and their contributions to the structure of the membrane–water interface (88–91). Deuteration of the methylene groups of lipid hydrocarbon chains results in ~730 cm^{-1} spectral downshift of the methylene stretching vibration, thus providing FTIR spectral resolution between unlabeled and deuterated lipids or between two chains of the same, selectively deuterated lipid. This effect was used to gain structural information on individual components of the membrane and the effect of membrane structure on membrane-bound enzyme activity (92–97). For example, lipid phase separation was found to promote protein kinase C activity (92). ATR-FTIR experiments on the action of PLA_2 on membranes composed of unlabeled and acyl chain deuterated lipids showed that (a) increased catalytic activity of PLA_2 toward negatively charged lipid membranes results from stronger membrane binding rather than selectivity for anionic lipids and (b) lipid hydrolysis is followed by preferential dissociation of the lyso-lipids and accumulation of free fatty acids in the membrane, which modulates the membrane charge and the morphology and thereby PLA_2 activity (ref. (13) and Fig. 8).

3.5. Conclusions and Perspectives

FTIR spectroscopy is a capable yet relatively easy biophysical technique to gain structural information on proteins and peptides. Studies on membrane proteins reconstituted in detergent micelles or lipid vesicles allow structural characterization of both the protein and the lipid, without any problems related to the molecular size or light scattering like in NMR or UV spectroscopy. While FTIR does not provide the atomic structures of molecules, it is highly sensitive to protein conformational changes, structural flexibility, and solvent accessibility. For example, differences between very similar structures such as α_I and α_{II} helices, which mainly differ by helical H-bonding strength, can be identified easily based on significant differences in vibrational frequencies of the two structures (6, 98). Polarized ATR-FTIR is especially well suited for studies on membrane-bound proteins and provides the protein's orientation with respect to the membrane in addition to protein secondary structure and the kinetics of amide HX. While significant progress has been made in site-directed structural analysis of proteins and peptides by isotope-edited FTIR, still both the theory and the experimental techniques need to be developed to a higher level. Complete understanding and structural interpretation of the spectra of isotope-labeled proteins and peptides requires a stringent theoretical framework of vibrational coupling between various amide modes and its relation to secondary and tertiary structure of the protein. Combination of protein engineering techniques, including peptide ligation, nonsense suppression, and other procedures, with FTIR spectroscopy, is one of the strategies that can break new ground in the whole field of structural biology.

4. Notes

In most studies, the secondary structures of proteins or peptides are estimated by curve-fitting of the amide I band assuming equal molar absorptivities (extinction coefficients) for all secondary structure types. While this may be an acceptable approximation, more accurate data may be obtained when the curve-fitting is conducted using the individual amide I extinction coefficients of various secondary structures. No consensus has been reached thus far regarding the quantitative determination of their values, however. Table 3 presents the values of extinction coefficients determined by Venyaminov and Kalnin (99), which are widely accepted and considered reliable. Jackson et al. (100) evaluated the following relative molar absorptivities of poly-L-lysine at different conformations: 1.00/1.16/0.88 for α-helix/β-sheet/random structure, respectively. De Jongh et al. (101) reported the relative extinction coefficients for α-helix (1.00), β-sheet (1.44–1.57), β-turns (0.48–0.75), and random structure (0.54–0.58). These estimates were derived from comparison of FTIR amide I bands of 15 proteins, deposited on a germanium plate from an H_2O buffer and dried, with their structures determined by X-ray crystallography or NMR. Vedantham et al. (3) reported the wavenumbers, relative extinction coefficients, and bandwidths at half height for proteins in H_2O solutions. Based on these data, the peak absorbance and integrated extinction coefficients are calculated using the α-helical extinction coefficients reported in ref. (99) (Table 6). The relative extinction coefficients of various secondary structures are sufficient to determine the relative contents of secondary structures in a protein, which is the most useful information derived from FTIR, implying that the absolute values of the extinction coefficients are not needed to gain this information. Still there are significant differences in reported relative absorptivities, such as $\varepsilon_\alpha/\varepsilon_\beta \approx 1.0/1.5$ according to Venyaminov and Kalnin (99) and de Jongh et al. (101) vs. $\varepsilon_\alpha/\varepsilon_\beta \approx 1.0/0.8$ according to Vedantham et al. (3). More systematic studies are needed to resolve these differences.

Amide I frequencies of various secondary structures can vary depending on numerous factors, including the polypeptide chain length, environment, rigidity, etc. For example, the amide I wavenumbers of poly-L-lysine in α-helical and β-sheet, and random conformations in D_2O were 1,638 cm^{-1}, 1,610 cm^{-1} plus a weaker component at 1,680 cm^{-1}, and 1,644 cm^{-1} (100). It is possible that homopolypeptides generate lower amide I vibrational frequencies than proteins (i.e., heteropolypeptides) because of more regular secondary structures, corresponding to stronger H-bonding and hence weaker C=O bonds. Also, there may be significant overlap between different structures in the amide I region, as seen from Table 2.

Acknowledgments

Experimental data presented in this article have been obtained and published elsewhere by Kathleen N. Nemec, Shan Qin, Abhay H. Pande, and Pranav Garg.

References

1. Krimm S, Bandekar J (1986) Vibrational spectroscopy and conformation of peptides, polypeptides and proteins. Adv Protein Chem 38:181–364
2. Brauner JW, Dugan C, Mendelsohn R (2000) ^{13}C isotope labeling of hydrophobic peptides. Origin of the anomalous intensity distribution in the infrared amide I spectral region of β-sheet structures. J Am Chem Soc 122:677–683
3. Vedantham G, Sparks HG, Sane SU, Tzannis S, Przybycien TM (2000) A holistic approach for protein secondary structure estimation from infrared spectra in H_2O solutions. Anal Biochem 285:33–49

4. Hsu SL, Moore WH, Krimm S (1976) Vibrational spectrum of the unordered polypeptide chain: a Raman study of feather keratin. Biopolymers 15:1513–1528
5. Naik VM (1992) Vibrational spectroscopic studies of L, D-alternating valine peptides. Vib Spectrosc 3:105–113
6. Barnett SM, Edwards CM, Butler IS, Levin IW (1997) Pressure-induced transmembrane α_{II}- to α_I-helical conversion in bacteriorhodopsin: an infrared spectroscopic study. J Phys Chem B 101:9421–9424
7. Tamm LK, Tatulian SA (1997) Infrared spectroscopy of proteins and peptides in lipid bilayers. Q Rev Biophys 30:365–429
8. Fringeli UP (1993) *In situ* infrared attenuated total reflection membrane spectroscopy. In: Mirabella FM Jr (ed) Internal reflection spectroscopy. Theory and applications. Marcel Dekker, New York, pp 255–324
9. Fringeli UP, Günthard HH (1981) Infrared membrane spectroscopy. In: Grell E (ed) Membrane spectroscopy. Springer, Berlin, pp 270–332
10. Venyaminov SY, Prendergast FG (1997) Water (H_2O and D_2O) molar absorptivity in the 1000–4000 cm^{-1} range and quantitative infrared spectroscopy of aqueous solutions. Anal Biochem 248:234–245
11. Arrondo JLR, Goñi FM (1993) Infrared spectroscopic studies of lipid-protein interactions in membranes. In: Watts A (ed) Protein-lipid interactions. Elsevier Science Publishers B.V, Amsterdam, pp 321–349
12. Tatulian SA (2003) Attenuated total reflection Fourier transform infrared spectroscopy: a method of choice for studying membrane proteins and lipids. Biochemistry 42:11898–11907
13. Tatulian SA (2001) Toward understanding interfacial activation of secretory phospholipase A_2 (PLA_2): Membrane surface properties and membrane-induced structural changes in the enzyme contribute synergistically to PLA_2 activation. Biophys J 80:789–800
14. Tatulian SA, Cortes DM, Perozo E (1998) Structural dynamics of the *Streptomyces lividans* K^+ channel (SKC1): secondary structure characterization from FTIR spectroscopy. FEBS Lett 423:205–212
15. Tatulian SA, Tamm LK (2000) Secondary structure, orientation, oligomerization, and lipid interactions of the transmembrane domain of influenza hemagglutinin. Biochemistry 39:496–507
16. Tatulian SA (2003) Structural effects of covalent inhibition of phospholipase A_2 suggest allosteric coupling between membrane binding and catalytic sites. Biophys J 84:1773–1783
17. Tatulian SA, Biltonen RL, Tamm LK (1997) Structural changes in a secretory phospholipase A_2 induced by membrane binding: a clue to interfacial activation? J Mol Biol 268:809–815
18. Pande AH, Moe D, Nemec KN, Qin S, Tan S, Tatulian SA (2004) Modulation of human 5-lipoxygenase activity by membrane lipids. Biochemistry 43:14653–14666
19. Tatulian SA, Qin S, Pande AH, He X (2005) Positioning membrane proteins by novel protein engineering and biophysical approaches. J Mol Biol 351:939–947
20. Glasoe PK, Long FA (1960) Use of glass electrodes to measure acidities in deuterium oxide. J Phys Chem 64:188–190
21. Makhatadze GI, Clore GM, Gronenborn AM (1995) Solvent isotope effect and protein stability. Nat Struct Biol 2:852–855
22. Nemec KN, Pande AH, Qin S, Bieber Urbauer RJ, Tan S, Moe D, Tatulian SA (2006) Structural and functional effects of tryptophans inserted into the membrane-binding and substrate-binding sites of human group IIA phospholipase A_2. Biochemistry 45:12448–12460
23. Marley J, Lu M, Bracken C (2001) A method for efficient isotope labeling of recombinant proteins. J Biomol NMR 20:71–75
24. Hill JM (2008) NMR screening for rapid protein characterization in structural proteomics. Methods Mol Biol 426:437–446
25. Kandori H, Nakamura H, Yamazaki Y, Mogi T (2005) Redox-induced protein structural changes in cytochrome *bo* revealed by Fourier transform infrared spectroscopy and [^{13}C]Tyr Labeling. J Biol Chem 280:32821–32826
26. Cheong JJ, Hwang I, Rhee S, Moon TW, Choi YD, Kwon HB (2007) Complementation of an *E. coli* cysteine auxotrophic mutant for the structural modification study of 3'(2'),5'-bisphosphate nucleotidase. Biotechnol Lett 29:913–918
27. Strømgaard A, Jensen AA, Strømgaard K (2004) Site-specific incorporation of unnatural amino acids into proteins. Chembiochem 5:909–916
28. Hendrickson TL, de Crécy-Lagard V, Schimmel P (2004) Incorporation of nonnatural amino acids into proteins. Annu Rev Biochem 73:147–176
29. Xie J, Schultz PG (2005) Adding amino acids to the genetic repertoire. Curr Opin Chem Biol 9:548–554

30. Seyedsayamdost MR, Stubbe J (2009) Replacement of Y_{730} and Y_{731} in the $\alpha 2$ subunit of *Escherichia coli* ribonucleotide reductase with 3-aminotyrosine using an evolved suppressor tRNA/tRNA-synthetase pair. Methods Enzymol 462:45–76
31. Liu CC, Schultz PG (2010) Adding new chemistries to the genetic code. Annu Rev Biochem 79:413–444
32. Anderson RD 3rd, Zhou J, Hecht SM (2002) Fluorescence resonance energy transfer between unnatural amino acids in a structurally modified dihydrofolate reductase. J Am Chem Soc 124:9674–9675
33. Sisido M, Ninomiya K, Ohtsuki T, Hohsaka T (2005) Four-base codon/anticodon strategy and non-enzymatic aminoacylation for protein engineering with non-natural amino acids. Methods 36:270–278
34. Qin S, Pande AH, Nemec KN, He X, Tatulian SA (2005) Evidence for the regulatory role of the N-terminal helix of secretory phospholipase A_2 from studies on native and chimeric proteins. J Biol Chem 280:36773–36783
35. Venyaminov SY, Hedstrom JF, Prendergast FG (2001) Analysis of the segmental stability of helical peptides by isotope-edited infrared spectroscopy. Proteins 45:81–89
36. Huang R, Kubelka J, Barber-Armstrong W, Silva RA, Decatur SM, Keiderling TA (2004) Nature of vibrational coupling in helical peptides: an isotopic labeling study. J Am Chem Soc 126:2346–2354
37. Petty SA, Decatur SM (2005) Intersheet rearrangement of polypeptides during nucleation of β-sheet aggregates. Proc Natl Acad Sci U S A 102:14272–14277
38. Ramajo AP, Petty SA, Starzyk A, Decatur SM, Volk M (2005) The α-helix folds more rapidly at the C-terminus than at the N-terminus. J Am Chem Soc 127:13784–13785
39. Lewis RN, McElhaney RN (2009) The physicochemical properties of cardiolipin bilayers and cardiolipin-containing lipid membranes. Biochim Biophys Acta 1788:2069–2079
40. Qin S, Pande AH, Nemec KN, Tatulian SA (2004) The N-terminal α-helix of pancreatic phospholipase A_2 determines productive-mode orientation of the enzyme at the membrane surface. J Mol Biol 344:71–89
41. Pande AH, Qin S, Tatulian SA (2005) Membrane fluidity is a key modulator of membrane binding, insertion, and activity of 5-lipoxygenase. Biophys J 88:4084–4094
42. Renthal R (2006) An unfolding story of helical transmembrane proteins. Biochemistry 45:14559–14566
43. Sanders CR, Sönnichsen F (2006) Solution NMR of membrane proteins: practice and challenges. Magn Reson Chem 44:S24–S40
44. Vinogradova O, Sönnichsen F, Sanders CR 2nd (1998) On choosing a detergent for solution NMR studies of membrane proteins. J Biomol NMR 11:381–386
45. Tatulian SA, Hinterdorfer P, Baber G, Tamm LK (1995) Influenza hemagglutinin assumes a tilted conformation during membrane fusion as determined by attenuated total reflection FTIR spectroscopy. EMBO J 14:5514–5523
46. Tatulian SA, Chen B, Li J, Negash S, Middaugh CR, Bigelow DJ, Squier TC (2002) The inhibitory action of phospholamban involves stabilization of α-helices within the Ca^{2+}-ATPase. Biochemistry 41:741–751
47. Axelsen PH, Citra MJ (1996) Orientational order determination by internal reflection infrared spectroscopy. Prog Biophys Mol Biol 66:227–253
48. Goormaghtigh E, Raussens V, Ruysschaert JM (1999) Attenuated total reflection infrared spectroscopy of proteins and lipids in biological membranes. Biochim Biophys Acta 1422:105–185
49. Silvestro L, Axelsen PH (2000) Membrane-induced folding of cecropin A. Biophys J 79:1465–1477
50. Citra MJ, Axelsen PH (1996) Determination of molecular order in supported lipid membranes by internal reflection Fourier transform infrared spectroscopy. Biophys J 71:1796–1805
51. Silvestro L, Axelsen PH (1999) Fourier transform infrared linked analysis of conformational changes in annexin V upon membrane binding. Biochemistry 38:113–121
52. Vigano C, Goormaghtigh E, Ruysschaert JM (2003) Detection of structural and functional asymmetries in P-glycoprotein by combining mutagenesis and H/D exchange measurements. Chem Phys Lipids 122:121–135
53. Beevers AJ, Kukol A (2006) The transmembrane domain of the oncogenic mutant ErbB-2 receptor: a structure obtained from site-specific infrared dichroism and molecular dynamics. J Mol Biol 361:945–953
54. Marsh D, Müller M, Schmitt F-J (2000) Orientation of the infrared transition moments for an α-helix. Biophys J 78:2499–2510
55. Marsh D, Páli T (2001) Infrared dichroism from the X-ray structure of bacteriorhodopsin. Biophys J 80:305–312

56. Páli T, Marsh D (2001) Tilt, twist, and coiling in β-barrel membrane proteins: relation to infrared dichroism. Biophys J 80:2789–2797
57. Harrick NJ (1987) Internal reflection spectroscopy. Harrick Scientific Corporation, Ossining, NY
58. Marsh D (2000) Infrared dichroism of twisted β-sheet barrels. The structure of *E. coli* outer membrane proteins. J Mol Biol 297:803–808
59. Marsh D (1999) Quantitation of secondary structure in ATR infrared spectroscopy. Biophys J 77:2630–2637
60. Arkin IT (2006) Isotope-edited IR spectroscopy for the study of membrane proteins. Curr Opin Chem Biol 10:394–401
61. Decatur SM (2006) Elucidation of residue-level structure and dynamics of polypeptides via isotope-edited infrared spectroscopy. Acc Chem Res 39:169–175
62. Flach CR, Cai P, Dieudonné D, Brauner JW, Keough KM, Stewart J, Mendelsohn R (2003) Location of structural transitions in an isotopically labeled lung surfactant SP-B peptide by IRRAS. Biophys J 85:340–349
63. Paul C, Wang J, Wimley WC, Hochstrasser RM, Axelsen PH (2004) Vibrational coupling, isotopic editing, and β-sheet structure in a membrane-bound polypeptide. J Am Chem Soc 126:5843–5850
64. Tatulian SA (2010) Structural analysis of proteins by isotope-edited FTIR spectroscopy. Spectroscopy Int J 24:37–43
65. Barber-Armstrong W, Donaldson T, Wijesooriya H, Silva RA, Decatur SM (2004) Empirical relationships between isotope-edited IR spectra and helix geometry in model peptides. J Am Chem Soc 126:2339–2345
66. Ludlam CF, Arkin IT, Liu XM, Rothman MS, Rath P, Aimoto S, Smith SO, Engelman DM, Rothschild KJ (1996) Fourier transform infrared spectroscopy and site-directed isotope labeling as a probe of local secondary structure in the transmembrane domain of phospholamban. Biophys J 70:1728–1736
67. Arkin IT, MacKenzie KR, Brünger AT (1997) Site-directed dichroism as a method for obtaining rotational and orientational constraints for oriented polymers. J Am Chem Soc 119:8973–8980
68. Marsh D (2004) Infrared dichroism of isotope-edited α-helices and β-sheets. J Mol Biol 338:353–367
69. Torres J, Adams PD, Arkin IT (2000) Use of a new label, $^{13}C{=}^{18}O$, in the determination of a structural model of phospholamban in a lipid bilayer. Spatial restraints resolve the ambiguity arising from interpretations of mutagenesis data. J Mol Biol 300:677–685
70. Torres J, Kukol A, Goodman JM, Arkin IT (2001) Site-specific examination of secondary structure and orientation determination in membrane proteins: the peptidic $^{13}C{=}^{18}O$ group as a novel infrared probe. Biopolymers 59:396–401
71. Kukol A, Torres J, Arkin IT (2002) A structure for the trimeric MHC class II-associated invariant chain transmembrane domain. J Mol Biol 320:1109–1117
72. Brewer SH, Song B, Raleigh DP, Dyer RB (2007) Residue specific resolution of protein folding dynamics using isotope-edited infrared temperature jump spectroscopy. Biochemistry 46:3279–3285
73. Amunson KE, Ackels L, Kubelka J (2008) Site-specific unfolding thermodynamics of a helix-turn-helix protein. J Am Chem Soc 130:8146–8147
74. Londergan CH, Wang J, Axelsen PH, Hochstrasser RM (2006) Two-dimensional infrared spectroscopy displays signatures of structural ordering in peptide aggregates. Biophys J 90:4672–4685
75. Kim YS, Liu L, Axelsen PH, Hochstrasser RM (2008) Two-dimensional infrared spectra of isotopically diluted amyloid fibrils from Abeta40. Proc Natl Acad Sci U S A 105:7720–7725
76. Kim YS, Liu L, Axelsen PH, Hochstrasser RM (2009) 2D IR provides evidence for mobile water molecules in beta-amyloid fibrils. Proc Natl Acad Sci U S A 106:17751–17756
77. Shim SH, Gupta R, Ling YL, Strasfeld DB, Raleigh DP, Zanni MT (2009) Two-dimensional IR spectroscopy and isotope labeling defines the pathway of amyloid formation with residue-specific resolution. Proc Natl Acad Sci U S A 106:6614–6619
78. Zanni MT, Hochstrasser RM (2001) Two-dimensional infrared spectroscopy: a promising new method for the time resolution of structures. Curr Opin Struct Biol 11:516–522
79. Haris PI, Robillard GT, van Dijk AA, Chapman D (1992) Potential of ^{13}C and ^{15}N labeling for studying protein-protein interactions using Fourier transform infrared spectroscopy. Biochemistry 31:6279–6284
80. Zhang M, Fabian H, Mantsch HH, Vogel HJ (1994) Isotope-edited Fourier transform infrared spectroscopy studies of calmodulin's interaction with its target peptides. Biochemistry 33:10883–10888

81. Das KP, Choo-Smith LP, Petrash JM, Surewicz WK (1999) Insight into the secondary structure of non-native proteins bound to a molecular chaperone α-crystallin. An isotope-edited infrared spectroscopic study. J Biol Chem 274:33209–33212
82. Haris PI (2010) Can infrared spectroscopy provide information on protein-protein interactions? Biochem Soc Trans 38:940–946
83. Tam JP, Yu Q, Miao Z (1999) Orthogonal ligation strategies for peptide and protein. Biopolymers 51:311–332
84. Camarero JA, Muir TW (2001) Native chemical ligation of polypeptides. Curr Protoc Protein Sci Ch. 18:Unit18.4
85. Muralidharan V, Muir TW (2006) Protein ligation: an enabling technology for the biophysical analysis of proteins. Nat Methods 3:429–438
86. Muir TW (2008) Studying protein structure and function using semisynthesis. Biopolymers 90:743–750
87. Tatulian SA (2008) Determination of helix orientations in proteins. Comput Biol Chem 32:370–374
88. Blume A, Hübner W, Messner G (1988) Fourier transform infrared spectroscopy of ^{13}C=O-labeled phospholipids hydrogen bonding to carbonyl groups. Biochemistry 27:8239–8249
89. Hübner W, Mantsch HH, Paltauf F, Hauser H (1994) Conformation of phosphatidylserine in bilayers as studied by Fourier transform infrared spectroscopy. Biochemistry 33:320–326
90. Lewis RN, McElhaney RN (1993) Studies of mixed-chain diacyl phosphatidylcholines with highly asymmetric acyl chains: a Fourier transform infrared spectroscopic study of interfacial hydration and hydrocarbon chain packing in the mixed interdigitated gel phase. Biophys J 65:1866–1877
91. Lewis RN, McElhaney RN, Pohle W, Mantsch HH (1994) Components of the carbonyl stretching band in the infrared spectra of hydrated 1,2-diacylglycerolipid bilayers: a reevaluation. Biophys J 67:2367–2375
92. Dibble AR, Hinderliter AK, Sando JJ, Biltonen RL (1996) Lipid lateral heterogeneity in phosphatidylcholine/phosphatidylserine/diacylglycerol vesicles and its influence on protein kinase C activation. Biophys J 71:1877–1890
93. Moore DJ, Gioioso S, Sills RH, Mendelsohn R (1999) Some relationships between membrane phospholipid domains, conformational order, and cell shape in intact human erythrocytes. Biochim Biophys Acta 1415:342–348
94. Binder H, Gawrisch K (2001) Dehydration induces lateral expansion of polyunsaturated 18:0–22:6 phosphatidylcholine in a new lamellar phase. Biophys J 81:969–982
95. Mimeault M, Bonenfant D (2002) FTIR spectroscopic analyses of the temperature and pH influences on stratum corneum lipid phase behaviors and interactions. Talanta 56:395–405
96. Fidorra M, Heimburg T, Seeger HM (2009) Melting of individual lipid components in binary lipid mixtures studied by FTIR spectroscopy, DSC and Monte Carlo simulations. Biochim Biophys Acta 1788:600–607
97. Gorcea M, Hadgraft J, Moore DJ, Lane ME (2011) Fourier transform infrared spectroscopy studies of lipid domain formation in normal and ceramide deficient stratum corneum lipid models. Int J Pharm. doi:10.1016/j.ijpharm.2011.11.004
98. Dwivedi AM, Krimm S (1984) Vibrational analysis of peptides, polypeptides, and proteins. XVIII. Conformational sensitivity of the alpha-helix spectrum: alpha I- and alpha II-poly(L-alanine). Biopolymers 23:923–943
99. Venyaminov SY, Kalnin NN (1990) Quantitative IR spectrophotometry of peptide compounds in water (H_2O) solutions. II. Amide absorption bands of polypeptides and fibrous proteins in α-, β-, and random coil conformations. Biopolymers 30:1259–1271
100. Jackson M, Haris PI, Chapman D (1989) Conformational transitions in poly(L-lysine): studies using Fourier transform infrared spectroscopy. Biochim Biophys Acta 998:75–79
101. de Jongh HH, Goormaghtigh E, Ruysschaert JM (1996) The different molar absorptivities of the secondary structure types in the amide I region: an attenuated total reflection infrared study on globular proteins. Anal Biochem 242:95–103
102. Stuart B (1997) Biological applications of infrared spectroscopy. Wiley, Chichester
103. Némethy G, Phillips DC, Leach SJ, Scheraga HA (1967) A second right-handed helical structure with the parameters of the Pauling-Corey α-helix. Nature 214:363–365
104. Heimburg T, Schuenemann J, Weber K, Geisler N (1996) Specific recognition of coiled coils by infrared spectroscopy: analysis of the three structural domains of type III intermediate filament proteins. Biochemistry 35:1375–1382
105. Reisdorf WC, Krimm S (1996) Infrared amide I' band of the coiled coil. Biochemistry 35:1383–1386

106. Kennedy DF, Crisma M, Toniolo C, Chapman D (1991) Studies of peptides forming 3_{10}- and α-helices and beta-bend ribbon structures in organic solution and in model biomembranes by Fourier transform infrared spectroscopy. Biochemistry 30:6541–6548
107. Martinez G, Millhauser G (1995) FTIR spectroscopy of alanine-based peptides: assignment of the amide I' modes for random coil and helix. J Struct Biol 114:23–27
108. Miick SM, Martinez GV, Fiori WR, Todd AP, Millhauser GL (1992) Short alanine-based peptides may form 3_{10}-helices and not α-helices in aqueous solution. Nature 359:653–655
109. Naik VM, Krimm S (1986) Vibrational analysis of the structure of gramicidin A. I. Normal mode analysis. Biophys J 49:1131–1145
110. Naik VM, Krimm S (1986) Vibrational analysis of the structure of gramicidin A. II. Vibrational spectra. Biophys J 49:1147–1154
111. Jackson M, Mantsch HH (1995) The use and misuse of FTIR spectroscopy in the determination of protein structure. Crit Rev Biochem Mol Biol 30:95–120
112. Papanikolopoulou K, Mills-Henry I, Thol SL, Wang Y, Gross AA, Kirschner DA, Decatur SM, King J (2008) Formation of amyloid fibrils in vitro by human γD-crystallin and its isolated domains. Mol Vis 14:81–89
113. Itkin A, Dupres V, Dufrêne YF, Bechinger B, Ruysschaert JM, Raussens V (2011) Calcium ions promote formation of amyloid β-peptide (1–40) oligomers causally implicated in neuronal toxicity of Alzheimer's disease. PLoS One 6:e18250
114. Karjalainen EL, Ravi HK, Barth A (2011) Simulation of the amide I absorption of stacked β-sheets. J Phys Chem B 115:749–757
115. Pauling L, Corey RB (1951) The pleated sheet, a new layer configuration of polypeptide chains. Proc Natl Acad Sci U S A 37:251–256
116. Armen RS, DeMarco ML, Alonso DO, Daggett V (2004) Pauling and Corey's α-pleated sheet structure may define the prefibrillar amyloidogenic intermediate in amyloid disease. Proc Natl Acad Sci U S A 101:11622–11627
117. Daggett V (2006) Alpha-sheet: the toxic conformer in amyloid diseases? Acc Chem Res 39:594–602
118. Wu H, Canfield A, Adhikari J, Huo S (2010) Quantum mechanical studies on model α-pleated sheets. J Comput Chem 31:1216–1223
119. Venyaminov SY, Kalnin NN (1990) Quantitative IR spectrophotometry of peptide compounds in water (H_2O) solutions. I. Spectral parameters of amino acid residue absorption bands. Biopolymers 30:1243–1257
120. Chirgadze YN, Fedorov OV, Trushina NP (1975) Estimation of amino acid residue side-chain absorption in the infrared spectra of protein solutions in heavy water. Biopolymers 14:679–694

Chapter 10

Förster Resonance Energy Transfer as a Tool for Quantification of Protein–Lipid Selectivity

Luís M.S. Loura, Manuel Prieto, and Fábio Fernandes

Abstract

This chapter addresses the determination of protein–lipid selectivity, here described as the preference of a protein for having a specific type of lipid in its vicinity (annular lipids), from Förster resonance energy transfer methodologies. These allow a quantification of the effect, i.e., the determination of the biasing in distribution of the lipid under study around the protein, as compared to its bulk membrane distribution, with advantages over established approaches that have been used for the same purpose, such as electron spin resonance spectroscopy. The experiment can be carried out with steady-state instrumentation, the formalisms are described in detail, and the model can be applied to a membrane protein of any size.

Key words: Fluorescence, FRET, Lipid/protein interaction, Membrane protein, Protein/lipid selectivity

1. Introduction

1.1. Protein–Lipid Selectivity

Membrane proteins are strongly influenced by the lipidic composition of the protein–lipid interface. Lipids with different charge, hydrophobic thickness, and inherent curvature can drastically promote or reduce protein activity (1, 2). Proteins exhibit preferential interactions with selected lipids, generating a driving force for the creation of lateral lipid heterogeneities in the membrane and these can potentially extend to several lipid shells around the protein. Several peripheral proteins reversibly interact with the membrane after binding to specific lipids, and temporal as well as spatial regulation of these events has been shown to be crucial for several biological functions, including signaling cascades (3).

On the other hand, hydrophobic matching constraints at the transmembrane protein–lipid interface induce enrichment of specific lipids around the protein. Enrichment occurs for lipids that upon interaction with the protein effectively prevent the exposure of hydrophobic protein residues and of the lipid hydrophobic

Jörg H. Kleinschmidt (ed.), *Lipid-Protein Interactions: Methods and Protocols*, Methods in Molecular Biology, vol. 974, DOI 10.1007/978-1-62703-275-9_10,

groups to the polar aqueous environment. The lipids in direct contact with the transmembrane protein surface embedded in the bilayer are known as annular lipids and typically display fast on–off rates for protein interaction. In some cases, proteins display hydrophobic pockets that correspond to binding sites with higher affinity for specific lipids. Binding of lipids to these sites is often essential for protein activity (4). Protein–lipid selectivity problems are commonly addressed by electron spin resonance (ESR) and fluorescence quenching methods. The former can discriminate between immobilized lipids in contact with the protein and mobile lipids away from the protein interface (5), while the latter relies on the measurement of fluorescence static or collisional quenching induced by contact between the protein and the lipid for a fluorophore in one of the molecules (6, 7). These techniques detect direct contact between protein and lipids but are insensitive to the creation of lateral lipid heterogeneities away from the protein–lipid interface. On the other hand, Förster resonance energy transfer (FRET) is sensitive to distances from 1 to 10 nm and is able to detect both direct binding and deviations from homogeneity of lipid distribution around the protein.

Additionally, FRET sensitivity, like all fluorescence methodologies, is unmatched by other spectroscopic techniques, and lower amounts of membrane proteins are necessary, and there is no restriction on the temperature range to carry out the experiment. Since membrane protein expression and/or purification can be challenging in some cases (particularly for integral proteins), the possibility of conducting experiments with low concentrations of materials is highly useful. Also, a large variety of fluorescent-labeled lipids with different structural features is commercially available allowing for great flexibility in the design of experiments. For these reasons, FRET is an excellent tool for protein–lipid selectivity studies. Here, we present an approach to the quantification of protein–lipid selectivity through FRET that has been successfully applied to different systems.

1.2. Characterization of Protein/Lipid Selectivity by FRET

FRET is a photophysical process by which an electronically excited fluorophore, the donor (D), is quenched by a nearby chromophore, termed acceptor (A), which in turn becomes excited. FRET does not involve molecular contact between the intervening species, but depends acutely on the D–A separation distance, ***d***. Its rate constant for a single D–A pair, k_T, is given by (8)

$$k_T = \frac{1}{\tau_0}\left(\frac{R_0}{d}\right)^6, \qquad (1)$$

where τ_0 is the fluorescence lifetime of D in absence of A, and R_0, the so-called Förster radius, is defined as the D–A distance for which k_T is half of the decay rate constant of D in absence of A (given by τ_0^{-1}).

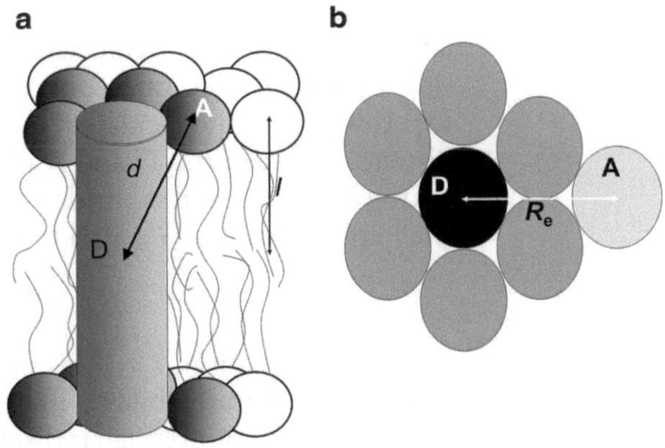

Fig. 1. Molecular model for the FRET analysis according to the model of Fernandes et al. (2004): (**a**) *side view* and (**b**) *top view*. In the model, protein–lipid organization presents a hexagonal geometry. The T36C mutant of M13 MCP was labeled with the FRET donor (coumarin) so that the fluorophore locates in the center of the bilayer, whereas the acceptors are distributed in the bilayer surface. Two different environments are available for the labeled lipids (acceptors): the annular shell surrounding the protein and the bulk lipid. Reprinted with permission from (10). Copyright 2004 Biophysical Society.

R_0 (typically in the 1–6 nm range) is a very important parameter as it measures the range of effective FRET interaction for a given pair. It can be calculated from spectral data, such as D's fluorescence quantum yield (Φ_D) and normalized emission spectrum ($I(\lambda)$) and A's molar absorption spectrum ($\varepsilon(\lambda)$), using the following equation (which assumes nm units for both λ and R_0):

$$R_0 = 0.02108\left[\kappa^2\Phi_D n^{-4}\int_0^\infty I(\lambda)\varepsilon(\lambda)\lambda^4 d\lambda\right]^{1/6}. \quad (2)$$

The other parameters needed for this purpose are the orientation factor κ^2 (see (9) for a detailed discussion) and the refractive index n. Experimentally, FRET is usually quantified by the FRET efficiency, E:

$$E = 1 - \tau/\tau_0 = 1 - I_{DA}/I_{D'} \quad (3)$$

where τ is the lifetime of D in the presence of A, and I_{DA}/I_D is the ratio between the steady-state fluorescence intensity of D in the presence (DA) and absence (D) of A.

Let us consider a membrane protein with approximate cylindrical symmetry, containing a D chromophore (Trp residue or extrinsic covalently attached label) in an axial position, at a given transverse distance. The protein is inserted in a bilayer composed of a single-component host lipid matrix and a small amount of an A lipid probe, surrounded by an annular layer of N closest lipid neighbors (Fig. 1a). N is ~12 (6 in each leaflet) for a protein with a single transmembrane (TM) segment (Fig. 1b) and obviously larger for proteins with multiple TM segments. Each annular site may be occupied by a host lipid or an A probe molecule. In the latter case, this A molecule acts as potential quencher of D's emission by FRET. The rate constant for this interaction is given by Eq. 1, with d easily calculated, once estimates for average transverse distance between the D and A planes (l) and the lateral separation between both D and A inside the transmembrane protein-annular shell lipids complex (R) are known:

$$d_i = \left(l_i^2 + R^2\right)^{1/2}. \quad (4)$$

In this equation, $i = 1{,}2$ denotes the possibility that D might not be located in the bilayer midplane, implying that there may be two distinct D–A interplanar distances l_i (and hence two distinct d_i values, and two corresponding k_{Ti} rate constants): one for A lipid probe molecules located on the top bilayer leaflet and another for those on the bottom leaflet. To compute the FRET contribution of annular A molecules to the decay of D, it is assumed that multiple acceptors (or none at all) may occupy the N available sites ($N/2$ in each leaflet), following a binomial distribution. This is done using the following equation (10):

$$\rho_{annular}(t) = \rho_{annular,1}(t)\rho_{annular,2}(t), \quad (5)$$

where

$$\rho_{annular,i} = \sum_{n=0}^{N/2} e^{-nk_{Ti}t} \binom{N/2}{n} \mu^n (1-\mu)^{N/2-n}. \quad (6)$$

μ represents the probability of each of the 12 annular sites to be occupied by an A molecule. It depends on the acceptor molar fraction and on a relative selectivity constant (K_S) which quantifies the relative affinity of the labeled and unlabeled phospholipids (whose concentrations are denoted below by [A] and [L], respectively):

$$\mu = K_S \frac{[A]}{[A] + [L]}. \quad (7)$$

Obviously, not all A molecules are located in the annular region. Some of them will be scattered in the bilayer, assumedly in a uniform distribution. The contribution of these A molecules can be calculated using the analytical solution for FRET in an infinite planar geometry (11, 12):

$$\rho_{ramdom}(t) = \exp\left(-\frac{t}{\tau_0}\right) \prod_{i=1}^{2} \exp\left\{-\pi R_0^2 n_2 \gamma \left[\frac{2}{3}, \left(\frac{R_0}{R_{ei}}\right)^6 \left(\frac{t}{\tau_0}\right)\right] \left(\frac{t}{\tau_0}\right)^{1/3}\right\}.$$
$$\exp\left\{\pi R_{ei}^2 n_2 \left(1 - \exp\left[-\left(\frac{R_0}{R_{ei}}\right)^6 \left(\frac{t}{\tau_0}\right)\right]\right)\right\}. \quad (8)$$

In this equation, γ is the incomplete gamma function, R_{ei} is the minimum distance (exclusion distance, Fig. 1b) between D and A molecules in leaflet i, and n_2 is the numerical concentration of A (molecules of each bilayer leaflet/unit area = total A molecules/unit area). The value n_2 must be corrected for the presence of labeled lipid in the annular region, which therefore is not part of the randomly distributed acceptors pool.

The overall D decay in presence of A ($i_{DA}(t)$) is now simply computed by multiplying that in absence of A ($i_D(t)$) by the annular and non-annular FRET contributions:

$$i_{DA}(t) = i_D(t)\rho_{annular}(t)\rho_{ramdom}(t). \tag{9}$$

From $i_{DA}(t)$ and $i_D(t)$, E can be computed according to

$$E = 1 - \int_0^\infty i_{DA}(t)dt \Big/ \int_0^\infty i_D(t)dt. \tag{10}$$

This allows comparison of experimental E values (Eq. 3) with theoretical expectations (Eq. 10), as well as retrieval of the model-fitting parameter values (K_S or μ), as illustrated below for the problem at hand.

2. Materials

1. Buffer I: 50 mM sodium cholate, 150 mM NaCl, 10 mM Tris–HCl, 1 mM EDTA, pH 8.
2. Buffer II: 150 mM NaCl, 10 mM Tris–HCl, 1 mM EDTA, pH 8.
3. Buffer III: 20 mM HEPES, 150 mM NaCl buffer, 0.008 % *N*-dodecyl-β-D-maltoside (DDM), pH 7.4.
4. Purified proteins containing D fluorophore: will vary according to the particular system under study. In the examples described in more detail below, these are the T36C mutant of the M13 major coat (purified from M13 bacteriophage and labeled with DCIA fluorophore (7-diethylamino-3((4′iodoacetyl)amino) phenyl-4-methylcoumarin), hereby named DCIA-M13 mcp (10, 13), and lactose permease (LacY) from *Escherichia coli*, which possesses a single tryptophan residue (Trp151) (14). DCIA-M13 mcp was stored in buffer I, while LacY was kept in buffer III.
5. Nonfluorescent lipids: will vary according to the particular system under study. In the examples described in more detail below, these are 1,2-dioleoyl-*sn*-glycero-3-phosphocholine (DOPC), 1,2-dierucoyl-*sn*-glycero-3-phosphocholine (DEuPC), and

1,2-dimyristoleoyl-*sn*-glycero-3-phosphocholine (DMoPC) for the DCIA-M13 mcp system and 1-palmitoyl-2-oleoyl-*sn*-glycero-3-phosphoethanolamine (POPE), 1-palmitoyl-2-oleoyl-*sn*-gly cero-3-phosphoglycerol (POPG), 1,2-dioleoyl-*sn*-gly cero-3-phosphoethanolamine (DOPE), 1,2-dipalmitoyl-*sn*-glycero-3-phosphoethanolamine (DPPE), 1′,3′-bis[1,2-dimyristoyl-*sn*-glycero-3-phospho]-*sn*-glycerol (myristoyl-CL), and 1′,3′-bis[1,2-dioleoyl-*sn*-glycero-3-phospho]-*sn*-glycerol (oleoyl-CL) for the LacY system. Prepare stock solutions in chloroform–methanol mixture (3:1, vol./vol.) and store at −20°C.

6. Lipids with linked A fluorophore: will vary according to the particular system under study. In the examples described inmore detail below, these are 1,2-dioleoyl-*sn*-glycero-3-phosphoethanolamine-*N*-(7-nitro-2-1,3-benzoxadiazol-4-yl) (*N*-NBD-DOPE), 1-oleoyl-2-[12-[(7-nitro-2-1,3-benzoxadiazol-4-yl)amino]dodecanoyl]-*sn*-glycero-Phosphocholine (NBD-PC), 1-oleoyl-2-[12-[(7-nitro-2-1,3-benzoxadiazol-4-yl)amino]dodecanoyl]-*sn*-Glycero-Phosphoethanolamine (NBD-PE), 1-oleoyl-2-[12-[(7-nitro-2-1,3-benzoxadiazol-4-yl)amino]dodeca noyl]-*sn*-glycero-phosphoserine (NBD-PS) (sodium salt), 1-oleoyl-2-[12-[(7-nitro-2-1,3-benzoxadiazol-4-yl)amino] dodecanoyl]-*sn*-glycero-phosphate (NBD-PA) (monosodium salt), and 1-oleoyl-2-[12-[(7-nitro-2-1,3-benzoxadiazol-4-yl)amino]dodecanoyl]-*sn*-glycero-3-[phospho-*rac*-(1-glycerol)] (NBD-PG) (sodium salt) for the DCIA-M13 mcp system, and 1-hexadecanoyl-2-(1-pyrenedecanoyl)-*sn*-glycero-3-phosphoglycerol ammonium salt (Pyr-PG) and 1-hexadecanoyl-2-(1-pyrenedecanoyl)-*sn*-glycero-3-phosphoethanolamine ammonium salt (Pyr-PE) for the LacY system. Prepare stock solutions in chloroform–methanol mixture (3:1, vol./vol.) and store at −20 °C.
7. Spectra/Por CE dialysis membrane (Spectrum Laboratories, Rancho Dominguez, CA) or polystyrene Bio-Beads SM-2, from Bio-Rad (Hercules, CA).

3. Methods

1. Mix adequate volumes of nonfluorescent lipids' stock solutions to prepare 1–10 μmol of total lipid in the desired proportion (see Note 1), to prepare samples without A (D samples).
2. Repeat previous step, also adding an adequate volume of fluorescent lipid stock solution (see Note 2), to prepare samples.
3. After thorough mixing of each sample, dry all samples under a gentle stream of nitrogen until complete evaporation.
4. Dry all samples further by leaving them in vacuum for 6 h.

5. Suspend the dried lipid film in buffer. In the described examples, buffer I is used for the DCIA-M13 mcp system, and buffer III for the LacY system. Liposomes may be extruded or sonicated at this stage to produce unilamellar vesicles.
6. Reconstitute the proteins in the lipid vesicles, in an adequate proportion to yield the desired protein/lipid ratio, which should be the same for all samples of a given system. The exact procedure will depend on the system under study (see Note 3).
7. Measure fluorescence of both D and DA samples. Either a time-resolved or a steady-state fluorescence setup may be used, depending on availability (see Note 4).
8. Calculate experimental FRET efficiencies, E, using Eq. 3 either with the steady-state fluorescence intensities I_D and I_{DA} for a steady-state setup or the lifetimes τ_D and τ_{DA} for a time-resolved setup (see Note 5).
9. If unknown, calculate the Förster distance, R_0, for the D/A pair in study, using Eq. 2. For this purpose, the D emission spectrum and the A absorption spectrum are required, as well as the fluorescence quantum yield of D. Integration is straightforwardly carried out in a spreadsheet, using the rectangle, trapezoidal, or Simpson rules (see Note 6).
10. Fit the model described in Subheading 2 to the experimental FRET efficiencies. For optimal fitting, inputs for all structural parameters (number N of annular sites, areas/host lipid molecule, interplanar distances) should be provided, as well as the R_0 value computed in the preceding step, so that only selectivity parameters are optimized (see Note 7). If a single A concentration is used, this simply equates to determining the value of the μ parameter that matches the experimental and theoretical FRET efficiencies, from which K_s may be calculated using Eq. 7. If (preferably) multiple A concentrations are used, then μ varies for each data point, but is determined by the acceptor mole fraction and the value of the selectivity constant, K_s. The latter is the sole fitting parameter in this case. The whole procedure may be carried out in a (large) spreadsheet (see Note 8).

FRET efficiencies obtained between DCIA-M13 mcp and NBD-labeled lipids are presented in Fig. 2 and Table 1. In order to determine selectivity of M13 mcp for particular phospholipid classes, PE, PC, PG, PS, and PA phospholipids labeled with NBD at the acyl chain were incorporated in proteoliposomes composed of DOPC. M13-mcp exhibits preferential interaction with anionic lipids, particularly for PA and PS. The protein presents a higher affinity for the labeled lipid than for the bulk lipid ($K_S > 1$), possibly as a result of electrostatic interactions with the NBD group. Different K_S values were also recovered for *N*-NBD-DOPE depending on the thickness of the proteoliposome

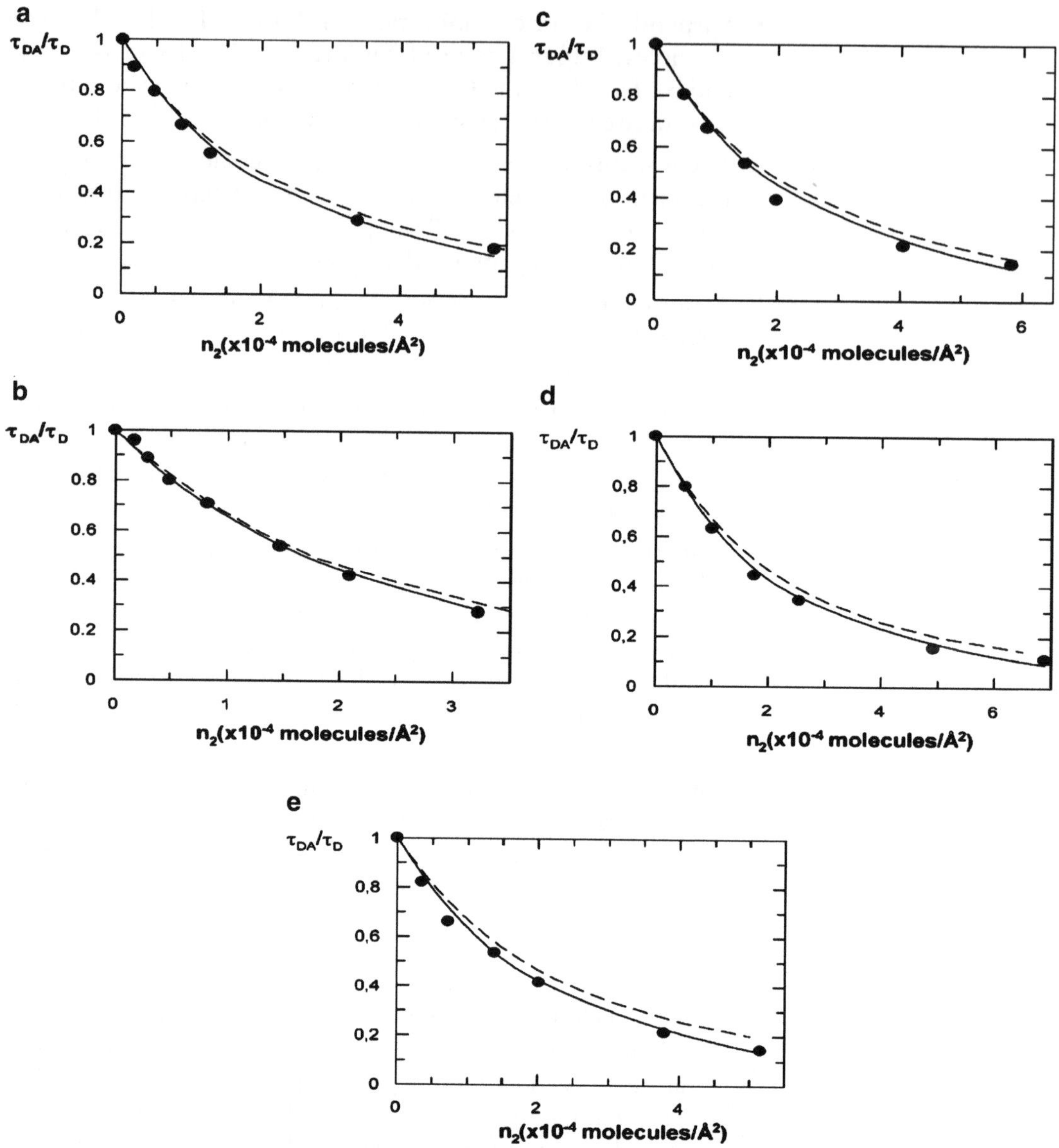

Fig. 2. Donor (coumarin-labeled protein) fluorescence quenching by energy transfer acceptor (18:1-(12:0-NBD)-PX, where X stands for the different headgroup structures) in pure DOPC bilayers. (*filled circle*) Experimental energy transfer efficiencies; (*solid line*) theoretical simulations obtained from the annular model for protein–lipid interaction using the fitted K_S; and (*dotted line*) simulations for random distribution of acceptors ($K_S = 1.0$). (**a**) PC-labeled phospholipid (fitted $K_S = 2.0$); (**b**) PE-labeled phospholipid (fitted $K_S = 2.0$); (**c**) PG-labeled phospholipid (fitted $K_S = 2.3$); (**d**) PS-labeled phospholipid (fitted $K_S = 2.7$); and (**e**) PA-labeled phospholipid (fitted $K_S = 3$). Reprinted with permission from (10). Copyright 2004 Biophysical Society.

membrane (Table 1). M13 mcp is shown to have an increased affinity for *N*-NBD-DOPE when there was considerable mismatch (both positive and negative) between the protein and the bulk lipid, reflecting an enrichment of the hydrophobically matching lipid in the protein interface.

Table 1
Labeled phospholipid relative association constants toward M13 major coat protein

Labeled phospholipid	Bilayer composition	K_S	$K_S/K_S(PC)^a$
N-DOPE-NBD	di(18:1)PC	1.4	–
N-DOPE-NBD	di(22:1)PC	2.1	–
N-DOPE-NBD	di(14:1)PC	2.9	–
(18:1-(12:0-NBD))-PE	di(18:1)PC	2.0	1.0
(18:1-(12:0-NBD))-PC	di(18:1)PC	2.0	1.0
(18:1-(12:0-NBD))-PG	di(18:1)PC	2.3	1.1
(18:1-(12:0-NBD))-PS	di(18:1)PC	2.7	1.3
(18:1-(12:0-NBD))-PA	di(18:1)PC	3.0	1.5

Reprinted with permission from (10). Copyright 2004 Biophysical Society
[a] K_S(PC) is the relative association constant of (18:1-(12:0-NBD))-PC

Figure 3 shows the FRET efficiencies measured between the Trp151 in LacY protein and either Pyr-PG or Pyr-PE as acceptors in the different lipid systems at 37 °C, together with the theoretical values, calculated using different values for the probability of finding a given phospholipid in the annular region (μ). Higher FRET efficiencies are obtained for transfer to Pyr-PE compared to Pyr-PG in both POPE/POPG and DOPE/POPG proteoliposomes, reflecting preferential interaction for PE in both systems. Application of the selectivity model described above to the experimental values indicates that PE concentration in the protein-annular region is approximately ~86 mol% for DOPE/POPG (μ(PE) = 0.86, μ(PG) = 0.14), and 100 % for POPE/POPG (μ(PE) = 1.00, μ(PG) = 0.00), whereas 75 mol% would be expected for a random distribution of both phospholipids. For proteoliposomes composed of DPPE/POPG with POPG-enriched fluid domains coexisting with DPPE-enriched gel-phase bilayer regions, an increase in the efficiency of FRET to Pyr-PG and a decrease in that to Pyr-PE are verified (Fig. 3), to an extent that the efficiency of FRET to Pyr-PG now clearly surpasses that to Pyr-PE. This result indicates that the protein is preferentially located in fluid domains where Pyr-PG is more abundant than Pyr-PE.

In order to test for the effects of cardiolipin (CL) in the annular region of LacY, myristoyl-CL and oleoyl-CL were incorporated in the POPE/POPG proteoliposomes. Addition of CL resulted always in a decrease of FRET efficiencies (Table 2) reflecting displacement of POPE and POPG lipids from the annular region of the protein. The effect was more significant when

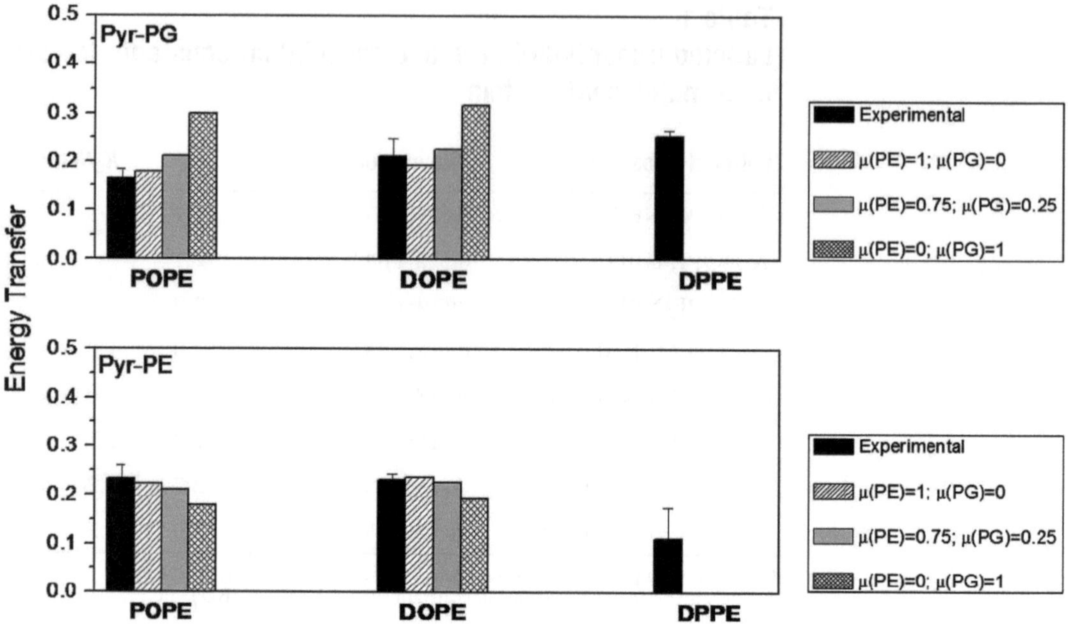

Fig. 3. Comparison of experimental and theoretical values of FRET efficiency between W151 and Pyr-PG (*top*) and Pyr-PE (*bottom*) at 37 °C in POPE:POPG (3:1, mol/mol) (*left*), DOPE:POPG (3:1, mol/mol) (*center*), and DPPE:POPG (3:1, mol/mol) (*right*) proteoliposomes (1.5 μM LacY). Reprinted with permission from (14). Copyright 2010 Elsevier.

Pyr-PG was used as the acceptor, likely due to the preference of the protein for PE species as described above. In fact, when the acceptor is Pyr-PG, even by imposing segregation of this probe from the first annular layer ($\mu(\mathrm{PG}) = 0$) in the FRET quantitative model, it is not possible to conciliate the theoretical (0.162) and the experimental values (0.143 for oleoyl-CL, 0.142 for myristoyl-CL); see Table 2. This is an indication that the changes in lateral lipid distribution induced by the presence of the protein extend beyond the first layer of lipids around it, as PG is still somewhat rarefied in this area. On the other hand, when A is Pyr-PE, a model matching to the experimental efficiencies (0.183 for oleo CL, 0.196 for myristoyl-CL) requires only partial replacement of PE by CL. When the CL lipid is oleoyl-CL, the retrieved composition of the annular layer is 40 mol% PE and 60 mol% CL. On the other hand, when the CL lipid is myristoyl-CL, the composition of the annular layer is 68 mol% PE and 32 mol% CL, indicating that in this case PE is kept in close proximity of the protein, in the same proportion as in the bulk. In fact, upon addition of myristoyl-CL to the system, CL enrichment around the protein is solely produced by replacing PG molecules. Unlike oleoyl-CL, myristoyl-CL is not able to displace PE from the area around the protein, probably due to the hydrophobic mismatch between the short myristoyl acyl chains and the protein.

Table 2
Comparison of experimental and theoretical FRET efficiencies between LacY W151 and Pyr-PE or Pyr-PG acceptors, for ternary mixtures PE:PG:CL 67:23:10 at 37°C

Lipid composition	Experimental	μ(PE) = 0.67, μ(PG) = 0.23, μ(CL) = 0.10 (all random)	Best fit	Parameter set for composition of first layer
67 POPE:22.75 POPG:10 oleo CL:0.25 Pyr-PG	$E = 0.1434$	$E = 0.1960$	$E = 0.1621$ (μ(PG) = 0.00)	μ(PE) = 0.40, μ(PG) = 0.00, μ(CL) = 0.60
66.75 POPE:23 POPG:10 oleo CL:0.25 Pyr-PE	$E = 0.1826$	$E = 0.1960$	$E = 0.1826$ (μ(PE) = 0.40)	
67 POPE:22.75 POPG:10 myr CL:0.25 Pyr-PG	$E = 0.1423$	$E = 0.1960$	$E = 0.1621$ (μ(PG) = 0.00)	μ(PE) = 0.68, μ(PG) = 0.00, μ(CL) = 0.32
66.75 POPE:23 POPG:10 myr CL:0.25 Pyr-PE	$E = 0.1963$	$E = 0.1960$	$E = 0.1964$ (μ(PE) = 0.68)	

4. Notes

1. In the examples that illustrate this chapter, one-component systems (DOPC, DEuPC, DMoPC) were addressed in the DCIA-M13 mcp system, whereas two- (POPE/POPG, DOPE/POPG, DPPE/POPG; 3:1 in all cases) and three-component (POPE/POPG/oleoyl-CL and POPE/POPG/myristoyl-CL, 67:23:10 in both cases) mixtures were explored in the LacY system.
2. The amount of added acceptor is somewhat arbitrary, but affected by the R_0 value. Lower R_0 values may require higher amounts of acceptor. In the examples that illustrate this chapter, acceptor concentration was varied between ~0.1 and 2.5 mol% of total lipid in the DCIA-M13 mcp system (for the DCIA/NBD FRET pair, $R_0 = 3.9$ nm) and fixed at 0.25 mol% in the LacY system (for the LacY/Pyr FRET pair, $R_0 = 3.0$ nm). Very high acceptor concentrations are not advised, as in these circumstances, FRET to non-bound A molecules dominates over the annular FRET component, rendering recovery of K_s or μ more difficult.

3. In the examples that illustrate this chapter, DCIA M13 mcp was reconstituted by a dialysis procedure. Lipid vesicles were mixed with the wild-type and labeled protein. Dialysis was carried out at room temperature and in the dark, with a 100-fold excess of buffer II, which was replaced five times every 12 h. On the other hand, for LacY reconstitution, lipid vesicle suspensions containing 0.2 % of DDM were incubated overnight at room temperature. Liposomes were subsequently mixed with the solubilized protein and incubated at 4°C for 30 min, with gentle agitation. DDM was extracted by addition of Bio-Beads SM-2.
4. This FRET methodology requires accurate knowledge of acceptor concentrations. It is therefore advisable to conduct a spectrophotometric determination of acceptor concentration and a quantification of phospholipid content (through the analysis of inorganic phosphate) after the reconstitution procedure. This is particularly important in case dialysis was performed, as a significant amount of material is lost during the procedure. Additionally, FRET measurements based on donor steady-state fluorescence intensities are more prone to error due to variability in donor concentration between different samples. Fluorescence lifetime measurements circumvent this problem as they are independent of concentration.
5. In the frequent case of multiexponentially decaying D, fluorescence lifetimes are replaced in Eq. 3 by the amplitude-weighted mean lifetime (15), defined by

$$\overline{\tau} = \sum_{i=1}^{m} \alpha_i \tau_i, \qquad (11)$$

where α_i and τ_i are the normalized amplitudes and lifetime values of the $m > 1$ components necessary for a statistically adequate description of the D decay (either in absence or in presence of acceptor).
6. Typical values for the refractive index n and the orientation factor κ^2 are 1.4 (16) and 2/3 (dynamic isotropic limit, good approximation in fluid bilayers; see (17) for a discussion), respectively.
7. In the examples that illustrate this chapter, area per lipid was taken as 0.72 nm^2 for the host lipids used in the DCIA M13 mcp system, and 0.56, 0.56, and 1.26 nm^2 for POPE, POPG, and CL, respectively, in the LacY system. As for the interplanar distances, in the DCIA M13 system, the D chromophore was known to reside in the bilayer midplane (18), and $l_1 = l_2$ values were 1.89, 1.98, 1.54, and 2.24 nm for *N*-NBD-DOPE in DOPC, all NBD acyl chain lipids in DOPC, *N*-NBD-DOPE in DMoOPC, and *N*-NBD-DOPE in DEuPC, respectively.

On the other hand, for the LacY system, it was assumed that the A chromophore (pyrene in acyl-chained labeled lipids) was located in the center of the bilayer, and $l_1 = l_2 = 1.2$ nm was considered. The number of annular sites was $N = 12$ for DCIA M13 mcp and $N = 46$ for LacY.

8. If the host lipid matrix is a mixture of two or more components, then the meaning of K_s is not so well defined, as in this instance it becomes dependent on the particular composition of the lipid mixture. For this reason, in the LacY case example, no K_s values are retrieved, and only μ is calculated for each system.

Acknowledgments

F.F. acknowledges a research grant (SFRH/BPD/64320/2009) from Fundacão para a Ciência e Tecnologia (FCT). F.F., M.P., and L.M.S.L. acknowledge funding by FEDER (COMPETE program), and by FCT (Fundação para a Ciência e a Tecnologia), project references PTDC/QUI-BIQ/112067/2009, PTDC/QUI-BIQ/099947/2008, and FCOMP-01-0124-FEDER-010787 (FCT PTDC/QUI-QUI/098198/2008).

References

1. Marsh D (2008) Electron spin resonance in membrane research: protein-lipid interactions. Methods 46:83–96
2. Nyholm TK, Ozdirekcan S, Killian JA (2007) How protein transmembrane segments sense the lipid environment. Biochemistry 46:1457–1465
3. Czech MP (2000) PIP2 and PIP3: complex roles at the cell surface. Cell 100:603–606
4. Lee AG (2003) Lipid–protein interactions in biological membranes: a structural perspective. Biochim Biophys Acta 1612:1–40
5. Marsh D, Horváth LI (1998) Structure, dynamics and composition of the lipid–protein interface. Perspectives from spin-labeling. Biochim Biophys Acta 1376:267–296
6. London E, Feigenson GW (1981) Fluorescence quenching in model membranes. 1. Characterization of quenching by a spin-labeled phospholipid. Biochemistry 20:1932–1938
7. Everett J, Zlotnick A, Tennyson J, Holloway PW (1986) Fluorescence quenching of cytochrome b5 in vesicles with an asymmetric transbilayer distribution of brominated phosphatidylcholine. J Biol Chem 261:6725–6729
8. Förster T (1949) Experimentelle und theoretische Untersuchung des Zwischenmolekularen übergangs von Elektrinenanregungsenergie. Z Naturforsch 4a:321–327
9. Van Der Meer B, Coker V III, Chen S-YS (1994) Resonance energy transfer: theory and data. VCH Publishers, New York
10. Fernandes F, Loura LMS, Koehorst R, Spruijt RB, Hemminga MA, Fedorov A, Prieto M (2004) Quantification of protein-lipid selectivity using FRET: application to the M13 major coat protein. Biophys J 87:344–352
11. Fung BK, Stryer L (1978) Surface density determination in membranes by fluorescence energy transfer. Biochemistry 17:5241–5248
12. Wolber PK, Hudson BS (1979) An analytical solution to the Förster energy transfer problem in two dimensions. Biophys J 28:197–210
13. Spruijt RB, Wolfs CJ, Verver JW, Hemminga MA (1996) Accessibility and environment probing using cysteine residues introduced along the putative transmembrane domain of

the major coat protein of bacteriophage M13. Biochemistry 35:10383–10391

14. Picas L, Suárez-Germà C, Montero MT, Vázquez-Ibar JL, Hernández-Borrell J, Prieto M, Loura LM (2010) Lactose permease lipid selectivity using Förster resonance energy transfer. Biochim Biophys Acta 1798:1707–1713
15. Lakowicz JR (2006) Principles of fluorescence spectroscopy, 3rd edn. Kluwer Academic/Plenum, New York
16. Davenport L et al (1985) Transverse location of the fluorescent probe 1,6-diphenyl-1,3,5-hexatriene in model lipid bilayer membrane systems by resonance energy transfer. Biochemistry 24:4097–4108
17. Loura LMS, Fedorov A, Prieto M (1996) Resonance energy transfer in a model system of membranes: application to gel and liquid crystalline phases. Biophys J 71:1823–1836
18. Koehorst RBM, Spruijt RB, Vergeldt FJ, Hemminga MA (2004) Lipid bilayer topology of the transmembrane α-helix of M13 major coat protein and bilayer polarity profile by site-directed fluorescence spectroscopy. Biophys J 87:1445–1455.

Chapter 11

A Guide to Tracking Single Transmembrane Proteins in Supported Lipid Bilayers

Kumud Raj Poudel, Jeffrey P. Jones, and James A. Brozik

Abstract

The purpose of this chapter is to serve as a guide for those who wish to carry out experiments tracking single transmembrane proteins in planar supported membrane biomimetics. This chapter describes, in detail, the construction of a simple single-molecule microscope, which includes (1) a parts list, (2) an alignment procedure, (3) a calibration procedure, and (4) a procedure for measuring the mechanical stability of the instrument. It also gives procedures for making planar supported POPC bilayers on hydrophilically treated borosilicate and quartz, POPC/PEG-PE cushioned bilayers on hydrophilically treated surfaces, and POPC/PEG-PE cushioned bilayers on BSA passivated substrates. The procedure for the detergent-mediated incorporation of the transmembrane protein $5HT_{3A}$ (a serotonin receptor) is also described and can be used as a starting point for other large non-self-inserting transmembrane proteins. The final experimental section of this chapter details different procedures for data analysis including (1) a quantitative analysis of mean displacements from individually tracked particles, (2) a Gaussian analysis of step-size distributions, (3) the Gaussian analysis of diffusion coefficients from ensembles of transmembrane proteins, and (4) a perspective associated with the interpretation of single-particle tracking data.

Key words: Single-molecule spectroscopy, Biomimetics, Protein diffusion, Transmembrane protein, In vitro protein studies, Single-molecule instrument design, Single-molecule tracking

1. Introduction

Transmembrane (TM) proteins are an important subcategory of membrane proteins whose main distinguishing structural characteristics are that they span a cellular membrane and often possess significant intra- and extracellular domains (1). Functionally, most TM proteins can be classified either as being (1) a transport protein, (2) an ion channel, or (3) an enzyme. The majority of TM proteins regulate their function through ligand binding, and it is this property that makes these proteins especially important as pharmaceutical targets (1–5). The local environment(s) surrounding these proteins are also very complex and often heterogeneous

Jörg H. Kleinschmidt (ed.), *Lipid-Protein Interactions: Methods and Protocols*, Methods in Molecular Biology, vol. 974, DOI 10.1007/978-1-62703-275-9_11,

(both chemically and structurally). Moreover, membrane proteins often interact with lipids, cholesterol, cytoskeletal proteins, peripheral proteins, sugars, fatty acids, small molecules and ions in the intra- and extracellular fluids, and many other chemically distinct species. This complexity and heterogeneity makes detailed in vivo biophysical studies of TM proteins extremely difficult to conduct under conditions that most biophysicists would describe as being controlled.

An attractive alternative to in vivo studies is the design of in vitro experiments that systematically introduces complexity in a well-controlled and well-characterized manner. This "divide and concur" approach is a well-proven strategy in biophysics, especially for determining the cause and effect relationships that can be attributed to adding or subtracting different components from a much larger system. In simplest terms, the strategy for studying TM proteins within artificial lipid membranes is to isolate a TM protein, incorporate it into a lipid bilayer (or biomimetic assembly), verify its functionality, and then increase membrane complexity one component at a time until the system has reached a level of sophistication suitable for the needs of the project. The drawback to such an experimental design is that no matter the sophistication, such membranes can only mimic naturally occurring cell membranes. Be that as it may, many in vitro studies and assays involving TM proteins reconstituted into supported artificial lipid membranes have been carried out, and protocols published (2–4).

1.1. Planar Supported Lipid Bilayers as Biomimetic Materials

When scientists talk about biomimetic materials, they are most often referring to materials that have structural and functional characteristics that parallel a material and/or property found in a naturally occurring living biological system. There are many biomimetic materials that can "act" as a cell membrane by keeping a TM protein intact and functional. These include detergents (e.g., polyoxyethylenes and alkyl glucosides in particular) (5–7), amphiphilic polymers (8, 9), liposomes (10), micelles (11, 12), and planar lipid monolayers (13). Emphasized in this chapter are protocols for tracking TM proteins in supported lipid bilayers of modestly increasing complexity, and from here on, only supported lipid bilayers will be discussed.

TM proteins often present unique challenges that can greatly affect the choice of biomimetic assembly. In this regard, there are two particular challenges that routinely affect the choice of an assembly and its preparation. The first involves protocols for the efficient integration of functional TM proteins into a lipid bilayer. This can be a delicate process in that TM proteins often do not self-insert (or do not efficiently self-insert) into lipid bilayers and require a detergent to mediate their incorporation. This is most often followed by a lengthy dialysis step or chromatography to remove the remaining detergent (2). The second tricky part is to minimize the severity of the interactions between intra- and extracellular domains of the TM protein and the underlying substrate. Currently there is no way to completely eliminate these interactions and so one has to resort to minimizing interactions and understanding their nature as well as their consequences to the protein itself. Below are the protocols for making three standard types of supported bilayer assemblies. The simplest is a planar supported bilayer on hydrophillically treated glass, the second is a planar supported bilayer on a polyethylene glycol (PEG) cushion (14, 15), and the third is a PEG cushioned bilayer on a bovine serum albumin (BSA) passivated glass substrate (16). Protein incorporation into these assemblies varies, and procedures can be found elsewhere (2, 4, 16, 17). Cushioned assemblies can be modified to control the interstitial separation between lipid bilayers and solid support, allowing for more room to accommodate intra- or extracellular TM protein domains. Passivating the solid supports modifies the nature of the physical interactions between the solid support and TM protein.

The main topics that will be covered in this chapter are (1) instrument design (for those who wish to build their own single-molecule fluorescence tracking instrument from scratch), (2) sample preparation including the preparation of solid supports and the formation of solid supported lipid bilayers suitable for single-molecule studies, (3) experimental details of single-particle tracking measurements, control experiments, and data analysis.

2. Materials

2.1. Instrument Design

Most laboratories use a commercial total internal reflection fluorescence microscope (TIRFM) together with an electron multiplying CCD (EMCCD) camera for single-particle tracking studies. A cost-effective alternative is to build your own instrument with off-the-shelf components readily available from a number of suppliers. The other advantage of a custom designed/built instrument is that it can be easily modified to accommodate multiple laser lines and multiple detectors for the simultaneous tracking of more than one type of membrane protein.

Below is a "how to" guide for building a sturdy single color "work horse" instrument. Included is a parts list (excluding bolts, washers, etc.), a description of the optical design, alignment tips, a calibration procedure, and a procedure to measure the mechanical stability of the instrument.

2.2. Parts List

1. Breadboard (24″ × 48″ × 1/2″) (MB2448, Thorlabs).
2. Breadboard (24″ × 24″ × 1/2″) (MB2424, Thorlabs).

3. 4× optical posts—10″ length, 1.5″ diameter (P10, Thorlabs).
4. Doubled cw-ND:YAG laser (LDCU3/5086, Power Technology).
5. EMCCD camera (Photonmax 512 7527-0001, Princeton Instruments).
6. Microscope objective (Olympus PlanApo 100X 1.45 TIRFM objective).
7. 532 nm ¼ waveplate (WPMQ10M-532, Thorlabs).
8. 532 nm laser line filter (FL532-1, Thorlabs).
9. 550 nm long pass filter (HQ550LP, Chroma Technologies Corp.).
10. Dichroic mirror (FF552-Di02-25x36, Semrock, Inc.).
11. 3× 0.5″ broadband dielectric mirror (BB05-E02—Ø1/2″ Thorlabs).
12. 3× motorized piezoelectric actuators (PZA12, Newport).
13. Shutter (SC10, Thorlabs).
14. 30 mm cage cube (CM1-G01, Thorlabs).
15. 2× 150 mm focal length achromatic lens (CAC254-150-A-ML, Thorlabs).
16. 45° broadband mirror (CM1-E02, Thorlabs).
17. Dichroic filter mount (FFM1, Thorlabs).
18. Neutral density filter (NDC-100S-4, Thorlabs).
19. 4× 0.5″ mini series cage assembly rods (SR05).
20. 4× 1″ mini series cage assembly rods (SR1).
21. 10× 4″ optical posts (TR4, Thorlabs).
22. SM1 adapter ring (SM1A1, Thorlabs).
23. 2× 2″ optical posts (TR2, Thorlabs).
24. Lens tube (SM1L30, Thorlabs).
25. 2× lens tube slip rings (SM1RC, Thorlabs).
26. 10× post holders (PH4, Thorlabs).
27. XYZ stainless steel stage (461-XYZ-LH-M, Newport).
28. 3× translation stages (5T1XY-A, Thorlabs).
29. XY stage for 30 mm cage assembly (Thorlabs 5T1XY-A).
30. 2× alignment iris (D36s, Thorlabs).
31. 1× 9.5″ post (P9.5, Thorlabs).
32. 3× 1″ diameter standard fixed optical mount (FMP1, Thorlabs).
33. 1× c-mount-to-SM1 adapter ring (SM1A9, Thorlabs).
34. 1× SM1 lens tube (SM1L0, Thorlabs).

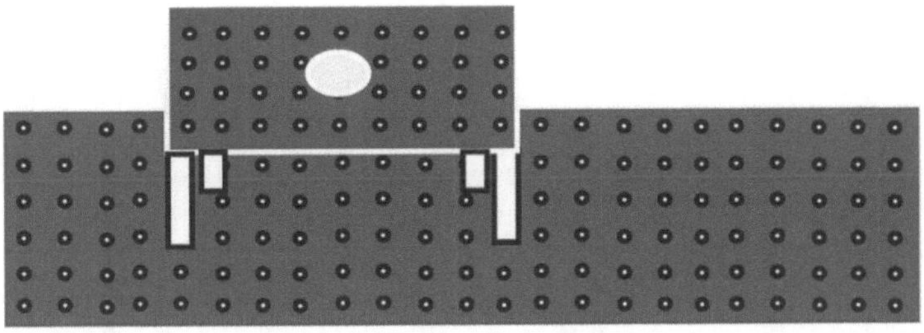

Fig. 1. Depiction of two breadboards connected together to construct a frame.

35. 2× 1.5″ optical posts (P2, Thorlabs).
36. 2× 1.5″ base adapter (PF175, Thorlabs).

2.3. Frame

If an optical bench is not available, a cost-effective and stable frame for a benchtop instrument can be constructed from two breadboards joined together (bolted) by four 10″ (1½″ diameter) stainless steel optical posts (Fig. 1). The larger breadboard (MB2448, Thorlabs) serves as the base in which the laser, laser optics, shutter, microscope optics, and EMCCD camera will be mounted. The smaller breadboard will serve as the top of the instrument and will support the sample stage. Before bolting the frame together, a 2½″ diameter hole must be drilled through the middle of the top breadboard (MB2424, Thorlabs) as shown in Fig. 1. If an optical bench is available, then it can serve as the base of the instrument.

2.4. Laser Optics

A schematic diagram of the laser optics is depicted in Fig. 2. Laser excitation is provided by a double cw-Nd:YAG laser (532 nm) mounted 7½″ above the base. The beam first passes through a 532 nm laser line filter, which is mounted in a 1″ diameter fixed optical mount (FMP1, Thorlabs). It is then passed through a 532 nm ¼ waveplate in order to produce a circularly polarized laser beam, also mounted in a 1″ diameter fixed optical mount (FMP1, Thorlabs). A neutral density filter (NDC-100S-4, Thorlabs) is then used to attenuate the power to 1 mW (see Note 1). The beam is then turned 90° using a broadband dielectric mirror (BB05-E02—Ø1/2″, Thorlabs) which sits on a translation stage used in the alignment procedure and passed through a shutter (Thorlabs SC10) for experimental control. The beam is then passed through an iris and turned 90° using another broadband dielectric mirror. This mirror is mounted on a translation stage as well. Finally the beam is passed through another alignment iris and a 150 mm focal length lens that focuses the beam into the microscope optics.

2.5. Microscope Optics

Figure 3a is a schematic of the microscope optics and 3(b) is a depiction of the mechanical setup. A 1.5″ diameter, 9.5″ high post (P9.5, Thorlabs) with a mounting plate (BA2, Thorlabs) and two 2″ posts with two 2″ post holders are used to mount an XY

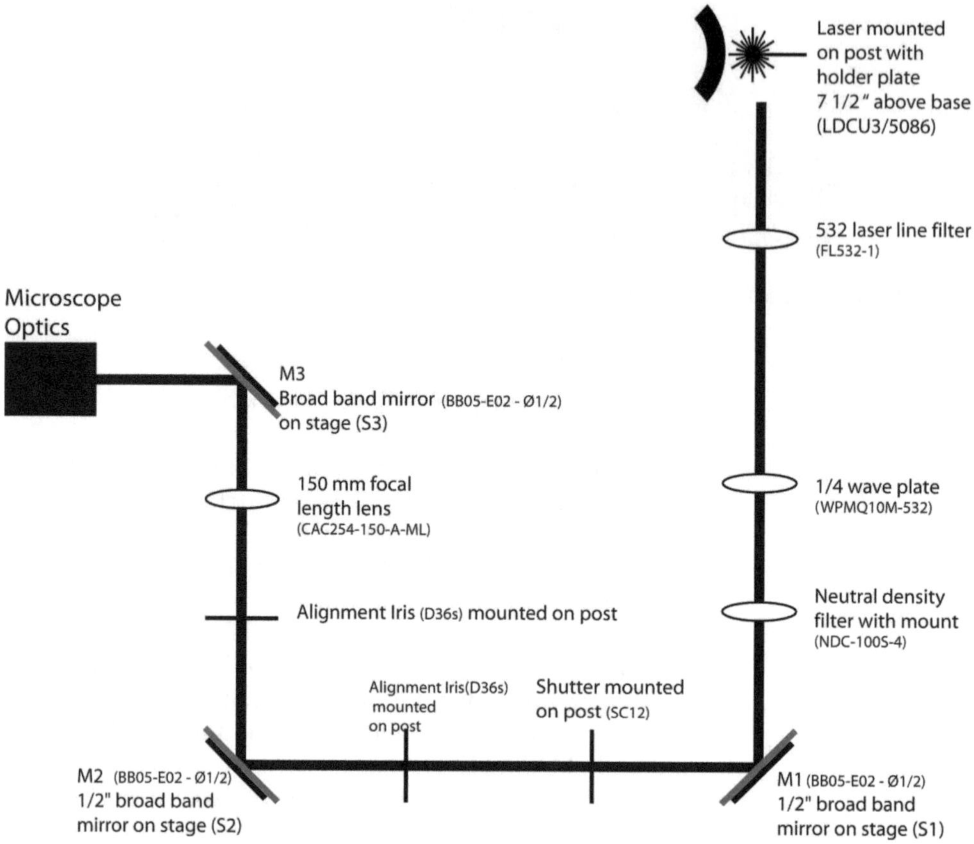

Fig. 2. Schematic for the laser optics; here a 532 nm laser beam is passed through a laser line filter to remove light from the pump diodes and fundamental of the cw-doubled-Nd:YAG laser, the beam is then passed through a ¼ waveplate to produce circular polarization to help relieve any probe orientation dependence on measured single-molecule fluorescence intensities, and finally through a lens that will focus the beam on the back edge of the microscope objective (this lens is optional; it is used to compensate for the divergence of the laser beam; if your laser has small divergence, then omit this lens).

stage horizontally as shown in Fig. 3b. The XY stage contains an SM1-to-microscope objective adapter ring in the center hole of the stage (SM1A1, Thorlabs). A microscope objective (1.45 N/A 100×, PlanApo, TIRFM Olympus, Inc.) is then securely screwed into the stage. Also attached to the bottom of stage are 4 cage rods, each 1″ long (SR1, Thorlabs). These cage rods are attached to a cube, which encloses the dichroic mirror (FF552-Di02-25x36, Semrock, Inc.) and held in place by a dichroic filter mount (FFM1, Thorlabs). This dichroic mirror accepts the attenuated laser beam and turns it 90° so it passes vertically through the microscope objective. The dichroic mirror reflects 532 nm laser excitation and passes all wavelengths longer than 552 nm. To the bottom of the cube are attached four 0.5″ cage rods (SR05, Thorlabs) which hold a pre-mounted 45° broadband dielectric mirror (CM1-E02, Thorlabs). This mirror receives the fluorescence that

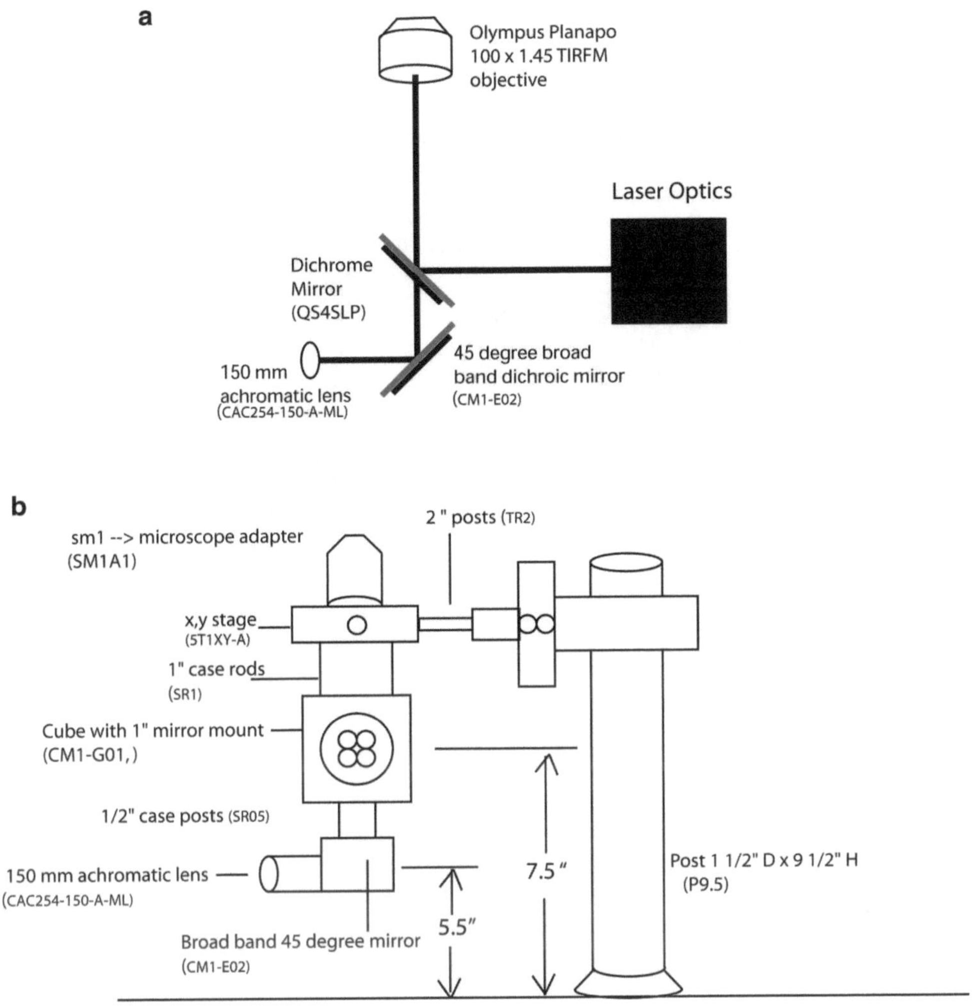

Fig. 3. (**a**) Schematic of the microscope optical path and (**b**) mechanical sketch of the setup for the microscope optics.

was passed through the dichroic mirror and turns it at 90° through a 150 mm achromatic lens (CAC254-150-A-ML, Thorlabs).

2.6. Sample Stage

The sample stage (Fig. 4) is comprised of three motorized actuators (NanoPZ Ultra-High Resolution Actuator, 12.5 mm Travel, Newport, Inc.) used to control the motion of the sample holder in the X, Y, and Z directions. The actuators area attached to a high precision XYZ optical stage (Model 461-XYZ-LH-M, Newport Corporation) which is positioned such that the sample is centered over the 2″ hole in the top breadboard. The actuators are connected to a motion controller (PZC200 Switchbox, Newport Inc.), which can be controlled via an application program provided with the hardware or a handheld controller (PZA12, Newport, Inc.). The sample holder itself is made from a 1″ lens tube (SM1L30, Thorlabs) attached to the XYZ stage via two small

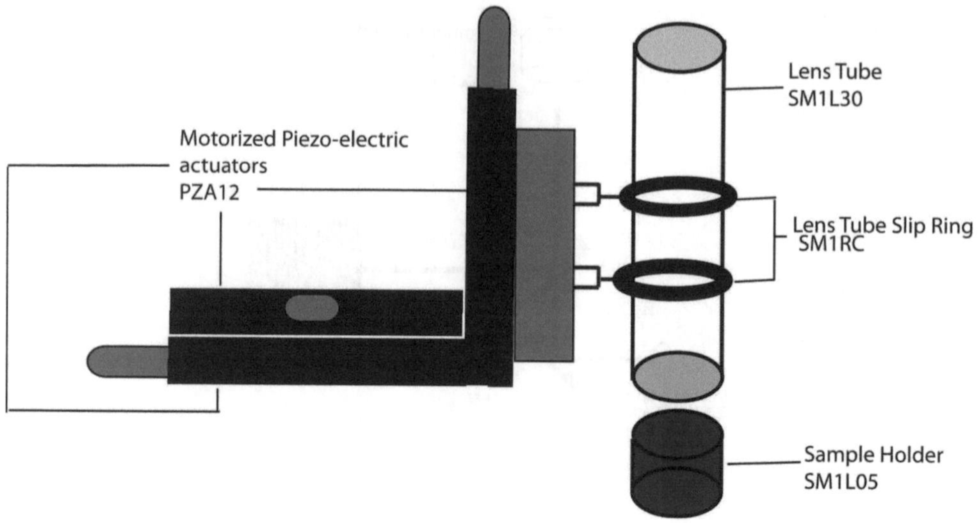

Fig. 4. Sketch of the sample stage.

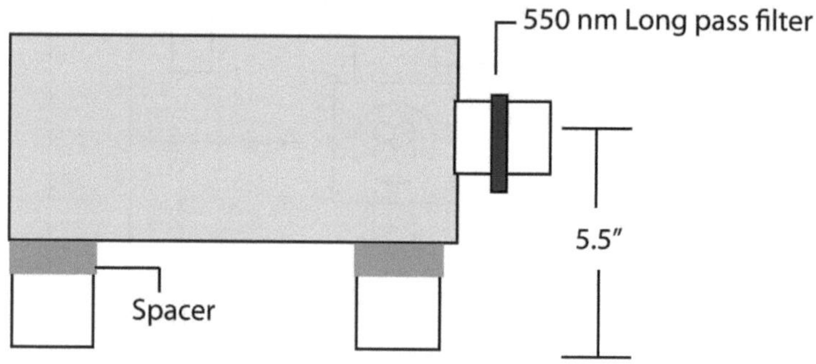

Fig. 5. Camera height and 550 nm long pass filter.

tube slip rings (SM1RC, Thorlabs) and two optical posts with two post holders. The sample can then be attached to the bottom of the sample holder via the threads in the lens tube.

2.7. Camera Placement

The camera is secured to the bottom breadboard by two 1.5″ diameter, 2″ high optical posts (P2, Thorlabs) such that its active area is 5½″ high and in line with the height of the 45° broadband mirror (Figs. 3 and 5). The active area of the camera is placed 150 mm from the imaging lens, and a 550 nm long pass filter (HQ550LP, Chroma Technologies Corp.) is placed before the camera via a lens tube attached directly to the camera via a c-mount-to-SM1 adapter ring (SM1A9, Thorlabs) and SM1 lens tube (SM1L0, Thorlabs).

2.8. Alignment Procedure

There are two optical trains that need to be aligned before the microscope is ready for use. They are:

2.8.1. Laser Optics Alignment

For laser optics alignment, first, visually ensure that the laser beam entering M1 (Fig. 2b) is approximately flat and 7½″ high. Then by moving translation stage S1 and mirror M1, walk the beam onto M2. The beam should be kept about 7½″ high and straight along the tapped holes of the breadboard. Next, adjust the height of the beam between M2 and M3 so that it is precisely 7½″ high. Adjusting M1 and M2 can do this. Next, the horizontal beam path can be adjusted with M2 and S2 such that it precisely travels down the taped holes in the breadboard that is between M2 and M3. The lens between M2 and M3 should then be installed such that the beam is not displaced in the X or Y direction. M3 needs to be installed such that the beam is turned 90° into the center of the dichroic mirror of the microscope optics (7½″ high). Finally, the irises (I1 and I2) can be installed and used as starting points for maintaining the alignment of the instrument.

2.8.2. Microscope Optics Alignment

When aligning the microscope optics, the objective can be taken out. Once the objective is taken out, the beam is turned vertically with the dichroic mirror. The cube mirror mount (for the dichroic mirror) contains adjustments screws, which can be used to straighten the beam such that it is perpendicular to the microscope base. Next, the microscope objective is replaced, and the XY objective stage is used to center the objective over the beam (epi-fluorescence mode; see Note 2). An evanescent field can be generated by first focusing on a sample under epi-illumination and then translating the objective in either the X or Y direction with the objective stage until a strong back reflection appears back through the microscope optics and onto M3 (laser optics).

2.9. Calibration

Calibration grids are commercially available and can be used to calibrate the instrument. For our setup, we use a custom-made grid of zero-mode waveguides (ZMW) that have been precisely aligned by electron beam lithography. Each ZMW is spaced 10 μm apart and the center of the ZMW can be found from its centroid. This is used to determine the calibration in μm/pixels. Each calibration grid should be verified by translating the grid with the piezoelectric actuators and plotting the known actuator translation (nm) vs. grid position (in pixels).

2.10. Mechanical Stability

Vibration free rooms provide excellent mechanical stability. Even with that, the mechanical stability of the microscope still needs to be measured under the conditions of the experiment (temperature, excitation, etc.). One way to do this is to dry a very dilute sample of quantum dots onto a coverslip. These QDs can then be imaged under the mechanical and temporal conditions of the experiment. The centroid of the QD can then be determined for each frame of a long movie (at least as long as the proposed experiments). For each frame, the deviation from its mean position in both the X and Y directions is determined, and histograms of these displacements should yield Gaussians centered at zero. The FWHM is a measure of the instruments frame-to-frame mechanical stability and should be the same in both X and Y directions.

3. Methods

3.1. Sample Preparation

3.1.1. Protein Purification and Labeling

Transmembrane protein purification (or isolation) and labeling schemes are usually quite specific to the protein of interest. On the other hand, procedures used to incorporate TM proteins into lipid bilayer assemblies are much more universal. Below is a robust protocol for reconstituting transmembrane proteins in liposomes. This procedure is a successful adaptation of published methods that has been used for a variety of lipid compositions and TM proteins (2, 18).

Purification and labeling of $5HT_{3A}$ (the main example in this protocol) is published elsewhere (2, 18).

3.1.2. Coverslip Preparation

A clean coverslip is key to ensuring samples with minimal background, and a hydrophilic treatment is necessary for efficiently fusing liposomes onto glass (or quartz substrates) as well as for producing a thin homogenous passivation layer of BSA for multi-cushioned lipid bilayers. There are different types of coverslips available. Borosilicate coverslips are fairly inexpensive, so can be used once and discarded but generally have increased background fluorescence compared to quartz. Quartz coverslips are fairly expensive and etching them can make them very fragile, which limits the number of times they can be used. We have noticed various degrees of contamination from one batch of coverslips to the next even though they might come from the same manufacturer or supplier. Below are two effective protocols to clean and hydrophilically treat coverslips (borosilicate and quartz; these protocols are not compatible with flint glass coverslips).

Technique 1:

1. Gently agitate coverslips in a solution of 20 mL 5 % NH_4OH and 20 mL 4 % H_2O_2 at 80–90°C for 15 min.
2. Decant the solution and rinse coverslips with a copious amount of nanopure water.
3. Next, hydrophobically treat the coverslips by agitating them in 20 mL of 0.4 M HCl and 20 mL of 8% H_2O_2 at 80–90°C for 15 min.
4. Individually rinse the coverslips with a copious amount of nanopure water and dry under a stream of nitrogen.
5. Place coverslips in a UV-ozone generator for 30 min prior to use.

Technique 2:

1. Boil the coverslips extensively in a mixture of concentrated nitric acid, hydrogen peroxide (30%), and nanopure water in a ratio of 1:1:1 for 20 min.
2. Make sure the coverslips are constantly agitated during this period, so they are well separated and do not stick to each other.
3. Wash coverslips with copious amounts of nanopure water.
4. Individually dry the coverslips under a stream of nitrogen.
5. Place coverslips in a UV-ozone generator for 30 min prior to use.

3.1.3. Liposome Preparation and Protein Reconstitution

There are many different ways to prepare liposomes and planar supported bilayers. Below is an established protocol that works well with detergent-mediated protein insertion followed by dialysis. A protocol for the formation of planar supported bilayers by vesicle fusion is also described. Please see Note 3 for tips on optimizing protein concentration.

For Planar POPC:

1. 910 μL of a 5 mM chloroform stock solution of 1-palmitoyl-2-oleoyl-sn-glycero-3-phosphocholine (POPC; Avanti Polar Lipids) is dried under a gentle stream of nitrogen for a minimum of 4 h.
2. Next, reconstitute the lipid cakes in 1 mL of 10 mM HEPES buffer (pH 7.4). Ensure the cake is completely reconstituted.
3. Sonicate the reconstituted lipid for 30 min at 60 °C or until the solution turns from turbid to translucent.

   For PEG Cushioned POPC Bilayer:
   For cushioned POPC liposomes, the same procedure as the planar POPC is followed but the initial lipid cocktail also contains a 1.4 mole percent PEG-PE (1,2-dioleoyl-*sn*-glycero-3-phosphoethanolamine-N-[methoxy(polyethyleneglycol)-2000] (ammonium salt), Avanti Polar Lipids).
   For self-inserting transmembrane proteins that do not require detergent-mediated insertion, skip to step 10. Steps 4–9 are for incorporating a TM protein into liposomes via a detergent.
4. Pretreat vesicles with 6 stepwise additions (16 μL each) of 24 mM C12E9 over the course of 1 h at 35°C.
5. The proteomicelles (purified/isolated protein of interest in detergent) are now ready to be added to the pretreated liposomes.
6. Add the solution of proteomicelles to the solution from step 4 and gently rock the mixture for 1 h.
7. Dialyze the solution (from step 6) overnight using a dialysis cassette (G-Biosciences, 3 kDa MWCO) in 10 mM HEPES

with 200 μL of a Bio-Beads® slurry (SM2- Adsorbent, 152-3920 Biorad) added to the dialysis buffer in order to remove the detergent.

8. Following dialysis, add the protein containing liposomes solution (proteoliposomes, from step 7) to an Eppendorf tube containing 15 μL of the Bio-Beads® slurry and then gently rock for 30 min.
9. Separate the proteoliposomes from the Bio-Beads® by centrifugation (16,000 × *g* for 5 min) and pipette the supernatant (containing the proteoliposomes) into a separate Eppendorf tube and store on ice until used.

Preparation of Planar Supported Bilayers:
If using negatively charged lipids in your preparation, please see Note 4.

10. Make a Parafilm® gasket such that it is a 1″ diameter circle with a 10 mm diameter circular hole cut through the center. This is most easily done with sharpened 1″ and 10 mm punches.
11. Place a previously cleaned coverslip in a 1″ SM1 lens tube.
12. On top of the coverslip, place the Parafilm® gasket.
13. Secure the Parafilm® gasket and coverslip using two mounting rings (SM1RR, Thorlabs).
14. Gently heat the holder/coverslip/gasket assembly with a heat gun to melt the Parafilm® to the coverslip.
15. After the assembly cools down, pipette 50 μL of the proteoliposome solution (step 9) into the center of the assembly.
16. Incubate the sample for 30 min.
17. Wash by gently pipetting 50 μL of the buffer onto the center of the assembly. Being careful not to touch the surface of the coverslip, take the buffer off with a pipette and discard.
18. Repeat step 17 six times and after the final rinse, gently place 50 μL of buffer onto the sample and attach to the microscope. The sample is now ready for single-particle tracking studies.

For PEG Cushioned Bilayers on a BSA Passivated Coverslips (Double Cushion):

For double-cushion systems, the same lipid cocktail and vesicle fusion procedures as the cushioned POPC are used. The only difference is that the coverslips are incubated with 0.1 mg/mL of bovine serum albumin (BSA) solution for 25 min prior to the vesicle fusion step. Before passivation, the BSA solution must first be centrifuged at 16,000 × *g* for 20 min. After the incubation, the BSA-coated surface is washed 5× with buffer (50 μL each-wash).

3.1.4. Preparation of Controls

Control experiments contain no fluorescent protein. These are conducted to systematically test for background contamination. Controls should be prepared in the exact same manner as outlined above. This includes the TM protein incorporation steps. The only difference is that non-labeled TM protein is substituted for the labeled TM protein.

Extemporaneous background signal in single-particle tracking experiments can originate from fluorescence contamination or scatter from particulates. To get rid of particulates, all the buffers and solutions should be filtered with a 0.250 μm filter and then centrifuged at 16,000 × *g* for 20 min. Ensuring that only the highest purity materials are used and all glassware rigorously cleaned prior to use can minimize fluorescence contamination. (Note: After rigorously cleaning, we rinse all glassware with concentrated sulfuric acid, followed by copious amounts of nanopure water.)

3.2. Data Analysis

Any quantitative or analytical conclusion for a set of single-particle tracking experiments requires a very large amount of data such that the results become statistically meaningful and direct observations can be generalized. A minimal guide on how to analyze data and interpret results is given below.

3.2.1. Data Tracking

Most automated single-molecule tracking algorithms are based on the work by Crocker and Grier (19) and are available as free downloads in a variety of formats including IDL, GDL (a free version of IDL), LabView, MatLab, and as stand-alone GUIs. We have used a modified MatLab scripts written by M. Kilfoil and coworkers at the University of Massachusetts Amherst (http://www.people.umass.edu/kilfoil/downloads.html). These scripts are very similar to the original IDL codes by Crocker and Grier.

3.2.2. Immobile vs. Mobile

Mobility is a relative measurement and the best anyone can do is to define what is meant by immobility. We suggest establishing a fiducial point by which immobility can be defined or by curve fitting the combined frame-to-frame displacement of all particles in data set (see below).

1. *Comparison vs. Mechanical Stability:* The easiest way to define a particle as being immobile is by comparison with the mechanical stability of the instrument. Usually, if a particle moves less than the full width at 1/10 maximum $FWTM = 2 \times \left(\sqrt{2 \times \ln(10)}\right) \times \sigma$, where σ is the standard deviation of the mechanical stability of the microscope), then we can consider the particle to be immobile. For example, the stability of our microscope is FWHM = 22 nm ($\sigma = 9.34$ nm). Therefore, under normal conditions, if a particle moves more than 40 nm in either the X or Y direction at anytime during the track of that particle, we would consider it to be mobile.

2. *Frame-to-Frame Image Analysis:* An internal fiducial point can be generated from the data itself. This can be done by fitting 1D frame-to-frame displacements of all particles in a data set to at least two Gaussian distributions and extracting the standard deviation for each population. The distribution with the smallest standard deviation can be considered to belong to the immobile fraction and used as the criterion to judge whether or not individual TM proteins are mobile (based on the FWTM cutoff described above).

This analysis can also be used to determine the percent mobility of TM proteins in the lipid assembly. This is done by simple integration of each Gaussian distribution. This result can then be compared to the individual tracking criterion. An example of such has been previously published (2).

3.2.3. Data Analysis

Once individual tracks have been inspected, diffusion coefficients can be determined for individual mobile TM proteins. The diffusion coefficient of a transmembrane protein in a biomimetic assembly can be determined by the analysis of (1) the mean square displacement of individual proteins vs. lag time (20) or (2) the frame-to-frame step-size distribution of individually tracked TM proteins. Individual squared displacements are given by

$$(\Delta r_{n\Delta t})^2 = (\Delta x_{n\Delta t})^2 + (\Delta y_{n\Delta t})^2 \tag{1}$$

where $n\Delta t$ is the time lag (n = number of frames during a time lag and Δt is the time between adjacent frames), and $\Delta x_{n\Delta t}$ and $\Delta y_{n\Delta t}$ are the spatial displacements in the x and y directions for a particular time lag. The mean square displacement (MSD) is the average of all discrete displacements corresponding to a time lag with a value of $n\Delta t$ for an individually tracked TM protein:

$$\langle \Delta r_{n\Delta t}^2 \rangle = \frac{1}{N} \sum_{i=1}^{N} \Delta r_{i,n\Delta t}^2 \tag{2}$$

There are several types of motion that can be distinguished by single-particle tracking. For TM proteins in biomimetic assemblies, there are three common types of diffusion, (1) normal diffusion, (2) corralled (or confined) diffusion, and (3) a type of hindered diffusion that is characterized by either hops or some sort of partial confinement. All tend to show Brownian diffusion characteristics at short time lags. Examples of these are shown in Fig. 6.

The upper part of Fig. 6 depicts example tracks of TM proteins diffusing in different biomimetic assemblies, and the bottom part of Fig. 6 depicts their corresponding MSD plots. Figure 6a is an example of the serotonin $5HT_{3A}$ receptor diffusing in a cushioned POPC lipid bilayer supported on a hydrophilically treated borosilicate coverslip. In this example, the TM protein is undergoing normal diffusion, which is characterized by a random trajectory with no discernable structure associated with its path. The MSD

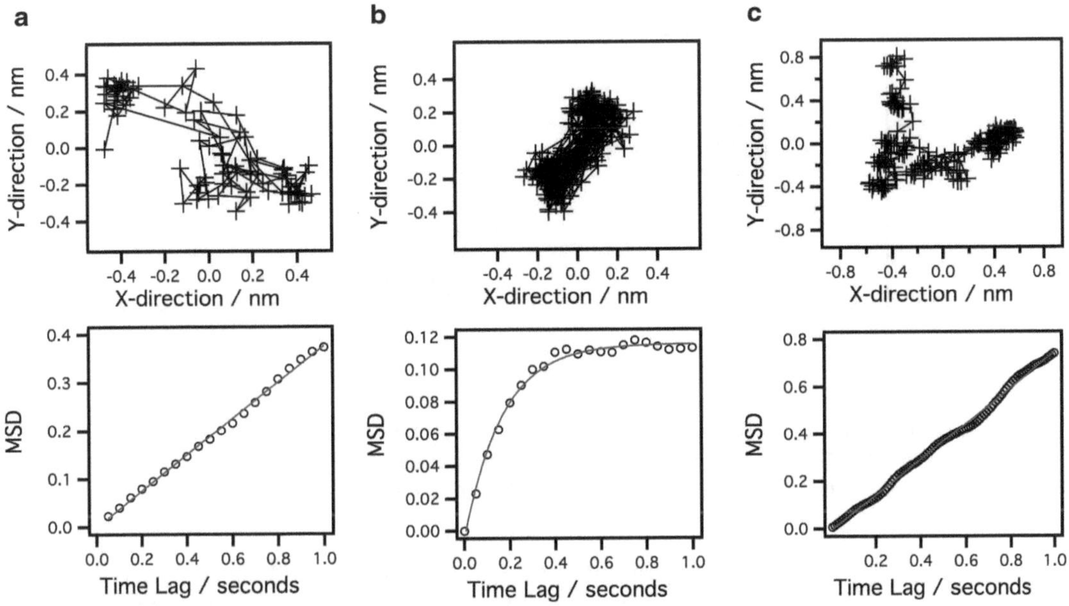

Fig. 6. Examples of different "types" of diffusion. In each example, the top figure is an actual trajectory of a TM protein in a membrane biomimetic, and the bottom figure is its corresponding MSD vs. lag time. (**a**) $5HT_{3A}$ in a POPC/PEG-PE membrane on hydrophilically treated borosilicate at 25° undergoes normal diffusion; time resolution is 50 ms. (**b**) $5HT_{3A}$ in a POPC/PEG-PE membrane on BSA passivated borosilicate at 25° displays marked corralled diffusion; time resolution is 50 ms. (**c**) Annexin V in a DMPC/PEG-PE/Brain PS on hydrophilically treated borosilicate at 23 °C displays hindered diffusion; time resolution is 10 ms.

vs. $n\Delta t$ plot for this track remains linear over longer time lags but, like all single-molecule tracks, will deviate severely at very large time lags. For normal diffusion, the MSD vs. time lag is related to its diffusion coefficient (D) by Eq. 3 below.

$$\langle r^2(n\Delta t)\rangle = 4D \times (n\Delta t) \tag{3}$$

Corralled TM proteins are easy to identify from their tracks. These tracks map out a defined area in which the protein is corralled and their MSD plots reflect this confined behavior by possessing a plateau at a maximum mean square displacement that is related to the size of the corral. Figure 6b is an example track of $5HT_{3A}$ diffusing in a cushioned POPC lipid bilayer assembly on a BSA passivated borosilicate surface. In this example, the bulk structure of the underlying BSA creates "hard" barriers, which encircle $5HT_{3A}$ and confine the TM protein to a small well-defined area in which it is free to diffuse. MSD vs. time lag plots for corralled motion can be fit to Eq. 4 below:

$$\langle r^2(n\Delta r)\rangle = \langle r_C^2\rangle \left[1 - A_1 \exp\left(\frac{-4A_2 D \times (n\Delta t)}{\langle r_C^2\rangle}\right)\right] \tag{4}$$

where D is the diffusion coefficient, $\langle r_C^2\rangle$ is the corral size, and A_1 and A_2 are constants determined by the shape of the corral.

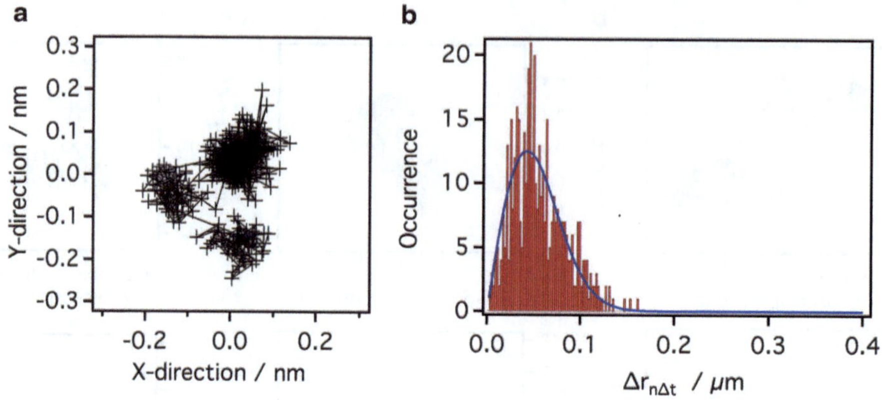

Fig. 7. $5HT_{3A}$ in double cushion POPC assembles (**a**) track showing discrete hop diffusion between corrals separated by ~150 nm and (**b**) SSD analysis of the track shows that the TM protein is undergoing normal Brownian diffusion and its SSD is related to its diffusion coefficient through Eqs. 11.5 and 11.6 (*solid blue line*).

The most common type of hindered diffusion observed for TM proteins in biomimetic lipid bilayers is intermediate between normal and corralled motion. In this case, a single TM protein can undergo normal diffusion during one part of the track only to be confined during a different part of the track or simply hop from one location to the next. This type of motion often leads to undulating MSD vs. $n\Delta t$ plots (see Fig. 6c bottom). Two examples of this type of motion are given in Figs. 6c and 7. The first is a track of annexin V in a cushioned DMPC/PEG-PE/Brain PS assembly at 23°C, which is close to the center of its liquid-to-gel phase transition and is dominated by heterogeneity caused by the coexistence of different phase domains. The second example (Fig. 7) is $5HT_{3A}$ in a cushioned POPC assembly on a BSA passivated surface in which the TM protein can hop from one corral to the next in a tight little cluster. There are a few choices in relating tracks like these to a diffusion coefficient(s). One way is to simply fit the short time lag MSD curve to a normal diffusion model and disregard the undulations. This will give you an "average" diffusion coefficient that is weighted by the part of the track in which the TM protein spends the most time. A second way is to relate the step-size distribution of the track to a quasi diffusion coefficient (see Subheading 3.2.4 below). Finally, if the track can be cleanly separated into normal regions and confined regions, then these different segments can be analyzed piecewise with Eqs. 3 and 4.

3.2.4. Step-Size Displacement

Heterogeneous environments in biomimetic assemblies (chemical domains, phase domains, physical barriers, etc.) can lead to pronounced hindered diffusion. When this happens, it may be more beneficial to address the mass transport properties of a TM protein from its step-size distribution (SSD) and determine a quasi diffusion coefficient rather than following its stepwise trajectory. Here, the step size, $\Delta r_{n\Delta t}$, is defined as the distance a single membrane protein traverses during a single time lag $n\Delta t$ ($\Delta r_{n\Delta t} = \sqrt{(\Delta x_{n\Delta t})^2 + (\Delta y_{n\Delta t})^2}$). In this analysis method, the previous position of the particle is not taken into account. This seems to be especially useful for estimating the diffusion coefficients of individual particles in heterogeneous environments. As long as the diffusion is Brownian in nature, the step-size distribution is expected to be Gaussian multiplied by a factor that accounts for the larger area as the size of $\Delta r_{n\Delta t}$ increases. The variance of the distribution is related to the time lag and a quasi diffusion coefficient. The SSD distribution is given by Eq. 5 below (21):

$$P(\Delta r_{n\Delta t}) = \left(\frac{1}{4\pi D_q n\Delta t}\right)\left[e^{\frac{-\Delta r^2_{n\Delta t}}{4D_q n\Delta t}}\right] \times 2\pi\Delta r_{n\Delta t} \quad (5)$$

where D_q is the quasi diffusion constant, $n\Delta t$ is the time between observations (time lag), and $\Delta r_{n\Delta t}$ is the step size for a given time lag. The histogram of step sizes is built by counting the number of steps within an interval ΔR and can be fit to Eq. 6 below:

$$F(\Delta R) = N \times P(\Delta r_{n\Delta t}) \times \Delta R \quad (6)$$

where N is the total number of steps and $P(\Delta r_{n\Delta t})$ is Eq. 5 above. An example is given in Fig. 7.

3.2.5. Histograms

If every TM protein in an ensemble were in exactly the same environment and if we were able to track each for an infinite amount of time and with infinitely fine time lag resolution, then every tracked TM protein would yield exactly same diffusion coefficient. The very fact that a distribution of diffusion coefficients can be measured suggests that there is heterogeneity in the sample and/or a fair amount of experimental variance (from limited time lag resolution, a finite tracking period, and mechanical instability). Be that as it may, histograms of diffusion coefficients can provide some insight about average diffusion coefficients and the number of "on-average" local environments present in the ensemble. With this perspective in mind, a histogram of diffusion coefficients that is more or less Gaussian indicates that the ensemble of TM proteins "sees" a more or less homogenous environment (see Note 5). On the contrary, if the histograms were to have a shape that was consistent with multiple distributions, this would indicate multiple populations of TM proteins with fundamentally different diffusion properties. Figure 8a depicts an example of a transmembrane protein ($5HT_{3A}$) in a "homogenous" environment. The histogram reveals a normal distribution with no reason to fit the data to anything more than a single Gaussian. In contrast, a sample containing gross heterogeneity or more than one type of mobile particle (biochemically distinct) can have a complex distribution. This can result in the distribution have a shoulder or may simply display

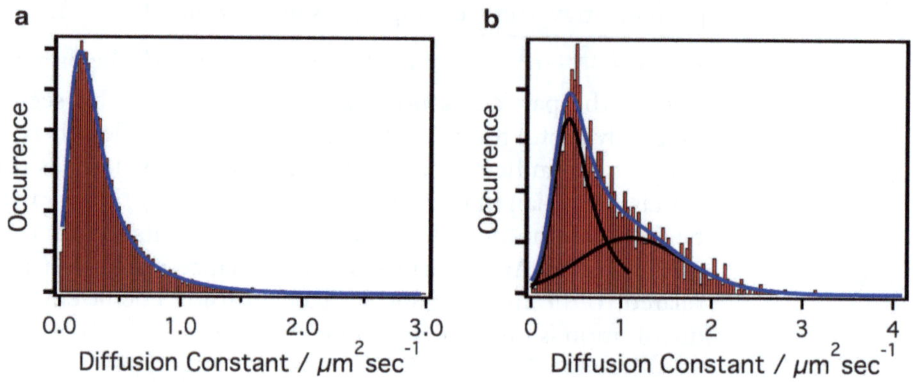

Fig. 8. Histograms (*red*) of diffusion coefficients measured from individually tracked TM protein. The *blue line* through the data is a multi-Gaussian fit of the data and the *black lines* are the individual Gaussians that make up the composite fitting function; (**a**) single Gaussian fit of diffusion coefficients from individually tracked $5HT_{3A}$ receptors in a cushioned POPC assembly at 25 °C and (**b**) diffusion coefficients of individual annexin V TM proteins fit to two Gaussians; data was collected in a DMPC/PEG-PE/Brain PS assemble at 23 °C.

an asymmetric distribution like the example in Fig. 8b, annexin V midway through a liquid-to-gel phase transition.

3.3. Conclusion and Perspectives

The mass transport properties of TM proteins are an essential part of their function, and single-particle tracking provides an essential insight into this process. The power to directly observe the motion of individual TM proteins cannot be overstated in this regard and has led to a much richer understanding of membrane biophysics that goes well beyond the classical fluid mosaic model of Singer and Nicolson (22) and the continuum fluid model of Saffman and Delbrück (23). Single-molecule tracking experiments have revealed complicated diffusion mechanisms in plasma membranes such as hop diffusion, corralling, and hindered diffusion. Moreover, these observations have been used to reconcile the differences between protein diffusion in plasma membranes vs. protein diffusion in liposomes (24). In membrane biomimetics, single-particle tracking has revealed new insight into how one can control TM protein mobility and transmembrane protein interactions with solid supports (2). These experiments are essential for the development of membrane biomimetics and will be used to gain a greater understanding of TM protein biophysics in these controlled, yet artificial, membranes.

4. Notes

1. Final Laser Power

   The final power of the laser beam should be adjusted to get the best compromise between signal-to-noise and the longevity of the fluorescence probe. 1 mW is a good place to start but should be optimized for each series of experiments.

2. Epi-Fluorescence

   It is sometimes difficult to determine when the laser is aligned in epi-fluorescence mode. You can check by placing a solution cell (a live imaging petri dish is good for this) filled with water and a drop of milk on top of the microscope objective and inspect the scatter from the focused laser beam. The scattered beam should look like a cone that is perfectly vertical (i.e., not slanted from side-to-side).

3. Protein Concentration

   The right range of concentration can make a significant impact on data analysis. Too high of a protein concentration leads to ambiguity between the individual tracks. That is, if there is too much crowding and if the membrane proteins move relatively long distances between frames, then it becomes impossible to discriminate between any two proteins. On the other hand, if there are too few membrane proteins, then the data collection can become prohibitively long and even minor contamination can overwhelm protein tracks.

4. Negatively Charged Lipids

   If you require negatively charged lipids such as phosphatidylserines or phosphoglycerols, you will need either a high ionic strength buffer or cations with a plus two charge in order to achieve vesicle fusion and lipid bilayer formation. We suggest 8 mM $CaCl_2$ in either HEPES or MES buffer. Note that Ca^{2+} will precipitate out of solution in PBS, and therefore, they are incompatible with one another.

5. Shape of Diffusion Coefficient Histogram

   In general if the protein can be tracked for a large number of frames (>60), then the distribution of diffusion coefficients from the linear analysis of early time lag MSD curves will give Gaussian distributions centered a $<D>$. Because of the limited photostability of most fluorescent probe molecules, it is often the case that a TM protein will only be tracked for a limited number of frames. This can make (1) the leading edge of the distribution modestly steeper than predicted by a Gaussian, (2) the most probable value (D_{MP}) will be slightly smaller than the average value ($<D>$), and (3) the trailing edge of the distribution will be longer. Regardless of modest apparent asymmetry in distributions of this sort, the $<D>$ value should remain unchanged by measuring a limited number of frames. In general, we will only consider tracks with 20 frames or higher and find that the distributions are described fairly well by a Gaussian, but the fit will slightly overestimate the leading edge of the data set. For a complete discussion, please see reference 25 by Saxton (25).

References

1. Thompson AJ, Lummis SCR (2007) The 5-HT3 receptor as a therapeutic target. Expert Opin Ther Targets 11:527–540
2. Poudel KR, Keller DJ, Brozik JA (2011) Single particle tracking reveals corralling of a transmembrane protein in a double-cushioned lipid bilayer assembly. Langmuir 27(1):320–327
3. Diaz JA, Albertorio F, Daniel S, Cremer SP (2008) Double cushions preserve transmembrane protein mobility in supported bilayer systems. Langmuir 24:6820–6826
4. Isas JM, Cartailler JP, Sokolov Y, Patel DR, Langen R, Luecke H, James E, Haigler HT (2000) Annexins V and XII insert into bilayers at mildly acidic pH and form ion channels. Biochemistry 39(11):3015–3022
5. Trubetskoy VS, Torchilin VP (1995) Use of polyoxyethylene-lipid conjugates as long-circulating carriers for delivery of therapeutic and diagnostic agents. Adv Drug Deliv Rev 16(2–3):311–320
6. Schwendener RA, Asanger M, Weder HG (1981) n-Alkyl-glucosides as detergents for the preparation of highly homogeneous bilayer liposomes of variable sizes (60–240 nm [phi]) applying defined rates of detergent removal by dialysis. Biochem Biophys Res Commun 100(3):1055–1062
7. Zhu J, Xue J, Guo Z, Zhang L, Marchant RE (2007) Biomimetic glycoliposomes as nanocarriers for targeting P-selectin on activated platelets. Bioconjug Chem 18(5):1366–1369
8. Tribet C, Audebert R, Popot JL (1996) Amphipols: polymers that keep membrane proteins soluble in aqueous solutions. Proc Natl Acad Sci 93(26):15047
9. Charvolin D, Perez JB, Rouvière F, Giusti F, Bazzacco P, Abdine A, Rappaport F, Martinez KL, Popot JL (2009) The use of amphipols as universal molecular adapters to immobilize membrane proteins onto solid supports. Proc Natl Acad Sci 106(2):405
10. Eytan GD (1982) Use of liposomes for reconstitution of biological functions. Biochimica et Biophysica Acta 694(2):185
11. Martinez KL, Gohon Y, Corringer PJ, Tribet C, Mérola F, Changeux JP, Popot JL (2002) Allosteric transitions of Torpedo acetylcholine receptor in lipids, detergent and amphipols: molecular interactions vs. physical constraints. FEBS Lett 528(1–3):251–256
12. Wallach DFH (1990) Lipid vesicles formed of surfactants and steroids. US Patent 4:452–895
13. Srinivasan MP, Ratto TV, Stroeve P, Longo ML (2001) Patterned supported bilayers on self-assembled monolayers: confinement of adjacent mobile bilayers. Langmuir 17(25):7951–7954
14. Wagner ML, Tamm LK (2000) Tethered polymer-supported planar lipid bilayers for the reconstitution of integral membrane proteins: silane-polyethyleneglycol-lipid as a cushion and covalent linker. Biophys J 79:1400–1414
15. Albertorio F, Diaz JA, Yang T, Chapa AV, Kataoka S, Castellana TE, Cremer SP (2005) Fluid and air stable lipopolymer membranes for biosensor applications. Langmuir 21:7476–7482
16. Diaz AJ, Albertorio F, Daniel S, Cremer PS (2008) Double cushions preserve transmembrane protein mobility in supported bilayer systems. Langmuir 24(13):6820–6826
17. Tamm LK, McConnell HM (1985) Supported phospholipid bilayers. Biophys J 47(1):105–113
18. Davis RW, Flores A, Barrick TA, Cox JM, Brozik SM, Lopez GP, Brozik JA (2007) Nanoporous microbead supported bilayers: stability, physical characterization, and incorporation of functional transmembrane proteins. Langmuir 23(7):3864–3872
19. Crocker JC, Grier DG (1996) Methods of digital video microscopy for colloidal studies. J Colloid Interface Sci 179(1):298–310
20. Saxton MJ, Jacobson K (1997) Single-particle tracking: applications to membrane dynamics. Annu Rev Biophys Biomol Struct 26(1):373–399
21. McCain KS, Hanley DC, Harris JM (2003) Single-molecule fluorescence trajectories for investigating molecular transport in thin silica sol–gel films. Anal Chem 75(17):4351–4359
22. Garner OB, Baum LG (2008) Galectin-glycan lattices regulate cell-surface glycoprotein organization and signalling. Biochem Soc Trans 36(Pt 6):1472
23. Wennmalm S, Simon SM (2007) Studying individual events in biology. Annu Rev Biochem 76:419–446
24. Kusumi A, Nakada C, Ritchie K, Murase K, Suzuki K, Murakoshi H, Kasai RS, Kondo J, Fujiwara T (2005) Paradigm shift of the plasma membrane concept from the two-dimensional continuum fluid to the partitioned fluid: high-speed single-molecule tracking of membrane molecules. Annu Rev Biophys Biomol Struct 34:351–378
25. Saxton MJ (1997) Single-particle tracking: the distribution of diffusion coefficients. Biophys J 72:1744–1753

Chapter 12

Fluorescence Correlation Spectroscopy to Examine Protein–Lipid Interactions in Membranes

Viktoria Betaneli and Petra Schwille

Abstract

Fluorescence correlation spectroscopy (FCS) is a versatile technique to study membrane dynamics and protein–lipid interactions. It can provide information about diffusion coefficients, concentrations, and molecular interactions of proteins and lipids in the membrane. These parameters allow the determination of protein partitioning into different lipid environments, the identification of lipid domains, and the detection of lipid–protein complexes on the membrane. During the last decade, FCS studies were successfully performed on model membrane systems as also on living cells, to characterize protein–lipid interactions. Recent developments of the method described here improved quantitative measurements on membranes and decreased the number of potential artifacts. The aim of this chapter is to provide the reader with the necessary information and some practical guidelines to perform FCS studies on artificial and cellular membranes.

Key words: Fluorescence correlation spectroscopy, FCS, Cross correlation, Diffusion, Model membranes, Giant unilamellar vesicles, Protein–lipid interactions, Lipid rafts

1. Introduction

Membranes are important and indispensible structural elements of the cell. Their major components are proteins and lipids. For a long time, lipids were primarily considered as the building blocks to provide boundaries and two-dimensional environments for membrane proteins—the active regulators of the cell membrane function. Nowadays, it becomes obvious that the physiological role of lipids was vastly underestimated, and they may be as important constituents of membranes as the proteins. Only collectively, lipids and proteins can fulfill many complex functions of the cell membrane: maintaining gradients of ions, pH, and electrical charge, accomplishing signal transduction, biochemical reactions, the recognition between cells, and helping the membrane to adopt certain shapes. In recent years, a large number of studies using different biochemical and biophysical methods have been performed to shed

Jörg H. Kleinschmidt (ed.), *Lipid-Protein Interactions: Methods and Protocols*, Methods in Molecular Biology, vol. 974, DOI 10.1007/978-1-62703-275-9_12, © Springer Science+Business Media New York 2013

more light on the specific interactions between proteins and lipids and their implications for the biological functions of the membrane. In this chapter, we will describe fluorescence correlation spectroscopy (FCS) as a powerful method to study protein–lipid interactions. This method allows the determination of diffusion coefficients, concentrations, molecular interactions, conformational changes, and many other parameters characteristic for the specific fluorescently labeled molecule. The listed parameters can be used to access different aspects of protein–lipid interactions. Measurements of lipid diffusion coefficients can provide information about lateral lipid heterogeneities in the membrane or so-called lipid "rafts." These dynamic sterol- and sphingolipid-enriched domains supposedly regulate in time- and space-specific protein-–protein and protein–lipid interactions in the membrane. FCS can yield information about protein partitioning into these domains, by measuring local concentrations and diffusion coefficients of fluorescently labeled proteins and lipids on the membrane. These studies can provide new insights into the regulation of protein localization and function by the lipids. Some FCS studies in this area have been successfully performed for membrane proteins like syntaxin and synaptobrevin (1), the mechanosensitive channel MscL, the secondary lactose transport protein LacS (2), and GPI-anchored protein PLAP (3, 4).

Other parameters, which can be accessed by FCS, are the affinity and specificity of protein binding to the membrane. Hydrophobic interactions of protein 4.1 with phosphatidylserine liposomes (SUVs) were measured based on the change in the diffusion time of the fast-diffusing protein upon its binding to slowly diffusing liposomes (5). FCS measurements to access spatio-temporal interaction of protein with the membrane were also performed on GUVs. FCS allowed to follow the incorporation of NaPi II cotransporter into the membrane by measuring diffusion rates of its fluorescent antibody (6). It is worth to mention that the exact amounts of these proteins in the membrane could not simply be obtained from the fluorescence images because of their poor resolution and contrast; therefore, FCS measurements were particularly useful in this work.

The FCS technique was also used to investigate how protein binding to the membrane can modulate structure and dynamics of the lipid membrane. For example, FCS measurements have shown that the binding of cholera toxin to the membrane, containing lipid GM1, alters the lateral diffusion of fluorescently labeled lipids not involved in the binding, which was interpreted as an alteration in the fraction of lipid in the gel phase upon protein binding (7).

Protein–protein interactions in general and protein oligomerization in particular can be regulated by lipids. Fluorescence correlation and cross-correlation spectroscopy (FCCS) are very useful to study molecular interactions. Oligomerization properties of bacteriorhodopsin (8), epidermal growth factor receptor (9), and interleukin-4 and erythropoietin receptors (10) were investigated by this method. The influence of lipid membranes on interactions between apoptotic proteins was also studied by FCCS (11).

By combining FCS and other techniques, one can study the effect of lipids on protein structure and function. For example, the combination of electrophysiology and FCS allowed a quantitative study on the influence of the lipid phase separation on the gramicidin ion channel function (12).

The above-mentioned research work was mostly conducted on model membrane systems. Among the most widely used systems are giant unilamellar vesicles (GUVs), small unilamellar vesicles (SUVs), supported lipid bilayers (SLBs), and black lipid membranes (BLMs). Preparation of some model membranes will also be discussed in this chapter. Lipid and protein composition of these artificial membranes can be accurately controlled, and FCS measurements can be easily performed, which allows distinguishing particular protein–lipid interactions. SUVs can be used to study protein binding to the membrane. They have sizes of 20–50 nm in diameter, and thus, the diffusion of entire SUVs can be followed by FCS in solution. In contrast, GUVs are of 1–300 μm in diameter, and their position can be fixed in the observation chamber, which allows conducting FCS measurements on a fixed spot in the membrane. Another important advantage of GUVs is the possibility to achieve a large-scale lipid phase separation and incorporate membrane proteins (13).

Optically resolved lipid domains can also be obtained in SLBs. In addition, these model membranes are very stable; however, their disadvantage is the potential interaction with the support. Because of this reason, diffusion coefficients measured on SLBs are usually lower compared to those measured on free-standing bilayers like GUVs. In addition, interaction with the support can create difficulties in incorporation of membrane proteins. Another alternative is to use BLMs as a model system. The main drawback is that they are not stable and might contain residual solvent what can also influence the values of diffusion coefficients of lipids and membrane proteins. Additionally, the equipment allowing to carry out FCS measurements in horizontal BLMs has just started to develop (12).

FCS is also a valuable technique to investigate molecular interactions in the membrane of living cells. FCS performed in vivo helped to, e.g., reveal the diffusion coefficient of the membrane-bound form of the Min proteins in *E. coli* (14), to investigate interaction of morphogen FGf8 and its receptors on the membrane (15), to suggest the existence of small heterogeneous membrane domains by detecting anomalous subdiffusion in cells (16), and to estimate raft partitioning of β-secretase inhibitors (17), Fas ligand (18), and AKT protein kinase (19).

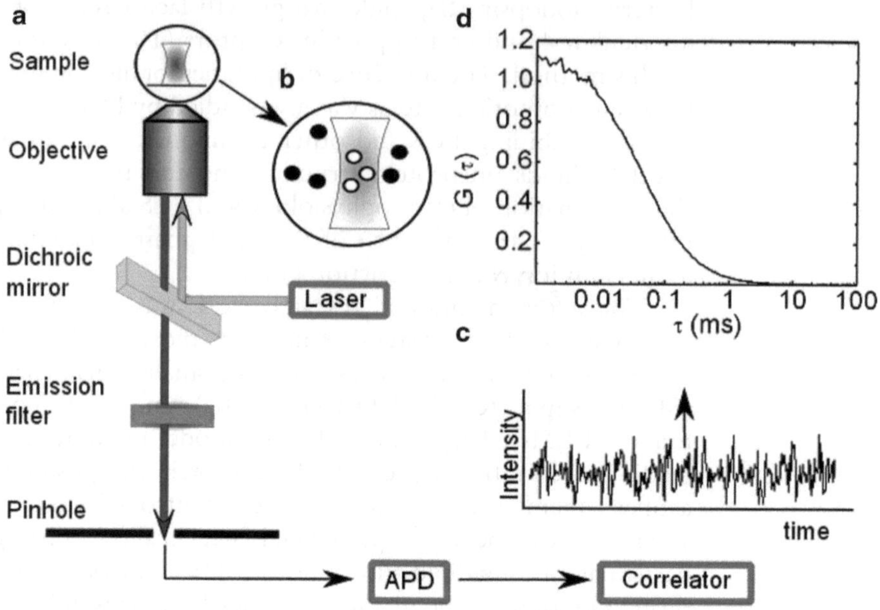

Fig. 1. Principle of FCS. (**a**) FCS setup. The laser light is reflected by a dichroic mirror into a high numerical aperture objective and is focused on the sample. Fluorescent molecules in the sample are excited and detected only if they diffuse through the confocal volume (**b**). The light emitted by the molecules is collected by the objective, and after passing the emission filter and the pinhole, it is detected by an APD, giving rise to a fluctuating intensity signal (**c**) which can be correlated in time to obtain the autocorrelation function (**d**).

The standard FCS setup consists of a confocal inverted microscope equipped with a high-numerical aperture objective. To apply FCS either in vivo or in vitro, the molecules of interest have to be fluorescently labeled. For FCS measurements, the laser is focused on the area of interest (a lipid bilayer or solution with labeled molecules), and fluorophores diffusing in and out of the detection volume induce fluctuations in the fluorescence intensity signal, which is collected by the microscope objective and detected by means of avalanche photodiodes (APD) giving rise to a fluctuating intensity trace $F(t)$ (Fig. 1). By temporally correlating this trace in time or, by other words, measuring the self-similarity of the obtained intensity signal, one can obtain the autocorrelation curve:

$$G(\tau) = \frac{\langle \delta F(t) \cdot \delta F(t + \tau) \rangle}{\langle F(t) \rangle^2}. \tag{1}$$

$\langle F(t) \rangle = (1/T) \int_0^T F(t)dt$ is the time average of $F(t)$ with measurement time T. τ is called "lag time." The experimental autocorrelation curve is fitted to a mathematical model, which is suitable for the given experimental situation (e.g., free diffusion or flow), and important quantities are obtained as fitting parameters. The basic mathematical models used to fit the autocorrelation curves for the molecules diffusing in 2D (membrane) or 3D (volume) are

$$G_{3D}(\tau) = \frac{1}{N} \cdot \frac{1}{(1 + \tau/\tau_d)} \cdot \frac{1}{\sqrt{1 + \tau/S^2 \cdot \tau_d}}, \tag{2}$$

$$G_{2D}(\tau) = \frac{1}{N} \cdot \frac{1}{(1 + \tau/\tau_d)}, \tag{3}$$

here $S = w_z/w_0$ is the so-called structural parameter which describes the ratio of axial w_z and radial w_0 dimensions of the detection volume (see Note 1). The decay time of the autocorrelation curve is the diffusion time τ_d—the time which the molecule spends in the detection volume (see Note 2). If the dimensions of the detection volume are known, one can obtain the diffusion coefficient D:

$$D = \frac{w_0^2}{4\tau_d}. \tag{4}$$

The number of particles N in the detection volume is inversely proportional to the amplitude of the autocorrelation curve:

$$G(0) = \frac{1}{N}. \tag{5}$$

The concentration of labeled molecules can be calculated as

$$C = \frac{N}{V_{eff}}, \tag{6}$$

where the effective detection volume V_{eff} for the 3D Gaussian is determined as

$$V_{eff} = \pi^{3/2} w_0^2 w_z. \tag{7}$$

Another important parameter in FCS is the effective molecular brightness ε_{eff}, which is measured in counts per particle per second (cpps) and is determined as the ratio of average fluorescence count rate and the number of particles in the detection volume:

$$\varepsilon_{eff} = \frac{\langle F(t) \rangle}{N} = \langle F(t) \rangle G(0). \tag{8}$$

This parameter is important not only to characterize the performance of fluorescent labels and particular FCS setups but also to sense the environment change of the fluorophore and to access the oligomerization state of the proteins. For example, if a dimer is formed, it is in principle twice as bright as the monomer.

The binding of molecules can also be reflected in the change of the diffusion time. However, only a substantial increase in particle mass of one binding partner will produce a significant effect on the diffusion time. The diffusion coefficient is connected to the hydrodynamic radius r of a particle by the Stokes–Einstein relation

$$D = \frac{k_B \cdot T}{6\pi \cdot \eta \cdot r}, \tag{9}$$

where k_B is the Boltzmann constant, T the temperature, and η the viscosity.

The hydrodynamic radius for globular particles usually scales with the cubic root of the molecular mass. A globular protein would therefore need to gain eight times its own mass to provide a change in the diffusion coefficient of a factor two, which is often not the case in protein–protein interactions in the fluid phase.

To overcome this problem, two-color fluorescence cross-correlation spectroscopy (FCCS) was developed to study molecular interactions. In this case, the interacting molecules should be labeled with spectrally separated dyes, and the fluorescence should be detected in two separate channels. Similar to the autocorrelation curve, the spectral cross-correlation curve can be obtained from the intensity traces $F_i(\tau)$ and $F_j(\tau)$ measured in different spectral channels:

$$G_x(\tau) = \frac{\langle \delta F_i(t) \cdot \delta F_j(t+\tau) \rangle}{\langle F_i(t) \rangle \cdot \langle F_j(t) \rangle}. \tag{10}$$

If the number of interacting molecules increases, the amplitude of the cross-correlation curve also increases. The degree of binding can be calculated from the amplitudes of auto- $(G_g(0), G_r(0))$ and cross-correlation $(G_{rg}(0))$ curves using the following equations:

$$\frac{C_{rg}}{C_{rg} + C_r} = \frac{G_{rg}(0)}{G_g(0)} \text{ or } \frac{C_{rg}}{C_{rg} + C_g} = \frac{G_{rg}(0)}{G_r(0)}, \tag{11}$$

where C_g, G_r, G_{rg} are concentrations of non-bound single-color molecules and molecules in complexes.

As it was mentioned, the determination of the concentration C and the diffusion coefficient D of the molecule requires knowledge of the dimensions of the focal volume V_{eff} or w_0. These values can be obtained from calibration measurements, using a dye with a known diffusion coefficient. Therefore, such FCS measurements will yield only relative values with respect to the calibration. FCS variants for calibration-free concentration and diffusion coefficient determinations were also developed. Among them are z-scan FCS and various other scanning FCS techniques as described below. In some cases, these methods can reduce the number of artifacts associated with the measurements on the membrane and therefore improve the quality of the data.

2. Materials

2.1. FCS Measurements

1. Confocal laser scanning microscope (see Note 3) equipped with:
   (a) Water immersion objective with numerical aperture >0.9 and correction ring (e.g., Zeiss 40× NA 1.2 UV-VIS-IR C-Apochromat water immersion).

   (b) Lasers of suitable wavelength such as argon ion laser (lines: 514, 488 nm) and helium–neon laser (lines: 543, 633 nm).
   (c) The main beam splitter to separate the excitation from the emission (e.g., 488/543/633 nm dichroic mirror).
   (d) The second beam splitter to split emission from different channels in case FCCS is performed (e.g., 570 nm dichroic mirror).
   (e) Different band-path filters (e.g., 580/60 nm, 700/75 nm, 520/40 nm, 530/60 nm).
   (f) Avalanche photodiodes (APDs) for the detection.
2. Hardware correlator (e.g., Flex 02-01D, correlator.com, Bridgewater, NJ, USA).
3. Chambers or coverslips of no. 1.5 thickness.
4. Nonfluorescent spectral grade immersion water.
5. Solutions for calibration measurements: e.g., Alexa 488, Atto 655, and Alexa 546 dissolved in water at a concentration of 10–100 nM.
6. Software to perform the fitting of correlation curves, e.g., Origin, Matlab, Zeiss ConfoCor, or Leica/ISS software packages.

2.2. Preparation of Giant Unilamellar Vesicles (GUVs) on ITO Coverslips (see Note 4)

1. Lipid mixtures: Lyophilized lipids dissolved in chloroform at a concentration of 10 mg/ml or proteoliposomes (see Note 5) at a concentration of 1–2 mg/ml.
2. Appropriate fluorescent labels (see Note 6).
3. Dark glass vials and parafilm.
4. Round coverslips no. 1.5 thickness and 25 mm in diameter custom coated from one side with indium thin oxide (ITO) (e.g., GeSIM, Grosserkmannsdorf, Germany).
5. Custom-made or commercial flow chambers for GUV formation (Fig. 2).
6. Material to assemble the flow chamber: Disposable syringes of 1 ml volume with appropriate needles, copper conductive wire adhesive on one side, polyethylene tubing matching the size of the inlets in the spacer of the chamber, and vacuum grease (e.g., Glisseal, Borer Chemie, Switzerland).
7. Sucrose solution, matching the osmolarity of the buffer outside the vesicles (300 mM in case PBS is used).
8. Buffer, e.g., PBS: pH 7.4, 3.2 mM Na_2HPO_4, 0.5 mM KH_2PO_4, 1.3 mM KCl, and 135 mM NaCl. The buffer should be nonfluorescent.
9. Sweep/function generator (e.g., Voltcraft 8202, Conrad Electronic SE, Germany) and multimeter.

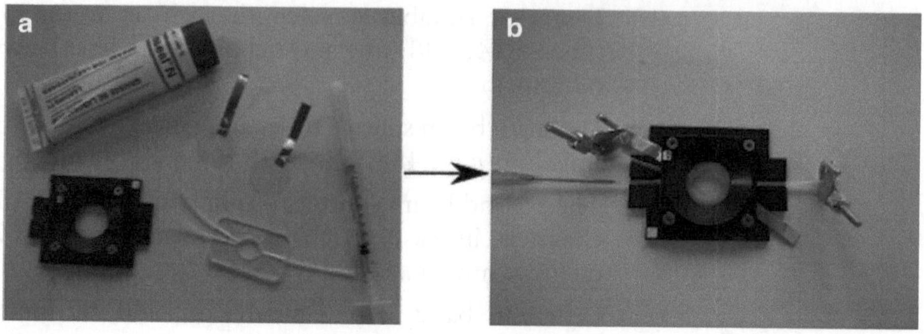

Fig. 2. ITO chamber for GUV formation. (**a**) Metal chamber, plastic spacer with polyethylene tubing, and ITO-covered cover slides with an attached copper wire. (**b**) Assembled chamber for GUV formation: ITO slides are attached to the plastic spacer, which is positioned between the two parts of the metal holder. The solution is injected using a syringe; other inlets are closed by metal clips.

10. Vacuum desiccator.
11. Osmometer.
12. Compressed argon gas with regulator.
13. Heating device.

2.3. Preparation of Supported Lipid Bilayers (SLBs)

1. Lipids dissolved in chloroform or proteoliposomes.
2. Appropriate fluorescent labels (see Note 6).
3. Coverslip no. 1.5 thickness, with the dimensions 24 × 50 mm.
4. Mica sheets and adhesive tape.
5. Eppendorfs with cut bottom, transparent glass vials.
6. SLB buffer: pH 7.4, 10 mM HEPES, and 150 mM NaCl.
7. Solution of 0.1 M $CaCl_2$.
8. Ultraviolet adhesive with refractive index 1.56 (e.g., Norland optical adhesive 63, Norland Products Incorporated, Cranbury, NJ).
9. Ultraviolet lamp.
10. Water-bath-type sonicator.
11. Vacuum desiccator.
12. Compressed argon and nitrogen gas with gas regulator and tube.
13. Heating device.

2.4. Preparation of Cells

1. Cells of interest in culture (see Note 7).
2. Minimum essential medium (check the required media and conditions for established cell lines at www.atcc.org).
3. Plasmids with GFP, mRFP, YFP, etc. fusion proteins of interest (see Note 8).
4. Stock solutions of fluorescent lipid analogs (see Note 6).

5. Observation chambers (cover glass no. 1.5 sealed at the bottom) (e.g., MatTek or LabTek, Nalge Nunc Intl., Rochester, New York, USA).
6. Imaging medium: Phenol red-free medium, e.g., PBS supplemented with Ca^{2+}/Mg^{2+}.
7. Transfection reagents (e.g., Lipofectamine, Invitrogen GmbH, Darmstadt, Germany).
8. Electroporator.
9. Heaters for chambers (e.g., heater/cooler for LabTek and MatTek chambers, Warner Instruments, LLC, Hamden, CT) or the objective (e.g., objective heater/cooler Bioptechs Inc., Butler, PA).

3. Methods

3.1. Preparation of GUVs

1. Measure osmolarity of the buffer, which will be used outside the vesicles by means of osmometer (osmolarity of PBS is 300 mOsm).
2. Prepare sucrose solution with the same osmolarity as the working buffer (1 mM of sucrose has osmolarity of 1 mOsm).
3. Prepare the lipid mixture at desired molar ratio, e.g., DOPC/DOPG 8/2 in chloroform. Add fluorescent lipid or lipid analog at 0.001–0.01 mol% (see Note 9).
4. Determine the conductive surface of ITO by measuring resistance with multimeter. Only ITO-coated side of the slide is conductive.
5. Attach the copper wire to the conductive side of ITO coverslips. Make sure that the copper wire will not be in touch with the buffer solution after assembling and filling the chamber.
6. Spread 5 μl of lipid mixture at concentration 10 mg/ml on the conductive surface of the ITO coverslip. In case proteoliposomes are used, deposit them on ITO coverslip at concentration 1–2 mg/ml of lipids, in the form of several 2 μl drops (0.1–1 mol% of labeled protein).
7. Dry the lipids in desiccator to allow complete evaporation of chloroform (see Note 10).
8. Assemble the flow chamber (Fig. 2). Put the ITO with dried lipids on the lower part of the chamber to be able to observe the vesicles with the inverted microscope. Seal the parts with vacuum grease.
9. Fill the chamber with degassed water or sucrose (see Note 11), using a syringe and polyethylene tubing. Avoid bubbles.

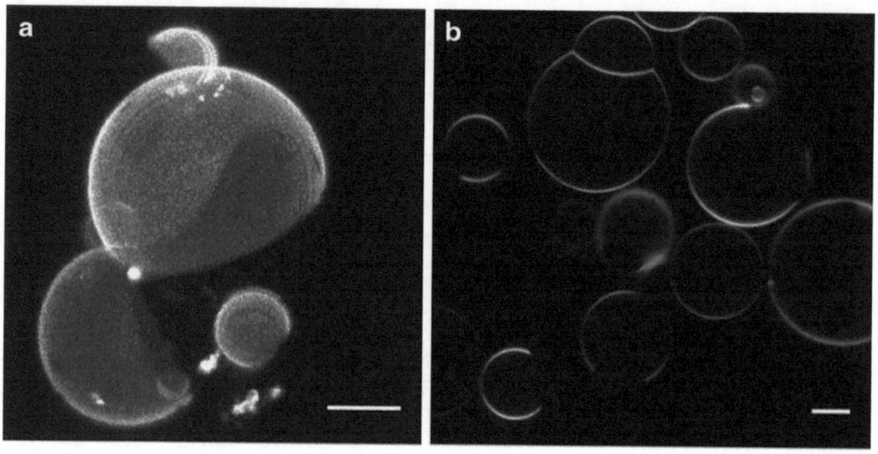

Fig. 3. Confocal fluorescence images of giant unilamellar vesicles with phase coexistence: (**a**) 3D image and (**b**) 2D image. Membrane composition: dioleoylphosphatidylcholine:sphingomyelin:cholesterol (2:2:1). The liquid-disordered phase is labeled with DiD. Scale bar 10 μm.

10. Incubate the chamber 10°C above the temperature of the highest melting temperature of the lipids in the mixture (electric heater can be used).
11. Connect copper wires to the function generator and apply a sinusoidal alternating current electric field 400 V/m at frequency 10 Hz for 1–2 h. If GUVs need to be detached from the coverslip and transferred into another chamber, change the frequency to 2 Hz for 30 min.
12. Turn off the function generator and let the chamber reach room temperature. Carefully exchange the sucrose in the chamber by the buffer using syringe and polyethylene tubing system or transfer GUVs into appropriate observation chamber. Vesicles of 20–100 μm in size should be obtained (Fig. 3).

3.2. Preparation of Supported Lipid Bilayers (SLBs)

1. Mix lipids and fluorescent probes in the required ratio in chloroform (see Note 9).
2. Dry lipids under the nitrogen flow to evaporate the solvent and then 1 h in vacuum desiccator, to remove chloroform traces.
3. Prepare multilamellar vesicles (MLVs): Dissolve dried lipids in SLB buffer at a concentration of 5 mg/ml; mix until lipids are completely dissolved and the solution becomes opaque. Prepare 20 μl aliquots. Aliquots can be kept at −20°C.
4. Add 130 μl of SLB buffer to an MLV aliquot and transfer the solution to the transparent glass vial. Add argon to the vial and warm it up to 37°C.

5. Prepare small unilamellar vesicles (SUVs): Sonicate the MLV solution in a water-bath sonicator for 15 min or until the solution becomes transparent.
6. Prepare the mica. Cleave the mica using adhesive tape; try to obtain a layer of mica as thin as possible.
7. Put the drop of UV adhesive on the coverslip and place the layer of mica on the glue.
8. Cut an Eppendorf tube into two pieces and glue the upper part with UV adhesive around the mica.
9. Place the prepared chamber under a UV lamp for 10 min, until the glue becomes solid.
10. Transfer the SUV solution into the chamber. Incubate the chamber and 3 ml of SLB buffer on the heater above the temperature of the highest melting temperature of the lipids in the mixture.
11. Add 3 μl of $CaCl_2$ 0.1 M to the SUV solution and incubate for 5 min.
12. Add 150 μl of SLB buffer and wait for 15 min. SUVs should fuse and form SLB.
13. Wash SLB with heated SLB buffer (carefully withdraw 75 μl and add 75 μl of fresh buffer). Cool down the sample. For the preparation of membranes with a phase separation, a slow cooling rate is preferable to ensure the formation of big (several micrometers in diameter) lipid domains.
14. To check if the bilayer is prepared properly and SUVs are fused, image the membrane using confocal microscope (Fig. 4), photobleach the sample using high laser powers and observe the recovery of fluorescence after photobleaching.

3.3. Preparation of Living Cells

1. After passaging, seed cells on chambers with cover glass thickness no. 1.5. If electroporation is needed, cells are seeded after electroporation, and measurements should be done 24 h after electroporation.
2. Incubate cells in the required media (check the conditions for established cell lines at www.atcc.org) for at least 24 h (depending on the cell can be longer) to reconstitute to the normal metabolism before transfection or staining. Confluency 40–60 % is optimal for FCS measurement.
3. Perform cell transfection using appropriate transfection reagents. The measurements should be done at least 24 h after the transfection.
4. If cells need to be stained, remove the medium culture, since the components in medium can bind to the stain, decreasing the efficiency of staining. Instead of the medium, 1× PBS

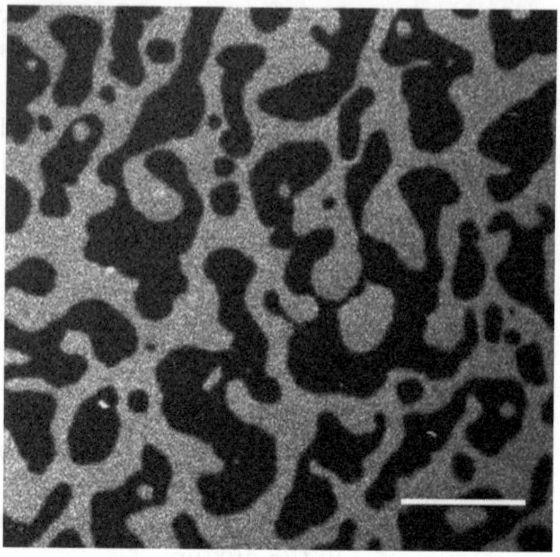

Fig. 4. The confocal fluorescence image of supported lipid bilayers shows phase coexistence in the membrane. Membrane composition: dioleoylphosphatidylcholine:sphingomyelin:cholesterol 1:1:0.7. The liquid-disordered phase is labeled with BODIPY-FL cholesterol. Scale bar 10 μm.

supplemented with Ca^{2+} and Mg^{2+} (these bivalent ions are necessary for attachment of the cells) can be used as staining buffer. Then, the stain (e.g., a fluorescent lipid analog) is added to the buffer at an optimal concentration for the measurement. In case of membrane staining, the procedure can be done at 4°C to avoid endocytosis. Otherwise, 37°C is optimal.

5. Transfer cells into imaging medium (phenol red-free medium, e.g., PBS with Ca^{2+}/Mg^{2+}).
6. The measurements should be carried out either at room temperature or at 37°C. To reach 37°C, the sample or the objective can be heated.

3.4. Alignment of the Setup and FCS Measurements in Solution

The system has to be aligned for all the channels, which will be used, and for all the methods listed below to ensure optimal conditions during data acquisition. If FCCS will be used, both channels have to be aligned simultaneously. Experiments should be performed in a dark room at a constant temperature.

1. Switch on the lasers at least 30 min before the measurements to stabilize the system.
2. Select the configuration of the microscope appropriate for the fluorophore under study. Choose the necessary dichroic mirrors. Select appropriate bandpass filters, broad enough to collect as much of the fluorescence emission of the used dye as possible, but, on the other hand, to exclude the excitation wavelengths. If FCCS is performed, narrower double-bandpass

filters are needed to reduce the cross talk between two spectral channels. Measure the temperature close to the objective or directly inside the sample. The temperature should be stable during the experiment.

3. Put the drop of the 10–100 nM dye solution with a known diffusion time and diffusion coefficient, e.g., Alexa 488 or (and) Atto 655, on the coverslip of the same thickness as the one, which will be used later in the experiment. The same dyes or dyes with similar spectral properties should be used for the alignment and measurements. For FCCS measurements, spectrally well-separated fluorophores with emission maxima separated by at least 30 nm should be used.
4. Place the coverslip directly on the drop of immersion water on the objective. Cover the sample to avoid evaporation. If the observation chamber is used instead of a coverslip for alignment or calibration measurements, place the focal volume well inside the dye solution, approximately 100 μm above the surface of the coverslip.
5. Illuminate the solution with the appropriate laser and start single-spot FCCS measurements. Select the minimum scanning speed. Detect the fluorescence signal with APDs and monitor the intensity trace. Use the same laser power as later in the experiment (see Note 12).
6. Choose the pinhole diameter 70–100 μm for objectives with numerical aperture >0.9. Align the pinhole in *x* and *y* directions to maximize fluorescence count rate on the detector or counts per molecule (molecular brightness).
7. Move the correction ring of the objective till the maximum count rate or maximum counts per molecule is obtained to correct for the cover glass thickness.
8. If FCCS will be used in systems with different pinholes for detection channels, maximize the overlap of red and green detection volumes:
   (a) Adjust green channel pinhole using short-wavelength excitation and green fluorescence dye.
   (b) Adjust red channel pinhole using long-wavelength excitation and red fluorescence dye; note pinhole position.
   (c) Adjust the red channel pinhole using short-wavelength excitation and red fluorescence dye; note the pinhole position.
   (d) Compare positions of the red channel pinhole obtained with the excitation of the short and long wavelengths; if they match, the overlap is optimal.

9. Record FCS curves. 100 s measurements are usually enough to obtain an average correlation curve. Fit the FCS curve by the formula in Eq. 2 (see Note 13).
10. Get the parameters S and τ_d from the fit. If the alignment was done properly, values for S should be between 4 and 6. The results should be reproducible and in agreement with the literature data.
11. If calibration is required for further experiments, the beam waist w_0 can be determined from Eq. 4. In this case, the diffusion coefficients of the used dyes should be known. The values for the diffusion coefficients of some fluorescent dyes can be found in the literature (24, 25). Diffusion coefficients depend on the viscosity according to Eq. 9. The viscosity of solvent and water depends on temperature (26); therefore, if the calibration experiment is not performed at 25°C, the tabulated values of the diffusion coefficients should be recalculated according to

$$D(T) = D(25^\circ C) \cdot \frac{t + 273.15}{\eta(T)} \cdot 2.985 \cdot 10^{-6}\, Pa \cdot s \cdot K^{-1}, \quad (12)$$

where $\eta(T)$ is the viscosity of the solvent and t is the temperature in °C.

3.5. Point FCS and FCCS on the Membrane

The advantage of point FCS is that it is a fast technique for routine measurements, which is relatively easy to perform. It can be easily purchased as a "build-in" or additional module in many commercial confocal microscopes, and therefore, no specialized software or equipment is needed for the measurements. Auto- and cross correlation in the microsecond range can be resolved by this method, and high-quality FCS curves can be obtained because photons are collected very fast. However, for determination of absolute values of diffusion coefficients and concentrations, calibration measurements are required. From these measurements, the size of the focal volume can be accessed if absolute values of diffusion coefficients and concentrations of used fluorescent probes are known. However, only for a few dyes, reliable diffusion coefficients are published, which in addition depend quite strongly on the temperature, what was often not considered in the literature. Other problems of point FCS are photobleaching and stable positioning of the focal spot on the membrane. The latter one can be particularly dramatic, if the membranes are not stable and fluctuate in z direction in the focus during the measurements (e.g., in case of cell membranes), what can significantly change diffusion times and introduce artifacts in cross-correlation values. However, for stable membranes like membranes of GUV or SLB, point FCS can be successfully performed.

1. Place the sample on the microscope; image GUV, cell, or SLB. In case of GUVs, focus on the upper pole. In case cells are used, focus on lower pole (membrane is more stable there).
2. Focus the laser beam onto the membrane plane, which corresponds to the highest fluorescence intensity in z position, using the "spot mode" in the laser scanning microscope (LSM) software (see Note 14). It is important to use small steps (200 nm) in z direction to detect the highest fluorescence intensity. Use a relatively low laser power to obtain counts per molecule below 10 kHz. For proteins with several labels, this-value has to be multiplied by the number of labels per molecule.
3. Perform several FCS measurements with the duration of 30–100 s (see Note 15).
4. Fit the FCS curve using Eq. 3. Determine the diffusion time τ_d and the number of molecules per focal area N.
5. The diffusion coefficient can be found using Eq. 4 if the waist w_0 was determined by calibration measurements (see Subheading 3.3, step 9).

   In case FCCS measurements are performed:
6. Record cross-correlation signal.
7. Fit auto- and cross-correlation curves by Eq. 3. Determine effective numbers of particles N_g and N_r from autocorrelation curves and the amplitude of the cross-correlation curve $G_{rg}(0)$. Take the inverse of N_g and N_r, what will be equal to amplitudes of the autocorrelations $G_g(0)$ and $G_r(0)$.
8. From the amplitudes of cross and autocorrelations, determine the degree of cross-correlation binding using Eq. 11. Compare the values with positive-control cross correlation (see Note 16).

3.6. z-Scan FCS

In the z-scan FCS method, autocorrelation curves are recorded at different positions along the z-axis of the excitation volume relative to the membrane plane. Compared to point FCS, this technique does not require extrinsic calibration. It allows the determination of diffusion coefficients and concentrations of the molecules of interest, along with the waist of the focal volume directly from the dependency of the diffusion time on the z position. The problem of correctly positioning of the focus on the membrane plane is thus solved here. The minimum values of the number of particles and of the diffusion time correspond with the position on the membrane plane.

The disadvantage of the technique is that it is time consuming for measurements and also depends on the optical properties of the system. Nevertheless, z-scan FCS was successfully performed on GUVs (27) and SLBs (28). However, it can be difficult to

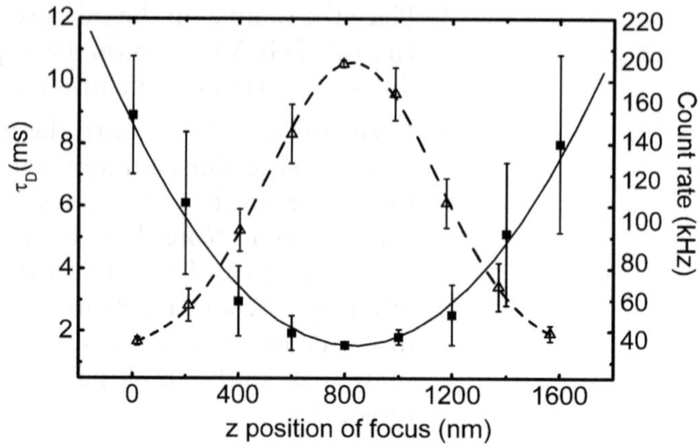

Fig. 5. Example of a z-scan FCS measurement. Dependences of the diffusion time τ_d (*squares*) and the fluorescence count rate (*triangles*) on the *z* position of the focal volume for a free-standing membrane. The data was kindly provided by Dr. Fabian Heinemann, TU Dresden.

perform the measurements on living cells because of the membrane movements.

1. Place the observation chamber with the sample on the microscope. Using the imaging mode of LSM, find GUV and try to focus on top of the vesicle. If the measurements will be performed on SLB, focus approximately on the membrane plane.
2. Find the membrane plane by determining the maximum fluorescence intensity along the *z*-axis.
3. Position the focus 1 μm below the membrane plane.
4. Move the focus up in 0.2 μm steps and record correlation curve for each step. For a reliable z-scan analysis, it is recommended to acquire data at a minimum of 7–9 different *z* positions and average over several z-scans.
5. Fit all obtained correlation curves using Eq. 3 and determine τ_d and N values for each focus position. Usually, a nonlinear least-square algorithm is used for fitting.
6. Plot obtained parameters τ_d and N against the focus position z (Fig. 5).
7. Fit the plotted dependencies using equations

$$\tau_d(\Delta z) = \frac{w_0^2}{4D} \cdot \left(1 + \frac{\lambda_0^2 \Delta z^2}{\pi^2 n^2 w_0^4}\right), \quad (13)$$

$$N(\Delta z) = \pi c w_0^2 \cdot \left(1 + \frac{\lambda_0^2 \cdot \Delta z^2}{\pi^2 n^2 w_0^4}\right), \quad (14)$$

where N is the particle number in the focal volume, c is area concentration of fluorophores in the waist of the focus w_0, λ_0 is the wavelength of the excitation light, and n is the refractive index of the medium.

Determine the diffusion coefficient D, the minimal radius of the laser beam w_0, and area concentration N.

3.7. One-Focus Scanning FCS Perpendicular to the Membrane Plane

Scanning FCS is a powerful method to accurately determine diffusion and binding parameters in membranes. The advantages of this technique are the reduced photobleaching and possibility to correct for membrane undulations and instabilities, as well as to perform long time measurements (up to half an hour) on the membranes, which allows studying slow diffusion. In addition, the method is insensitive to focus mispositioning on the membrane. However, compared to point FCS, the measurement times are longer, and a movable detection volume is required to perform measurements. Also, auto- and cross-correlation curves cannot be resolved below the microsecond range and therefore do not capture free solution diffusion or photophysical dynamics. The quality of FCS curves is also lower compared to point FCS because the number of obtained points to generate the FCS curve is limited here by the scanning speed, and the fluorescence counts per molecule are lower because of the bigger detection area. One-focus scanning FCS is not a calibration-free technique; however it allows to implement an alternating excitation scheme, which can be important to reduce the cross talk in FCCS measurements if the fluorophores with overlapping emission spectra cannot be avoided. Here the method developed by Ries et al. (20) is described (Fig. 6).

1. Acquire a fluorescence image of the cell or GUV (SLBs cannot be used for these measurements). Focus on the equatorial plane of the vesicle or cell.
2. Select the linear scan path perpendicular to the membrane with a length of approximately 15 μm and scan it repeatedly in a "line scan mode" of the LSM with a repetition rate of 1.3 kHz. Try to position only the short part of the scanned line inside the membrane to avoid photobleaching.
3. Detect the emitted fluorescence and acquire the photon arrival times using the photon mode of the hardware correlator.
4. Using mathematical software, like Matlab, bin the data in bins of 2 μs and perform a fast Fourier transform. To determine the exact repetition rate of the line scan, arrange the binned data stream in a matrix with rows corresponding to one line scan (see Note 17). The matrix can be displayed as an image, where membrane movements can be visualized as a deviation of the image trace from the straight line (Fig. 6).

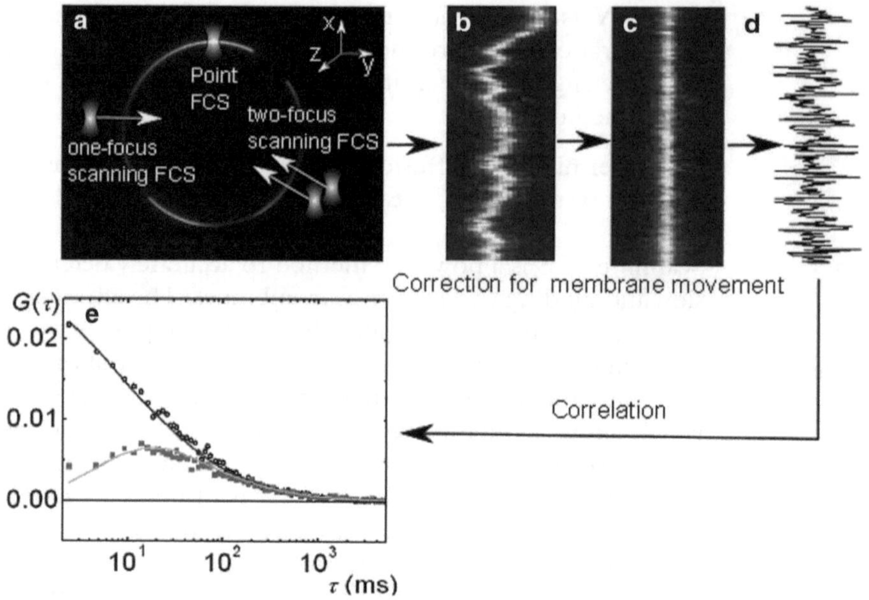

Fig. 6. Scanning FCS perpendicular to the membrane plane. (**a**) For one-focus scanning FCS, a single line is scanned repeatedly through the membrane; for two-focus scanning FCS, two lines separated by a known distance are scanned through the membrane. (**b**) The data obtained for each focus is arranged in a matrix with rows corresponding to one line scan. (**c**) The maxima signals are shifted relative to each other along one line to correct for membrane movements. (**d**) Photons for each line scan are summed up to obtain an intensity value. (**e**) The intensity trace is correlated to obtain the autocorrelation curve (*circles*) and one cross-correlation curve (*squares*) (in case two-focus FCS was used).

5. To correct for membrane instabilities, evaluate the membrane positions in windows of several hundred scans by finding the position of the maximum signal. Then, all line scans should be shifted relative to each other to align all maxima in one line.
6. Estimate the width of the membrane trace from the obtained matrix. Collect photons in a window significantly larger than this width and sum them up for each line scan to obtain intensity F (see Note 18).
7. Calculate the correlation curve $G(\tau_i)$ with the multiple-tau correlation algorithm (see Note 19).
8. Fit the correlation curve using equation

$$G(\tau) = \frac{1}{C\pi w_0 w_z} \cdot \left(1 + \frac{4D\tau}{w_0^2}\right)^{-(1/2)} \cdot \left(1 + \frac{4D\tau}{w_z^2}\right)^{-(1/2)}$$
$$= \frac{1}{N}\left(1 + \frac{\tau}{\tau_d}\right)^{-(1/2)} \cdot \left(1 + \frac{\tau}{\tau_d S^2}\right)^{-(1/2)}. \quad (15)$$

Determine N, τ_d, and S (see Note 20).

3.8. Two-Focus Scanning FCS and FCCS Perpendicular to the Membrane Plane

Two-focus scanning FCS is an extension of the scanning FCS technique. In this method, two lines parallel to each other and separated by the known distance are scanned through the membrane (Fig. 6). Absolute values of the concentration, the diffusion coefficient, and the focal volume size can be determined by two-focus scanning FCS without any calibration measurements, which was not possible with one-focus scanning FCS. The implementation of a dual-focus and dual-color FCS scheme using two spectral and two spatial channels allows circumventing the calibration of the detection volumes and their overlap, as well as false-positive cross correlation due to membrane movements (29).

1. Calibrate the distance d between two scanned lines. Prepare a film of a dried fluorophore on the coverslip. Scan two lines on a well-defined distance: choose multitrack mode in the LSM software; select bidirectional scanning mode with the lasers switched on only during the back scanning. Scan the frame with the size $N \times 2$ pixels. Bleach traces on the film of the dried fluorophore using these settings (see Note 21). Determine the distance d between the lines.
2. Acquire a fluorescence image of GUV or cell (SLBs cannot be used for these measurements). Focus on the equatorial plane of the vesicle.
3. Scan two lines separated by the distance d perpendicular to the membrane with the settings noted in step 1. In case FCCS is performed, both lasers implemented for cross correlation should be on for the back scanning.
4. Repeat steps 3–7 of the Subheading 3.7. with the difference that now two lines corresponding to two spatial channels should be analyzed in case of one-color measurements and four lines corresponding to two spectral and two spatial channels in case of two-color measurements. One should mention, however, that photons in the two foci are collected with a delay t_d, which is given by the scan period. Therefore, cross-correlation curves are shifted relative to autocorrelation curves according to equation

$$G_{cc}^{d}(\tau) = G_{cc}(\tau + t_d) \propto F_1(t)F_2(t + t_d + \tau)dt. \qquad (16)$$

5. Globally (simultaneously) fit autocorrelation curves using Eq. 15. Fit two-focus single-color red (r) and green (g) cross-correlation curves and two-focus two-color (red–green, rg) cross-correlation curves using the following equations:

$$G_n = \eta_r \eta_g C_n \frac{2}{\pi} a(\tau)^2 b(\tau) \exp[-2d_{ij}^2 b(\tau)^2 - 2d^2 a(\tau)^2], \qquad (17)$$

where $a(\tau) = (8D_i\tau + w_i^2 + w_j^2)^{-(1/2)}$, $b(\tau) = (8D_i\tau + w_i^2 S_i^2 + w_j^2 S_j^2)^{-(1/2)}$, $i, j = r, g$, $i \neq j$, $n = r, g, rg$.

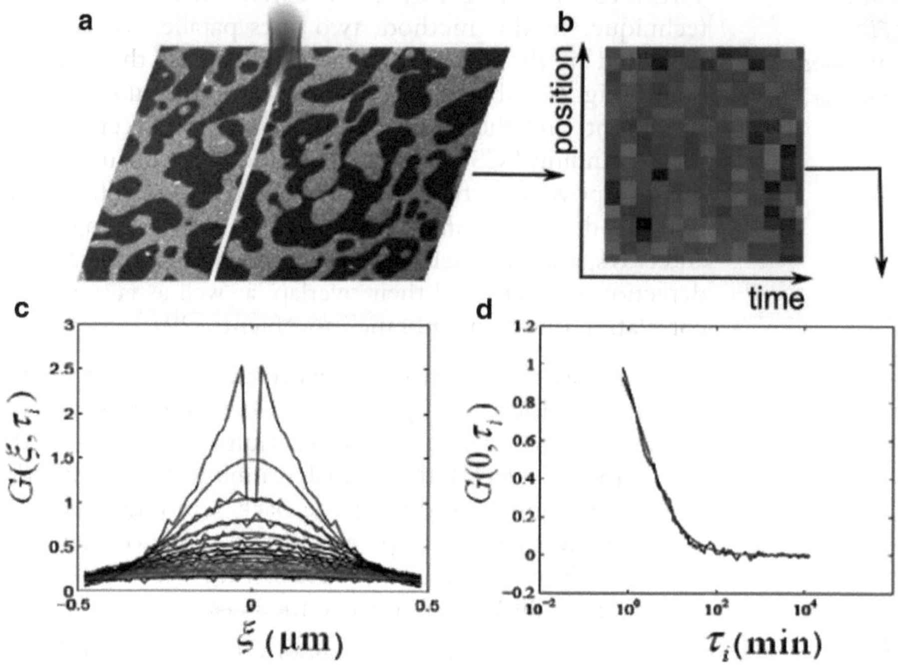

Fig. 7. Principle of line scan FCS in the plane of the membrane. (**a**) The fluorescence intensity signal along the line scan path is collected. (**b**) For each line scan, *i* intensity traces are arranged according to position (x_i) and time (τ_i) to form a pseudo-image. (**c**) Spatiotemporal correlation curves $G(\xi, \tau_i)$ are calculated for different τ_i values and are presented along with a fit to Eq. 18. (**d**) Temporal correlation curve $G(0, \tau_i)$ with a fit of Eq. 19.

For two-focus single-color cross-correlation curves, assume $d_{ij} = w_j = 0$.

In these equations, d_{ij} describes overlap between red and green detection volumes, d is the distance between two lines, $\eta_r = (F_r - kF_g)/(C_r + C_{rg})$ and $\eta_g = F_g/(C_g + C_{rg})$ are channel brightnesses, F_g and F_r are measured mean intensities which can be corrected for the background, and $k = F_r/F_g$ is cross talk coefficient from the green fluorophore to the red channel.

6. Determine diffusion coefficients; waists w_r, w_g; and concentrations C_r, C_g, C_{rg} (see Note 22). Use Eq. 11 to determine the degree of binding.

3.9. Line Scan FCS Parallel to the Membrane

Line scan FCS, which belongs to the class of image correlation techniques, is also calibration-free and allows accurate and fast determination of diffusion coefficients and concentrations in lipid membranes. In contrast to the scanning FCS method presented above, the measurements are performed on the membrane parallel to the membrane plane (Fig. 7). This allows carrying out measurements not only on GUVs and cells but also on supported lipid membranes (30). Unfortunately, no cross-correlation scheme was introduced so far to this technique.

1. Acquire a fluorescent image of a GUV, cell, or SLB. If measurements are performed on SLB, the correction ring has to be realigned to correct for the thickness of the mica. For this, scan a frame (e.g., 32 × 32 pixels) repeatedly on SLB while recording the intensity trace and aligning the correction ring to get the maximum intensity.
2. Focus on the top of the GUV, on the plane of the SLB, or on the bottom of the cell or the supported plasma membrane patches (30). Try to find the position of the membrane by finding the maximum intensity.
3. Select the linear scan path in the LSM operation software. The length of the scan should be between 10 and 30 μm.
4. Choose the maximum repetition rate (1.3 kHz). Note the nominal scan speed v displayed by the LSM software.
5. Detect the emitted fluorescence and acquire the photon arrival times using the photon mode of the hardware correlator with a time resolution of 0.15 μs.
6. Arrange the intensity traces of the line scans i vertically to form the pseudo-image $F(x, t_i)$, where $t_i = iT$ (T is scanning period). Due to scanning, the intensity $F(x, t_i)$ will be acquired at the time $t_i + x/v$. The horizontal axis on this image will denote the position x in the sample and the vertical axis time t_i.
7. Use the obtained pseudo-image to calculate the spatiotemporal correlation curve $G(\xi, \tau_i)$ and the temporal correlation curve $G(0, \tau_i)$:

$$G(\xi, \tau_i) = \frac{\langle \delta F(x, t_i) \cdot \delta F(x + \xi, t_i + \tau_i) \rangle}{\langle F(x, t_i) \rangle^2}, \tag{18}$$

where $\delta F(x, t_i) = F(x, t_i) - \langle F(x, t_i) \rangle$, ξ is a spatial variable which can be restricted to $\xi = -1.5\,\mu m \ldots + 1.5\,\mu m$ and $\tau_i = iT$ is discrete lag time, which can be restricted to $\tau_i = \tau_1 \ldots \tau_{50}$ (see Note 23).
8. Globally fit the obtained correlation curve to the following equations:

$$G(\xi, \tau_i) = \frac{1}{C\pi\omega_0^2} \cdot \exp\left(-\frac{\xi^2}{w_0^2 + 4D(\tau_i + \xi/v)}\right) \cdot \left(1 + \frac{4D}{w_0^2}\left(\tau_i + \frac{\xi}{v}\right)\right)^{-1}, \tag{19}$$

$$G(0, \tau_i) = \frac{1}{C\pi S^2}\left(\sqrt{\pi}\mu_i erf(\mu_i) + e^{-\mu_i^2} - 1\right), \tag{20}$$

with $\mu_i = \dfrac{S}{\sqrt{w_0^2 + 4D\tau_i}}$.
9. Determine the diffusion coefficient D, the waist w_0, and the concentration C from the fit.

4. Notes

1. At the high laser powers applied for FCS, the probability of the transition between the first excited singlet and triplet state increases in most fluorescent dyes. The excited state lifetime for the triplet state is in the microsecond range.

   Fluorophores start to blink on and off because of intersystem crossing. Intersystem crossing changes the quantum efficiency of the system. The fluorescent molecules, in this case, can be divided on two fractions: the nonfluorescent fraction T and the fluorescent fraction:

$$X_{triplet}(\tau) = \frac{1 - T + T \cdot e^{-(\tau/\tau_{triplet})}}{1 - T}.$$

   The correlation function must be multiplied by this fraction; then, the expression for correlation function will be

$$G_{total}(\tau) = X_{triplet}(\tau) \cdot G_{3D}(\tau) \; or \; G_{total}(\tau) = X_{triplet}(\tau) \cdot G_{2D}(\tau).$$

2. In living cell membranes, diffusion can deviate from normal Brownian diffusion. These deviations might be due to the confinement of the particles in lipid domains or interactions with other cellular structures. This phenomenon is called anomalous diffusion. In case of anomalous diffusion, the mean square displacement $\langle r^2 \rangle$ is no longer directly proportional to time t, but depends on t^{α}, so in Eqs. 2 and 3, τ/τ_d has to be replaced by $(\tau/\tau_{anom})^{\alpha}$.
3. A commercial microscope (such as Zeiss Confocor 2) can be used for point FCS and z-scan FCS, but the detection volume cannot be moved in the standard FCS mode. This problem was solved for Confocor 3 or FCS2 from Leica, where scanning FCS measurements can be performed. Otherwise, a home-built setup based on a Zeiss LSM 510 meta with a constructed FCS detection unit on the optical output can be implemented for scanning FCS (20).
4. GUVs can be prepared not only on ITOs but also on platinum wires. In this case, the custom-made chambers with two platinum wires are used, the lipid solution is dried on the wires, and an electric field between the wires is applied. GUVs can be then mechanically transferred into the observation chamber.
5. Proteoliposomes can be prepared by different methods. In general, purified protein in detergent is mixed with detergent-destabilized SUVs (lipid/protein ratio 40/1, w/w) and incubated for 30 min. Then the detergent is removed by BioBeads or by means of size exclusion chromatography for several hours (21). Proteoliposomes can be kept at –20°C.

6. The selection of fluorophores is an important issue for FCS. Fluorophores with high quantum yield, low photobleaching, and no or little blinking and triplet population should be chosen. As membrane dyes, fluorescent lipid analogs DiD, DiO, and DiI or fluorescent lipids like BODIPY-cholesterol or Alexa647-DOPE can be used. Purified proteins can be labeled with cysteine- or amine-reactive dyes like Atto and Alexa. For in vivo measurements, fusion fluorescent proteins like eGFP, mCherry, or mRFP are employed. It should be considered that fluorescent labeling of proteins or lipids can change their physicochemical properties (e.g., the partitioning into Lo or Ld domain in model membranes) (for review see ref. 22). If dyes will be used for cross correlation, spectrally well-separated dyes should be chosen.
7. If possible, for FCS studies, cell lines with less flexible membranes are preferable, e.g., human embryonic kidney 293 cells (HEK 293 cells), because membrane undulations and movements can create serious artifacts in FCS measurements.
8. An appropriate plasmid has to be chosen, depending on the protein of interest, to ensure proper protein folding and reduce the number of aggregates in the cell, as the appearance of aggregates disturbs correlation curves. Also, promoters for high expression (e.g., CMV promoter) should be avoided, because very high concentrations are not recommended for FCS measurements.
9. Save lipids in dark glass vials, sealed with parafilm, in argon atmosphere at −20°C (to avoid evaporation of the solvent and lipid oxidation). Let the lipid stock reach room temperature before opening the vial, to avoid water condensation inside the vial. Try to minimize the time that the vial is open, to avoid the evaporation of the solvent and a change in lipid concentration. Avoid contact of chloroform with plastic, wear protective gloves, and work under the hood.
10. To evaporate organic solvents from the lipid solution, it should be dried 5–10 min in a desiccator. To prepare the lipid film from proteoliposomes, longer times usually are required. The drying procedure should be optimized for the particular protein. Solutions can be dried from 10 min to 8 h at room temperature or at 4°C. Sometimes, sugars or the vapor pressure of a saturated NaCl solution are recommended for use during the drying procedure, to maintain the protein functionality (1, 2, 21).
11. GUVs can also be prepared in buffer; in this case, a modified protocol should be used (23).
12. Usually, laser powers of 1–10 μW are used. In general, it is better to check for the used sample, which is the highest laser power

that can be used without photobleaching. Photobleaching can be evidenced as a decrease in the diffusion time of the autocorrelation curve at higher laser powers.

If APDs are used for the detection, count rates below 300 kHz are recommended.

13. Data can be fitted using Zeiss Confocor 2 (3) Leica software. Alternatively, FCS curves can be exported to Origin or other analysis programs providing a nonlinear least-squares fitting algorithm.
14. Correct positioning of the confocal volume on the membrane is very important for obtaining a correct FCS curve. The fluorescence intensity should be checked after the measurements, to ensure the correct focus positioning on the membrane. Displacement in z direction of the focus from the membrane plane will lead to illumination of a bigger area at the membrane and therefore increase in the diffusion time.
15. In general, the measurement time should be 10,000 times longer than the diffusion time of the molecules of interest. However, the movement of the membrane and photobleaching do not allow very long measurements with single-point FCS. In this case, many repetitive scans have to be done.
16. Usually, a double-labeled sample is used as an FCCS standard. The dyes used for the labeling should be photostable and lack cross talk or FRET. So far, the best cross correlation standards are approximately 40 base pairs long (to avoid FRET) DNA oligonucleotides, labeled with red and green fluorescent dyes, both at $3'$ or $5'$ prime ends at a 1:1 ratio.
17. The software in commercial setups like Zeiss Confocor 3 allows the direct determination of the matrix.
18. To correct the intensity trace $F(t)$ for photobleaching, the following empirical formula can be used (20):

$$F_c(t) = \frac{F(t)}{\sqrt{f(t)/f(0)}} + f(0)(1 - \sqrt{f(t)/f(0)}),$$

where $f(t)$ is an analytical function describing the decrease in the intensity trace $F(t)$. For the restricted reservoir, this decrease is approximated by the exponential function $f(t) = f_0 e^{-t/\tau_b}$, whereas, if the reservoir is connected weakly to a larger reservoir, a multi-exponential function is used:

$$f(t) = f_1 e^{-t/\tau_1} + \ldots + f_n e^{-t/\tau_n}.$$

19. It is recommended to correlate parts of the intensity trace (around 50 s) manually to exclude distorted correlation curves, especially in the systems where fluorophores tend to aggregate.

20. Triplet contribution can be neglected in scanning FCS, because triplet lifetimes cannot be resolved by this technique.

21. The distance between two lines depends on the scan direction and is also not constant along the scan. Therefore it is recommended to use the same magnification and scanning direction during the measurements (e.g., zoom 12 and frame size 32 × 2) and the position of the scan path relative to the GUV membrane.

22. If there are too many fitting parameters, the data becomes less reliable. It is better to reduce the number of fitting parameters by fixing the parameters, which can be determined in calibration measurements. For example, the structural parameters S_r and S_g can be fixed during the fit, as they were already determined during the alignment of the setup. Diffusion coefficients D_r, D_g, and D_{rg} can be assumed approximately equal for the entities in the membrane, the diffusion of which follows Saffman–Delbrueck equation (Jonas comprehensive).

23. To correct the intensity trace $F(x, t_i)$ for depletion, use the following formula:

$$F_c(x, t_i) = \frac{F(x, t_i)}{\sqrt{f(t_i)/f(0))}} + f(0) \cdot (1 - \sqrt{f(t_i)/f(0)}).$$

To correct for inhomogeneities, correct $\delta F(x, t_i)$ in formula (18) using the following equation:

$$\delta F^c(x, t_i) = F^c(x, t_i) - \langle F^c(x, t_i) \rangle_{t_i}.$$

Acknowledgments

We thank Dr. Fabian Heinemann (BIOTEC, TU Dresden) for the critical reading of this chapter and Erdinc Sezgin (BIOTEC, TU Dresden) for the help with the protocol describing the preparation of living cells for FCS. This work was supported by Dresden International Graduate School for Biomedicine and Bioengineering.

References

1. Bacia K, Schuette CG, Kahya N et al (2004) SNAREs prefer liquid-disordered over "raft" (liquid-ordered) domains when reconstituted into giant unilamellar vesicles. J Biol Chem 279:37951–37955
2. Doeven MK, Folgering JHA, Krasnikov V et al (2005) Distribution, lateral mobility and function of membrane proteins incorporated into giant unilamellar vesicles. Biophys J 88:1134–1142
3. Chiantia S, Ries J, Kahya N et al (2006) Combined AFM and two-focus SFCS study of raft-exhibiting model membranes. Chemphyschem 7:2409–2418

4. Kahya N, Brown DA, Schwille P (2005) Raft partitioning and dynamic behavior of human placental alkaline phosphatase in giant unilamellar vesicles. Biochemistry 44:7479–7489
5. Takakuwa Y, Pack CG, An XL et al (1999) Fluorescence correlation spectroscopy analysis of the hydrophobic interactions of protein 4.1 with phosphatidyl serine liposomes. Biophys Chem 82:149–155
6. Ruan Q, Cheng MA, Levi M et al (2004) Spatial–temporal studies of membrane dynamics: scanning fluorescence correlation spectroscopy (SFCS). Biophys J 87:1260–1267
7. Forstner MB, Yee CK, Parikh AN et al (2006) Lipid lateral mobility and membrane phase structure modulation by protein binding. J Am Chem Soc 128:15221–15227
8. Kahya N, Wiersma DA, Poolman B et al (2002) Spatial organization of bacteriorhodopsin in model membranes. Light-induced mobility changes. J Biol Chem 277:39304–39311
9. Liu P, Sudhaharan T, Koh RML et al (2007) Investigation of the dimerization of proteins from the epidermal growth factor receptor family by single wavelength fluorescence cross-correlation spectroscopy. Biophys J 93:684–698
10. Worch R, Bökel C, Höfinger S et al (2010) Focus on composition and interaction potential of single-pass transmembrane domains. Proteomics 10:4196–4208
11. García-Sáez AJ, Ries J, Orzáez M et al (2009) Membrane promotes tBID interaction with BCL(XL). Nat Struct Mol Biol 16:1178–1185
12. Honigmann A, Walter C, Erdmann F et al (2010) Characterization of horizontal lipid bilayers as a model system to study lipid phase separation. Biophys J 98:2886–2894
13. Schwille P, Diez S (2009) Synthetic biology of minimal systems. Crit Rev Biochem Mol Biol 44:223–242
14. Meacci G, Ries J, Fischer-Friedrich E et al (2006) Mobility of Min-proteins in *Escherichia coli* measured by fluorescence correlation spectroscopy. Phys Biol 3:255–263
15. Ries J, Yu SR, Burkhardt M et al (2009) Modular scanning FCS quantifies receptor-ligand interactions in living multicellular organisms. Nat Methods 6:643–645
16. Schwille P, Korlach J, Webb WW (1999) Fluorescence correlation spectroscopy with single-molecule sensitivity on cell and model membranes. Cytometry 36:176–182
17. Rajendran L, Schneider A, Schlechtingen G et al (2008) Efficient inhibition of the Alzheimer's disease beta-secretase by membrane targeting. Science 320:520–523
18. Cahuzac N, Baum W, Kirkin V et al (2006) Fas ligand is localized to membrane rafts, where it displays increased cell death-inducing activity. Blood 107:2384–2391
19. Lasserre R, Guo X-J, Conchonaud F et al (2008) Raft nanodomains contribute to Akt/PKB plasma membrane recruitment and activation. Nat Chem Biol 4:538–547
20. Ries J (2008) Advanced fluorescence correlation techniques to study membrane dynamics. Technical University Dresden
21. Girard P, Pécréaux J, Lenoir G et al (2004) A new method for the reconstitution of membrane proteins into giant unilamellar vesicles. Biophys J 87:419–429
22. Kahya N (2010) Light on fluorescent lipids in rafts: a lesson from model membranes. Biochem J 430:e7–e9
23. Méléard P, Bagatolli LA, Pott T (2009) Giant unilamellar vesicle electroformation from lipid mixtures to native membranes under physiological conditions. Methods Enzymol 465:161–176
24. Dertinger T, Pacheco V, von der Hocht I et al (2007) Two-focus fluorescence correlation spectroscopy: a new tool for accurate and absolute diffusion measurements. Chemphyschem 8:433–443
25. Korlann Y, Dertinger T, Michalet X et al (2008) Measuring diffusion with polarization–modulation dual-focus fluorescence correlation spectroscopy. Opt Exp 16:14609–14616
26. Korson L, Drost-Hansen W, Miller FJ (1969) Viscosity of water at various temperatures. J Phys Chem 73:34–39
27. Guo L, Har JY, Sankaran J et al (2008) Molecular diffusion measurement in lipid bilayers over wide concentration ranges: a comparative study. Chemphyschem 9:721–728
28. Machán R, Hof M (2010) Lipid diffusion in planar membranes investigated by fluorescence correlation spectroscopy. Biochim Biophys Acta 1798:1377–1391
29. Ries J, Petrasek Z, Garcia-Saez AJ et al (2010) A comprehensive framework for fluorescence cross-correlation spectroscopy. New J Phys 12:113009–113041
30. Ries J, Chiantia S, Schwille P (2009) Accurate determination of membrane dynamics with line-scan FCS. Biophys J 96:1999–2008

Chapter 13

Analyzing Transmembrane Protein and Hydrophobic Helix Topography by Dual Fluorescence Quenching

Gregory A. Caputo and Erwin London

Abstract

The location of fluorescent groups relative to the lipid bilayer can be evaluated using fluorescence quenchers embedded in the membrane and/or dissolved in aqueous solution. Quenching can be used to define the membrane topography of membrane proteins and individual membrane-embedded hydrophobic helices by combining it with the placement of fluorescent groups, including Trp, at defined sequence positions. This chapter briefly discusses various quenching methods for studies of membrane protein topography and provides detailed protocols for dual quencher analysis, a rapid, highly sensitive, and experimentally flexible approach in which the information gained from both a membrane-embedded and aqueous quencher is combined.

Key words: Fluorescence spectroscopy, Fluorescence quenching, Nitroxides, Spin-labels, 10-Doxyl nonadecane, DQA, Transmembrane helices

1. Introduction

Recent years have seen a tremendous increase in the understanding of the biochemical, structural, and biophysical properties of cellular membranes and the integral membrane proteins incorporated within those membranes. A number of high-resolution structures solved by X-ray crystallography and NMR methods have yielded essential insights into the specific interactions that are likely drivers of membrane protein folding and function. Nevertheless, the number of solved structures lags behind soluble proteins by a significant margin, which is even more evident when the structures are clustered by structural class. As a result, much biochemical and biophysical information regarding membrane protein structure and architecture has relied upon spectroscopic methodologies. In addition, spectroscopic methods are valuable for studies of membrane protein dynamics, including studies of how physiological state alters conformational equilibria and the kinetics of conformational

Jörg H. Kleinschmidt (ed.), *Lipid-Protein Interactions: Methods and Protocols*, Methods in Molecular Biology, vol. 974,
DOI 10.1007/978-1-62703-275-9_13, © Springer Science+Business Media New York 2013

interconversion between different states. This type of information is often not available from high-resolution studies.

Two spectroscopic methods have predominated in studies aimed at site-by-site characterization of protein topography: fluorescence and electron spin resonance (ESR). The simplest fluorescence method is to introduce a single fluorescent label (most commonly either a unique Trp or an exogenous fluorescent group attached to a unique Cys residue) with fluorescence properties, such as emission λ_{max}, that respond to local polarity. A label on a residue that is deeply buried in the membrane will exhibit blue-shifted fluorescence while the one exposed to the aqueous milieu will have more red-shifted fluorescence. This method has the drawback that emission λ_{max} responds to more than membrane depth. An alternative is to measure exposure of the fluorescent group either to a quencher molecule localized to aqueous solution or to a membrane-associated quenching molecule. The greater the quenching, the closer the proximity of the fluorescent group to the location of the quencher. The problem with single quencher strategies is that the amount of quenching also can respond to more than just the depth of burial of the fluorescent group within a lipid bilayer. For example, consider a fluorescent group equally deeply buried in the membrane in terms of distance from the membrane surface in two different proteins, but in one case also buried within the center of the protein. In the latter case, the level of quenching observed using a membrane-embedded or aqueous quencher would be weaker than in the former case because the fluorescent group is buried within the protein.

To minimize this and similar problems, the ratio of quenching levels using two or more quenchers that simply differ in their location relative to the membrane can be measured. Use of a quenching ratio tends to cancel out factors that influence quenching other than the depth because they tend to equally affect accessibility to quencher and/or overall sensitivity to quenching. The first method developed for such studies was to compare the level of quenching by a series of nitroxide (spin)-labeled fatty acids in which nitroxide groups are attached at different positions along the fatty acyl chain, and thus in which the nitroxide resides at different depths in a membrane. The principle of the analysis using nitroxide-labeled fatty acids is that the amount of quenching is maximal when the depth of the fluorescent group matched that of the nitroxide group (1). Nitroxides are short-range quenchers which require near-contact in order for quenching (fluorophore–quencher distances <15 Å) and are able to quench a wide variety of fluorescent groups (2–8). Their short quenching range makes the degree of quenching very sensitive to the difference in location between the fluorophore and quencher. On the other hand, although similar FRET methods have been proposed (9), FRET interaction radii (R_o) can be too large for good depth resolution.

However, use of labeled fatty acids is not ideal. Variations in the amount of fatty acid binding to membranes, the need for 4 or 5 different probes to define depth to good resolution, and the need to introduce a large amount of fatty acid to a membrane to obtain quenching are variables that complicate such studies. A variation has been to use model membranes containing nitroxide-labeled (i. e., spin-labeled) phospholipids or brominated phospholipids (2–6, 10–13). This circumvents the binding issue and can be used with mathematical analyses (either parallax analysis or distribution analysis (5, 13–16)) which minimize how many different quenchers are needed to define fluorophore depth. A comparison of these analyses has been presented elsewhere (16).

A number of factors limit methods in which the quenchers are located at fixed depths. For example, the level of quenching can be affected by bilayer width, which alters the positions of the quenchers relative to the bilayer center. This is an issue when studying the effect of bilayer width, i.e., hydrophobic mismatch, upon the conformation and topography of hydrophobic helices. In addition, quenching can be affected by specific interactions between the lipid carrying the quencher and the fluorescent molecules, especially when lipids and fluorescent molecules with a net charge are under study. The most desirable situation would be that the quencher lipids and unlabeled lipids have the same structure, i.e., the same acyl chain length and headgroup structure. This would require synthesis of a series of quenchers, each series itself composed of a series of molecules with quenching groups at different positions. In an attempt to circumvent these limitations, variations of fluorescence quenching methods have been developed that parallel an approach that has proven powerful in spin-labeling studies. In the site-directed spin-labeling method the extent to which the ESR signal of a spin-labeled group attached to a specific amino acid residue is perturbed by a membrane-embedded probe (oxygen) is compared to that induced by an aqueous paramagnetic probe (in which nickel is complexed with ethylenediamine-N,N′-diacetic acid (NiEDDA: Nickel (II))) (17) to define the depth of the spin-label. The ratio of the oxygen/ion ESR perturbation is dependent upon the depth of the spin-label within the membrane. This method had to be calibrated with primary spin-label standards at known depths, unlike fluorescence quenching methods in which quencher positions are known and relatively fixed, but this has not proven to be a major limitation.

In the analogous fluorescence methods two quenchers are used (Fig. 1), one in aqueous solution and one embedded in the membrane bilayer (18–22). The most commonly used aqueous quenchers used are iodide, which quenches the fluorescence many different molecules, and acrylamide, which is an especially good quencher of Trp fluorescence. The membrane-embedded quencher is usually a nitroxide-labeled molecule. In dual quencher analysis (DQA), the most widely used approach (19, 20, 23–38), the membrane-embedded quencher is

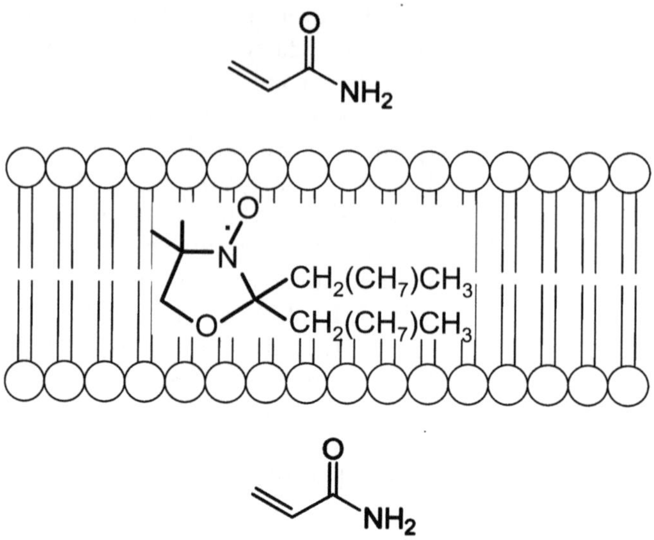

Fig. 1. Quencher structures and their locations relative to the lipid bilayer. Acrylamide predominantly resides in aqueous solution, although it can pass through membranes quickly. 10-Doxylnonadecane is dissolved in the bilayer core.

most often 10-doxylnonadecane (10-DN) although other membrane-embedded nitroxide-bearing quenchers have been used. DQA using acrylamide and 10-DN as quenchers has been calibrated for the relationship between Trp quenching and Trp depth (20). DQA has also been used with other fluorophores (28–30, 36–38), but standard quenching vs. depth curves for exactly calibrating the relationship between quenching and depth of the fluorescent group are not yet available for fluorophores other than Trp.

Acrylamide (Fig. 1) is useful as it is a small, water-soluble molecule which can easily cross lipid bilayers and, importantly, does not exhibit high quenching efficiency while in the membrane. As a result acrylamide acts as an "aqueous" or "shallow" quencher, that is, a strong quencher of Trp fluorophores which are exposed to the aqueous environment in the sample and hence not buried in the membrane interior. Since acrylamide can cross the bilayer rapidly, its addition to a solution does not induce significant osmotic pressure changes across the membrane bilayer. Combined these properties make exogenous addition of acrylamide to preformed lipid vesicles containing the peptide or protein of interest a practical experimental approach. Furthermore, acrylamide is a common biochemical reagent that allows quenching experiments to be carried out economically.

The nitroxide-labeled aliphatic molecule 10-DN is very hydrophobic and resides in the core of the bilayer (Fig. 1). We find that it is best mixed with lipids prior to vesicle formation (see below). By comparison to quenching with lipids that contain nitroxides at fixed positions, studies show that the position of the quenching-active nitroxide moiety of the 10-DN molecule has an average position between the 5 and 12 carbons (i.e., near the middle) of the lipid acyl chains, as expected if it is simply spread throughout the hydrophobic core of the lipid bilayer. This is consistent with the fact that 10-DN lacks a highly polar group that might anchor it to a specific location within a bilayer. The likelihood that 10-DN is spread throughout the bilayer minimizes complications in interpretation when bilayer width is varied. In addition, the 10-DN molecule was shown not to significantly alter bilayer properties such as membrane thickness or permeability to small ions (20).

Defining detailed membrane protein structure from spectroscopic information can be laborious because it often entails the use of a single spectroscopic probe attached to one specific site on the protein of interest. To gain a detailed picture of structure, the site of the spectroscopic probe must be varied. In principle, one would need to scan the entire membrane protein sequence residue by residue, with appropriate controls to rule out perturbation by the label. However, in practice, crucial information on conformational state and topography can be defined with a small number of strategically located labeling sites. In the case of single-pass membrane proteins or single hydrophobic helices inserted in a bilayer, as described below, topography with respect to the bilayer can even be determined through careful choice of the position of a single Trp position in the peptide primary sequence (19, 23, 24, 39). Single hydrophobic helices are of interest because they have proven to be a highly useful experimental system for understanding the basic properties of membrane-inserted/transmembrane helices and how they depend upon sequence and lipid composition (40–43).

The DQA method has been proven to be particularly successful when applied to synthetic helix-forming hydrophobic peptides which contain a single fluorescent group, e.g., Trp, as a depth reporter (19, 24–27, 44). For example, consider a fluorescent group placed at the center of a hydrophobic helix flanked by polar juxtamembrane residues. As shown in Fig. 2, the formation of the transmembrane state would show a characteristic location of the labeled group at the exact center of the bilayer while in a surface-bound state it would be in the vicinity of the much more polar environment of the lipid headgroups. Thus, these two configurations are easily distinguished from one another by quenching. It is also possible to detect if transmembrane helix has undergone a transverse displacement that maintains the transmembrane state, but alters what residues are membrane buried, from the change such a shift induces in the location of the fluorescent group (Fig. 2). By combining λ_{max} and depth measurements DQA can also distinguish between fluorophores with a single average intermediate depth in the membrane and an intermediate depth due to a mixture of populations in which some molecules form a transmembrane orientation with a deep fluorophore depth, and other molecules

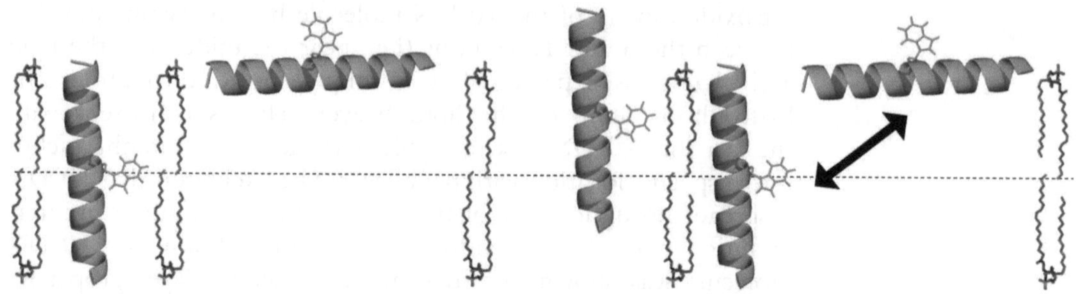

Fig. 2. Schematic illustration of potential topographies of monomeric helical peptides associated with lipid membranes. Peptides are drawn as *cartoon-ribbons*, and a centrally located tryptophan is depicted in *stick* representation to illustrate differences in Trp depth as a function of peptide topography. A *dashed line* is included to represent the bilayer center. Illustrated from *left* to *right* are a transmembrane helix with a Trp at the bilayer center, a surface helix with a shallow Trp depth, a transmembrane helix with a shifted transverse position and intermediate Trp depth, and coexisting/equilibrating transmembrane and surface helix populations with an intermediate average Trp depth. See Table 1 for how fluorescence λ_{max} and quenching would differ in each of these cases.

Table 1
How combining dual quenching and emission λ_{max} measurements can distinguish different hydrophobic helix topographies in membranes.

	Location of fluorescent group within the bilayer			
Parameter measured	**Deep = close to bilayer center**	**Surface**	**Intermediate depth**	**Mixture of deep and surface depths**
λ_{max}	Blue shifted	Red shifted	Intermediate	Intermediate
Q-Ratio value	Low	High	Intermediate	Intermediate
$\Delta\lambda_{max}$	Small	Small	Small	Large

$\Delta\lambda_{max}$ is the absolute value of the difference between λ_{max} in the presence of acrylamide and 10-DN

reside near the membrane surface, and so contain a shallowly located fluorophore (Fig. 2). In a mixed population, the presence of two populations with different λ_{max} results in quenching that is accompanied by large λ_{max} shifts which reflect whether the deep or shallow population is being quenched (Table 1). These shifts are observed because the aqueous quencher will more strongly quench the shallowly located fluorescent groups and the membrane-embedded quencher will more strongly quench the deeply located fluorescent groups.

Below, the DQA protocols used to define Trp depth are given. Variations for other fluorescent groups are described in the appropriate reports (2, 29, 30, 36–38).

2. Materials and Equipment

2.1. Lipid Stock Solutions

1. Lipids (Avanti Polar Lipids, Alabaster, AL) lyophilized or dissolved in ethanol or chloroform. Stored in glass vials at either −20 °C or −70 to −80 °C.
2. 10-DN (Avanti Polar Lipids, Alabaster, AL). Dissolved and stored in ethanol at −20 °C.
3. Acrylamide (Sigma-Aldrich). Acrylamide should be dissolved in deionized water (at 4 M) and stored in glass (see Note 1).
4. Aluminum foil. To shield samples from light (see Note 2).
5. A balance accurate to 0.01 mg. A Cahn C-33 microbalance was used for our studies. An ordinary balance can be used but will require higher amounts of lipid. It should be noted that higher amounts of lipids increase the risk of incomplete drying which impacts proper calculation of concentration.
6. Nitrogen gas source with regulator capable of producing a gentle stream (~2 psi).
7. Vacuum desiccator.

2.2. Sample Preparation

1. Pipettors which are organic solvent compatible (for example, Drummond positive-displacement pipettors with disposable glass bores). Changing bores between samples is essential to avoid cross-contamination.
2. Nitrogen gas source with regulator capable of generating a gentle stream (~2 psi).
3. Vortexing mixer.
4. A dark place to store samples before measurement.

2.3. Fluorescence Measurements

1. Fluorometer.
2. Quartz cuvettes (see Note 3).
3. Standard air-displacement pipettors.

3. Methods

3.1. Experimental Outline

The experiments described use two sets of samples to measure topography of TM helices based on the differential accessibility of a fluorophore to an aqueous or membrane-embedded quencher (acrylamide and 10-DN, respectively). The experiments can also be adapted to investigate the depth of fluorescent groups on proteins that spontaneously insert into membranes (2, 29, 30, 36–38).

As noted above, acrylamide can easily cross lipid bilayers and can be exogenously added to preformed lipid vesicles containing the peptide or protein of interest. In contrast 10-DN is mixed with lipids prior to vesicle formation. This is necessary due to the low solubility of 10-DN in water. In the 10-DN containing sample sets the phospholipid concentration is reduced by an amount equal to the amount of 10-DN added to account for the incorporation of the hydrophobic quencher in the bilayer. In previous studies, 10-DN samples in DOPC bilayers contained 10 mol% of the quencher and phospholipid concentration reduced by 10 %. When bilayer width is varied, 10-DN concentration is varied to keep it a constant value relative to the volume of the core of the bilayer. This is based upon the idea that 10-DN is spread throughout the hydrophobic core of the bilayer and that the volume of the bilayer core is proportional to acyl chain length. For example we used 12 mol% 10-DN for di(C22:1)PC (dierucoyl-PC) (19).

Two independent sets of samples are prepared in at least triplicate, the first containing lipid and fluorescent species (i.e., peptide) and the second containing lipid, fluorescent species, and 10-DN. Single background samples lacking fluorescent molecule are prepared with and without 10-DN. The fluorescence intensities from the lipid + peptide sample give the "F_o" or initial, unquenched fluorescence values. These intensities are recorded at two sets of excitation/emission wavelengths: 280 nm/330 nm and 295 nm/340 nm. The former data is used for calculation of quenching by 10-DN and the latter for acrylamide. This is done to minimize the significant inner filter effect (see Note 8) from acrylamide at 280 nm. Fluorescence intensities are then recorded for the 10-DN samples referred to as "$F_{10\text{-DN}}$." Upon completion of the measurements, an aliquot of acrylamide is added to the "F_o" samples (final concentration of 0.235 M acrylamide) which are then referred to as "F_{acryl}." After ~5 min of equilibration, the fluorescence emission in these samples is also measured. The same sets of measurements are carried out upon the background samples, which lack the fluorescent species, i.e., the peptide.

3.2. Lipid Stock Solutions

In order to determine a meaningful membrane depth lipid concentration must be accurately controlled. This point is of particular importance for sequences that may exhibit the tendency to form homo-oligomers, have behavior that depends upon the composition of lipid headgroups in the sample, or exist in equilibrium between soluble and membrane-bound conformations. As a way of controlling for potential lot-to-lot inaccuracies in lipid concentrations reported by the manufacturer and changes in lipid concentration due to evaporation in stock solutions we periodically measure the concentration of lipid in a stock solution by drying a known volume of the lipid stock and precisely weighing the dried material on a milligram-sensitive or sub-milligram-sensitive microbalance.

1. Lipids are purchased as lyophilized solids or dissolved in chloroform or ethanol. Solids are dissolved in ethanol. Lipid solutions are stored in glass vials at −20 °C or −70 to −80 °C.
2. Aluminum foil is cut into squares, approximately 15 mm to a side, numbered on the exterior surface, and then shaped into a liquid-tight container. Alternatively, the caps of microcentrifuge tubes can be cut off and used if their weights are within the measurement range of the microbalance.
3. Using forceps, the aluminum or plastic caps are weighed on a microbalance to the closest microgram, and the weights are recorded (see Note 4).
4. After warming lipids to room temperature and mixing well to redissolve any precipitated lipid, 50 μl of lipid solution is pipetted into the cup. For best accuracy, 3–5 such samples are prepared.
5. The solvent is dried under a gentle stream of nitrogen (<2 psi) for 10 min. It is absolutely essential that drying is complete.
6. The cups are placed in a high vacuum for at least 45 min to remove the remaining solvent.
7. The cups are then reweighed, and the difference in weight is used to calculate the concentration of the solution (see Notes 5 and 6).

3.3. 10-DN Stock Solution

Accurate quencher concentration is important for insuring the reproducibility of results across batches of 10-DN. Quantification of 10-DN concentration can be most easily performed using absorbance spectroscopy (see Notes 13 and 14).

1. 10-DN is purchased as yellow-orange oil. A known volume of ethanol is added directly to the container in which the 10-DN is shipped and lightly vortexed to mix. Typical final concentration of 10-DN based on the estimated amount of 10-DN in the container should be between 5 and 10 mM. This stock solution is stored in glass vials at −70 to −80°C.
2. The stock solution is transferred to a 1 cm quartz cuvette, and an absorbance spectrum is recorded over the range 200–500 nm.
3. 10-DN concentration is calculated using the molar extinction coefficients (ε) at spectral peaks in the ultraviolet and visible. We have used values of 4,480 M^{-1} cm^{-1} at 231 nm or 12 M^{-1} cm^{-1} at 422 nm but have found the 422 nm value to be more reliable (20, 23, 36).

3.4. Sample Preparation

Lipid vesicles, with or without peptides incorporated, are formed using the ethanol dilution method. To summarize, the dried lipid–peptide film is dissolved in a small volume of absolute ethanol and buffer is added rapidly while vortexing. This rapid buffer addition to ethanolic lipids promotes the formation of small unilamellar vesicles (SUVs) (45). In more detail:

1. The appropriate amount of each lipid stock solution needed in the sample is calculated using the stock concentrations. The lipids are sequentially added to a glass test tube using the organic-compatible pipettor. 10-DN should also be added at this point to the "$F_{10\text{-DN}}$" samples (see Note 7). Both the F_o and the $F_{10\text{-DN}}$ samples are prepared at least in triplicate. A single background sample for each F_o and $F_{10\text{-DN}}$ is also prepared with the omission of the peptide.
2. If the peptide sequence to be analyzed is highly hydrophobic, the peptide of interest is then added to all samples (except the backgrounds) from a stock solution in ethanol or isopropanol/water. (Alternatively, if the membrane-binding peptide/protein is truly soluble in aqueous solution it can be added after formation of vesicles.) In any case, controls to demonstrate that the peptide topography reaches a topographic equilibrium are required (see Note 9). A modified sample preparation protocol would be required for membrane proteins that must be incorporated into membranes by detergent dialysis, but we have not yet developed appropriate protocols (see Note 15).
3. Solvents are then evaporated using a gentle stream (~2 psi) of pure nitrogen for 10 min or until the samples are dry. Nitrogen can be replaced with any other inert gas, such as argon. Gas flow should be strictly controlled to minimize splashing of the peptide/lipid mixture onto the walls of the test tube.
4. The lipid films are placed in a desiccator jar which is then evacuated and held under vacuum for 1–2 h.
5. The dried lipid–peptide film is dissolved in 10 μl of ethanol and swirled. Complete dissolution of the lipid–peptide film is essential for vesicles to form with the proper components and component ratios.
6. Upon dissolution of the film, 790 μl of buffer (for example phosphate-buffered saline composed of 10 mM NaH_2PO_4, 10 mM Na_2HPO_4, 150 mM NaCl at pH 7.1) is added and vortexed to mix.
7. Samples are allowed to equilibrate at room temperature, shielded from light, before beginning the fluorescence measurements (see Note 2).
8. Samples are vortexed prior to transfer to cuvettes.

3.5. Fluorescence Measurements

1. Samples are placed into semimicro quartz fluorescence cuvettes (10 mm excitation path length, 4 mm emission path length) and placed into the cuvette holder in the sample compartment.

2. The excitation and emission slits are then set (see Note 10).
3. Initial emission spectra of all samples are recorded in order to determine the emission λ_{max} in the absence of quencher and presence of 10-DN. The excitation monochromator is set to 280 nm, and fluorescence emission is scanned from 300 to 400 nm (also see Note 11).
4. Upon measuring emission spectra for all F_o and $F_{10\text{-}DN}$ samples, separate fluorescence intensity measurements are made at fixed wavelengths. The excitation monochromator is set to 280 nm, and the emission monochromator is set to 330 nm. Slit sizes may have to be reset to suitable values.
5. Ideally, an instrument setting should be used in which the excitation shutter is only open during measurement. Typically 3–10 readings are averaged for each sample.
6. For the F_o set of samples (but not the $F_{10\text{-}DN}$ set of samples) intensity is then remeasured with the excitation monochromator set to 295 nm and the emission monochromator set to 340 nm (see Note 11).
7. Acrylamide is added to each of the F_o samples, mixed by pipetting or with a small magnetic stir bar, and allowed to equilibrate for 10 min. We have typically used a 4 M stock solution of acrylamide to create samples with a final acrylamide concentration of 0.235 M (50 μl acrylamide added to 800 μl samples for a final volume of 850 μl). These samples are designated F_{acryl} samples after acrylamide addition.
8. Using the fixed wavelengths, the F_{acryl} sample intensities are then remeasured with the excitation monochromator set to 295 nm and the emission monochromator set to 340 nm.
9. Final emission spectra of the F_{acryl} samples are recorded. The excitation monochromator is set to 280 nm, and fluorescence emission is scanned from 300 to 400 nm in order to determine emission λ_{max} in the presence of acrylamide.

3.6. Backgrounds and Controls

Background samples lacking the peptide, but containing all other components, should be prepared with and without quencher. The apparent fluorescence intensity in these samples is subtracted from the intensity of the samples containing the fluorescent peptide molecule in order to determine the true fluorescence intensity and spectral properties of the peptide of interest. When emission spectra are being recorded at an excitation wavelength of 280 nm or 295 nm, a significant peak at ~310 nm or ~325 nm, respectively, arising from Raman scattering from water should be expected.

It is highly desirable that peptides containing Trp at known positions should be employed as shallow and deep Trp control samples. This can insure that the quenching experiment is being properly performed. We have successfully employed synthetic poly-Leu peptides with Lys flanking juxtamembrane residues and a single Trp as such controls. These peptides adopt stable transbilayer α-helical structures, unless the Leu sequence is very short (20). The Trp position is designed so that when the peptide is in the transmembrane orientation the Trp is located either at/near the center of the bilayer or near the polar headgroup region. Alternatively, short, amphiphilic peptides which lie on the membrane surface also can be employed as a shallow control. If a transmembrane peptide is employed as a shallow control, the binding affinity for the membrane should be determined to ensure that the peptide is associated with the membrane and not existing in equilibrium with an aqueous population.

Other controls may be necessary in the cases in which chemical reaction between fluorophore and nitroxide is possible (see Note 16).

3.7. Data Analysis

All sample intensities are corrected for background fluorescence from the lipid or lipid + 10-DN by subtraction. Acrylamide-quenched samples are also corrected for dilution of the fluorescent molecules by the acrylamide solution and the loss of fluorescence due to the inner filter effect at the excitation wavelength of 295 nm (acrylamide does not exhibit any significant inner filter effect at the emission wavelength of 340 nm; see Note 8). The corrected emission intensities are then used to calculate the F_o/F values (F_o/F_{acryl} or $F_o/F_{10\text{-}DN}$) which increase as quenching increases. F_o/F can be calculated using any standard spreadsheet software. For acrylamide individual F_o/F_{acryl} values are then averaged. Since, unlike the acrylamide samples, where F_o and F_{acryl} measurements can be made on a single sample, the 10-DN quenching samples require independent F_o and F samples, an average of the three emission intensities from three F_o samples and from the three F samples is calculated prior to calculation of F_o/F. The final Q-ratio is calculated from a ratio of the individual F_o/F ratios for acrylamide and 10-DN. Standard formulas for calculating standard deviations, and for calculating standard deviations for ratios in which the standard deviation of the numerator and denominator are known, are used. Determination of λ_{max} for use in spectral shift ($\Delta\lambda_{max}$) experiments can be performed visually by inspection of emission intensity vs. λ (smoothing spectra mathematically if necessary).

The above-described quenching ratios are then used to calculate the Q-ratio, which is basically a ratio of ratios:

$$Q\text{-ratio} = \frac{(F_o/F_{acryl}) - 1}{(F_o/F_{10\text{-}DN}) - 1}, \tag{1}$$

where F_o is the fluorescence of the sample in the absence of quencher, F_{acryl} is the fluorescence emission intensity after addition of acrylamide, and $F_{10\text{-}DN}$ is the fluorescence emission intensity in the 10-DN-loaded sample. The *Q*-ratio will be low (~0.1) if the Trp is located at or near the bilayer center and will increase (above 1.0) for Trp at membrane surface. A value of 1 is observed when the Trp is at the polar/nonpolar interface, and values up to 3–4 have been observed for more shallowly located Trp (20, 24, 25, 27). Even higher values are sometimes measured, but these likely reflect cases in which peptides are only partly membrane bound (27).

If the *Q*-ratio is found to be an intermediate value between the deep and shallow controls then the effect of quenchers upon λ_{max} is also analyzed. As noted above, an intermediate depth (and thus intermediate *Q*-ratio) can arise from either (1) a single population of peptides with a Trp at an average intermediate depth or (2) a heterogeneous population of peptides in which some Trp are located at the bilayer center and others are located at or near the bilayer surface (Fig. 2). In cases of intermediate *Q*-ratio, a comparison of emission λ_{max} in the unquenched (F_o) and quenched (F_{acryl} and $F_{10\text{-}DN}$) samples can be employed to effectively identify heterogeneous populations with very different depths. If two populations with very different depths exist the unquenched emission λ_{max} will be an aggregate of emission spectra arising from the deep subpopulation and the shallow subpopulation. Upon quenching with acrylamide, the shallow conformation will be strongly quenched, so the aggregate emission will be dominated by that of the *deep*, unquenched population. This will generally result in an overall blue shift of observed emission λ_{max}. Conversely, samples quenched with 10-DN will have the deep population more strongly quenched, such that the aggregate emission is predominantly reflecting the *shallow*, unquenched population resulting in an overall red shift of emission λ_{max}.

In practice, the differences in λ_{max} between the F_{acryl} or $F_{10\text{-}DN}$ samples are calculated and the absolute value of the difference is defined as $\Delta\lambda_{max}$ term. A relatively homogeneous population will show very small $\Delta\lambda_{max}$ (<4 nm). A large $\Delta\lambda_{max}$ (>5 nm) indicates the presence of multiple populations with very different emission spectral properties (see Note 12).

4. Notes

1. Acrylamide is highly toxic. Do not breathe in acrylamide dust, and do not allow it or solutions containing acrylamide to contact the skin.
2. After samples are prepared, they are stored in a dark location (e.g., a cardboard box) until fluorescence is measured in order to limit potential photodamage to the fluorophore.

3. High-quality quartz cuvettes are required. The excitation and emission wavelengths are in the UV region, and the cuvettes must be transparent in this wavelength region. Either a single cuvette or a matched set of cuvettes is used, meaning that a solution placed in different cuvettes would have the same fluorescence intensity. (When F values are measured in different cuvettes than F_o measurements, incorrect F_o/F values will be obtained if the cuvettes are not matched).
4. After weighing, care should be given not to touch the cups to minimize the transfer of hand oils. Precautions include wearing gloves and handling aluminum cups with forceps.
5. Care should be taken when transporting the aluminum cups to avoid tipping.
6. Three to five separate samples for dry weight measurements are prepared and weighted for each solution, and an average is calculated.
7. Typically a positive displacement pipettor is used as both lipids and 10-DN are dissolved in organic solvents. Such solvents cannot be pipetted accurately with a standard air-displacement micropipettor.
8. The inner filter effect arises from molecules in the sample that can absorb the excitation radiation and/or the emitted fluorescence radiation. Samples containing inner-filter molecules will exhibit abnormally low fluorescence emission due to less excitation energy reaching the fluorophores (excitation radiation absorbed), less emitted fluorescence reaching the detector (emission radiation absorbed), or both. The equation for calculating the inner filter effects using a 1 cm path length fluorescence cuvette is

$$F_{corr} = F_{meas}{}^{*} \, antilog[(A_{EX}(1\ cm)^{*} n) + (A_{EM}(1\ cm)^{*} m)], \quad (2)$$

where F_{corr} is the corrected fluorescence of the sample, F_{meas} is the fluorescence emission intensity measured in the instrument, n and m are the path lengths of light in the cuvette in cm, A_{EX} (1 cm) is the absorbance of the sample at the excitation wavelength in a 1 cm cuvette, and A_{EM} (1 cm) is the absorbance of the sample at the emission wavelength in a 1 cm cuvette. Notice that the values of absorbance must be appropriate for cuvette path lengths. For an ordinary 1 cm path length cuvette in which the detected light is collected from the center of the cuvette, the appropriate path lengths for both excitation and emission are 0.5 cm, and thus the A value used for excitation and emission would be half that measured in a 1 cm cuvette in an absorbance instrument. Hence n and $m = 0.5$. If semimicro or micro cuvettes are used, this factor must be adjusted accordingly.

9. Controls should be performed to ensure that samples have achieved equilibrium with respect to topographic distributions. If there are dynamic rearrangements that occur within the samples on the timescale of the experiment, quenching results will be skewed in response to the changing topographic distribution. Additionally, these controls ensure that the topography detected does not reflect the peptide or the protein being kinetically trapped in a nonequilibrium conformation. Such controls are necessarily sample dependent. Our studies to insure that equilibrium has been achieved have used various approaches, including showing that hydrophobic helices can change their topography after incorporation into vesicles by in situ variation of bilayer thickness using decane (19) or by changing pH (39).
10. The width settings of the fluorometer slits will be dependent on your instrument. You may have to adjust the slit widths to get a usable signal.
11. Recording entire emission spectra is only strictly *required* if spectral shift ($\Delta\lambda_{max}$) analysis is being investigated. However, we have often found it informative to have spectral information to ensure comparable behavior in samples from day to day as λ_{max} is more consistent than raw intensity values.
12. Alternatively, spectral barycenter can be calculated for each sample, and the total change in barycenter (ΔBc) can also be used to measure heterogeneity (Caputo, unpublished observations).
13. Determination of "active" nitroxide groups can be obtained by ESR spectroscopy (46).
14. Using control samples we have found that quenching efficiency by 10-DN does not diminish over the course of several months presuming that the stock solutions are handled and stored properly. We recommend keeping a laboratory stock of one or more peptides with known topographic characteristics to use as controls for quenching efficiency upon creation of each new 10-DN stock.
15. The formation of lipid vesicles with membrane proteins embedded in the bilayer has routinely been performed by creation of mixed micelles of detergent, lipid, and protein which are subjected to dialysis to remove the detergent resulting in formation of lipid–protein vesicles. To use this protocol for dual quenching experiments the mixed micelles would also contain 10-DN. If there is any 10-DN loss during dialysis, a method to quantify the remaining 10-DN in the vesicles, e.g., ESR spectroscopy, would be necessary. It may turn out that using a nitroxide-bearing quencher that can be added to

preformed vesicles from solution, such as tempo (18), will be better for such experiments.

16. It should be kept in mind that some fluorescent molecules (but not Trp as far as we know) will undergo reaction with nitroxides. Appropriate controls can be carried out to rule out such reactions or correct for them (7).

Acknowledgments

This work was supported by NSF grant MCB 1019986 to E.L.

References

1. London E (1982) Mol Cell Biochem 45:181–188
2. Abrams FS, Chattopadhyay A, London E (1992) Biochemistry 31:5322–5327
3. Asuncion-Punzalan E, Kachel K, London E (1998) Biochemistry 37:4603–4611
4. Asuncion-Punzalan E, London E (1995) Biochemistry 34:11460–11466
5. Chattopadhyay A, London E (1987) Biochemistry 26:39–45
6. Kachel K, Asuncion-Punzalan E, London E (1995) Biochemistry 34:15475–15479
7. Kachel K, Asuncion-Punzalan E, London E (1998) Biochim Biophys Acta 1374:63–76
8. Kaiser RD, London E (1998) Biochemistry 37:8180–8190
9. Kleinfeld AM, Lukacovic MF (1985) Biochemistry 24:1883–1890
10. Ladokhin AS, Wang L, Steggles AW, Holloway PW (1991) Biochemistry 30:10200–10206
11. Bolen EJ, Holloway PW (1990) Biochemistry 29:9638–9643
12. Everett J, Zlotnick A, Tennyson J, Holloway PW (1986) J Biol Chem 261:6725–6729
13. Abrams FS, London E (1992) Biochemistry 31:5312–5322
14. Ren J, Lew S, Wang Z, London E (1997) Biochemistry 36:10213–10220
15. Ladokhin AS (1999) Biophys J 76:946–955
16. London E, Ladokhin AS (2002) Measuring the depth of amino acid residues in membrane-bound peptides by fluorescence quenching. In: Simon SA, McIntosh TJ (eds) Peptide–lipid interactions. Academic, San Diego, CA, pp 89–115
17. Altenbach C, Greenhalgh DA, Khorana HG, Hubbell WL (1994) Proc Natl Acad Sci U S A 91:1667–1671
18. Jittikoon J, East JM, Lee AG (2007) Biochemistry 46:10950–10959
19. Caputo GA, London E (2003) Biochemistry 42:3275–3285
20. Caputo GA, London E (2003) Biochemistry 42:3265–3274
21. Shatursky O, Heuck AP, Shepard LA, Rossjohn J, Parker MW, Johnson AE, Tweten RK (1999) Cell 99:293–299
22. Powl AM, Wright JN, East JM, Lee AG (2005) Biochemistry 44:5713–5721
23. Caputo GA, London E (2004) Biochemistry 43:8794–8806
24. Krishnakumar SS, London E (2007) J Mol Biol 374:1251–1269
25. Krishnakumar SS, London E (2007) J Mol Biol 374:671–687
26. Shahidullah K, Krishnakumar SS, London E (2010) J Mol Biol 396:209–220
27. Shahidullah K, London E (2008) J Mol Biol 379:704–718
28. Musse AA, Wang J, Deleon GP, Prentice GA, London E, Merrill AR (2006) J Biol Chem 281:885–895
29. Wei Z, White D, Wang J, Musse AA, Merrill AR (2007) Biochemistry 46:6074–6085
30. White D, Musse AA, Wang J, London E, Merrill AR (2006) J Biol Chem 281:32375–32384
31. Ho D, Merrill AR (2009) Biochemistry 48:1369–1380
32. Antollini SS, Xu Y, Jiang H, Barrantes FJ (2005) Mol Membr Biol 22:471–483

33. Chattopadhyay A, Rawat SS, Greathouse DV, Kelkar DA, Koeppe RE 2nd (2008) Biophys J 95:166–175
34. Kelkar DA, Chattopadhyay A (2006) Biochem Biophys Res Commun 343:483–488
35. Pande AH, Qin S, Nemec KN, He X, Tatulian SA (2006) Biochemistry 45:12436–12447
36. Hayashibara M, London E (2005) Biochemistry 44:2183–2196
37. Lai B, Zhao G, London E (2008) Biochemistry 47:4565–4574
38. Zhao G, London E (2005) Biochemistry 44:4488–4498
39. Lew S, Ren J, London E (2000) Biochemistry 39:9632–9640
40. Nyholm TK, Ozdirekcan S, Killian JA (2007) Biochemistry 46:1457–1465
41. Killian JA, Nyholm TK (2006) Curr Opin Struct Biol 16:473–479
42. London E, Shahidullah K (2009) Curr Opin Struct Biol 19:464–472
43. London E (2007) J Gen Physiol 130:229–232
44. Fujita K, Krishnakumar SS, Franco D, Paul AV, London E, Wimmer E (2007) Biochemistry 46:5185–5199
45. Kremer JM, Esker MW, Pathmamanoharan C, Wiersema PH (1977) Biochemistry 16:3932–3935
46. Abrams FS, London E (1993) Biochemistry 32:10826–10831

Chapter 14

Studying Lipid–Protein Interactions with Electron Paramagnetic Resonance Spectroscopy of Spin-Labeled Lipids

Tibor Páli and Zoltán Kóta

Abstract

Spin label electron paramagnetic resonance (EPR) of lipid–protein interactions reveals crucial features of the structure and assembly of integral membrane proteins. Spin label EPR spectroscopy is the technique of choice to characterize the protein-solvating lipid shell in its highly dynamic nature, because the EPR spectra of lipids that are spin labeled close to the terminal methyl end of their acyl chains display two spectral components, those corresponding to lipids directly contacting the protein and those corresponding to lipids in the bulk fluid bilayer regions of the membrane. In this chapter, typical spin label EPR procedures are presented that allow determination of the stoichiometry of interaction of spin-labeled lipids with the intra-membranous region of membrane proteins or polypeptides, as well as the association constant of the spin-labeled lipid with respect to the host lipid. The lipids giving rise to the so-called immobile spectral component in the EPR spectrum of such samples are identified as the motionally restricted first-shell lipids solvating membrane proteins in biomembranes. Stoichiometry and selectivity are directly related to the structure of the intra-membranous sections of membrane-associated proteins or polypeptides and can be used to study the state of assembly of such proteins in the membrane. Since these characteristics of lipid–protein interactions are discussed in detail in the literature [see Marsh (Eur Biophys J 39:513–525, 2010) for a most recent review], here we focus more on how to spin label model and biomembranes and how to measure and analyze the two-component EPR spectra of spin-labeled lipids in phospholipid bilayers that contain proteins or polypeptides. After a description of how to prepare spin-labeled model and native biological membranes, we present the reader with computational procedures for determining the molar fraction of motionally restricted lipids when both, one, or none of the pure isolated—mobile or immobile—spectral components are available. With these topics, this chapter complements a recent methodological paper [Marsh (Methods 46:83–96, 2008)]. The interpretation of the data is discussed briefly, as well as other relevant and recent spin label EPR techniques for studying lipid–protein interactions, not only from the point of view of lipid chain dynamics.

Key words: Spin label, Spin-labeled, Electron paramagnetic resonance (EPR), Electron spin resonance (ESR), Integral membrane proteins, Lipid–protein interactions, Immobile and mobile membrane lipids, Phospholipids, Annular, solvation, motionally restricted, and bulk lipids, Two-component spectrum

Jörg H. Kleinschmidt (ed.), *Lipid-Protein Interactions: Methods and Protocols*, Methods in Molecular Biology, vol. 974,
DOI 10.1007/978-1-62703-275-9_14,

1. Introduction

Membrane lipids and proteins possess very different rotational and translational mobility mainly because of the big difference in their mass, giving rise to frictional forces at their colliding molecular interface. The lipid bilayer maintains a continuous permeability barrier also around membrane proteins, via tight molecular packing, because fluid lipids can instantaneously match any packing irregularities arising during lateral and rotational movements as well as conformational rearrangements of proteins. The lipid–protein interface (Fig. 1) not only seals proteins in the membrane and motionally couples them with the dynamic bulk lipid environment required for related life processes but is also a very important site of specific modulation of protein function (1–7). Therefore, the structural, energetic, and dynamic aspects of lipid–protein interactions are key to understanding biochemical processes related to various membrane proteins. The lateral mobility gradient interfaces to the solvation shell of lipids around integral membrane proteins: lipids suffer motional restriction by interaction with the protein, which has been explored mostly by spin label electron paramagnetic resonance (EPR) spectroscopy.

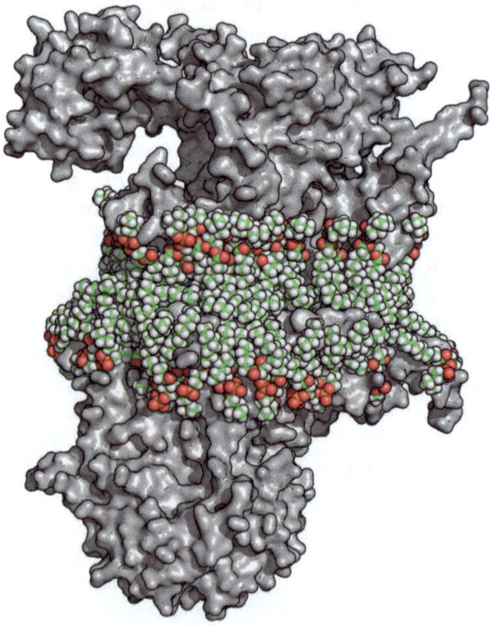

Fig. 1. The first shell of energy-minimized 1,2-dimyristoyl-*sn*-glycero-3-phosphocholine (DMPC) lipids surrounding the Van der Waals surface from the crystal structure of the monomeric form of bovine cytochrome c reductase (PDB code 1L0L) (unpublished figure; see ref. (26) for details on how the lipid shell was modeled).

Spin labeling is a technique of introducing stable nitroxide radicals into biological samples at low, non-perturbing concentrations (8–11). In general proteins are labeled covalently, whereas membranes are labeled with chemically modified lipids, which freely mix with the host lipids. Except for rare cases (when, e.g., studying paramagnetic membrane proteins) no EPR spectrum can be detected from intact biomembranes or artificial, model membranes. Model membranes are aqueous dispersions of bilayer-forming lipids that can also host membrane proteins or polypeptides (12, 13). Even if paramagnetic molecules are present in the bio- or model membrane under study, they usually do not disturb the spectrum of nitroxide spin labels at ambient temperatures (see, e.g., ref. 14). In this chapter, under EPR spectroscopy we mean conventional EPR spectroscopy, that is, continuous-wave detection of the first-harmonic absorption signal under field-modulation at ~100 kHz and microwave radiation at ~9 GHz (i.e., the X band frequency region). Spin label EPR spectroscopy is sensitive to the molecular mobility of membrane components over a wide and biologically very relevant timescale. Since the first observation of boundary lipids, which have more restricted mobility than fluid bulk lipids (6, 16), around membrane proteins (15) (conventional) EPR of spin-labeled lipids has developed into the main and most versatile technique to study lipid–protein interactions (6, 17–26). The physical bases for this success are the facts that (a) the time window of spin label EPR, ~10^{-11}–10^{-7} s, covers a most important region of rotational and translational correlation times of lipids and proteins in biological and model membranes (27–32); (b) the rate of exchange between the mobile and motionally restricted lipid populations is slower than the spectral separation, in frequency units, of their corresponding spectrum components (33, 34) (see also Note 7); and (c) spin–spin interactions can be quantitated by nonlinear EPR techniques and used to measure interactions and proximity relations between spin-labeled lipid–lipid, lipid–protein, and protein–protein molecular regions (14, 35, 36). The sensitivity of spin label EPR is remarkable too: One can detect stable nitroxide radicals down to sub-nM concentrations in 10–50 μl volume. Spin-labeled lipids can carry their nitroxide group on their headgroup or on different positions of their *sn*-2 chain (typically from the fifth to the sixteenth carbon) as dictated by the purpose of the study, i.e., whether dynamics, selectivity, or depth profiles (of chain-flexibility or polarity) at the lipid–-protein interface are to be investigated. In the most frequent application of spin-labeled lipids in lipid–protein interaction studies, with membrane-embedded or membrane-penetrating proteins, changes in lipid chain dynamics are used to characterize and quantitate the interaction. In such studies the spin label is attached close to the terminal methyl end of the acyl chain, the C14-atom of

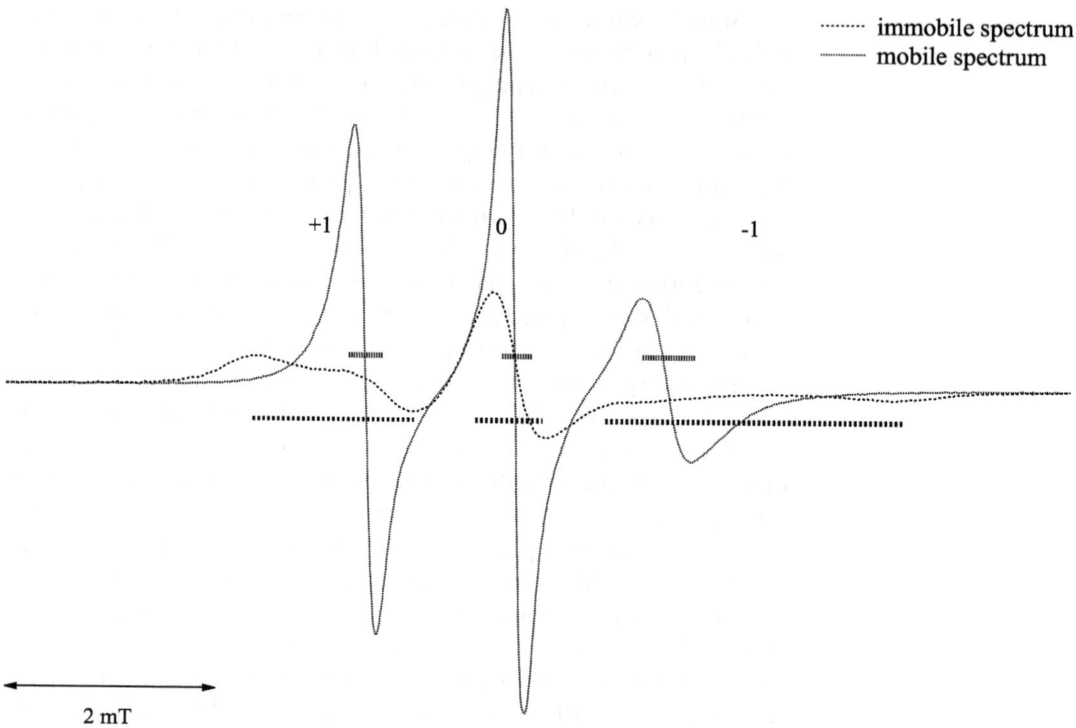

Fig. 2. Representative conventional (X-band) electron paramagnetic resonance (EPR) spectra of spin-labeled lipid (in this case DMPC) in gel (*dashed line*)- and fluid (*dotted line*)-phase membranes. Of the total lipids 1 % is labeled on their carbon 14 (C14) position with a doxyl group carrying a stable nitroxide radical. In the gel and fluid membrane phases, the spin labels have mobility that is slow (immobile) and fast (mobile) on the EPR timescale (10^{-11}–10^{-7} s) of spin label EPR. These spectra are often indistinguishable from those of protein-solvating first-shell and bulk lipids, respectively, and hence can serve as single-component spectra in spectral analysis. The spectra are normalized to represent the same number of spins. The *horizontal bars* indicate the spectral regions of the +1, 0, and −1 hyperfine lines, originating from electron-nuclear spin–spin interaction in the three nuclear spin states of the ^{14}N nucleus of the NO bond, which is responsible for the largest anisotropy in X-band spin label EPR spectra.

stearic acid, or that of the *sn*-2 chain of phospholipids, being the most common labeling sites.

Due to the hyperfine interaction between the electron and the nuclear (^{14}N) spins of the nitroxide, the spin label EPR spectrum consists of three lines (Fig. 2). The exact position, spread, and width of these hyperfine lines depend on, among other factors, the rate and amplitude of the rotation of the spin-labeled segment of the lipid molecule (27–29, 31). In a fluid bilayer, acyl chain rotations about the individual C–C bonds give rise to large-amplitude angular fluctuations of the spin-labeled chain segment, which average the orientational anisotropy in the spin label EPR spectrum (see the mobile component in Fig. 2). The optimal sensitivity of spin label EPR to the rotational dynamics of the doxyl nitroxide makes it possible to separate spectrally C14-position spin-labeled lipids in direct contact with, but almost never covalently bound to, a transmembrane protein from those in the bulk regions of the membrane (23). Lipids in contact with the membrane protein are hindered in their chain motion and are therefore detected in the outer wings of the spectrum (see the immobile component in Fig. 2). The interpretation of the origin of the immobile and mobile spectral components as protein-associated spin labels and those in the bulk, protein-free fluid phase, respectively, is verified (a) by the predicted and observed linear dependence of the mobile/immobile ratio (of the corresponding integrated intensities) on the total lipid–protein ratio (6, 23, 37) (see Subheading 3.2.6); (b) by the fact that the weight of the two components is invariant to different spin label configurations (see, e.g., ref. 38); and (c) by nonconventional EPR spectroscopic techniques (33, 39, 40). In addition, it has been proven for all known membrane proteins that have been studied with this spin label technique and have known X-ray structure that the first-shell lipids around membrane proteins are the ones identified in spin label EPR spectroscopy as motionally restricted lipids in lipid bilayers containing integral membrane proteins (25, 26) (see boundary lipids in Fig. 1 and how they are manifested in the EPR spectrum of a spin-labeled membrane in Fig. 2). It follows from the above arguments that endogenous lipids strongly bound to membrane proteins (25, 41) cannot be detected with spin label EPR techniques, because these lipids do not exchange with the exogenous spin label. Nevertheless they are important for the structure or the function of the protein concerned (see, e.g., refs. 5, 24). In the literature, membrane protein-solvating, motionally restricted first-shell lipids are also called as immobile, boundary, solvation, or annular lipids, whereas the bulk lipids are also called as free or mobile lipids. It should be noted that the same or similar techniques described here can also be used with other spin-labeled membrane-active compounds, e.g., local anesthetics or inhibitors, to study their interaction with membrane proteins (42–44).

There are other characteristics of spin label EPR that make it the most powerful technique for studying lipid–protein interactions in biological membranes (35, 36, 45–50), but in this chapter we focus on lipid–protein interactions as "seen" by lipids, more specifically how to measure and analyze the two-component EPR spectra of spin-labeled lipids in phospholipid bilayers that contain proteins or polypeptides. The examples and figures are illustrations rather than reproductions of data from our experience. Interaction of lipids with water-soluble proteins can also be studied with spin label EPR (see, e.g., refs. 51, 52), but this special topic is not considered in this chapter. The use of spin labels in proteins to study proximity relations, dynamics, and how they are embedded in membranes, also by microwave saturation and modern pulsed EPR techniques, is subject of the chapter "EPR techniques to probe insertion and conformation of spin-labeled proteins in lipid bilayers" of this book.

2. Materials

1. *Spin-labeled lipids.* The structure, synthesis, chemical characteristics, advantages, and special uses of spin-labeled lipids are discussed in great detail in a recent paper (17). Spin-labeled lipids, in which the nitroxide ring is incorporated as an integral part of the hydrocarbon chain, such as *n*-(4,4-dimethyloxazolidine-*N*-oxy)stearic acid spin label (*n*-SASL) or *n*-1-acyl-2-[*n*-(4,4-dimethyloxazolidinyl-*N*-oxyl)stearoyl]-*sn*-glycero-3-X spin label (*n*-PXSL, with X = C, E, G, S, I, and A for phosphocholine, phosphoethanolamine, phosphoglycerol, phosphoserine, phospho-*sn*-1-myo-inositol, and phosphate, respectively), cause the smallest perturbation in a phospholipid bilayer (38, 53), and by far the most frequently used spin labels are the ones with the doxyl ring on their C14 position, e.g., 14-SASL and 14-PXSL (with X = A, C, E, G, I, or S). Some of the spin-labeled lipids are available commercially (from, e.g., Avanti Polar Lipids, Sigma-Aldrich); others need to be synthesized according to standard procedures (17, 54, 55). Of the commercially available *stearic-acid derivatives*, that labeled on the C16-atom (16-SASL) is most suitable. Phosphatidylcholine spin-labeled on the C14- or C16-atom of the *sn*-2 chain, 14- or 16-1-acyl-2-[*n*-(4,4-dimethyloxazolidinyl-*N*-oxyl)stearoyl]-*sn*-glycero-3-phosphocholine (14- or 16-PCSL), respectively, can be purchased from Avanti Polar Lipids.
2. *Lipids* for model membrane preparations are available commercially (from, e.g., Sigma-Aldrich, Avanti Polar Lipids, Larodan Fine Chemicals AB).
3. *ORG*anic solvents (used in Subheading 3.1.1 for lipid–polypeptide membranes), e.g., hexafluoroisopropanol, trifluoroethanol, and chloroform/methanol mixture. One often needs to experiment with different solvents to find the best one for a given lipid, polypeptide, and spin label system.
4. *Buffer solutions.* There is no special restriction on the buffer solutions used for membrane dispersions from the spin labeling point of view, except that the pK_a of *n*-SASL is around 7–8 and the protonated and deprotonated *n*-SASL has a slightly different vertical location in the membrane, which results in slightly different hyperfine splitting (56, 57). The buffer solutions used depend largely on the actual lipid–protein system to be studied.
   (a) *HEN* (used in Subheading 3.1.2 for gap junction membranes), 10 mM *N*-(2-hydroxyethyl)piperazine-*N'*-2-

ethanesulphonic acid (HEPES), 10 mM ethylenediamine-tetraacetic acid (EDTA), and 10 mM NaCl (pH 7.8).

(b) *SORBI* (used in Subheading 3.1.2 for photosynthetic membranes), 0.4 M sorbitol, 10 mM NaCl, 5 mM $MgCl_2$, 1 mM $MnCl_2$, 1 mM phenylmethylsulfonyl fluoride (PMSF, toxic!), and 50 mM MES (pH 6.5).

(c) *OSMO* (used in Subheading 3.1.2 for photosynthetic membranes), 40 mM sucrose, 10 mM NaCl, 5 mM $MgCl_2$, 1 mM PMSF, and 20 mM tricine–KOH buffer (pH 7.8).

3. Methods

3.1. Sample Preparation and Instrument Setup

For studying lipid–protein interactions there are membrane preparations with two levels of complexity: (a) the original intact membrane or vesicles, liposomes formed from its membrane fragments (see some examples in refs. 4, 7, 36, 58, 59), and (b) artificial or model membranes made from the total lipid, or a subclass of that, extracted from the original membrane system or from synthesized lipids, which contain the target protein(s) (14, 24, 60–65). While the former, more complex structures resemble the properties of the original biomembrane closely, model membranes provide a physically and chemically better defined system (12, 13, 17), and therefore the interpretation of the data is more straightforward. For this latter reason there are many more lipid–protein interaction studies in model than in intact or native biomembranes (see ref. 37 for a recent overview of the topic).

More complete data can be obtained if the experiments are done at different lipid–protein molar ratios (N_t), which is an important input parameter in the calculation of lipid–protein stoichiometry, and different spin labels (see Subheading 3.2.6). N_t can be easily varied in reconstituted model membranes (see Subheading 3.1.1), but it is much more complicated with biological membranes (Subheading 3.1.2). In the latter case progressive delipidation or lipid enrichment might work (1, 15, 66, 67). Phospholipid and protein content can be determined by lipid phosphate (68, 69) and protein assays (70–72), respectively. The purity of lipids and the lipid composition of the membrane can be checked with thin layer or gas chromatography (69, 73–75). These assays can be done before or after the EPR measurement. It might be necessary to record the composite spectra at different temperatures in order to find the best separation of the two components. In addition, the (secondary) structure and the biological function of the target protein or polypeptide should be checked if possible. Since such structure and activity measurements depend largely on the actual system their discussion is beyond the scope of this chapter. Keep in mind to always check whether the compounds, especially the organic solvents to be used for dissolving lipids and spin labels, are hazardous, and follow the instructions for how to work with them safely.

Optimally, the membranes should contain 0.5–1 mol% spin label with respect to total lipid in the sample. This gives a strong EPR signal with 0.5–2 mg total lipid. Much less material than this might result in a spectrum that is too weak, and much more cannot be loaded fully into the EPR sample capillary. One should aim at 100 μl membrane dispersion for one sample (see Subheading 3.1.3 for details). The mobile and immobile spectral components are best resolved in fluid membranes, i.e., above the phase transition temperature of the host lipid. However, at too high temperatures the spin label might degrade. In general 60°C is still safe for most spin-labeled lipids.

3.1.1. Spin Labeling Model Membranes

Spin-labeled membranes of synthetic lipids can be prepared in two ways: (1) The most common technique is to mix the spin-labeled lipids with the host lipid in organic solvent, followed by complete removal of the solvent, and then by hydration of the mixture. (2) It is also possible to make the lipid bilayer first and label the membrane suspension as in the case of biological membranes (see Subheading 3.1.2). This latter process is preferable if parallel studies are carried out on biomembranes, but it might result in an inhomogenous distribution of the label if the vesicles have a multi-bilayer structure. In the case of artificial or model membranes a general requirement is that the lipid must be fully hydrated. For safety, water content in the final sample above 30 wt% (it is much higher in practice) is "excess," and it will result in a fully hydrated membrane (12, 13). The procedure to be chosen for preparing model membranes, with or without labels, depends on the lipids and the desired form of the membrane vesicles (i.e., uni- or multilamellar, small, large, or giant vesicles) (see, e.g., ref. 13). The procedures for the reconstitution of polypeptides and proteins depend largely on the chemical and physical characteristics of the actual biomolecules to be studied. Therefore, two simple examples from our experience are presented below.

If the subject of study is a small polypeptide or protein that can be co-dissolved with the host lipid, it might fold correctly once the membrane bilayers are formed, which is a very convenient situation (76, 77).

Example, lipid–protein interaction with transmembrane and surface-bound polypeptides (based on refs. 24, 62, 65):

1. Dissolve the polypeptide, at the desired molar ratio to the lipid, plus phospholipid and 1 mol% spin-labeled lipid (1–2 mg of lipid at 1–10 mg/ml, depending on the solubilities) in a suitable organic solvent (see Subheading 2 item 3).

2. Dry down the lipid–polypeptide mixture on the bottom of a glass test tube under nitrogen gas stream and then incubate under vacuum overnight.
3. Hydrate the dry sample with excess buffer by vortex mixing for about a minute at a temperature above the chain-melting temperature of the lipid (but below 60°C).
4. Load the dispersion into an EPR sample cell (flat cell or capillary), and go to Subheading 3.1.3.

Although there are some exceptions (see, e.g., refs. 78–82), for large membrane proteins, especially for those with multiple subunits, refolding into a membrane is not a practical approach. Instead, as a result of purification, membrane proteins are solubilized in a semi-native form in detergent, from which they are transferred into membrane vesicles. The most common technique is dialysis, in which detergent molecules are removed from the aqueous dispersion of membrane + detergent–protein system and the protein incorporates into the membrane (see, e.g., refs. 3, 13, 39, 78, 79, 83–86)

Example–cytochrome c oxidase (based on refs. 14, 61):

1. Cytochrome c oxidase can be purified from, e.g., beef heart in a lipidated form (87, 88).
2. The native lipid can be substituted with the desired one according to (89). Briefly, dissolve cytochrome c oxidase membranes with a 500-fold molar excess of the target lipid in the selected buffer solution for 1 h at room temperature. The buffer solution should contain 0.2% sodium cholate to facilitate the lipid exchange. Cool the solution on ice, and precipitate the enzyme by adding cold concentrated ammonium sulfate solution to a final concentration of 35%. Remove the supernatant after centrifugation at 4°C, and repeat the lipid incubation and ammonium sulfate-precipitation steps a second time. The enzyme is then taken up in a buffer containing 1% sodium cholate and centrifuged at 66,000 × *g* (4°C).
3. The supernatant should be assayed for the protein and lipid content (68–70).
4. Reconstitution: Divide the cytochrome c oxidase-containing solution into several aliquots, and add the chosen lipid to each aliquot to get the desired molar ratios of lipid per monomer enzyme. To remove cholate dialyze samples against the same buffer as above but also containing 1 M KCl at 4°C. Exchange the dialysis buffer after 8–12 h a total of three times.
5. Check the homogeneity of the samples, with respect to lipid-to-protein ratio, in a continuous sucrose density gradient.
6. Check enzyme activity (90).

7. Add the selected spin-labeled lipid in an ethanol solution, which should be prepared such that the total amount of ethanol does not exceed 1% of the sample volume and the final spin label concentration is 1 mol% of the lipid.
8. Pellet the membrane samples in a benchtop centrifuge at 4°C, and take them up in 100 μl of buffer, transfer them into EPR capillaries, and go to Subheading 3.1.3.

3.1.2. Spin Labeling Biological Membranes

The procedures for biomembrane preparation vary from membrane to membrane. For biomembrane samples, and often for membrane-protein samples obtained with dialysis, the spin-labeled lipid must be added exogenously to the aqueous membrane dispersion as a small volume of concentrated solution in organic solvent, during vortexing (gentle shaking or sonicating, depending on how sensitive the material is to mechanical stress), in such a small volume that is harmless to the biomembrane. The incubation time depends on the labeling efficiency (but also see Note 1). The spin label incorporates more readily in fluid biomembranes; therefore, the temperature should be higher than their phase transition temperature (but below 60°C). Ethanol-mediated labeling has the advantage that it can be controlled very conveniently and quantitatively. If one has to avoid the use of ethanol (or other organic solvents) then the partially water-soluble spin label has to be dried at the bottom of a tube (under vacuum) and the membrane suspension has to be layered onto it during vortexing. The sample must then be freed from unincorporated spin label and the organic solvent by sedimentation or gradient centrifugation followed by resuspension in label- and solvent-free buffer, several times if needed (but see also Notes 2 and 3). In addition, if the integrity or biological activity of the biological membrane can be measured, it must be ensured that the organic solvent used for spin labeling does not harm the biomembrane under the labeling conditions. See, e.g., refs. 4, 5, 7, 36, 58, 59, 91–95 for detailed procedures of labeling different biomembranes. Two examples are presented below.

Example—gap junction membranes from the hepatopancreas of Nephrops norvegicus (based on refs. 36, 58):

1. Membranes containing the 16-kDa channel proteolipid can be isolated for instance from the hepatopancreas of the decapod *Nephrops norvegicus* either by detergent extraction or by extraction with NaOH. The procedure is lengthy, and therefore we only refer to the relevant publications (58, 96–101).
2. In order to determine the lipid–protein ratio, membrane lipids need to be extracted (102), and the type of phospholipids in the native membranes can be determined with thin-layer chromatography (69, 73–75). Lipid phosphate and protein content should be determined by routine procedures (68, 70, 71).

3. Wash the membranes in HEN (see Subheading 2 item 4(a)), and add the spin-labeled lipid with vortex mixing from a stock solution in ethanol (ca. 1 mg/ml) at a level of 0.5–1 mol% relative to the total lipid such that the resulting ethanol concentration is no greater than 1%.
4. Incubate the mixture at room temperature overnight, and then pellet the membranes by centrifugation in a benchtop centrifuge.
5. Resuspend the pellets (of ca. 2 mg total protein each) in 60 μl HEN, load them into EPR sample cells, and go to Subheading 3.1.3.

Example—photosynthetic membranes (based on (59), interestingly, in this study the lipid–protein ratio and the degree of aggregation of membrane proteins varied during the biological process of the development of the photosynthetic membrane):

1. Plastids can be isolated in dim light at 4°C in SORBI (see Subheading 2 item 4(b)).
2. Disrupt plastids by osmotic shock in OSMO (see Subheading 2 item 4(c)).
3. Determine the fatty acid composition by gas chromatography (69, 73–75) after extracting total lipids (103, 104). The total protein content can be determined according to (72).
4. Add 5 μl of *n*-SASL solution in ethanol (1 mg/ml) to 100 μl thylakoid suspension (corresponding to 1–2 mg total protein) during vortexing.
5. Incubate for 10 min in the dark at room temperature, and vortex repeatedly.
6. Fill the material into an EPR sample cell, and go to Subheading 3.1.3.

3.1.3. EPR Sample Cells

The most common sample cells for membrane preparations are thin glass or quartz capillaries with ~1 mm internal diameter (see, e.g., ref. 17). Glass capillaries can have some microwave loss, but they have the advantage that they can be easily flame-sealed, and their background signal is usually very weak and broad and does not disturb the spin label spectrum. Typically 0.5–2 mg of membrane material needs to be concentrated in the bottom section of the capillaries with low-speed centrifugation. Excess supernatant, and, if needed, even enough membrane material from the pellet, is then removed with a micropipet to obtain an ideal sample length of ~5 mm. Water absorbs microwave and thereby decreases the quality factor of the resonator that results in a weaker signal. Removal of excess supernatant is especially important in the case of single-chain spin-labeled lipids, e.g., fatty acids, which have some solubility in water, hence establish a partition equilibrium (see Note 3). In this case the sample height is usually smaller than 5 mm. The sample capillary can be immersed in a thicker quartz tube (with an internal diameter of about 3–5 mm) loaded up to a height of 1–3 cm with silicone oil in order to improve the thermal stability of the sample.

3.1.4. Instrument Settings and Other Experimental Conditions

Most of the commercial EPR spectrometers work in the microwave X band (~9 GHz) frequency region and are suitable for detecting stable nitroxide free radicals, such as membrane spin labels (105). Though there are spectrometers with different microwave regions in use, as well as pulsed techniques, all with specific advantages and disadvantages, here we consider only the most common conventional continuous-wave X-band instruments. Standard microwave cavities serve spin label studies very well. Sample temperature can be controlled by thermostatted gas flow into the microwave cavity, normally through a quartz Dewar. Continuous-wave spectrometers use field modulation (normally at 100 kHz) in combination with phase-sensitive detection of the absorption signal, which results in first-harmonic EPR spectra that, under normal conditions, are identical within error with the first derivative of the absorption signal with respect to the magnetic field scan. Consequently, the absorption spectrum is the first integral of the EPR signal, whereas the total area of the absorption spectrum is the second integral (see Subheading 3.2.2).

Spectrometer settings should be optimized for the actual sample. For the nitroxide spin labels a scan width of 10–12 mT centered in the g-factor = 2 region, i.e., around 325 mT, is typical. The exact center of the spectrum depends on the type of the resonator and the sample geometry. The modulation amplitude should be about 30% of the width of the sharpest peak. For smaller modulation amplitude the signal decreases linearly; for higher field modulation the signal gets distorted. In practice, the modulation amplitude varies between 0.05 and 0.125 mT. The signal amplitude is proportional to the square root of the microwave power for low values, but at higher powers the signal saturates, i.e., it broadens and the amplitude increases less than the linear dependence or even decreases. The saturation properties vary from sample to sample even for the same spin label. One should set the microwave power such that it is still in the linear region, typically 5–10 mW. A low-pass filter is usually applied to improve the signal-to-noise ratio. The response time should be set to a value lower than 20–30% of the time needed to scan the sharpest peak. This means that longer time constants require longer scans. Typical settings are 4 min, 0.25 s or 2 min, 0.128 s for the scan time and time constant, respectively. Sample temperature can be measured with a fine-wire thermocouple positioned close to the sample into the silicone oil. The temperature at which the spectrum is recorded is adjusted to optimize the resolution of the mobile and immobile components, and this depends on the spin label and on the fluidity of

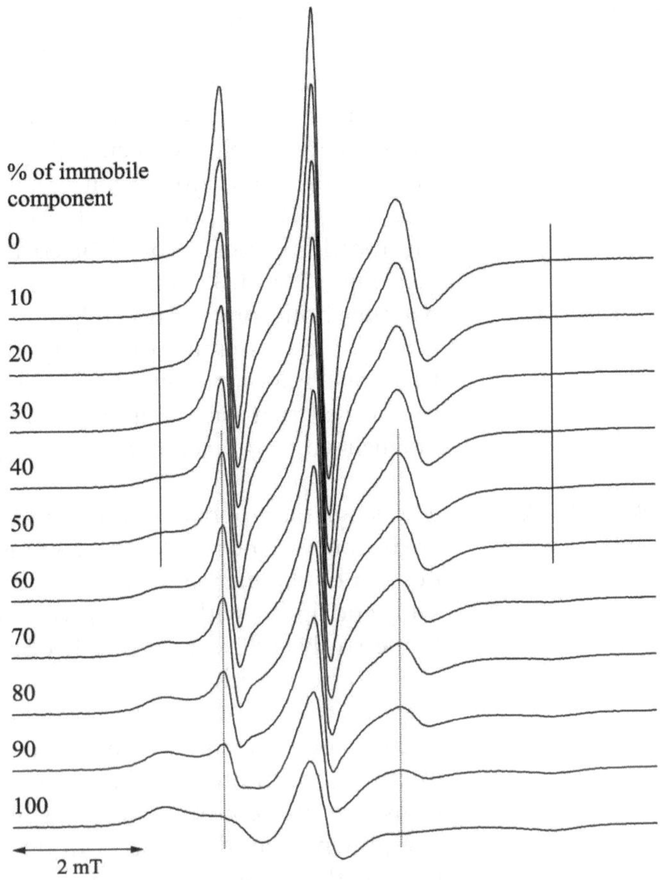

Fig. 3. Illustrative two-component EPR spectra of spin-labeled membrane lipids with varying contribution from immobile and mobile components. The spectra are normalized to represent the same number of spins. The vertical lines indicate spectral regions where the two components differ most. The researcher should look for signs of the presence of immobile (*solid vertical lines*) or mobile (*dotted vertical lines*) component in predominantly mobile or immobile spectra, respectively, in these regions.

the particular membrane system. A trivial requirement is that the bulk lipids must be in a fluid state; otherwise the separation becomes very difficult. Therefore, the spectra are recorded above the mean gel–fluid temperature of the host lipid. In the case of 1,2-dimyristoyl-*sn*-glycero-3-phosphocholine (DMPC), for instance, 30°C is common to record two-component spectra. The molar fraction of the motionally restricted lipids, and hence the corresponding spectral component, displays a systematic dependence on the lipid–protein ratio, on the number of lipid-binding sites per monomer protein, and on factors that affect the selectivity of the interaction. A series of typical two-component spectra are illustrated in Fig. 3 for the molar fraction of motionally restricted lipids varying from 0 to 100%.

3.2. Spectrum Analysis

The contribution of the mobile and immobile components to the total integrated intensity of the EPR spectrum of the spin-labeled lipid is a direct measure of the spin labels in the bulk phase and in the first (solvation) shell around the intra-membranous surfaces of the studied protein, respectively. The two components are quantified by computer subtraction or addition of the digitized spectra (see, e.g., refs. 17, 62, 106–113). Spectra from pure mobile and motionally restricted spin-labeled lipids are also called endpoints (see Figs. 2 and 3). The spectrum analysis differs depending on whether both, one, or no endpoint spectrum is available. We consider all these cases in Subheadings 3.2.3–3.2.6, in addition to how to obtain endpoint spectra (see Subheading 3.2.1). For simplicity, it is assumed that the composite, mobile and immobile spectra are normalized to the same second integral (see Subheading 3.2.2).

3.2.1. Obtaining Mobile and Immobile Components

The single-component spectrum corresponding to the fluid lipid environment (mobile component) is usually recorded from a dispersion of the extracted membrane lipids or from a dispersion of synthetic lipids (without the protein) in the case of a model membrane study (see, e.g., refs. 24, 65, 94, 102, 114). These spectra need to be recorded at closely spaced temperature intervals (1°C steps are typical) in order to have a library from which to choose the one that best matches the mobile component. For protein–DMPC mixtures, spectra recorded from sonicated DMPC above the phase transition may constitute a better library for matching the mobile component than do those from un-sonicated DMPC. The advantage of the sonicated system is that it exhibits a wider range of linewidths. The best single-component spectrum corresponding to the protein-interacting environment (the immobile component) would be the spectrum from a sample in which the protein/lipid ratio is so high that there are no bulk mobile lipids so that all lipids contact the protein. However, it is usually difficult or impossible to prepare such a sample, because the lipid–protein dispersion might not form a membrane at high protein/lipid ratio. In the case of biological membranes, the low-temperature spectra recorded from the bilayer membranes of the extracted lipids may serve as the library for the immobile component. As a practical alternative, the immobile component can be obtained from sonicated, small unilamellar vesicles of DMPC in the gel phase. Again a spectral library is recorded at closely spaced temperature intervals in the range from 0°C (or even a few degrees lower) to 20°C. Note that using a spectrum library from the temperature dependence of a spin-labeled lipid bilayer implies that the protein reduces the lipid chain mobility in a similar way as decreased temperature. Of course one can test experimental or even simulated libraries obtained by varying other physical parameters (e.g., viscosity, rate, or amplitude of rotation diffusion), also by other laboratories.

In practice, spectra of spin-labeled lipids in gel and fluid membranes work very well as immobile and mobile components, respectively. This is because (a) two-component analysis is done in membranes where the bulk lipids are in the fluid phase (mobile component), on the one hand, and because (b) the tight packing of lipid chains in the gel phase reduces both the rate and amplitude of rotational motion of lipid molecules in similar way as the intramembranous hydrophobic surface of membrane proteins do (with respect to the resolution of the EPR technique to these quantities of rotational dynamics of spin labels), on the other hand. The above efforts yield a set of mobile and immobile spectra, the library, from which the best matching single mobile and immobile spectra can be identified in an iterative process that differs whether only one, both, or none of the components are available.

3.2.2. Normalizing EPR Spectra

For a given spin label, sample composition, geometry, and instrument setting, the second integral of the conventional EPR spectrum is proportional to the number of spins in the sample. Because of variation in labeling efficiency, total mass of spin-labeled membrane, sample volume, and, to a lesser extent, instrument sensibility, differences in spin label content need to be considered when comparing spectral amplitudes and scaling factors between different samples and experiments. A practical approach is to rescale spectrum amplitudes, before any further processing, so that they have the same second integral A_0, which can be unity. This is done by multiplying each spectrum, C_i, with A_0/A_i, where A_i is the second integral of spectrum C_i. In this case, the scaling factors obtained from subtractions or additions are the true fractions of their respective spectral components in the composite spectrum. The A_i second integrals are calculated in the following way:

1. Check if the given EPR spectrum starts and ends at zero amplitude. Deviation from this situation can come because (a) the scan range is not optimized to the spectrum (too narrow or the spectrum is not in its center) or (b) there is a background signal under the EPR spectrum of the spin label. The former case can be easily corrected by changing the width and center of the field scan. In the latter case, however, the spectrum of the same sample (also the same capillary, geometry, instrument settings, etc.) has to be recorded under exactly the same conditions but without adding any spin label. The spectrum so obtained can be used to remove the background signal by spectral subtraction. The subtraction factor is to be determined by the appearance of resulting spectrum—it should satisfy the above criterion—and the baseline. If the background signal is very broad one can use a linear baseline fitted to the low- and high-field wings of the spectrum, and then this line is subtracted from the whole spectrum.

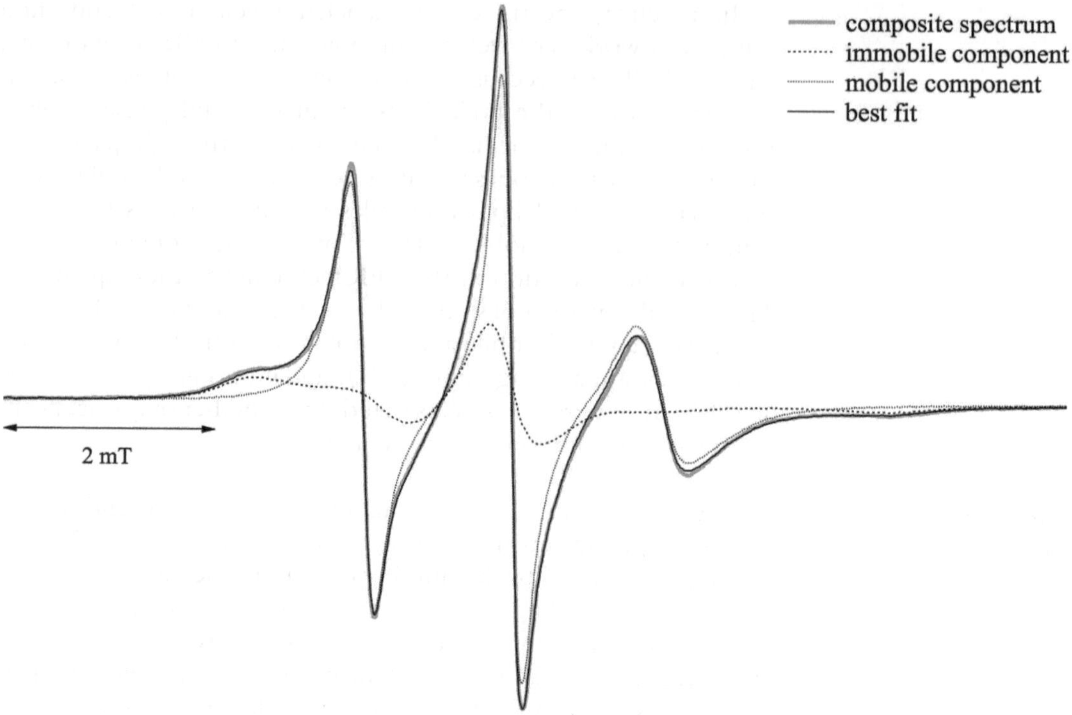

Fig. 4. Illustration of the approach of least-squares fitting of a linear combination of the mobile and immobile spectra (endpoints) to a composite (two-component) experimental spectrum. The two components are scaled to show their actual contribution to the composite spectrum as obtained from the fit.

2. If the baseline of the normal (first harmonic) spectrum is correct, its first integral, which is the absorption spectrum, must start and return to zero. If this is not the case, the most likely reason is a small amount of dispersion admixture to the absorption signal. If the problem is not severe, one can try to subtract again a linear baseline; otherwise, some instruments allow changing the phase between dispersion and absorption. In addition, some programs, e.g., ref. 115, can correct for dispersion admixture to the absorption signal.
3. Finally, the integration of the absorption spectrum gives the second integral A_i: It is the difference of the *y*-axis values at the high- and low-field (horizontal) ends of the doubly integrated spectrum.

3.2.3. Linear Combination of Mobile and Immobile Components: Least-Squares Fit

The easiest situation is, and the most accurate analysis can be done, when both the mobile and immobile components are available. In this case, the best match of the linear combination (weighted sum) of the components can be established by least-squares minimization of the difference from the original composite experimental spectrum (see, e.g., refs. 24, 62, 65 and Fig. 4). The composite spectrum C can be written as $C = f_1 \times I + f_2 \times M$, where I and M are the immobile and mobile spectra, respectively, and f_1 and f_2 are the scaling factors of the immobile and mobile spectra (fitting parameters) providing best fit in the least-squares optimization. If all the spectra in the analysis, i.e., the two components and the composite spectra, are normalized to the same second integral and the fit is good, then fraction of the immobile component $f = f_1$. If the components are close to the true ones, $f_1 + f_2$ should be very close to 1. The absolute difference $|(f_1 + f_2) - 1|$ measures the goodness of the selection of the components and can be used as an additional control parameter, in addition to the chi-square of the fit, in the iterative process of selecting the best mobile and immobile components: Select different components, repeat the fit, and note the least-squares and $|(f_1 + \mathrm{f}_2) - 1|$. Repeat the above procedure until you get the best fit (also see Note 8). It is advisable to test the goodness of fits over the whole spectral scan region and also only over the low-field (i.e., nuclear spin state = +1) hyperfine manifold, which is the spectral region most sensitive to the presence of motionally restricted lipids.

3.2.4. Subtracting One Component

If only a set of one type of components, mobile or immobile, is available, that component must be subtracted from the composite spectrum to get the "other" component as the resulting endpoint. The goodness of the subtraction must be judged by the appearance of the result; hence some experience is needed. As is evident from Fig. 5, it is much easier to identify a clean immobile endpoint than a clean mobile endpoint. Subtracting an immobile component can be prone to large errors; hence for optimizing the endpoint in subtracting the motionally restricted component, the difference spectrum should be displayed with vertical expansion. If the composite and the known component spectra are normalized, the factor f, with which the known (mobile or immobile) component is multiplied and subtracted from the composite spectrum [to result in the pure "other" (immobile or mobile) component], is the molar fraction of the known component in the composite spectrum. Consequently, $1 - f$ is that of the "other" component. The procedure should be repeated with different known component spectra from the library in an iterative process in order to find the best single component. This is especially important if there is a set of composite spectra from samples with different lipid–protein ratios. Note that the spectral result of this procedure is the missing "other" component. This means that, after normalizing the new component, one can go to Subheading 3.2.3 in order to make least-squares fits and check errors.

3.2.5. Cross-Subtraction

The spectrum analysis is most difficult when there are no immobile and mobile library spectra suitable for the given set of composite spectra. However, analysis is still possible if the composite spectra display different contributions from the same immobile and mobile

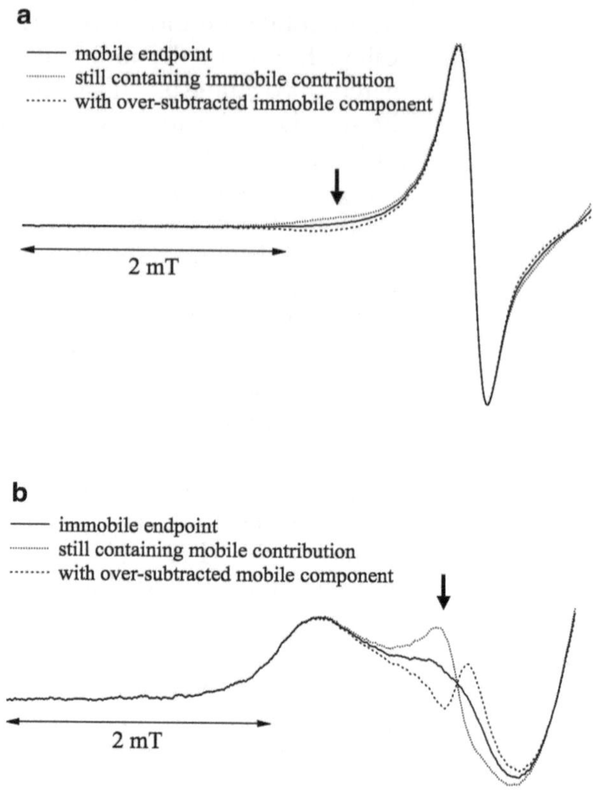

Fig. 5. Single-component subtractions. (**a**) Subtracting the immobile component from a composite spectrum in order to get a clean mobile endpoint. (**b**) Subtracting the mobile component from a composite spectrum in order to get a clean immobile endpoint. Only the low-field, i.e., ^{14}N nuclear spin state $= +1$, region of the spectra is shown.

components. In other words, the situation is such that one has a set of composite spectra without the isolated mobile and immobile components (as if the top and bottom spectra, respectively, of the stack plot in Fig. 3 were missing). This could be the case with spin-labeled native biological membranes (1, 4, 7, 15, 66, 67) if it is possible to vary the lipid–protein ratio or if it varies due to the biological phenomenon studied (see, e.g., ref. 59), but neither a deproteinated (with a clean mobile spectrum) nor a delipidated (with a clean immobile spectrum) sample can be prepared. The procedure is to generate the endpoints in a process that we call cross-subtraction. In the following, it is assumed that there is a series of (normalized) composite spectra, C_i, composed of different proportions of the same immobile, I, and mobile, M, components. Let us take spectrum C_A with more mobile component than in spectrum C_B. Consequently, spectrum C_B has more immobile component than spectrum C_A. If f_A and f_B are the fractions of the immobile component in spectrum C_A and C_B, respectively, then

$$C_A = f_A \times I + (1 - f_A) \times M,$$

$$C_B = f_B \times I + (1 - f_B) \times M,$$

and $f_B > f_A$. So, subtract spectrum C_B from spectrum C_A, aiming at a pure mobile component as the result, by removing the immobile component (by visual inspection as in Subheading 3.2.4). Similarly, subtract spectrum C_A from spectrum C_B, aiming at a pure immobile component as the result, by removing the mobile component (again by visual inspection). Let us denote the corresponding scaling factors, which are dimensionless quantities, as s_B and s_A, respectively. Their meaning is the following: Spectrum C_B is multiplied by s_B and subtracted from C_A to get a pure mobile spectrum. Similarly, spectrum C_A is multiplied by s_A and subtracted from C_B to get a pure immobile spectrum. Since C_A and C_B are normalized, the proper scaling factors in the subtractions are

$$s_B = \frac{f_A}{f_B},$$

$$s_A = \frac{1 - f_B}{1 - f_A}.$$

Solving this for f_A and f_B yields

$$f_A = s_B \times \frac{1 - s_A}{1 - s_A \times s_B},$$

$$f_B = \frac{1 - s_A}{1 - s_A \times s_B}.$$

The other possibility to determine the fractions is that, if the new endpoints are satisfactory, one can use I and M for optimized addition and perform the analysis under Subheading 3.2.3 after normalizing them to the same second integral (they are not normalized because of the cross-subtraction). The above procedure should be repeated with different pairs from the C_i set of the composite spectra. If the above assumptions are valid, one will find the same M and I spectra from different C_i pairs. If they are significantly different, then the spectra are not composed of the same invariant I and M components and/or they are not two-component spectra.

3.2.6. Interpretation of the Data

One of the main observations from spin label EPR on lipid–protein titrations in reconstituted membranes is that the number of motionally restricted lipids or lipid-"binding" sites per monomer protein (N_b) is constant for a given protein and is independent of the total amount of lipids per monomer protein in the membrane; hence it is a characteristic feature of that protein in its native structural state in the membrane (see, e.g., refs. 2, 23, 53, 62, 78, 107, 116–119 and 37 for a recent review). It has been proven (26) that N_b counts the number of first-shell lipids around proteins, polypeptides, or their aggregates (oligomers) (see, e.g., refs. 24, 58, 59, 65 for applications in biomembranes). Consequently, N_b depends on the structure, number, and arrangement of the transmembrane segments of the protein and also on the degree of protein oligomerization or aggregation. Therefore, this information can be used to build or verify models describing how the target protein is embedded in the lipid bilayer (see, e.g., refs. 24, 26, 58, 59, 62, 64, 65). The values of N_b have been determined for a wide range of integral membrane proteins by using spin label EPR under various conditions (6, 17, 23, 37).

In an optimal, truly two-component experimental system (see Notes 4 and 5 for when there are doubts), a complete study yields the immobile and mobile spectral components (if they were missing), the fractions of the corresponding motionally restricted and mobile lipids (f and $1 - f$, respectively) for a series of different lipid–protein molar ratios, N_t, the total number of lipids per monomer protein (determined in lipid phosphate and protein assays), and possibly for different spin-labeled lipids and experimental conditions. The molar fraction of motionally restricted lipids, f, provides information on both the stoichiometry and the selectivity of the lipid–protein interaction. N_b can be obtained from the equation for equilibrium lipid exchange association with the protein, which is given by

$$\frac{1-f}{f} = (1/K_r) \times (N_t - N_b)/(N_b),$$

where K_r is the association constant of the spin-labeled lipid probe with the protein relative to that of the unlabeled background host lipid (23, 53, 117). Therefore, the plot $(1 - f)/f$ vs. N_t is a straight line, which crosses the N_t and $(1 - f)/f$ axes at N_b and $-1/K_r$, respectively, and has a slope of $1/(K_r \times N_b)$. For such lipid–protein titrations it must hold that $N_t > N_b$; otherwise lipids are excluded from nonspecific protein–protein interfaces. Obviously, N_b and K_r are important parameters of lipid–protein stoichiometries and specificities both in intact biological (5, 42, 58, 59, 109) and reconstituted membranes (6, 17, 23–26, 62, 64, 65).

Even without a lipid–protein titration experiment, the fraction of immobile and mobile components (f and $1 - f$, respectively) can be determined from a composite spectrum of known lipid–protein molar ratio N_t. As shown in several studies (reviewed in refs. 23, 37), $K_r = 1$ for a spin-labeled analogue of the host lipid, because spin labeling of the lipid chain does not disturb lipid–protein interaction. Therefore, N_b can be calculated at a single lipid–protein ratio, since $N_b = f \times N_t$ in this case. It should be noted that two-component spectra of chain-labeled lipids can also be used to detect the penetration of peripheral or surface-binding proteins into the hydrophobic membrane core. In this case N_b measures the extent of membrane-penetration of the protein (see, e.g., refs. 62, 120–122).

The relative association constant K_r measures how much a specific spin-labeled lipid is preferred by the protein over the spin-labeled bulk lipid; therefore it defines the specificity of interaction of different lipids with a given membrane protein. It should be noted that this relative specificity is averaged over all binding sites of the protein (17). Apart from lipid–protein titration experiments, K_r can also be determined with different spin-labeled lipids in samples of a fixed composition and lipid–protein ratio, i.e., if N_t and N_b are constant. It follows from the above equation that in this case $K_r = K_{r0} \times f \times (1 - f_0)/(f_0 \times (1 - f))$, where K_{r0} and f_0 are the known and K_r and f are the unknown relative association constants and immobile fractions, respectively. In most studies phosphatidylcholine is the reference lipid (i.e., $K_r = 1$ for this lipid) because it is the least selective one, and largest variation in lipid specificity is observed between different lipid headgroups caused in most cases by electrostatic interactions with charged amino acid side chains in the vicinity of the phospholipid headgroups (5, 23, 37, 58). This information is important when considering which amino acid residues are close to the membrane–water interface (see, e.g., ref. 123). For further details on how to use lipid–protein stoichiometry and specificity of membrane proteins, based on spin label EPR studies, there are plenty of original studies and reviews in the literature; see, e.g., refs. 6, 17, 23, 24, 26, 36, 37, 62, 65, and also see Note 6. Proteins or polypeptides that do not penetrate into the hydrophobic core of the membrane might not give rise to a separable immobile component. In this case one might study their effect on lipid lateral diffusion (see Note 9). Nonlinear EPR and other advanced spin label EPR techniques are mentioned in Note 10.

4. Notes

Potential Pitfalls

1. Reducing agents like ascorbate can reduce the spin label and, similarly, strong oxidants might oxidize it and quench the signal. Therefore, strong reducing and oxidizing agents should be removed from the membrane suspension. For the same reason, active electron transport chains or redox-active proteins should be blocked in functioning biomembranes before spin labeling or the spectra need to be recorded faster than the rate of reduction of the nitroxide radical (4, 7).
2. It is not recommended to try to improve signal intensity by increasing the labeling level, because spin–spin interactions will broaden the lines so that the position and especially amplitudes will get distorted and the intensity will not increase as much as expected. In addition, various spin-labeled lipid analogues

incorporate into membranes to a different extent so that the labeling level varies. This is not a problem as long as the samples are magnetically dilute (i.e., label/lipid ratio is less than about 1%), but it will cause changes in the linewidth above that. It is recommended to test the labeling efficiency by using different amount of labels and compare their EPR intensity to that of standard samples in order to find the optimal labeling condition if quantitative analysis is required.

3. The doxylstearic acid spin label (*n*-SASL) partitions between the membrane and the aqueous buffer. This means that repeated centrifugation and washing of the labeled material can remove the label not only from the buffer but also from the membrane. In addition, it is advisable to minimize sample volume in the EPR capillary to reduce the disturbing signal originating from label molecules free in the buffer. If label is present in the buffer special care must be taken because its EPR signal might overlap with that of the mobile component.
4. In highly heterogenous membranes the interpretation of the spectrum can be difficult because of many different proteins, lipid domains, etc. Still, one can very often describe the spectrum with two components (15, 42, 58, 59, 66, 67, 109). In this case the fraction of motionally restricted lipids must be understood as the sum of lipids bound on the surface of all intra-membranous sections of all membrane-penetrating proteins in the membrane.
5. Theoretically, some anisotropic rotational modes of nitroxides might produce single-component spectra that, for the unexperienced eyes, look like two-component ones (see some examples in, e.g., ref. 124), although we have never seen such spectra from spin-labeled lipid bilayers. On the other hand, spin-labeled lipids with intermediate mobilities might remain unresolved (125). In addition, the presence of lateral domains with different lipid mobilities in phospholipid membranes can produce multicomponent spectra that, however, look like two-component ones at first glance (126–128). Conversely, fast physical exchange (see Note 7) could make a two-component system display a single-component spectrum (62). In general, if there are any doubts about the two-component nature of the spectrum, verification is advisable: (a) The lipid–protein titration should follow the dependence described in Subheading 3.2.6. (b) The goodness of fit with two-component vs. other models (e.g., continuous distribution over mobility parameters or a detailed simulation assuming a single component) should be checked. (c) The dependence on physical parameters, such as temperature, could discriminate between different models. For instance, the mobile–immobile spectral contribution should not depend on the temperature if the membrane is in a fluid state (58). If the spectrum has indeed two

components but the library spectra are not optimal one has to try different experimental or simulated libraries.

6. *Lipid Specificity of Membrane Proteins.* In addition to providing hints about electrostatic interactions between lipid headgroups and amino acids near the membrane–water interface, lipid selectivities (K_r) of integral and peripheral proteins can also be considered in a broader context, namely, the energetics of lipid–protein interactions, hydrophobic matching between the bilayer and the membrane-spanning regions of the protein, dependence on structure, ionic strength and pH, as well as profiles, along the membrane normal, of the lateral pressure in phospholipid bilayers (see refs. 6, 23, 37, 108 for reviews).
7. *Two-Site Exchange.* The mechanism by which equilibrium between protein-solvating (motionally restricted) and bulk (mobile) lipids is established is lateral diffusion of lipids in the membrane plane. Lateral diffusion (45, 129) exchanges lipids at the intra-membranous protein surface and the bulk phase. This is a physical exchange between two sites (130), and since the exchange is relatively slow in the spin label EPR timescale (33, 34), it has been considered in lineshape simulations of the two-component spectra based on the so-called exchange-coupled Bloch equations (see, e.g., refs. 39, 58, 78, 107, 123, 131). In such lineshape simulations, not only the molar fraction of mobile and motionally restricted lipids can be determined but also the on- and off-rates at which spin labels arrive and leave the protein surface, respectively. It follows from the two-site exchange model, and in agreement with experimental observations (78, 108, 119, 123, 132), that the off-rates are inversely proportional to the corresponding relative association constants (K_r), which means that lipids displaying higher selectivity than the reference lipid reside longer at the protein interface than the reference lipid. For lipids that do not display a selectivity for the protein, the off-rates for exchange are in the order of 1–10 MHz, which is at least an order of magnitude lower than the rates of diffusive exchange in fluid lipid membranes (6, 23, 39, 45, 129, 133). Lipids with an affinity towards the protein, relative to the background lipid, have off-rates for leaving the protein that are correspondingly slower. For further details, see refs. 23, 37, 78, 107, 108, 131, 133.
8. *Other Computational Procedures.* It should be routine to try various combinations of mobile–immobile component pairs from a given spectrum library in any of the Subheadings 3.2.1–3.2.5. Optimally, the linear combination of the very same mobile, M, and immobile, I, components should be satisfactory to fit all the composite spectra (recorded at different N_t). This procedure can be accelerated in an iterative global fit: Select a pair of I and M components, and fit all the

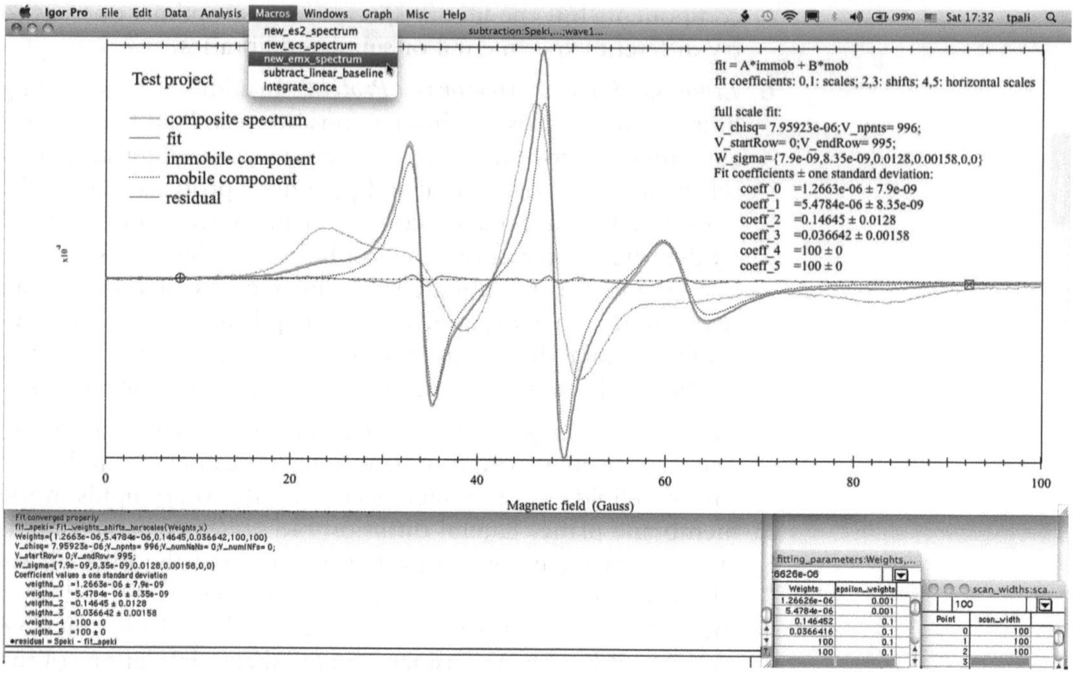

Fig. 6. Snapshot from the program "Optimized subtraction" for dealing with two-component EPR spectra as implemented in the Igor software environment in order to determine the mobile and immobile content in composite spectra. The program can import EPR spectra in all major data formats (new formats can be added easily) and can be used for integration and baseline corrections, least-squares fitting of two components to a composite spectrum, subtracting single components, and corrections for changes in polarity and microwave frequency between the spectra. The program is available from the authors upon request.

composite spectra in a single step. Then repeat the global fit with a different I and M pair until finding the best pair and fit.

Our program "Optimized subtraction" (Fig. 6) implemented in Igor (Wavemetrics, Lake Oswego) can be used for all of the computational tasks mentioned in Subheadings 3.2.1–3.2.5, except for computing and removing dispersion admixture (this feature is planned), and it is being developed further in order to implement and automate all of the above tasks, including global fits and distribution search routines. The program is available upon request from the authors.

9. *Lipid Lateral Diffusion*. Proteins that bind to the membrane surface but do not penetrate into the hydrophobic core do not give rise to easily separable immobile EPR spectral components of the spin-labeled lipid because they cause only a small perturbation in the mobility of the acyl chains and in the polarity around the doxyl nitroxide. Although such perturbations can be detected by detailed lineshape analysis (7, 62, 120, 134, 135), these methods are not widely used. An alternative approach might be to study the effect of surface-bound proteins on lipid lateral diffusion as changes in spin–spin interactions (45, 129, 136) or direct imaging (137–139).

10. *Other Advanced Spin Label EPR Techniques.* Nonlinear EPR, which is related to conventional and nonconventional detection modes at saturation of the spectrum at high microwave radiation intensities (14, 47, 50), is optimally sensitive to dynamics on the timescale of spin–lattice relaxation of nitroxide spin labels, i.e., the microsecond regime (140), and it is also sensitive to spin–spin interactions, either within the spin label system (45, 129, 141), including spin-labeled proteins (36, 50), or with an "external" spin system, i.e., paramagnetic centers in proteins (14) or quenching agents (36, 46, 48, 142, 143). Nonlinear EPR techniques on spin-labeled lipids can be used to study further details of lipid–protein interactions. For instance, both progressive saturation and saturation transfer EPR experiments provide definitive evidence that lipids at the protein interface are exchanging on this timescale (37). The exchange rate between the protein-associated and bulk fluid states can be determined more accurately because these recording modes are very sensitive to spin–lattice relaxation rates, which are optimally matched in timescale to lipid exchange processes (14, 36, 39, 144). The penetration of a spin-labeled surface-bound or integral protein into the bilayer can also be measured by spin–spin interactions with spin-labeled lipids with their doxyl nitroxide located at different positions along their acyl chain (36, 94) or with paramagnetic quenching agents (46, 48, 50, 94, 140, 142). Spin–spin interaction of a spin-labeled lipid chain segment can be detected with a paramagnetic metal center in a protein (14) too. The use of nonlinear EPR for spin–spin interactions to study lipid–protein interactions, in addition to proximity relations, in membranes is discussed in refs. 35, 47, 50, 140. Pulsed and high-field EPR techniques can also be used with spin-labeled lipids (see, e.g., refs. 34, 52, 145–148), but their full utilization for studying lipid–protein interactions remains mainly to be explored. The applications of some of these techniques for spin-labeled membrane proteins are described in the chapter "Spin labeling techniques to probe insertion and conformation of proteins in lipid bilayers by EPR" of this book (see also, e.g., refs. 149–152).

Acknowledgment

The library spectra used in this chapter for illustrations were recorded, and the optimized subtraction routines were started to be developed (by T.P.) in Derek Marsh's laboratory in Göttingen, Germany. We thank him also for reading the manuscript. We acknowledge financial support from the Hungarian National Science Fund (OTKA K68804 and K101633).

References

1. Hidalgo C, Ikemoto N, Gergely J (1976) Role of phospholipids in the calcium-dependent ATPase of the sarcoplasmic reticulum. J Biol Chem 251:4224–4232
2. Knowles PF, Watts A, Marsh D (1979) Spin-label studies of lipid immobilization in dimyristoylphosphatidylcholine-substituted cytochrome oxidase. Biochemistry 18:4480–4487
3. Fajer P, Knowles PF, Marsh D (1989) Rotational motion of yeast cytochrome oxidase in phosphatidylcholine complexes studied by saturation-transfer electron spin resonance. Biochemistry 28:5634–5643
4. Horvath I, Glatz A, Varvasovszki V, Torok Z, Pali T, Balogh G, Kovacs E, Nadasdi L, Benko S, Joo F, Vigh L (1998) Membrane physical state controls the signaling mechanism of the heat shock response in Synechocystis PCC 6803: Identification of hsp17 as a "fluidity gene". Proc Natl Acad Sci USA 95:3513–3518
5. Pali T, Garab G, Horvath LI, Kota Z (2003) Functional significance of the lipid–protein interface in photosynthetic membranes. Cell Mol Life Sci 60:1591–1606
6. Marsh D (2008) Protein modulation of lipids, and vice-versa, in membranes. Biochim Biophys Acta 1778:1545–1575
7. Balogh G, Maulucci G, Gombos I, Horvath I, Torok Z, Peter M, Fodor E, Pali T, Benko S, Parasassi T, De Spirito M, Harwood JL, Vigh L (2011) Heat stress causes spatially-distinct membrane re-modelling in K562 leukemia cells. PLoS One 6:e21182
8. Berliner LJ (1976) Spin labeling. Theory and applications, molecular biology. An international series of monographs and textbooks 1. Academic Press, New York
9. Berliner LJ (1979) Spin labeling II. Theory and applications, molecular biology. An international series of monographs and textbooks 2. Academic Press, New York
10. Berliner LJ, Reuben J (1989) Spin labeling. Theory and applications, biological magnetic resonance 8. Plenum Press, New York and London
11. Berliner LJ (1998) Spin labeling. The next millennium, biological magnetic resonance 14. Plenum Press, New York and London
12. Cevc G, Marsh D (1987) Phospholipid bilayers. Physical principles and models, cell biology: a series of monographs 5. John Wiley and Sons, New York
13. Lasic DD (1993) Liposomes: from physics to applications. Elsevier, Amsterdam
14. Pali T, Kleinschmidt JH, Powell GL, Marsh D (2000) Nonlinear electron paramagnetic resonance studies of the interaction of cytochrome oxidase with spin-labeled lipids in gel-phase membranes. Biochemistry 39:2355–2361
15. Jost PC, Griffith OH, Capaldi RA, Vanderkoi G (1973) Evidence for boundary lipid in membranes. Proc Natl Acad Sci USA 70:480–484
16. Meier P, Sachse J-H, Brophy PJ, Marsh D, Kothe G (1987) Integral membrane proteins significantly decrease the molecular motion in lipid bilayers: a deuteron NMR relaxation study of membranes containing myelin proteolipid apoprotein. Proc Natl Acad Sci USA 84:3704–3708
17. Marsh D (2008) Electron spin resonance in membrane research: protein–lipid interactions. Methods 46:83–96
18. Marsh D, Watts A, Pates RD, Uhl R, Knowles PF, Esmann M (1982) ESR spin label studies of lipid–protein interactions in membranes. Biophys J 37:265–274
19. Marsh D (1983) Spin-label answers to lipid--protein interactions. Trends Biochem Sci 8:330–333
20. Marsh D (1987) Selectivity of lipid–protein interactions. J Bioenerg Biomembr 19:677–689
21. Marsh D (1990) Lipid–protein interactions in membranes. FEBS Lett 268:371–375
22. Knowles PF, Marsh D (1991) Magnetic resonance of membranes. Biochem J 274:625–641
23. Marsh D, Horvath LI (1998) Structure, dynamics and composition of the lipid–protein interface. Perspectives from spin-labelling. Biochim Biophys Acta 1376:267–296
24. Kota Z, Pali T, Marsh D (2004) Orientation and lipid-peptide interactions of gramicidin A in lipid membranes: polarized attenuated total reflection infrared spectroscopy and spin-label electron spin resonance. Biophys J 86:1521–1531
25. Marsh D, Pali T (2004) The protein–lipid interface: perspectives from magnetic resonance and crystal structures. Biochim Biophys Acta 1666:118–141
26. Pali T, Bashtovyy D, Marsh D (2006) Stoichiometry of lipid interactions with

transmembrane proteins – deduced from the 3D structures. Protein Sci 15:1153–1161

27. Griffith OH, Jost PC (1976) Lipid spin labels in biological membranes. In: Berliner LJ (ed) Spin labeling. Theory and applications. Academic Press, New York, pp 453–523
28. McConnell HM (1976) Molecular motion in biological membranes. In: Berliner LJ (ed) Spin labeling. Theory and applications. Academic Press, New York, pp 525–560
29. Seelig J (1976) Anisotropic motion in liquid crystalline structures. In: Berliner LJ (ed) Spin labeling. Theory and applications. Academic Press, New York, pp 373–409
30. Lange A, Marsh D, Wassmer KH, Meier P, Kothe G (1985) Electron spin resonance study of phospholipid membranes employing a comprehensive line-shape model. Biochemistry 24:4383–4392
31. Marsh D (1989) Experimental methods in spin-label spectral analysis. In: Berliner LJ, Reuben J (eds) Spin labeling. Theory and applications. Plenum Press, New York and London, pp 255–303
32. Moser M, Marsh D, Meier P, Wassmer K-H, Kothe G (1989) Chain configuration and flexibility gradient in phospholipid membranes. Comparison between spin-label electron spin resonance and deuteron nuclear magnetic resonance, and identification of new conformations. Biophys J 55:111–123
33. Horvath LI, Brophy PJ, Marsh D (1994) Microwave frequency dependence of ESR spectra from spin labels undergoing two-site exchange in myelin proteolipid membranes. J Magn Reson B 105:120–128
34. Marsh D, Kurad D, Livshits VA (2002) High-field electron spin resonance of spin labels in membranes. Chem Phys Lipids 116:93–114-PII S0009-3084(02)00022-1
35. Marsh D, Pali T, Horvath LI (1998) Progressive saturation and saturation transfer EPR for measuring exchange processes and proximity relations in membranes. In: Berliner LJ (ed) Spin labeling, the next millennium. Plenum Press, New York and London, pp 23–82
36. Pali T, Finbow ME, Marsh D (1999) Membrane assembly of the 16-kDa proteolipid channel from *Nephrops norvegicus* studied by relaxation enhancements in spin-label ESR. Biochemistry 38:14311–14319
37. Marsh D (2010) Electron spin resonance in membrane research: protein–lipid interactions from challenging beginnings to state of the art. Eur Biophys J 39:513–525
38. Esmann M, Hideg K, Marsh D (1988) Novel spin-labels for the study of lipid–protein interactions. Application to (Na+, K+)-ATPase membranes. Biochemistry 27:3913–3917
39. Horvath LI, Brophy PJ, Marsh D (1993) Exchange rates at the lipid–protein interface of the myelin proteolipid protein determined by saturation transfer electron spin resonance and continuous wave saturation studies. Biophys J 64:622–631
40. Horvath LI, Brophy PJ, Marsh D (1993) Spin label saturation transfer EPR determinations of the stoichiometry and selectivity of lipid–-protein interactions in the gel phase. Biochim Biophys Acta 1147:277–280
41. Marsh D, Pali T (2006) Lipid conformation in crystalline bilayers and in crystals of transmembrane proteins. Chem Phys Lipids 141:48–65
42. Horvath LI, Arias HR, Hankovszky HO, Hideg K, Barrantes FJ, Marsh D (1990) Association of spin-labeled local anaesthetics at the hydrophobic surface of acetylcholine receptor in native membranes from *Torpedo marmorata*. Biochemistry 29:8707–8713
43. Dixon N, Pali T, Kee TP, Marsh D (2004) Spin-labelled vacuolar-ATPase inhibitors in lipid membranes. Biochim Biophys Acta Biomembr 1665:177–183
44. Dixon N, Pali T, Kee TP, Ball S, Harrison MA, Findlay JBC, Nyman J, Vaananen K, Finbow ME, Marsh D (2008) Interaction of spin-labeled inhibitors of the vacuolar H+–ATPase with the transmembrane Vo-sector. Biophys J 94:506–514
45. Pali T, Horvath LI (1989) Restricted lateral diffusion of acidic lipids in phospholipid vesicles aggregated by myelin basic protein. Biochim Biophys Acta 984:128–134
46. Stopar D, Jansen KAJ, Pali T, Marsh D, Hemminga MA (1997) Membrane location of spin-labeled M13 major coat protein mutants determined by paramagnetic relaxation agents. Biochemistry 36:8261–8268
47. Marsh D, Livshits VA, Pali T, Gaffney BJ (1999) Recent development in biological spin-label spectroscopy. In: Greve J, Puppels GJ, Otto C (eds) Spectroscopy of biological molecules: new directions. Kluwer Academic Publishers, Dordrecht, Boston, London, pp 647–650
48. Kostrzewa A, Pali T, Froncisz W, Marsh D (2000) Membrane location of spin-labeled cytochrome determined by paramagnetic relaxation agents. Biochemistry 39:6066–6074
49. Kota Z, Szalontai B, Droppa M, Horvath G, Pali T (2002) The formation of an inverted hexagonal phase from thylakoid membranes upon heating. Cell Mol Biol Lett 7:126–128

50. Pali T, Marsh D (2002) Structural studies on membrane proteins using non-linear spin label EPR spectroscopy. Cell Mol Biol Lett 7:87–91
51. Livshits VA, Marsh D (2000) Fatty acid binding sites of serum albumin probed by non-linear spin-label EPR. Biochim Biophys Acta 1466:350–360
52. De Simone F, Guzzi R, Sportelli L, Marsh D, Bartucci R (2007) Electron spin-echo studies of spin-labelled lipid membranes and free fatty acids interacting with human serum albumin. Biochim Biophys Acta 1768:1541–1549
53. Brotherus JR, Griffith OH, Brotherus MO, Jost PC, Silvius JR, Hokin LE (1981) Lipid–protein multiple binding equilibriums in membranes. Biochemistry 20:5261–5267
54. McConnell HM, Hubbell WL (1971) Molecular motion in spin-labeled phospholipids and membranes. J Am Chem Soc 93:314–326
55. Marsh D, Watts A (1982) [90] Diffusible spin labels used to study lipid–protein interactions with rhodopsin and bacteriorhodopsin. Methods Enzymol 88:762–772
56. Esmann M, Marsh D (1985) Spin-label studies on the origin of the specificity of lipid–protein interactions in Na+, K+–ATPase membranes from *Squalus acanthias*. Biochemistry 24:3572–3578
57. Sankaram MB, Brophy PJ, Jordi W, Marsh D (1990) Fatty acid pH titration and the selectivity of interaction with extrinsic proteins in dimyristoylphosphatidylglycerol dispersions. Spin label ESR studies. Biochim Biophys Acta 1021:63–69
58. Pali T, Finbow ME, Holzenburg A, Findlay JBC, Marsh D (1995) Lipid–protein interactions and assembly of the 16-kDa channel polypeptide from *Nephrops norvegicus*. Studies with spin-label electron spin resonance spectroscopy and electron microscopy. Biochemistry 34:9211–9218
59. Kota Z, Horvath LI, Droppa M, Horvath G, Farkas T, Pali T (2002) Protein assembly and heat stability in developing thylakoid membranes during greening. Proc Natl Acad Sci USA 99:12149–12154
60. Kleinschmidt JH, Marsh D (1997) Spin-label electron spin resonance studies on the interactions of lysine peptides with phospholipid membranes. Biophys J 73:2546–2555
61. Kleinschmidt JH, Powell GL, Marsh D (1998) Cytochrome c-induced increase of motionally restricted lipid in reconstituted cytochrome c oxidase membranes, revealed by spin-label ESR spectroscopy. Biochemistry 37:11579–11585
62. Kiricsi M, Horvath LI, Dux L, Pali T (2001) Spin label EPR studies of the effect of gramicidin S on lipid chain dynamics. J Mol Struct 563:469–475
63. Marsh D, Horvath LI, Swamy MJ, Mantripragada S, Kleinschmidt JH (2002) Interaction of membrane-spanning proteins with peripheral and lipid-anchored membrane proteins: perspectives from protein–lipid interactions. Mol Membr Biol 19:247–255
64. Anbazhagan V, Qu J, Kleinschmidt JH, Marsh D (2008) Incorporation of outer membrane protein OmpG in lipid membranes: protein–lipid interactions and beta-barrel orientation. Biochemistry 47:6189–6198
65. Kota Z, Pali T, Dixon N, Kee TP, Harrison MA, Findlay JB, Finbow ME, Marsh D (2008) Incorporation of transmembrane peptides from the vacuolar H(+)-ATPase in phospholipid membranes: spin-label electron paramagnetic resonance and polarized infrared spectroscopy. Biochemistry 47:3937–3949
66. Watts A, Davoust J, Marsh D, Devaux PF (1981) Distinct states of lipid mobility in bovine rod outer segment membranes. Resolution of spin label results. Biochim Biophys Acta 643:673–676
67. Ivancich A, Horvath LI, Droppa M, Horvath G, Farkas T (1994) Spin label ESR study of lipid solvation of supramolecular photosynthetic protein complexes in thylakoids. Biochim Biophys Acta Biomembr 1196:51–56
68. Eibl H, Lands WE (1969) A new, sensitive determination of phosphate. Anal Biochem 30:51–57
69. Rouser G, Fleischer S, Yamamoto A (1970) Two dimensional thin layer chromatographic separation of polar lipids and determination of phospholipids by phosphorous analysis of spots. Lipids 5:494–496
70. Lowry OH, Rosebrough NJ, Farr L, Randall RJ (1951) Protein measurement with the Folin phenol reagent. J Biol Chem 193:265–275
71. Peterson GL (1977) A simplification of the protein assay method of Lowry et al. which is more generally applicable. Anal Biochem 83:346–356
72. Markwell MAK, Haas SM, Tolbert NE, Bieber LL (1981) Protein determination in membrane and lipoprotein samples: manual and automated procedures. Methods Enzymol 72:296–303
73. Gasser A, Raddatz S, Radunz A, Schmid G (1999) Comparative immunological and chemical analysis of lipids and carotenoids of

the D1-peptide and of the light-harvesting-complex of photosystem II of *Nicotiana tabacum.* Zeitsch Naturforsch C J Biosci 54:199–208

74. Kruse O, Hankamer B, Konczak C, Gerle C, Morris E, Radunz A, Schmid G, Barber J (2000) Phosphatidylglycerol is involved in the dimerization of photosystem II. J Biol Chem 275:6509–6514
75. Mongrand S, Badoc A, Patouille B, Lacomblez C, Chavent M, Cassagne C, Bessoule J (2001) Taxonomy of gymnospermae: multivariate analyses of leaf fatty acid composition. Phytochemistry 58:101–115
76. Perez-Gil J, Casals C, Marsh D (1995) Interactions of hydrophobic lung surfactant proteins SP-B and SP-C with dipalmitoylphosphatidylcholine and dipalmitoylphosphatidylglycerol bilayers studied by electron spin resonance spectroscopy. Biochemistry 34:3964–3971
77. Cruz A, Casals C, Plasencia I, Marsh D, Perez-Gil J (1998) Depth profiles of pulmonary surfactant protein B in phosphatidylcholine bilayers, studied by fluorescence and electron spin resonance. Biochemistry 37:9488–9496
78. Peelen SJCJ, Sanders JC, Hemminga MA, Marsh D (1992) Stoichiometry, selectivity, and exchange dynamics of lipid–protein interaction with bacteriophage M13 coat protein studied by spin label electron spin resonance. Effects of protein secondary structure. Biochemistry 31:2670–2677
79. Aggeli A, Boden N, Cheng Y-L, Findlay JBC, Knowles PF, Kovatchev P, Turnbull JHP (1996) Peptides modeled on the transmembrane region of the slow voltage-gated IsK potassium channel: structural characterization of peptide assemblies in the beta-strand conformation. Biochemistry 35:16213–16221
80. Kleinschmidt JH, Tamm LK (1996) Folding intermediates of a beta-barrel membrane protein. Kinetic evidence for a multi-step membrane insertion mechanism. Biochemistry 35:12993–13000
81. Wolkers WF, Spruijt RB, Kaan A, Konings RNH, Hemminga MA (1997) Conventional and saturation-transfer EPR of spin-labeled mutant bacteriophage M13 coat protein in phospholipid bilayers. Biochim Biophys Acta 1327:5–16
82. Kleinschmidt JH, den Blaauwen T, Driessen AJ, Tamm LK (1999) Outer membrane protein A of *Escherichia coli* inserts and folds into lipid bilayers by a concerted mechanism. Biochemistry 38:5006–5016
83. Bosterling B, Stier A, Hildebrandt AG, Dawson JH, Trudell JR (1979) Reconstitution of cytochrome P-450 and cytochrome P-450 reductase into phosphatidylcholine-phosphatidylethanolamine bilayers: characterization of structure and metabolic activity. Mol Pharmacol 16:332–342
84. Powell GL, Knowles PF, Marsh D (1987) Spin label studies on the specificity of interaction of cardiolipin with beef heart cytochrome oxidase. Biochemistry 26:8138–8145
85. Horvath LI, Heimburg T, Kovachev P, Findlay JBC, Hideg K, Marsh D (1995) Integration of a K+ channel-associated peptide in a lipid bilayer: conformation. Lipid–protein interactions, and rotational diffusion. Biochemistry 34:3893–3898
86. Yang Q, Liu X, Hara M, Lundahl P, Miyake J (2000) Quantitative affinity chromatographic studies of mitochondrial cytochrome c binding to bacterial photosynthetic reaction center, reconstituted in liposome membranes and immobilized by detergent dialysis and avidin-biotin binding. Anal Biochem 280:94–102
87. Yu C, Yu L, King TE (1975) Interactions of the cytochrome oxidase protein with phospholipids and cytochrome c. J Biol Chem 250:1383–1392
88. Cable MB, Powell GL (1980) Spin-labeled cardiolipin: preferential segregation in the boundary layer of cytochrome c oxidase. Biochemistry 19:5679–5686
89. Watts A, Marsh D, Knowles PF (1978) Lipid-substituted cytochrome oxidase: no absolute requirement of cardiolipin for activity. Biochem Biophys Res Commun 81:403–409
90. Yonetani T (1966) Cytochrome oxidase from beef heart muscle. In: Maehly AC (ed) Biochemical preparations, vol 11. John Wiley & Sons, New York, pp 14–20
91. Gombos Z, Kis M, Pali T, Vigh L (1987) Nitrate starvation induces homeoviscous regulation of lipids in the cell envelope of the blue-green alga, *Anacystis nidulans.* Eur J Biochem 165:461–465
92. Horvath I, Vigh L, Pali T, Thompson GA Jr (1989) Effect of catalytic hydrogenation of Tetrahymena ciliary phospholipid fatty acids on ciliary phospholipase A activity. Biochim Biophys Acta 1002:409–412
93. Pali T, Pesti M (1996) Chapter V: phase transition of membrane lipids. In: Prasad R (ed) Manual on membrane lipids. Springer-Verlag, Berlin, Heidelberg, New York, pp 80–93
94. Pali T, Finbow ME, Marsh D (1997) Membrane assembly of the 16-kDa V-ATPase proteolipid subunit from spin–lattice relaxation

enhancements in spin label ESR. Biophys J 72:TUAM7
95. Belagyi J, Pas M, Rasport P, Pesti M, Pali T (1999) Effect of hexavalent chromium on eukaryotic plasma membrane studied by EPR spectroscopy. Biochim Biophys Acta Biomembr 1421:175–182
96. Finbow ME, Eldridge T, Buultjens J, Lane NJ, Shuttleworth J, Pitts JD (1984) Isolation and characterisation of arthropod gap-junctions. EMBO J 3:2271–2278
97. Hertzberg EL (1984) A detergent-independent procedure for the isolation of gap-junctions from rat-liver. J Biol Chem 259:9936–9943
98. Buultjens TEJ, Finbow ME, Lane NJ, Pitts JD (1988) Tissue and species conservation of the vertebrate and arthropod forms of the low-molecular weight (16–18000) proteins of gap-junctions. Cell Tissue Res 251:571–580
99. Leitch B, Finbow ME (1990) The gap junction-like form of a vacuolar proton channel component appears not to be an artifact of isolation – an immunocytochemical localization study. Exp Cell Res 190:218–226
100. Finbow ME, Eliopoulos EE, Jackson PJ, Keen JN, Meagher L, Thomson P, Jones P, Findlay JBC (1992) Structure of a 16 kDa integral membrane protein that has identity to the putative proton channel of the vacuolar H +–ATPase. Protein Eng 5:7–15
101. Finbow ME, Meagher L (1992) Connexins and the vacuolar proteolipid-like 16-kDa protein are not directly associated with each other but may be components of similar or the same gap junctional complexes. Exp Cell Res 203:280–284
102. Bligh EG, Dyer WJ (1959) A rapid method of total lipid extraction and purification. Can J Biochem Physiol 37:911–917
103. Folch JM, Lees M, Sloane SGH (1957) A simple method for the isolation and purification of total lipids from animal tissues. J Biol Chem 226:497–509
104. Horvath G, Droppa M, Hideg E, Rozsa Z (1989) The role of phospholipids in regulating photosynthetic electron transport activities: treatment of chloroplasts with phospholipase A2. J Photochem Photobiol B Biol 3:515–527
105. Jost PC, Griffith OH (1976) Instrumental aspects of spin labeling. In: Berliner LJ (ed) Spin labeling. Theory and applications. Academic Press, New York, pp 251–272
106. Datema KP, Wolfs CJAM, Marsh D, Watts A, Hemminga MA (1987) Spin-label electron spin resonance study of bacteriophage M13 coat protein incorporation into mixed lipid bilayers. Biochemistry 26:7571–7574
107. Ryba NJP, Horvath LI, Watts A, Marsh D (1987) Molecular exchange at the lipid-rhodopsin interface: spin-label electron spin resonance studies of rhodopsin-dimyristoyl phosphatidylcholine recombinants. Biochemistry 26:3234–3240
108. Horvath LI, Brophy PJ, Marsh D (1988) Influence of lipid headgroup on the specificity and exchange dynamics in lipid–protein interactions. A spin label study of myelin proteolipid apoprotein-phospholipid complexes. Biochemistry 27:5296–5304
109. Li G, Knowles PF, Murphy DJ, Nishida I, Marsh D (1989) Spin-label ESR studies of lipid–protein interactions in thylakoid membranes. Biochemistry 28:7446–7452
110. Li G, Knowles PF, Murphy DJ, Marsh D (1990) Lipid–protein interactions in thylakoid membranes of chilling-resistant and -sensitive plants studied by spin label electron spin resonance spectroscopy. J Biol Chem 265:16867–16872
111. Marsh D (1996) Membrane assembly studied by spin-label electron spin resonance. Braz J Med Biol Res 29:863–871
112. Mihailescu D, Horvath LI (1999) Molecular dynamics of lipid association at the hydrophobic interface of gramicidin S. Eur Biophys J 28:216–221
113. Szakonyi G, Dux L, Horvath LI (1999) Rearrangement of boundary, stand-by, and fluid lipids during the formation of two-dimensional crystals of Ca2+ –ATPase. J Mol Struct 483:253–256
114. Marsh D (1982) Electron spin resonance: spin label probes. In: Metcalfe JC, Hesketh TR (eds) Techniques in lipid and membrane biochemistry. Elsevier, Amsterdam, p B426/1–B426/44
115. Stoll S, Schweiger A (2006) EasySpin, a comprehensive software package for spectral simulation and analysis in EPR. J Magn Reson 178:42–55
116. Brophy PJ, Horvath LI, Marsh D (1984) Stoichiometry and specificity of lipid–protein Interaction with myelin proteolipid protein studied by spin-label electron spin resonance. Biochemistry 23:860–865
117. Marsh D (1985) ESR spin label studies of lipid–protein interactions, vol 1, Chapter 4. In: Watts A, Pont JJHHM (eds) Progress in protein–lipid interactions. Elsevier, Amsterdam, pp 143–172
118. Marsh D, Horvath LI (1989) Spin-label studies of the structure dynamics, of lipids

proteins. In: Hoff AJ (ed) Advanced EPR. Applications in biology and biochemistry. Elsevier, Amsterdam, pp 707–752

119. Horvath LI, Drees M, Beyer K, Klingenberg M, Marsh D (1990) Lipid–protein Interactions in ADP-ATP carrier/egg phosphatidylcholine recombinants studied by spin-label ESR spectroscopy. Biochemistry 29:10664–10669
120. Goerrissen H, Marsh D, Rietveld A, de Kruijff B (1986) Apocytochrome c binding to negatively charged lipid dispersions studied by spin-label electron spin resonance. Biochemistry 25:2904–2910
121. Jordi W, de Kruijff B, Marsh D (1989) Specificity of the interaction of amino- and carboxy-terminal fragments of the mitochondrial precursor protein apocytochrome c with negatively charged phospholipids. A spin-label electron spin resonance study. Biochemistry 28:8998–9005
122. Montich GG, Montecucco C, Papini E, Marsh D (1995) Insertion of diphtheria toxin in lipid bilayers studied by spin label ESR. Biochemistry 34:11561–11567
123. Horvath LI, Brophy PJ, Marsh D (1990) Influence of polar residue deletions on lipid--protein interactions with the myelin proteolipid protein. Spin-label ESR studies with DM-20/lipid recombinants. Biochemistry 29:2635–2638
124. Freed JH (1976) Theory of slow tumbling ESR spectra of nitroxides. In: Berliner LJ (ed) Spin labeling. Theory and applications. Academic Press, New York, pp 53–132
125. Marsh D, Watts A, Maschke W, Knowles PF (1978) Protein-immobilized lipid in dimyristoylphosphatidylcholine-substituted cytochrome oxidase: evidence for both boundary and trapped-bilayer lipid. Biochem Biophys Res Commun 81:397–402
126. Arsov Z, Schara M, Zorko M, Strancar J (2004) The membrane lateral domain approach in the studies of lipid–protein interaction of GPI-anchored bovine erythrocyte acetylcholinesterase. Eur Biophys J 33:715–725
127. Arsov Z, Strancar J (2005) Determination of partition coefficient of spin probe between different lipid membrane phases. J Chem Inf Model 45:1662–1671
128. Strancar J, Schara M, Pecar S (2003) New EPR method for cellular surface characterization. J Membr Biol 193:15–22
129. Pali T, Horvath LI (1989) Lipid lateral diffusion measurements in model membranes. Acta Physiol Hung 74:311–314
130. Davoust J, Devaux PF (1982) Simulation of electron spin resonance spectra of spin-labeled fatty acids covalently attached to the boundary of an intrinsic membrane protein. A chemical exchange model. J Magn Reson 48:475–494
131. Horvath LI, Brophy PJ, Marsh D (1988) Exchange rates at the lipid–protein interface of myelin proteolipid protein studied by spin-label electron spin resonance. Biochemistry 27:46–52
132. Wolfs CJAM, Horvath LI, Marsh D, Watts A, Hemminga MA (1989) Spin-label ESR of bacteriophage M13 coat protein in mixed lipid bilayers. Characterization of molecular selectivity of charged phospholipids for the bacteriophage coat protein in lipid bilayers. Biochemistry 28:9995–10001
133. Ryba N, Horvath LI, Watts A, Marsh D (1987) Exchange at the lipid–protein interface. Spin label ESR studies of rhodopsin-dimyristoyl phosphatidylcholine recombinants. Biochemistry 26:3234–3240
134. Sankaram MB, de Kruijff B, Marsh D (1989) Selectivity of interaction of spin-labelled lipids with peripheral proteins bound to dimyristoylphosphatidylglycerol bilayers, as determined by ESR spectroscopy. Biochim Biophys Acta 986:315–320
135. Sankaram MB, Brophy PJ, Marsh D (1989) Interaction of two complementary fragments of the bovine spinal cord myelin basic protein with phospholipid bilayers. An ESR spin label study. Biochemistry 28:9692–9698
136. Pali T, Tigyi G, Horvath LI (1988) Constrained lateral diffusion of membrane lipids as a consequence of myelin basic protein/lipid interaction. Acta Physiol Hung 71:146
137. Pali T, Ebert B, Horvath LI (1987) ESR imaging of myelin basic protein induced vesicle aggregation. Biochim Biophys Acta Biomembr 904:346–352
138. Pali T, Sass L, Ebert B, Horvath LI (1990) 2D ESR image reconstruction from 1D projections using the modulated field gradient method. J Magn Reson 86:338–345
139. Pali T, Ebert B, Horvath LI (1992) Dynamic imaging and spatially localized ESR spectroscopy of oriented phospholipid multilayers. J Magn Reson 96:491–500
140. Marsh D, Livshits VA, Pali T (1997) Nonlinear, continuous-wave EPR spectroscopy and spin–lattice relaxation: spin-label EPR methods for structure and dynamics. J Chem Soc Perkin Trans 2:2545–2548
141. Pali T, Bartucci R, Horvath LI, Marsh D (1993) Kinetics and dynamics of annealing

during sub-gel phase formation in phospholipid bilayers. A saturation transfer electron spin resonance study. Biophys J 64:1781–1788

142. Pali T, Bartucci R, Horvath LI, Marsh D (1992) Distance measurements using paramagnetic ion-induced relaxation in the saturation transfer electron spin resonance of spin-labeled biomolecules. Biophys J 61:1595–1602

143. Pali T, Finbow ME, Marsh D (2006) A divalent-ion binding site on the 16-kDa proton channel from *Nephrops norvegicus*-revealed by EPR spectroscopy. Biochim Biophys Acta Biomembr 1758:206–212

144. Arora A, Esmann M, Marsh D (1999) Microsecond motions of the lipids associated with trypsinized Na, K-ATPase membranes. Progressive saturation spin-label electron spin resonance studies. Biochemistry 38:10084–10091

145. Gaffney BJ, Marsh D (1998) High-frequency, spin-label EPR of nonaxial lipid ordering and motion in cholesterol-containing membranes. Proc Natl Acad Sci USA 95:12940–12943

146. Livshits VA, Marsh D (2000) Simulation studies of high-field EPR spectra of spin-labeled lipids in membranes. J Magn Reson 147:59–67

147. Bartucci R, Gambacorta A, Gliozzi A, Marsh D, Sportelli L (2005) Bipolar tetraether lipids: chain flexibility and membrane polarity gradients from spin-label electron spin resonance. Biochemistry 44:15017–15023

148. Bartucci R, Erilov DA, Guzzi R, Sportelli L, Dzuba SA, Marsh D (2006) Time-resolved electron spin resonance studies of spin-labelled lipids in membranes. Chem Phys Lipids 141:142–157

149. Hilger D, Polyhach Y, Padan E, Jung H, Jeschke G (2007) High-resolution structure of a Na+/H+ antiporter dimer obtained by pulsed electron paramagnetic resonance distance measurements. Biophys J 93:3675–3683

150. Volkov A, Dockter C, Bund T, Paulsen H, Jeschke G (2009) Pulsed EPR determination of water accessibility to spin-labeled amino acid residues in LHCIIb. Biophys J 96:1124–1141

151. Dockter C, Volkov A, Bauer C, Polyhach Y, Joly-Lopez Z, Jeschke G, Paulsen H (2009) Refolding of the integral membrane protein light-harvesting complex II monitored by pulse EPR. Proc Nat Acad Sci USA 106:18485–18490

152. Dockter C, Mueller AH, Dietz C, Volkov A, Polyhach Y, Jeschke G, Paulsen H (2011) Rigid core and flexible terminus. Structure of solubilized light-harvesting chlorophyll a/b complex (LHCII) measured by electron paramagnetic resonance (EPR). J Biol Chem 287:2915–2926

Chapter 15

EPR Techniques to Probe Insertion and Conformation of Spin-Labeled Proteins in Lipid Bilayers

Enrica Bordignon and Yevhen Polyhach

Abstract

Electron paramagnetic resonance (EPR) spectroscopy of spin-labeled membrane proteins is a valuable biophysical technique to study structural details and conformational transitions of proteins close to their physiological environment, e.g., in liposomes, membrane bilayers, and nanodiscs. Unlike in nuclear magnetic resonance spectroscopy, having only one or few specific side chains labeled at a time with paramagnetic probes makes the size of the object under investigation irrelevant in terms of technique sensitivity. As a drawback, extensive site-directed mutagenesis is required in order to analyze the properties of the protein under investigation. EPR can provide detailed information on side chain dynamics of large membrane proteins or protein complexes embedded in membranes with an exquisite sensitivity for flexible regions and on water accessibility profiles across the membrane bilayer. Moreover, distances between the two spin-labeled side chains in membrane proteins can be detected with high precision in the 1.5–6 nm range at cryogenic temperatures. The application of EPR to membrane proteins still presents some challenges in terms of sample preparation, sensitivity, and data interpretation; thus no ready-to-go methodological recipes can be given. However this chapter describes the state of the art in the application of nitroxide-based site-directed spin labeling EPR to membrane proteins, with specific focus on the different types of information which can be obtained with continuous wave and pulsed techniques and on the challenges in sample preparation and data analysis for functional and structural membrane protein studies.

Key words: Site-directed spin labeling, Nitroxide, EPR, Membrane proteins, Bilayer, Accessibility, High field, Mobility, Distances, Double electron electron resonance (DEER)

1. Introduction

Probing insertion and conformational transitions of proteins in lipid bilayers is a formidable task for any biophysical technique both in vivo and in vitro. Magnetic resonance techniques as liquid- and solid-state nuclear magnetic resonance (NMR) can be combined to obtain information in the two tumbling regimes (from the aqueous solution to the membrane bilayer), but the bigger the size of the protein under study, the more challenging is to follow

Jörg H. Kleinschmidt (ed.), *Lipid-Protein Interactions: Methods and Protocols*, Methods in Molecular Biology, vol. 974,
DOI 10.1007/978-1-62703-275-9_15, © Springer Science+Business Media New York 2013

insertion into membranes or conformational changes. X-ray crystallography is one of the techniques which can give structural snapshots of membrane proteins in action in a membrane-like milieu (most crystals are obtained in detergent micelles); however, the challenges in this field are still numerous. Site-directed spin labeling electron paramagnetic resonance (EPR), in contrast, can be performed with similar sensitivity in both aqueous and lipid environment, thus allowing to follow insertion and conformational changes of membrane proteins in the same sample. Clearly, the drawback of this advantage is the need to genetically introduce cysteines, and subsequently EPR-active paramagnetic centers which give information at the molecular level on the protein.

Once a spin label is inserted in a protein (Subheading 3.1), it is possible to extract information about the dynamics of the spin-labeled side chain and of the backbone associated with it (Subheading 3.2). Moreover water accessibility to that particular site can be addressed by several techniques (Subheading 3.3) and the topology of the protein with respect to the membrane–water interface can be investigated. By placing two spin labels in the protein of interest, one can also detect very precisely the mean distance and the distance distribution between the two spins (Subheading 3.4), allowing for detection of conformational changes in proteins during their functioning.

Subheading 3 gives detailed guidelines to perform the experiments on membrane proteins and to extract the information described, above with particular focus on the pitfalls which may be encountered both in data recording and processing. Selected examples from the literature will be used in order to convey the potential of the technique to obtain insights into different aspects of membrane protein structural details.

2. Materials

2.1. Sample Preparation and Instrumentation

In order to be able to perform the experiments outlined in this chapter one needs a basic lab equipment to treat membrane protein samples. Additionally, one needs:

1. 1.4-dithio-DL-threitol (DTT) (Sigma) to pre-reduce all cysteines prior to the spin labeling.
2. (1-Oxyl-2,2,5,5-tetramethylpyrroline-3-methyl)-methanethiosulfonate (spin label, MTSSL, Fig. 1). The reagent is available from Toronto Research Chemicals, Ontario, Canada. A 100 mM stock solution of the spin label in acetonitrile or DMSO is prepared and stored at −80°C in the dark.
3. 3-(2-Iodoacetamido)-PROXYL is available from Sigma.

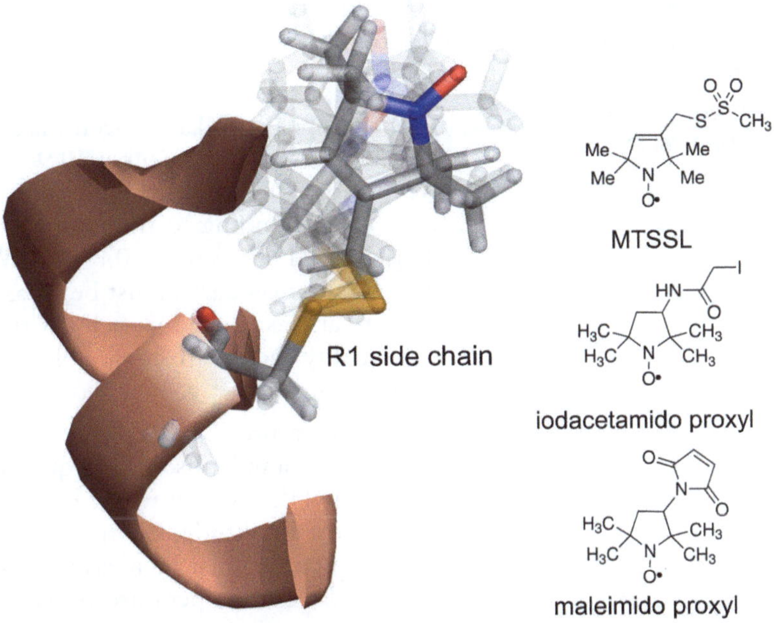

Fig. 1. Pyrroline-type spin labels. The three spin labels most commonly used to label cysteine residues are presented. The simulated MTSSL side chain rotamers attached to an alpha-helix in a protein with the program MMM (http://www.epr.ethz.ch/software/index) highlight the intrinsic flexibility of the spin label.

4. Liquid nitrogen.
5. Deuterated water (Sigma) for the low-temperature pulse experiments.
6. Deuterated glycerol (Sigma) for the low-temperature pulse experiments.
7. TPX (polymethylpentene) capillaries for accessibility measurements (Bruker BioSpin, Karlsruhe, Germany; Molecular Specialities, Inc. Milwaukee, WI, USA).
8. Glass tubes (0.9 mm outer diameter) for continuous-wave room-temperature measurements (for this purpose use 50 μl micropipettes closed with a Bunsen burner from Brand GMBH, Wertheim, Germany).
9. Quartz tubes for low-temperature measurements (Wilmad-Labglass, Vineland, NJ, USA or Aachener Quarz-Glas Technologie Heinrich, Aachen, Germany).
10. X- and W-band continuous wave EPR spectrometers (Jeol Ltd., Tokyo, Japan; Magnettech GmbH, Berlin, Germany; Bruker BioSpin, Karlsruhe, Germany).
11. X- and Q-band pulse EPR spectrometer (Bruker BioSpin, Karlsruhe, Germany).

2.2. Technical Requirements for Continuous-Wave and Pulse EPR Measurements

Sufficient conditions for the detection of reliable continuous-wave (CW) nitroxide spectra either at room or at low temperature are the following:

1. Room-temperature EPR spectra are detected on 10–20 μl of sample inserted in glass tubes (0.9 mm outer diameter) with a sweep width of 15 mT, a 0.15 mT maximal modulation amplitude (corresponding to the natural linewidth of a freely tumbling MTSSL in water), 0.5–1 mW microwave power (the power saturation curve must be measured at the working temperature at least once for each cavity in use). Resonators with high-quality factor are preferable (e.g., super high Q cavity from Bruker).
2. Low-temperature EPR spectra are detected at 160 K on 20–30 μl of sample inserted in quartz tubes (3–3.8 mm outer diameter) with a sweep width of 25 mT, a 0.25 mT maximal modulation amplitude, 0.2 mW microwave power (the power saturation curve must be measured at least once for the cavity in use). For low-temperature W-band spectra, 1 μl of sample is inserted in quartz capillaries (0.9 mm outer diameter, 0.5 mm inner diameter), and the spectra are detected with a sweep width of 40 mT, 0.25 mT modulation amplitude (due to the broad spectral linewidth), 5 μW microwave power (to avoid saturation at 160 K, the power saturation curve must be measured to avoid line-shape artifacts). Resonators with high-quality factor are preferable as well (e.g., super high Q cavity from Bruker for X band and standard cavity for W band).

Pulse measurements require specific characteristics from the hardware. Quite general, broad bandwidth and high microwave power are the most essential requirements. Additionally, the filling factor of the resonator (the amount of sample relative to the active region of the resonator) is important as it directly influences the experimental sensitivity.

1. For distance measurements with the Double Electron Electron Resonance (DEER) experiment (see Subheading 3.4.2), an X-band (preferably a Q-band) spectrometer equipped with two frequency sources and able to deliver hard (optimum 12 ns) π-pulses is required. At X band, commercial hardware (Elexsys E580, Bruker) with sufficient characteristics (1 kW microwave amplifier, two frequency sources) is readily available. The best performance in DEER experiments is achieved when at least four independent (in terms of power and phase) pulse channels are available: one for the pump pulse (ELDOR channel) and the rest three for the observer pulses. DEER is still possible when two phase-locked stripline channels and one ELDOR channel are only available. Among the commercial

resonators, the best sensitivity is obtained with the split-ring MS3 probehead (Bruker) adjusted for 3 mm o.d. sample tubes with 40–50 μl of material. Q-band commercially available hardware normally has a lack of power so that hard microwave pulses are not possible and DEER experiment has to be done with the soft excitation. But the field is rapidly developing, a high power upgraded Q-band spectrometer has recently become commercially available (Bruker). Finally, a cryostat for the low-temperature operation (down to 50 K) is required.

2. For Electron Spin Echo Envelope Modulation (ESEEM) measurements (Subheading 3.3) a pulse spectrometer operating at a single microwave frequency suffices. Requirements to the resonators are not that strict as those for DEER. One still achieves optimum performance with the Bruker split-ring MS3, but dielectric MD4 (max. tube diameter 4 mm) or MD5 (max. tube diameter 5 mm) resonators are suitable as well. A cryostat for the low-temperature operation is required.

3. Methods

3.1. Spin Labeling

The most commonly used spin labels for membrane proteins are commercially available pyrroline-type nitroxide radicals (Fig. 1): iodoacetamido proxyl label, maleimido proxyl, and MTSSL (also known as MTSL) (1).

Pyrroline-type nitroxide radicals are more resistant to reducing stress than the pyrrolidine-type analogous systems (2, 3), which can also be found commercially. MTSSL binds specifically and reversibly to cysteines (addition of reducing agents will cleave the disulfide bond, removing the label from the protein). Iodoacetamido and maleimido labels are irreversibly bound to cysteines, and if present in limiting quantities and slightly alkaline pH, cysteine modification will be the exclusive reaction. However, iodoacetamido spin labels in non-buffered solutions can also alkylate amines (lysine, N-termini), thioethers (methionine), imidazoles (histidine), and carboxylates (aspartate, glutamate, acidic C-termini), and maleimido spin labels were found to be able to bind to lysines. Iodoacetamido and MTSSL are flexible and do not affect protein structures at the majority of sites. In contrast, the maleimido group is bulkier and may cause problems at particular site in proteins (see Note 1).

To disentangle the overall motion of the label from the protein backbone motion, more rigid spin labels can be employed. As an example the new bifunctional MTSSL reagent which binds to two consecutive cys residues at position i, $i + 3$ (or $i + 4$) in a helix has been shown to report on the backbone dynamics of α-helical proteins (4) (see Note 2). The most commonly used spin label for membrane proteins is MTSSL due to its extreme specificity for the free thiol of cysteines, a stoichiometric reaction at most sites, and the relatively small size and flexibility of the modified side chain. The labeling strategies for membrane proteins are similar to those applicable to water-soluble proteins. However, more care must be taken in identifying sites which are accessible for the spin label. In particular, sites located in the middle of the transmembrane region of the protein may not be labeled at all, or once labeled may influence the protein stability (see Note 3).

In the following, a recipe for spin labeling membrane proteins with MTSSL is given, which can be however used as guideline for the labeling with other spin labels as well.

1. Experimentally, the spin labeling procedure is carried out incubating overnight at 4°C detergent-solubilized proteins (10–30 μM) with tenfold molar excess of MTSSL under gentle shaking (see Note 4). To avoid biradical formation the MTSSL concentration must be <500 μM in aqueous buffer (see Note 5).
2. To maximize the reaction yield, in some cases the protein needs to be pre-incubated for 1 h at room temperature with 10 mM DTT to reduce eventually oxidized cysteines. DTT must be removed before starting the labeling procedure (max. 1 μM final residual concentration) to avoid nitroxide reduction and to allow the formation of the disulfide bond between MTSSL and the cys residue. Immediately after DTT removal, MTSSL is added.
3. After the incubation period, remove the excess free label in the buffer. Different methods are possible: (a) dialysis: slow but efficient, requires at least 500 ml of detergent-containing buffer, which could be expensive; (b) centrifugal filter devices: fast and efficient, but some proteins precipitate on the membrane after centrifugation; (c) desalting columns: very fast and efficient, performed by centrifugation or gravity, requires a larger volume than centrifugal devices.
4. Once the free label is removed, the protein is concentrated to the desired final concentration, and the spin labeling efficiency is calculated based on the second integral of the derivative spectrum according to the method described in ref. 5. The spin-labeled protein is ready to be reconstituted in liposomes or nanodiscs (reconstitution procedures vary for different proteins and must be optimized accordingly).

3.2. Spin Label Dynamics Extracted at Room Temperature

EPR spectra of spin labels attached at protein sites are sensitive to even minor changes in the reorientational motion of the spin label with respect to the external magnetic field. Reorientational motions

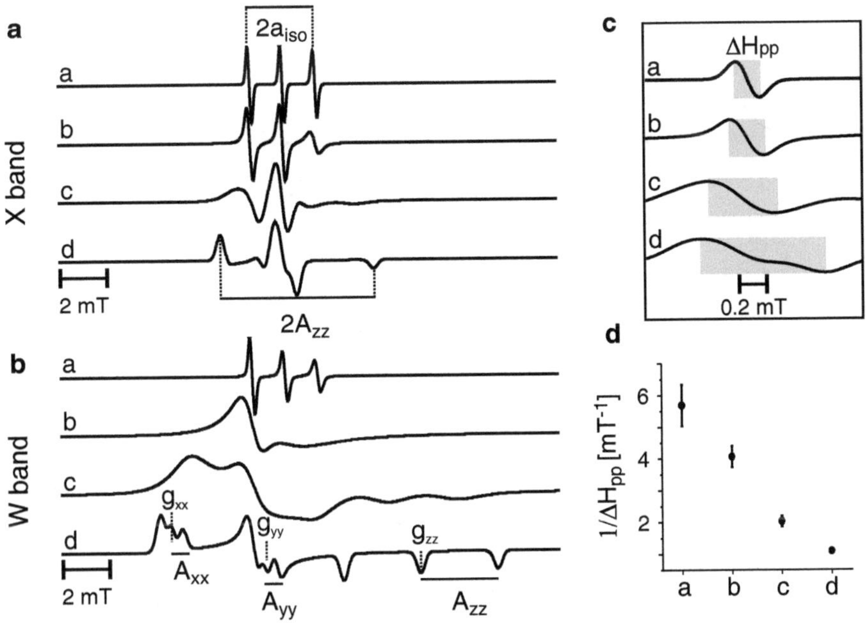

Fig. 2. Dynamics of the spin-labeled side chain encoded in the EPR spectra. (**a**) X-band EPR spectra (first derivative) simulated with Easyspin (a, routine "garlic," 0.1 ns rotational correlation time; b, routine "chili," 1 ns; c, routine "chili," 5 ns; d, routine "pepper," rigid limit). The relation between a_{iso} and A_{zz} and the outer peaks separation is presented in spectra *a* and *d*. (**b**) Analogous W-band EPR spectra simulated with Easyspin. The extractable parameters of the *g* and *A* tensors are depicted on the spectra. (**c**) Extraction of the ΔH_{pp} from the central line of the X-band spectra (denoted by the *grey bar*). (**d**) Plot of the inverse of the ΔH_{pp} for the four spectra simulated. Experimentally, a 10% error in the determination of ΔHpp is to be expected, which propagates into mobility parameter uncertainties.

of spin labels that are fast on the EPR timescale effectively average the anisotropic magnetic interactions leading to motionally narrowed spectral features. At X band, this corresponds to rotational correlation times less than 1 ns. Slower motions (5–100 ns) which are characteristic of buried sites in proteins and of global tumbling of proteins in membranes lead to increased apparent hyperfine splittings and broader spectral linewidths. In intermediate regimes the partial averaging of the anisotropic properties leads to a gradual change in the spectral features (Fig. 2) (see Note 6).

The reorientation of the spin label in the protein is strictly anisotropic being a complex function of the spin label molecular structure (e.g., MTSSL is characterized by five rotatable bonds, see Fig. 1) and the primary, secondary, tertiary, and eventually quaternary structure of the protein under investigation. Often complex distributions of motional states can be observed in spectra exhibiting more than one spectral component.

For a semiquantitative description of spin label dynamics, a simple empirical "mobility" parameter is extracted from the spectra at room temperature. This parameter is useful to obtain insights in the protein secondary structure, based on the available correlation between mobility and protein structure obtained from a large number of studies, mainly on α-helical proteins (see Note 7).

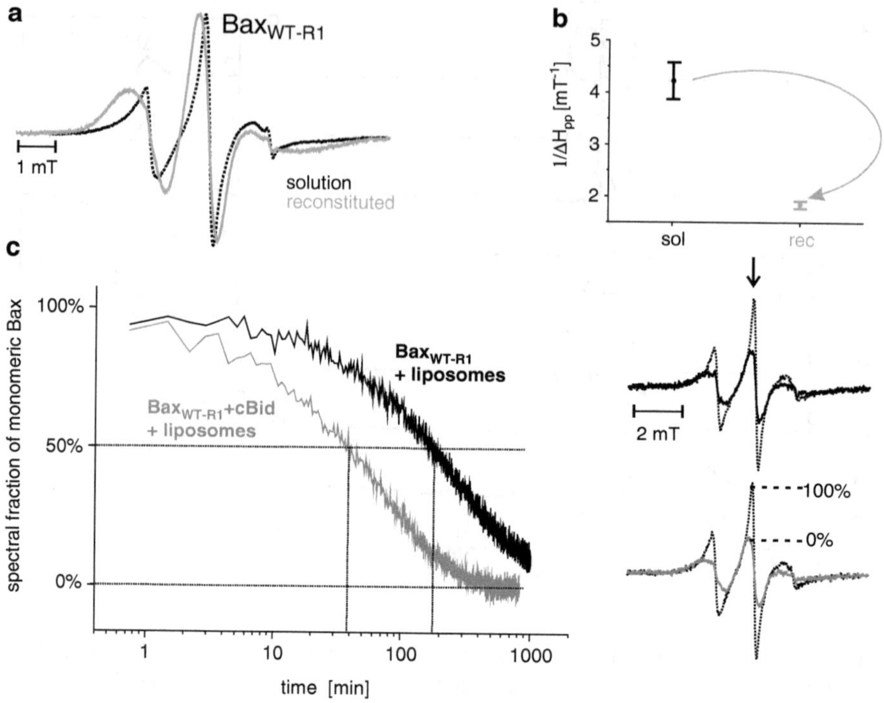

Fig. 3. Bax membrane insertion followed by changes in spin-labeled side chain dynamics. (**a**) X-band intensity-normalized derivative spectra of spin-labeled wild-type Bax in solution (*black*) and reconstituted in liposomes (*grey*). (**b**) Mobility parameters extracted from the two spectra presented in *panel A*. (**c**) Kinetics of Bax conformational changes at 37°C. Spectra were recorded each 43 s at 37°C with (*grey*) and without (*black*) cBid as triggering agent. The intensity of the central EPR line versus incubation time (logarithmic scale) is plotted. The *inset* in the right shows the spectra at time zero (*black dotted*) and at the end of the incubation time.

The most commonly used mobility parameter is the inverse of the central EPR linewidth ($\Delta H_{pp}{}^{-1}$), which is an increasing function of the mobility of the spin label (6) (Fig. 2c, d) (see Note 8).

One example of the use of the mobility parameter to address the conformational change of the human apoptotic Bcl-2 protein Bax from the inactive water-soluble to the oligomeric active membrane-embedded state is shown in Fig. 3 (7). Wild-type Bax contains two accessible natural cysteines at positions 62 and 126, which were labeled without additional mutagenesis, with a spin labeling efficiency close to 90%. Figure 3a shows the intensity-normalized EPR spectra detected at room temperature in the two protein conformations (see Note 9).

The two mobility parameters extracted from the spectra are shown in Fig. 3b. The time course of the slow conformational changes induced by the membrane could be followed at 37°C by detecting the EPR spectra over time (e.g., using the 2D CW experiment in the Bruker software Xepr). The decrease of the spectral amplitude over time denotes the broadening of the spectrum due to the protein conformational change. The intensity of the central line plotted versus time is presented in Fig. 3c. The spectra detected at time zero (just after insertion of the tube in the spectrometer), characteristic of 100% of the protein in the inactive soluble state, and after 6 h of incubation at 37°C, characteristic of the membrane-embedded state, are presented on the right panel. Analysis of the intensity decay gives the half time of the membrane insertion and conformational change of the protein. Addition of cleaved Bid, a Bcl-2 protein which triggers Bax insertion, resulted in a faster insertion of spin-labeled Bax, which could clearly be detected by EPR.

3.3. Water Accessibility: Topology of Membrane Proteins

The protein topology with respect to the water-membrane interface is an important aspect in membrane protein studies, and it is not directly addressable by X-ray crystallography. The three different types of possible side chain environments in a membrane protein, namely, the protein neighboring residues, the bulk water, or the lipid alkyl chains, can be addressed selectively by accessibility measurements in EPR. Several methods are available to obtain such information on membrane proteins both at room and cryogenic temperatures. This section presents such methods, with a particular focus on those, which require the use of cryogenic temperatures.

The most common EPR technique to obtain information on side chain accessibility at room temperature is power saturation CW EPR (8, 9), which consists in detecting spectra at increasing microwave power to induce saturation, a method which was already detailed in ref. 10. The amplitude of the central line is plotted versus the square root of the incident microwave power, and the saturation curve is analyzed to obtain the empirical $P_{1/2}$ value which is inversely proportional to the product of the effective relaxation times T_1 and T_2. The saturation behavior is characteristic of the local environment of the spin label and can be modified by addition of paramagnetic species soluble in water (chromium oxalate CrOX or NiEDDA) or preferentially partitioning in membrane bilayers (O_2) (see Note 10).

A second technique consists in the direct measure of the changes induced by paramagnetic agents in the spin–lattice relaxation T_1 by saturation recovery, a method detailed in ref. 11. In this case an intense saturation pulse of microwave is delivered at room temperature at a frequency corresponding to the central EPR line, and the return of the spectral intensity is monitored with a weak CW observing microwave field. The advantage of this method is the possibility to selectively measure T_1 at room temperature on different spectral components, which are usually present in the spectra of spin-labeled membrane proteins (see Note 11). An example of the application of saturation recovery to proteins to extract accessibility information can be found in ref. 12.

A third approach allows extraction of the polarity (electric fields created by charges or electric dipole moments of the water molecules) and proticity (propensity to form H-bonds) of the nitroxide microenvironment by analysis of the EPR spectra in the rigid limit (13–15). For membrane proteins in water–glycerol mixture, a temperature of 160 K, although not being exactly the rigid limit for nitroxides, can be safely used. The A_{zz} principal value of the ^{14}N hyperfine tensor (half of the splitting between the positive low-field peak and the negative high-field peak of the X-band CW spectrum, Fig. 2) is proportional to the polarity of the nitroxide microenvironment (for MTSSL the A_{zz} value can change from 3.3 to 3.7 mT going from an apolar to a polar environment). The A_{zz} parameter can be already extracted from X band CW EPR spectra. This information can be complemented by detection of the propensity of the nitroxide to form H-bond (proticity) via the g_{xx} principal value of the g tensor (for MTSSL the g_{xx} value can change from 2.0089 to 2.0083 mT going from an apolar-aprotic to a polar-protic environment) (Fig. 2). Measurements of g_{xx} require the continuous wave detection of the low-temperature nitroxide spectrum at W band (3.4 T/95 GHz).

A detailed description of the method to detect W-band spectra and extract the polarity parameters is given in the following:

1. Singly labeled membrane proteins at an optimal concentration of 100 μM (in detergent or liposomes) containing 10% v/v of glycerol as cryoprotectant are prepared, and 1 μl is inserted in W-band quartz capillaries (0.9 mm outer, 0.5 mm inner diameter) using, for example, a thin Pasteur pipette made with a Bunsen burner. The height of the sample in the tube is about 5 mm, enough to fill the W-band cavity.
2. Prepare the spectrometer (e.g., Bruker Elexsys E680 spectrometer equipped with a W-band probehead) in "tune mode" and position the microwave dip in the center of the screen.
3. The capillary is shock frozen in liquid nitrogen and inserted in the W-band cavity precooled at 160 K in a helium- or nitrogen-flow cryostat. After insertion of the capillary, the resonant frequency moves to lower values. Follow it and position the dip again in the middle of the screen (see Note 12).
4. The temperature of the cavity is allowed to equilibrate for at least 30 min. Reposition eventually the dip in the center of the screen. The equilibrium can also be judged by the invariance of the dip position and shape (see Note 13).
5. Before recording the spectrum, a power saturation curve must be measured at 160 K at least for one mutant of the spin-labeled protein under investigation to obtain the optimal microwave power. Additionally, the phase has to be properly adjusted. Usually 5 μW incident microwave power, 100 kHz

field modulation, 0.25 mT modulation amplitude, and 40 mT sweep width are good experimental parameters.

6. A first estimation of the g_{xx} and A_{zz} values, diagnostic for polarity and water penetration into the membrane, can be directly obtained from the spectra (Fig. 2). The fit of the spectra can provide more exact values of the parameters (e.g., using the Easyspin routine "pepper" (16) and esfit, see Note 14). A starting g-tensor for MTSSL is (2.0085, 2.0061, 2.0021), and starting A-tensor in MHz is (14, 14, 98). Calibration of the B field is mandatory. The g_{xx} region of the spectrum is the most complicated to fit. Due to g_{XX}-strain, there can be significant broadening of the spectral features, which needs to be taken into account in the simulation. Moreover, heterogeneities in the H-bonding may lead to the appearance of shoulders. In this case the fit is performed with two (max. three) g_{xx} components. A detailed study of this heterogeneity can be found in ref. 17.
7. The values obtained can be inserted in a plot of g_{xx} versus A_{zz} to obtain information on the relative location of each spin-labeled site with respect to known values. A comprehensive g_{xx} versus A_{zz} plot containing data from spin labels in organic solvents and from different proteins spin labeled with MTSSL can be found in refs. 14, 17.

An example of a polarity–proticity profile detected in the membrane protein bacteriorhodopsin is presented in Fig. 4a. Spin labels were placed in different position through the proton channel from the periplasmic to the cytoplasmic side of the protein. A clear water penetration barrier in the middle of the protein could be detected by both A_{zz} and g_{xx} (highlighted in red). This method is used to unravel relative polarities of different spin-labeled sites in a membrane protein or to detect changes in the water accessibility due to protein conformational changes.

A fourth approach, which is not yet established as a conventional method, is to detect the accessibility of the nitroxide towards deuterated water molecules by comparing the amplitude of the 3-pulse ESEEM oscillations detected at X-band at cryogenic temperatures (18). For deuterium ESEEM measurements on spin-labeled membrane protein embedded in liposomes or solubilized in detergent, one needs to deuterate the buffer up to more than 90%. Relative deuterium accessibility parameters for different side chains in a protein can be obtained using the ESEEM amplitude of a nitroxide in the deuterated buffer solution as a reference. An example of the water accessibility derived by 3-pulse ESEEM for two spin-labeled side chains in the light harvesting complex II compared to a water-soluble reference is presented in Fig. 4b (18). The advantage of the method is that deuterium ESEEM

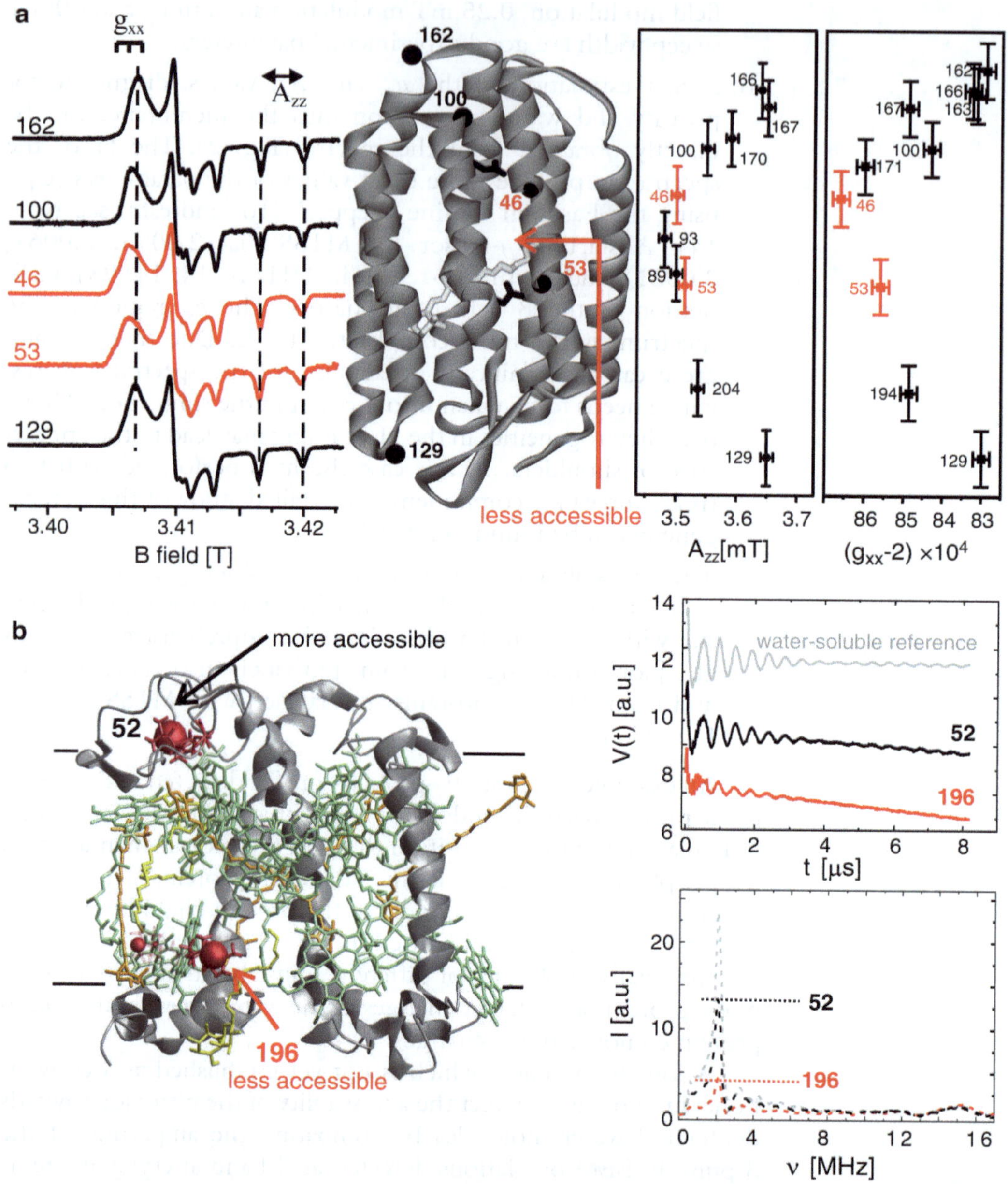

Fig. 4. Low-temperature methods to extract polarity and water accessibility. (**a**) W-band rigid-limit spectra of BR (PDB 1MOL) spin labeled in the proton channel. The changes in g_{xx} and A_{zz} are highlighted by *vertical lines*. Right panel, plots of g_{xx} and A_{zz} versus nitroxide position showing the hydrophobic barrier in the channel. (**b**) Structure of LHCIIb (PDB 2BHW) with two iodoacetamido spin labels at positions 52 and 196 calculated with MMM. Right panel, ESEEM traces detected on a water-soluble standard and on the two spin-labeled positions. Bottom, ESEEM modulation depths correlated with the water accessibility.

modulations provide a direct measure of the accessibility of the spin label side chain to the aqueous buffer. In addition, data are intrinsically well reproducible as the depth of the modulations is not very sensitive to the microwave field conditions once resonator is overcoupled. The disadvantages are the use of cryogenic temperatures and the difficulty to distinguish protein-buried from membrane-embedded sites and to disentangle the effects arising from deuterium atoms from water, exchangeable protons, or deuterated glycerol (the latter is used as cryoprotectant).

3.4. Methods Based on Interspin Distance Determination

Conformational changes of membrane proteins can be followed by monitoring distances between selected labeled sites on a protein (19). The existing CW and pulse EPR methods rely on extracting the dipolar coupling between two electron spins to obtain the distance distributions. There are common limitations for the CW and pulse EPR methods of distance determination. Generally, in addition to the dipolar (anisotropic) part, coupling between the spins contains an isotropic contribution known as Heisenberg exchange coupling. Since no general quantitative treatment of the distance dependence of the exchange coupling exists yet, distances at which exchange coupling is significant (<1 nm) cannot be reliably treated at present. Another limitation is imposed by the purely anisotropic nature of the dipolar interaction. In frozen solutions, one deals with a *static* ensemble of protein (and spin label) conformations which enables the experimental observation of the *full* dipolar tensor. In contrast, in the limiting case of fast tumbling (e.g., at room or higher temperatures), the dipolar interaction is completely averaged on the EPR timescale; thus it vanishes from the experimental spectra. In the intermediate situations (spin labels with restricted molecular motions on membrane proteins) only residual dipolar interaction is visible. Therefore, determination of distances from room-temperature spectra ignoring motional effects may be erroneous.

3.4.1. Interspin Distances at Cryogenic Temperatures Detected by CW EPR

This method is used when the magnitude of the dipolar coupling exceeds the natural linewidth of the nitroxide CW EPR spectrum and can therefore be separated from other interactions by lineshape analysis. Usually this is the case at interspin distances <2 nm. The distance extraction is based on the possibility to represent the spectrum with a sum of coupled and non-coupled species. The experimental labeling efficiency determines the fraction of the non-coupled (singly labeled) species and must be measured prior to the distance analysis. For the non-coupled spectrum, singly-labeled proteins are prepared and measured independently. The coupled dipolar broadened spectrum can be approximated with that of the non-coupled species with an appropriate dipolar broadening convolution function. In general, simplification and speeding up of the convolution method is achieved when one common coupling regime is used for all spin pairs, either strong coupling regime (e.g., DipFit from Steinhoff's group (20), Short-Distances from Altenbach (21), http://sites.google.com/site/

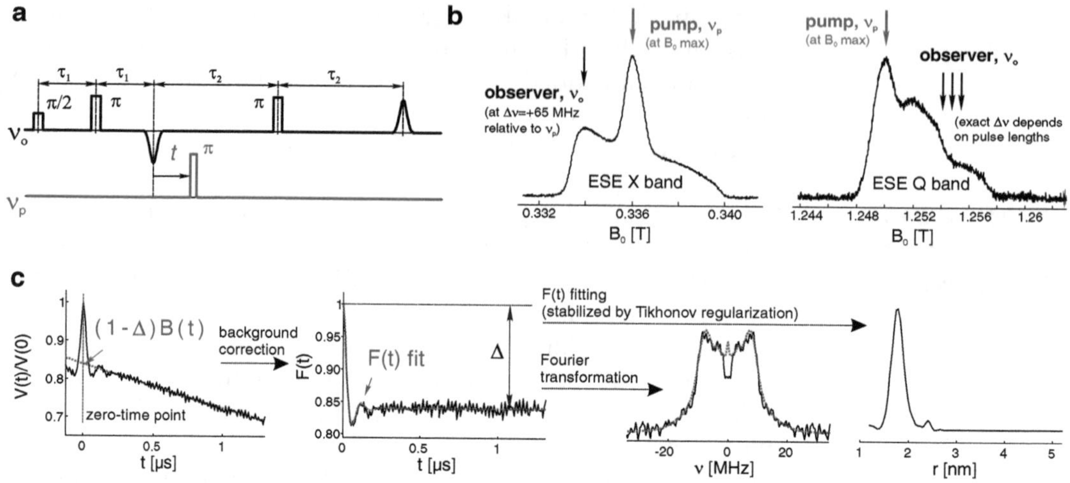

Fig. 5. DEER setup and data analysis. (**a**) DEER pulse sequence (**b**) Electron spin echo (ESE)-detected spectra from a pair of coupled nitroxides at X and Q bands with the observer and the pump positions indicated with arrows. (**c**) DEER data analysis flow (left–right): normalized and phase-corrected raw experimental trace $V(t)/V(0)$; DEER form-factor $F(t)$ obtained after background correction; dipolar (frequency-domain) spectrum; distance-distribution (also known as a $P(r)$ function).

altenbach/labview-programs/epr-programs/short-distances) or weak coupling regime (DipCon from Timmel's group (22)). In both cases a sum of Pake patterns corresponding to a Gaussian distribution of distances is used. Due to the approximations used in this method, the reliability of distance determination decreases if (a) the fraction of the singly labeled species is more than 30% or if this parameter is let free in the fitting and (b) the distance distribution is multimodal. At X-band in a range of 1.4–2.0 nm, such a fast distance extraction method gives generally consistent results on mean interspin distance and on the distribution width (22, 23) and can be used to complement the more precise DEER analysis which is presented in the next section. While sharing the common principles highlighted above, the exact protocols to extract distances vary from program to program.

3.4.2. Interspin Distance Determination by Double Electron Electron Resonance

Pulse EPR methods for distance determination provide much better resolution and sensitivity towards longer distances than the CW EPR lineshape analysis. Among various modern pulse EPR techniques for distance determination in membrane proteins, the DEER experiment [also known under the PELDOR (Pulsed ELectron DOuble Resonance) acronym, Fig. 5a] is presently the most used technique in the 1.5–6 nm range due to its precision combined with the relative simplicity in setting up, data acquisition, and data analysis (24). In the following, key theoretical points necessary for understanding and successfully performing DEER experiments are sketched, protocols for setting up DEER at standard X-band frequencies are given, and DEER data analysis is described.

Due to the about 20-fold sensitivity enhancement of Q-band DEER over the standard X-band operation (25), remarks about the Q-band setup are also given.

DEER Prerequisites

DEER on spin-labeled membrane proteins is a low-temperature method. Optimum sensitivity for proteins in water/glycerol mixture is achieved at 50–60 K (25, 26), but 80 K is also an option if cooling by liquid He is not available. Importantly, the DEER experiment is routinely applied to *weakly* coupled spins, meaning that not only the exchange coupling is zero, but also the dipolar coupling between the spins is small with respect to a resonance frequency difference of the coupled spins. In the DEER experiment this condition is satisfied for interspin distances >1.5 nm when $|\nu_p - \nu_o| > 50$ MHz (Fig. 5a). The pulses should be strong enough to excite the dipolar doublet of the spins (12 ns pulses are enough to measure down to 1.5 nm distance). Using softer pump π-pulses (>24 ns) not only leads to sensitivity losses but also increases the shortest measurable distances.

The DEER signal from a pair of spins residing on a single molecule (called $F(t)$ or *form-factor*) is given by

$$F(t, \theta, r) = 1 - \Delta(1 - \cos(\omega(r, \theta)t)),$$

in which Δ is the signal modulation depth ranging from 0 to 1 (equal for a two-spin system to the inversion efficiency λ), and $\omega(r,\theta)$ is the dipolar frequency for a particular distance r and a particular angle θ between the dipolar axis and the static magnetic field B_0. For a pair of nitroxide labels on a protein the modulation depth never reaches 1 due to incomplete excitation of the nitroxide spectrum by microwave pulses in DEER experiment (see Note 15). In frozen solutions all orientations of the interspin vector with respect to B_0 are present; therefore for a single distance one observes a weighted average of the individual orientation contributions known as a Pake pattern (see Note 16). In addition to the intramolecular part $F(t)$, the experimental DEER signal contains also *intermolecular* contributions $B(t)$ (i.e., dipolar couplings between spins residing on *different* protein molecules). The total DEER signal is then a product between the $F(t)$ and $B(t)$. Separation between intermolecular and intramolecular parts can readily be done if $B(t)$ (called *background* in such cases) has a simple analytical form, which is the case for diluted proteins (<500 μM spin concentration).

The optimal setup of the DEER pulse sequence to obtain the best experimental sensitivity can be derived from the absorption spectrum of the spins in the selected microwave band. The general principles are simple: (a) the stronger (i.e., shorter) the pump pulse, the larger the DEER modulation depth; (b) the stronger the observer pulses, the stronger the DEER echo (lower noise level);

(c) excitation bandwidths of the observer and pump pulses should not overlap; otherwise the DEER effect is lost (the modulation depth is reduced and artifacts, originating from the single-frequency excitation, may enter the DEER trace). The optimal X-band pulse setup for nitroxide spin labels is the following: 12 ns pump pulse (ν_p), placed in the middle of the resonator mode and on the global maximum of the nitroxide absorption spectrum; 32 ns observer pulses (ν_o), placed at the low-field shoulder of the nitroxide absorption spectrum (i.e., +65 MHz with respect to ν_p) (see Notes 17 and 18).

Scheme for DEER Measurements on Membrane Proteins

1. *Sample Insertion.* The resonator has to be precooled to the target temperature. The sample is shock frozen in liquid nitrogen prior to insertion and then transferred into the critically coupled resonator (spectrometer is in CW mode). After insertion, and temperature stabilization the resonator is overcoupled for the pulse operation (see Note 19).
2. *Acquiring the ESE-Detected Spectrum.* The Electron-Spin-Echo (ESE)-detected spectrum is acquired by sweeping the magnetic field B_0 using a primary (Hahn) echo sequence $\pi/2 - \pi -$ echo. Pulse lengths close to those for observer pulses can be used (see Note 20) (Fig. 5). The spectrum is primarily needed to determine the precise positions of the DEER pulses (see Note 21). By setting up the pulses for the ESE-detected spectrum, the optimum shot repetition time (srt) is determined: if increasing the srt by 1 ms brings less than 10% gain in the echo intensity, the previous srt value can be used (see Note 22).
3. *Measuring T_2 Relaxation by 2-Pulse ESEEM (Optional Step).* The primary echo pulse sequence (set on the previous step) can be used to measure the transverse relaxation time T_2 (with the pulse separation (τ_1) incremented normally in steps of 4–8 ns). The 2-pulse ESEEM is detected plotting the echo integral versus $2\tau_1$. The ESEEM trace can be fitted to obtain T_2, but it can serve directly as an estimation of the maximal length of the DEER trace, which is related to the longest detectable distance.
4. *Setting Up the DEER Pulse Sequence.* Move the B field to the maximum of the ESE-detected spectrum. Set the ELDOR frequency to the spectrometer frequency and optimize the pump π-pulse (see Note 23). Move the spectrometer frequency 65 MHz higher than the ν_p (X band, Fig. 5b) and optimize all observer pulses by Hahn echo detection. In order to avoid errors in the determination of the modulation depth related to hardware detection offsets, it is advisable to do a one-step phase cycling on the first ($\pi/2$) pulse of the DEER sequence; thus it is necessary to optimize also this channel.

5. After the setup is complete, load the DEER sequence with a standing pump pulse placed exactly at the position of the first echo (Fig. 5a) and adjust the phase of the DEER echo using the main phase of the spectrometer. Set the integration window around the DEER echo with the width equal to the length of the observer pulses (e.g., 32 ns for the optimal X-band DEER excitation).
6. *Choose the Dipolar Evolution Time* (τ_2). This is a crucial parameter which decides on the longest achievable distances and reliability of the background correction (i.e., reliability of the DEER form-factor $F(t)$). Basic principles: (a) frequency determination from the $F(t)$ is unambiguous if about two periods of the dipolar oscillation are observed; (b) background correction is reliable when at least half of the DEER trace (after dipolar oscillations have decayed) can be used for the background fit. Due to usually fast (T_2) relaxation rates in membrane proteins such criteria are rarely met. One can use the following algorithm instead: choose first a short τ_2 of about 1.2–1.5 μs and acquire a trial DEER trace. In case of short distances (<2.5 nm) this may be already sufficient. If long distances are present, τ_2 has to be increased bearing in mind that after 24 h of measuring time, the quality of the trace cannot be further increased much in a reasonable time, as the signal-to-noise ratio increases only with the square root of the measurement time.
7. *DEER Acquisition*. The DEER trace is acquired by moving the pump pulse through the dipolar evolution time usually with an 8 ns step (Fig. 5a). Due to residual bandwidth overlap between the observer and the pump pulses, nuclear (ESEEM) modulations may enter the DEER trace and contribute either to the noise or to the experimental frequencies if the DEER modulation depth is very low (~0.1). Main sources for the nuclear modulation effects are protons and deuterons. At X band, the modulations are suppressed by varying the τ_1 time of the DEER sequence in steps of 8 ns (protons) or 56 ns (deuterons) and subsequently summing up the traces. The acquisition of eight steps per cycle suffices (see Note 24). The stability of the hardware as well as of the experimental conditions is very important during data acquisition. Temperature oscillations, microwave mode instabilities, etc. may cause uncorrectable phase drifts, and measurements will have to be repeated. It is advisable therefore to pause the acquisition from time to time, control the phase, and make a trial distance extraction. Such a procedure helps to avoid unnecessary long measurements or to stop the measurement too early.

DEER Data Analysis

There are several software packages for DEER data processing which are freely available (e.g., DeerAnalysis, http://www.epr.ethz.ch/software/index; DeFit, http://www.sb.fsu.edu/~fajer/Programs/DEFit/defit.html). The steps of the DEER data analysis are described in relation to DeerAnalysis (current version 2011) (27).

1. Raw experimental data ($V(t)$) are phase corrected and normalized, and the zero time of the DEER trace is determined (Fig. 5c). This time corresponds to the exact point in which the pump pulse crosses the first echo. Knowing the exact zero time is particularly important if short distances are present. Due to finite pulse lengths it cannot be determined from the pulse sequence timings, but rather directly from the experimental data. The procedure gets less precise if DEER data are noisy. In such cases the zero time can be taken from the high-quality DEER records taken with analogous experimental parameters.
2. Normalized $V(t)$ is background corrected in order to obtain the DEER form-factor $F(t)$. If the sample is not over-concentrated (<500 μM spin concentration) and no aggregation occurred, $F(t)$ is obtained by division of the $V(t)$ by the background function $B(t)$ which is given by $B(t) = \exp(kt^{D/3})$ The parameter D reflects the dimensionality of the *homogeneous* distribution of the spins in a sample, and k is a concentration-dependent decay rate. The most commonly used D values are 3 for the 3D distribution (soluble proteins, membrane protein prepared with high lipid/protein ratios, e.g., >500) and 2 for the 2D distribution (membrane proteins with low lipid/protein ratio) (see Note 25). If possible, detect a DEER trace on a singly labeled mutant of the protein under investigation to extract precisely the D value which best fits the background decay. If D is known and the DEER trace is long enough (Fig. 5c) then the decay rate and background correction starting time are fitted to obtain an $F(t)$ trace oscillating about an horizontal line after background correction. However, DEER traces on membrane proteins are usually too short for such an ideal background correction procedure; decisions on particular background corrections require experience and are often made on an individual basis. DeerAnalysis provides a validation tool with which one can systematically check various background models for doubtful cases and try to estimate reliability of the extraction of the distance distributions. If sensitivity allows, increasing lipid/protein ratios or decreasing the concentration of detergent-solubilized proteins also help in background correction.

3. Extracting distance distributions from the DEER form-factors $F(t)$ is in mathematical terms an *ill-posed* problem, i.e., transition from $F(t)$ to distance distributions is not unique when experimental data are not perfect (26). In DeerAnalysis, the problem is solved by generating trial distance distributions, transforming them into form-factors and fitting to the experimental $F(t)$. Further stabilization is achieved by demanding distribution smoothness (Tikhonov regularization feature in DeerAnalysis, optimal regularization parameter is determined from the L-curve). Unbiased distance extraction from the experimental data is always preferred when measurement quality is good. That means that one should avoid model-based distance extractions as this may result in generating erroneous solutions.

Accounting for Spin Label Flexibility During Distance Analysis: A Rotamer Library Approach

Spin labels are relatively long (5–6 dihedral angles) and flexible objects; therefore judging on protein dynamics without taking into account the size of the labels is not a good practice as it reduces strongly both the precision and the reliability of the conclusions derived. A simple and quite reliable way to introduce the spin label conformational freedom into the EPR analysis is a rotamer library approach (28). Very briefly, the spin label is represented by a discrete set of rotamers whose interaction with the (immobilized) protein backbone is calculated with a certain algorithm. In a recent version of the rotamer library approach, sets of MTSSL and iodoacetamido-proxyl rotamers were derived from long molecular dynamics trajectories while rotamer–backbone interaction is approximated by softened Lennard-Jones potential. The approach is implemented in the freely available package MMM operating under Matlab (http://www.epr.ethz.ch/software/index), and it allows for prediction of labeling sites, simulating DEER traces (including multi-spin cases) on chosen protein models (PDB templates are necessary).

Studying Membrane Protein Conformations with DEER

Here we provide two examples of the application of DEER to membrane proteins. The first example deals with the detection of the conformational changes of a membrane-embedded ABC transporter during its function, and the second with the determination of the structure of the transmembrane helix of a membrane-embedded symporter PutP using only distance restraints.

Gate Movements in the Type II ABC Importer BtuCD-F During Nucleotide Cycle

Here DEER and CW measurements were combined to extract information about the motion of the periplasmic and cytoplasmic gates of BtuCD-F, the vitamin B_{12} importer from *E. coli* which belongs to the type II ABC importers. ATP-binding cassette (ABC) transporters are ubiquitous integral membrane proteins translocating substrates across cell membranes. The alternating access of their

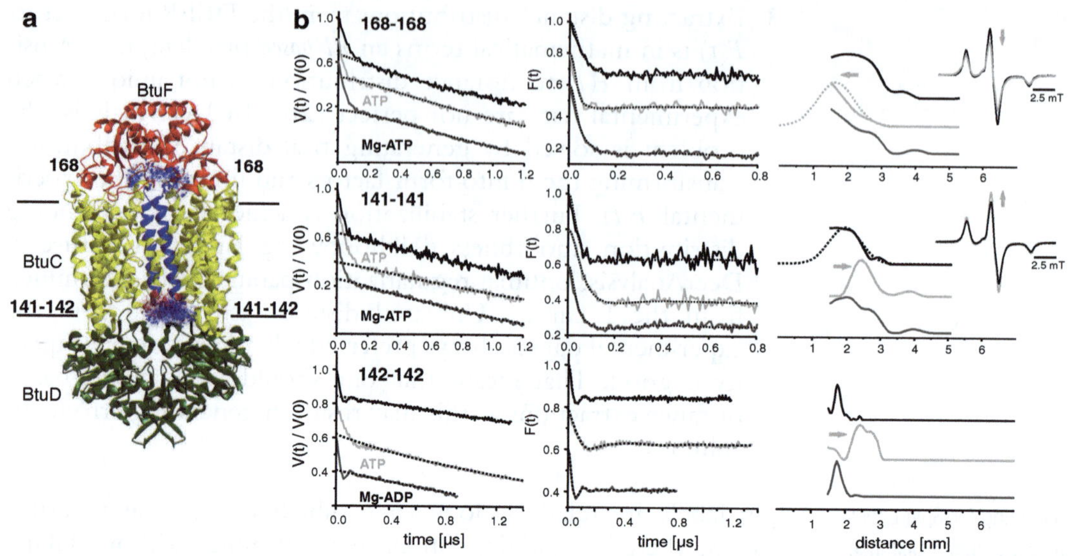

Fig. 6. Gate movements in BtuCD-F detected by EPR. (**a**) X-ray structure of BtuCD-F (PDB 2QI9) showing in ball and stick representation the calculated spin label rotamers attached to the engineered cysteines S141C, T142C, and T168C with the software MMM. (**b**) Normalized experimental data $V(t)/V(0)$, $F(t)$ and distance distributions obtained with DeerAnalysis 2011. DEER traces were recorded at X band for positions 141 and 168, at Q band for position 142 in the absence of nucleotides (*black*), in the presence of 5 mM ATP (*light grey*), and in the presence of Mg-ATP (after incubation at 37°C) or Mg-ADP, representing the post-hydrolytic state (*dark grey*). The *inset* shows the spin-normalized CW EPR spectra detected at 160 K characterized by different dipolar broadening for specific states during the nucleotide cycle. The Gaussian distance distributions obtained by DIPFIT (position 168: ATP-state, *grey dotted*; position 141: apo-state, *black dotted*) are superimposed to the DEER-derived distributions.

transmembrane domains to opposite sides of the membrane powered by the closure and reopening of the nucleotide-binding domains is proposed to drive the translocation events. Despite clear structural similarities, evidence for considerable mechanistic diversity starts to accumulate within the importers subfamily. The mechanism of substrate translocation for the type II importers is still poorly understood, and only X-ray structures of some nucleotide-free states are currently available (29–31). First insights into the gating mechanism of BtuCD-F were recently obtained by EPR (32). The distance changes measured at key positions in the translocation gates in the nucleotide-free, ATP- and ADP-bound conformations of the transporter reconstituted in liposomes are presented in Fig. 6. The translocation gates of the BtuCD-F complex undergo conformational changes in line with a "two-state" alternating-access model but DEER and lineshape analysis of low-temperature spectra provided the first direct evidence that, in contrast to type I importer binding of ATP, drives the gates to an inward-facing conformation, with the cytoplasmic gate opening and the periplasmic gate closing. The recently solved X-ray structure of the AMP-PNP-bound state of BtuCD-F confirmed the EPR findings (33).

Shape of the Transmembrane Helix IX of PutP

Here DEER measurements together with molecular modeling were used to determine from the scratch and without any additional information the backbone structure of transmembrane helix (TM) IX of the Na^+/proline transporter PutP (34) whose 3D structure was (and still remains) unknown. Interest in studying that particular helix was driven by biological evidences of its possible vicinity to the translocation pathway of the protein. By systematically mutating and labeling with MTSSL many positions on the TMIX, 16 double-mutants of PutP reconstituted in liposomes were prepared. DEER measurements taken at X band with the commercial hardware from Bruker (and using optimal DEER excitation conditions) were used to generate modeling constraints. A "helix–loop–helix" basic construct was chosen for the helix during modeling (Fig. 7a) while the beginning and the length of the loop (in terms of the number of residues) were fitting parameters. Primary DEER data (time-domain traces, phase corrected) were used as the only fitting constraints, whereas comparison between the simulated and the experimental distance distributions served for a visual control (Fig. 7b). As a result, a rather compact (~2.0 Å rmsd per residue) and moderately kinked backbone shape was found for TMIX (Fig. 7a). The relevance of the shape was supported by the solvent accessibility data: the beginning of the kinked region obtained by modeling correlates with the rapid increase of the solvent accessibility (35).

4. Notes

1. Unnatural amino acids can also be engineered when functional cysteines are present in the protein, for example carrying a keto group which can be functionalized with an hydroxylamine reagent carrying a paramagnetic center (36); however this approach has not been done yet on membrane proteins.
2. TOAC is an unnatural amino acid introduced by solid-state synthesis (a procedure not really feasible for large membrane proteins) and which can also be employed as a rigid spin label to obtain information on protein backbone dynamics, having only two degrees of freedom (37).
3. The freely downloadable software MMM based on a rotamer library approach (28) (http://www.epr.ethz.ch/software/index) can be used to predict the accessibility towards spin labels (currently MTSSL or iodoacetamido proxyl label) for each site of the protein prior to the site-specific mutagenesis if a structural template of the protein exists.
4. Very accessible sites can be labeled already after few minutes of incubation. In general, the labeling efficiency increases with

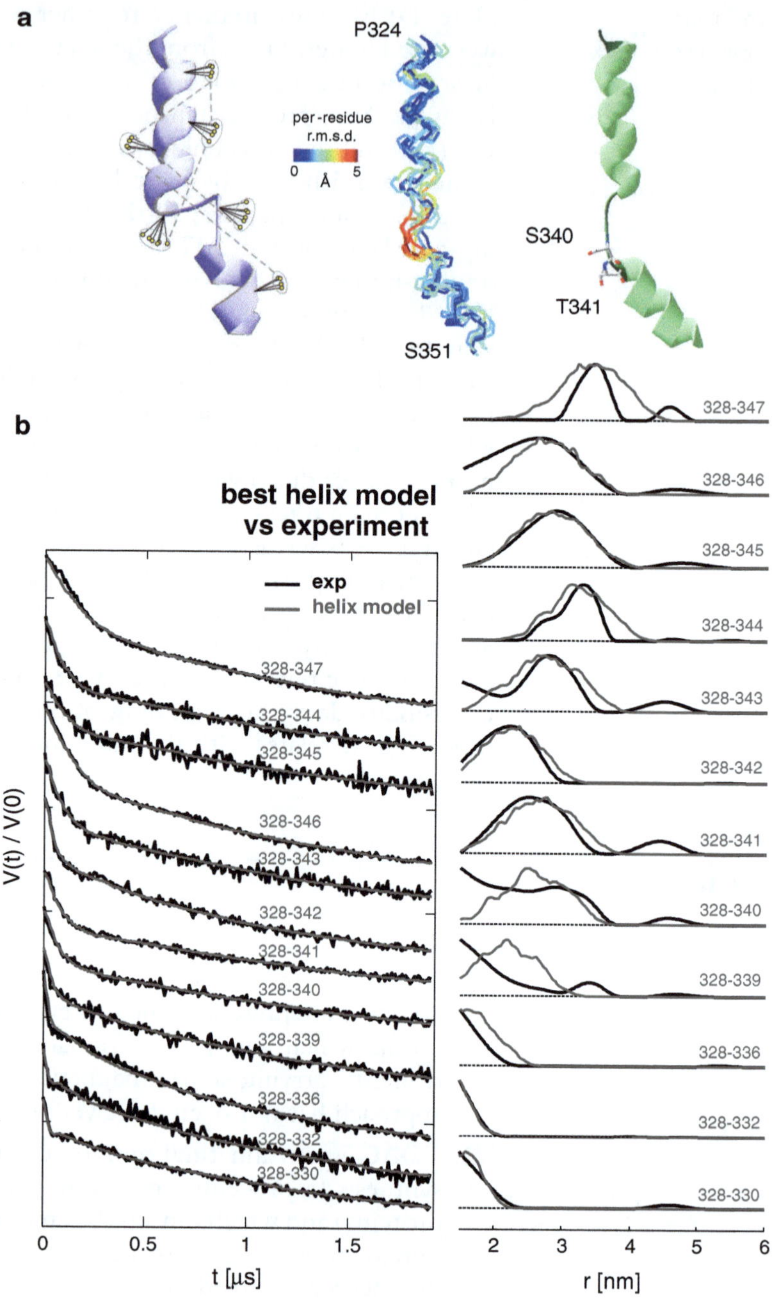

Fig. 7. Determining the backbone structure of the TMIX of PutP. (**a**) From the modeling prototype to experimental helix shape: "Helix–loop–helix" construct (*left*); final structure ensemble of the TMIX (*center*); final structure of the TMIX in ribbon representation based on the best-fitting model (*right*). (**b**) Fit performance during modeling: Primary DEER data versus best-fitting model (*left*); experimental versus model distance distributions (*right*).

increasing the label-to-protein molar excess (e.g., up to 40-fold) and increasing the temperature (from 4 to 37°C), depending on the protein stability.

5. In few cases the spin labels may bind unspecifically to the cys-less protein, or to the detergent micelles. It is then good practice to spin label prior to the SDSL study the cys-less protein and in the case of doubt also the empty micelles, to verify that the non-covalent unspecific labeling is negligible. Labels trapped in the micelles are removed by the reconstitution procedure, unless the protein is measured in detergent-solubilized form.

6. High-frequency spectra are more sensitive to fast motions due to the higher Zeeman spectral resolution. On the contrary, slow motions, which may be already close to the rigid-limit case at high frequencies, can be still detected at lower frequencies. The multifrequency approach allows extracting more information about the dynamics of the spin labels, and a global fitting can be performed to obtain the details of the anisotropic motion, for example with the SRLS program from Freed.

7. Although the semiquantitative analysis is usually enough to draw conclusions on secondary structure and conformational changes, interpretative problems may arise if spectra show complicated features with multiple components, which in the literature are often identified as "immobile" and "mobile" components.

8. The inverse of the spectral second moment can also be used as an increasing function of the spin label mobility and can be calculated to complement the mobility data extracted from the central linewidth. In the presence of two or more spectral components characterized by different dynamics, $\Delta H_{pp}{}^{-1}$ can overestimate the mobility due to the fact that the more mobile spectral component provides the leading fraction in the central linewidth. In contrast, the second moment, or spectra breath, is biased towards the low- and high-field regions of the spectrum which are mainly due to the immobile component, characterized by larger pseudo hyperfine splitting.

9. The intensity-normalized spectra, which are normalized to one at the maximum of the spectral amplitude, are used to highlight the spectral differences. (In contrast, the spin-normalized spectra, which are normalized by the second integral, show the broadening of the EPR lines due to restricted spin label motion as a decrease in spectral amplitude).

10. This method allows fast measurements on commercial CW spectrometers on a little amount of sample. However, indirect effects of paramagnetic quenchers are measured instead of the direct effects of water, and in the case of multiple spectral

components only averaged accessibility can be obtained. The use of the neutral paramagnetic water-soluble reagent NiEDDA (usually 20 mM) instead of the charged CrOX (usually 5 mM) is advisable in order to avoid possible electrostatic effects between the charged membrane headgroups and the paramagnetic reagent in membrane-embedded proteins. To increase the reproducibility, it is also advisable to prepare a large stock solution of concentrated NiEDDA (e.g., 200 mM) to be used for all measurements in a protein. When a new stock solution is prepared, control experiments to verify the reproducibility of previous measurements must be done. By this method, one does not measure relaxation but rather estimates the $P_{1/2}$ parameter. Since this estimation is sensitive to resonator tuning conditions, special care should be taken for data consistency (e.g., reproducible sample insertion and tuning conditions). Loop gap resonators are more suitable for these purposes.

11. With this method the resonator tuning conditions are less problematic in terms of signal reproducibility, because the T_1 relaxation time is measured directly. However, the setup for saturation recovery is more complex than for the CW power saturation method. Since a few years, the Bruker Elexsys bridge has the option for saturation recovery, which made the technique more widely available.
12. For commercial Bruker setups: The sample should be placed inside the resonator in two steps: the shock-frozen sample in the holder is introduced in the cryostat and fixed at about 1 cm before the entrance of the resonator. After about 15 min one can proceed with placing sample fully inside. This is because the very long holder of the commercial resonator shrinks at low temperatures affecting the sample position.
13. The linearity of the spectral baseline can be strongly influenced by temperature drifts during spectra acquisition; thus it is necessary to carefully optimize the nitrogen or helium flux and wait until the temperature in the cavity is perfectly stable. Strong helium/nitrogen fluxes should be avoided.
14. The exact fit of the whole spectrum is complicated by the multi-component nature of the fitting algorithm and by the non-perfect adjustment of the phase, which specially affect the central region of the spectrum.
15. If more than two spin labels are present in a protein complex, DEER signal will contain not only pairwise contributions but also combination (sum and difference) of interaction frequencies between all spins in the system. The total modulation depth of such system is larger than that for a simple pair of spins under the same experimental conditions (38).

16. In the absence of the orientation selection, each single distance contributes a full Pake pattern into a total DEER spectrum irrespective of the excitation bandwidth of the pulses. We do not consider here treatment of the orientation selection effects as their manifestation depends on the sample geometry and thus the treatment should be adopted on a case-to-case basis. Orientation selection is usually very rare in proteins (particularly, membrane proteins), especially at standard X-band frequencies.
17. In spectrometers equipped only with stripline pulse channels, it is not possible to achieve all observer pulses with the same length as well as to have a hard pump pulse. Sensitivity can be significantly lower in such cases.
18. At Q band, the same considerations apply: Pump pulse can be set at a global maximum of the nitroxide spectrum while observer pulses on the high-field side of the nitroxide absorption spectrum. Since absorption spectrum is broader than at X band, all pulses can be chosen hard (10–12 ns for instance) if enough microwave power is available.
19. At X band, resonators are usually always overcoupled. The degree of overcoupling varies from instrument to instrument, but generally speaking ringing should be avoided. In the case there is a lack of power (often the case at Q band) the DEER experiment is done with critically or slightly overcoupled resonators.
20. It is advisable to organize all steps of the DEER setup in one program controlling the experiment (known as PulseSPEL program in Bruker spectrometers).
21. The appearance of strongly broadened ESE-detected spectra points to the presence of short distances, which can be due to close vicinity of the two spin labels or to protein aggregation.
22. Srt depends on the longitudinal relaxation time (T_1). It decides how many DEER averages per time can be done without losing signal intensity. Generally, a 2–3 ms srt is used for spin-labeled proteins in detergents or liposomes. Srt may vary from <1 to 4 ms depending on the location of the labeled site (solvent-exposed site—long srt's when solvent deuteration was performed; facing bilayer site—short srt's due to relaxation enhancement by alkyl protons and dissolved oxygen).
23. The microwave power of the pump π-pulse can be optimized using an inversion recovery sequence in which the detection is done with a $\pi/2 - \pi -$ echo sequence set on the channel used for the ESE-detection. A nutation experiment can also be done to achieve a better precision.

24. At Q band, high-frequency proton modulations (~45 MHz) are practically not seen in the Q-band DEER traces. For deuterons, an 8-step averaging cycle performed with the step of 16 ns can be used.

25. DeerAnalysis provides exponential functions for homogeneous distribution of spins (2D, 3D, 4D, and higher dimensionality cases available), polynomial functions for inhomogeneous cases (requires particular experience), as well as the possibility to use experimental background function. This is particularly useful in cases when deducing background form from the DEER trace is not possible (far too short trace, high noise) or singly labeled mutants are used as background standards.

References

1. Berliner LJ, Reuben J (eds) (1989) Spin labeling theory and applications. Plenum Press, New York
2. Azarkh M, Okle O, Eyring P et al (2011) Evaluation of spin labels for in-cell EPR by analysis of nitroxide reduction in cell extract of *Xenopus laevis* oocytes. J Magn Reson 212:450–454
3. Krstic I, Hansel R, Romainczyk O et al (2011) Long-range distance measurements on nucleic acids in cells by pulsed EPR spectroscopy. Angew Chem Int Ed Engl 50:5070–5074
4. Fleissner MR, Bridges MD, Brooks EK et al (2011) Structure and dynamics of a conformationally constrained nitroxide side chain and applications in EPR spectroscopy. Proc Natl Acad Sci USA 108:16241–16246
5. Cooke JA, Brown LJ (2011) Distance measurements by continuous wave EPR spectroscopy to monitor protein folding. Methods Mol Biol 752:73–96
6. Hubbell WL, Mchaourab HS, Altenbach C et al (1996) Watching proteins move using site-directed spin labeling. Structure 4:779–783
7. Bleicken S, Classen M, Padmavathi PV et al (2010) Molecular details of Bax activation, oligomerization, and membrane insertion. J Biol Chem 285:6636–6647
8. Altenbach C, Greenhalgh DA, Khorana HG et al (1994) A collision gradient-method to determine the immersion depth of nitroxides in lipid bilayers – application to spin-labeled mutants of bacteriorhodopsin. Proc Natl Acad Sci USA 91:1667–1671
9. Altenbach C, Froncisz W, Hemker R et al (2005) Accessibility of nitroxide side chains: absolute Heisenberg exchange rates from power saturation EPR. Biophys J 89:2103–2112
10. Oh KJ, Altenbach C, Collier RJ et al (2000) Site-directed spin labeling of proteins. Applications to diphtheria toxin. Methods Mol Biol 145:147–169
11. Subczynski WK, Widomska J, Wisniewska A et al (2007) Saturation-recovery electron paramagnetic resonance discrimination by oxygen transport (DOT) method for characterizing membrane domains. Methods Mol Biol 398:143–157
12. Pyka J, Ilnicki J, Altenbach C et al (2005) Accessibility and dynamics of nitroxide side chains in T4 lysozyme measured by saturation recovery EPR. Biophys J 89:2059–2068
13. Kurad D, Jeschke G, Marsh D (2003) Lipid membrane polarity profiles by high-field EPR. Biophys J 85:1025–1033
14. Plato M, Steinhoff HJ, Wegener C et al (2002) Molecular orbital study of polarity and hydrogen bonding effects on the g and hyperfine tensors of site directed NO spin labelled bacteriorhodopsin. Mol Phys 100:3711–3721
15. Steinhoff HJ, Savitsky A, Wegener C et al (2000) High-field EPR studies of the structure and conformational changes of site-directed spin labeled bacteriorhodopsin. Biochim Biophys Acta Bioenerg 1457:253–262
16. Stoll S, Schweiger A (2006) EasySpin, a comprehensive software package for spectral simulation and analysis in EPR. J Magn Reson 178:42–55
17. Bordignon E, Brutlach H, Urban L et al (2010) Heterogeneity in the nitroxide microenvironment: polarity and proticity effects in spin-labeled proteins studied by multi-frequency EPR. Appl Magn Reson 37:391–403
18. Volkov A, Dockter C, Bund T et al (2009) Pulsed EPR determination of water

accessibility to spin-labeled amino acid residues in LHCIIb. Biophys J 96:1124–1141

19. Berliner LJ, Eaton SS, Eaton GR (eds) (2000) Distance measurements in biological systems by EPR. Kluwer Academic/Plenum, New York
20. Steinhoff HJ, Radzwill N, Thevis W et al (1997) Determination of interspin distances between spin labels attached to insulin: comparison of electron paramagnetic resonance data with the x-ray structure. Biophys J 73:3287–3298
21. Altenbach C, Oh KJ, Trabanino RJ et al (2001) Estimation of inter-residue distances in spin labeled proteins at physiological temperatures: experimental strategies and practical limitations. Biochemistry 40:15471–15482
22. Banham JE, Baker CM, Ceola S et al (2008) Distance measurements in the borderline region of applicability of CW EPR and DEER: a model study on a homologous series of spin-labelled peptides. J Magn Reson 191:202–218
23. Grote M, Bordignon E, Polyhach Y et al (2008) A comparative electron paramagnetic resonance study of the nucleotide-binding domains' catalytic cycle in the assembled maltose ATP-binding cassette importer. Biophys J 95:2924–2938
24. Pannier M, Veit S, Godt A et al (2000) Dead-time free measurement of dipole–dipole interactions between electron spins. J Magn Reson 142:331–340
25. Polyhach Y, Bordignon E, Tschaggelar R et al (2012) High sensitivity and versatility of the DEER experiment on nitroxide radical pairs at Q-band frequencies. Phys Chem Chem Phys 14:10762–10773
26. Jeschke G, Polyhach Y (2007) Distance measurements on spin-labelled biomacromolecules by pulsed electron paramagnetic resonance. Phys Chem Chem Phys 9:1895–1910
27. Jeschke G, Chechik V, Ionita P et al (2006) DeerAnalysis2006 – a comprehensive software package for analyzing pulsed ELDOR data. Appl Magn Reson 30:473–498
28. Polyhach Y, Bordignon E, Jeschke G (2011) Rotamer libraries of spin labelled cysteines for protein studies. Phys Chem Chem Phys 13:2356–2366
29. Hvorup RN, Goetz BA, Niederer M et al (2007) Asymmetry in the structure of the ABC transporter-binding protein complex BtuCD-BtuF. Science 317:1387–1390
30. Locher KP, Lee AT, Rees DC (2002) The *E. coli* BtuCD structure: a framework for ABC transporter architecture and mechanism. Science 296:1091–1098
32. Pinkett HW, Lee AT, Lum P et al (2007) An inward-facing conformation of a putative metal-chelate-type ABC transporter. Science 315:373–377
32. Joseph B, Jeschke G, Goetz BA et al (2011) Transmembrane gate movements in the type II ATP-binding cassette (ABC) importer BtuCD-F during nucleotide cycle. J Biol Chem 286:41008–41017
33. Korkhov VM, Mireku SA, Locher KP (2012) Structure of AMP-PNP-bound vitamin B12 transporter BtuCD-F. Nature 490:367–372
34. Hilger D, Polyhach Y, Jung H et al (2009) Backbone structure of transmembrane domain IX of the Na+/proline transporter PutP of *Escherichia coli*. Biophys J 96:217–225
35. Raba M, Baumgartner T, Hilger D et al (2008) Function of transmembrane domain IX in the Na+/proline transporter PutP. J Mol Biol 382:884–893
36. Fleissner MR, Brustad EM, Kálai T et al (2009) Site-directed spin labeling of a genetically encoded unnatural amino acid. Proc Natl Acad Sci 106:21637–21642
37. Toniolo C, Valente E, Formaggio F et al (1995) Synthesis and conformational studies of peptides containing TOAC, a spin-labelled Cα, α-disubstituted glycine. J Pept Sci 1:45–57
38. Jeschke G, Sajid M, Schulte M et al (2009) Three-spin correlations in double electron–electron resonance. Phys Chem Chem Phys 11:6580–6591

Chapter 16

Solid-State NMR Approaches to Study Protein Structure and Protein–Lipid Interactions

Christopher Aisenbrey, Matthias Michalek, Evgeniy S. Salnikov, and Burkhard Bechinger

Abstract

Solid-state NMR spectroscopy has been developed for the investigation of membrane-associated polypeptides and remains one of the few techniques to reveal high-resolution structural information in liquid-disordered phospholipid bilayers. In particular, oriented samples have been used to investigate the structure, dynamics, and topology of membrane polypeptides. Much of the previous solid-state NMR work has been developed and performed on peptides, but the technique is constantly expanding towards larger membrane proteins. Here, a number of protocols are presented describing among other the reconstitution of membrane proteins into oriented membranes, monitoring membrane alignment by ^{31}P solid-state NMR spectroscopy; investigations of the protein by one- and two-dimensional ^{15}N solid-state NMR; and measurements of the lipid order parameters using ^{2}H solid-state NMR spectroscopy. Using such methods solid-state NMR spectroscopy has revealed a detailed picture of the ensemble of both lipids and proteins and their mutual interdependence in the bilayer environment.

Key words: Membrane reconstitution, Oriented bilayer, Helix topology, Membrane protein structure, Detergent, Surface alignment, Transmembrane orientation, Protein insertion, Channel, PISEMA, Separated local field spectroscopy, Cross polarization, Lee-Goldburg decoupling

1. Introduction

Membrane proteins constitute about 30% of the expressed reading frames (1, 2) and are extremely important in cellular function and of major importance in biomedical and pharmaceutical research. However, their structural investigation lags much behind that of soluble proteins due to problems of obtaining the quantitative amounts of proteins needed for structural investigations and their limited stability in the absence of lipid bilayers. In addition, among those systems that can be made available many are too large for a detailed solution NMR spectroscopic analysis and/or they are

Jörg H. Kleinschmidt (ed.), *Lipid-Protein Interactions: Methods and Protocols*, Methods in Molecular Biology, vol. 974,
DOI 10.1007/978-1-62703-275-9_16,

difficult to crystallize in their native conformation, a requirement for high-resolution X-ray diffraction techniques. In particular, very few crystal structures of G-protein-coupled receptors (GPCR) could so far be solved (e.g., (3–5)) although this family of proteins is of major importance in pharmaceutical and medical research.

Solid-state NMR spectroscopy provides an interesting alternative route for protein structure determination also within large complexes including membrane polypeptides, and provides interesting complementary information when combined with other structural approaches (6–9). Whereas in solution fast diffusion and motional averaging assure that the NMR spectra exhibit narrow lines and high resolution, the anisotropy of NMR interactions remains apparent in the spectra taken from solids or from molecules or complexes whose size prevents efficient isotropic averaging. Therefore in static solid or semisolid samples the chemical shift, dipolar interactions, and quadrupolar splittings are all dependent on the molecular alignment relative to the magnetic field direction, and the anisotropy of nuclear interactions is a predominant feature of the NMR spectra also of membrane-bound proteins (10). Whereas under magic angle spinning conditions the spectral appearance is related to the isotropic chemical shift spectra observed in solution, here we present the analysis of oriented samples, which contain valuable structural information as the chemical shifts (e.g., ^{15}N, ^{13}C, ^{1}H), quadrupolar (e.g., ^{2}H, ^{14}N), and dipolar couplings (e.g., ^{15}N–^{13}C, ^{1}H–^{15}N) all depend on the alignment of bonds and molecules relative to the magnetic field direction. This approach requires that polypeptides are reconstituted into macroscopically oriented membranes and the resulting samples inserted in the magnetic field of the NMR spectrometer, in the simplest case with the bilayer-normal parallel to the magnetic field direction.

Clearly the reconstitution of the proteins into phospholipid bilayers is a critical step before the analysis by NMR spectroscopic methods can take place. Therefore, Subheading 3.1 provides a detailed step-by-step protocol for the membrane reconstitution of proteins that are soluble in aqueous buffer but also capable to insert into the membrane. A number of alternative pathways are described including the use of detergents thus that the approach can be extended to other membrane proteins. In particular, Notes 2, 5, and 9 (Subheading 4), respectively, give indications for how to proceed with other lipid bilayers and buffers, or which detergents are commonly used to extract and solubilize membrane proteins.

From such samples solid-state NMR spectra are obtained that provide orientation-dependent information by revealing ^{15}N and ^{31}P chemical shifts as well as ^{15}N–^{1}H dipolar couplings (Subheadings 3.2.1–3.2.3). This chapter therefore complements previous protocols where the investigation of synthetic peptides by ^{15}N and ^{2}H solid-state NMR spectroscopy was described (11).

In Subheading 3.2.2 the procedure to record oriented ^{15}N chemical shift solid-state NMR spectra is presented in detail. These spectra allow one to simply "read out" the membrane alignment of α-helical peptides that have been labelled with ^{15}N at one of the amide bonds (12, 13). At membrane alignments where the normal is parallel to the magnetic field direction transmembrane α-helical peptides exhibit ^{15}N resonances in the 200 ppm range whereas they resonate at <100 ppm when oriented parallel to the membrane surface (12). In a related manner, the dipolar coupling is $K \times (3\cos^2\Theta - 1)$, where Θ is the angle between the N–H vector and the magnetic field direction (parallel to the bilayer normal if oriented correspondingly) and for the amide nitrogen K is typically −10 kHz (12). As the N–H vectors are close to collinear with the helix long axis values around 20 kHz (ignoring the negative sign, which cannot be determined easily) indicate transmembrane alignments, whereas the splitting is zero at the magic angle and close to 10 kHz in planar alignments. Therefore, this parameter provides related information to ^{15}N chemical shift measurements from the same helix (9). When both values are correlated from a large number of residues of the same helix the resulting pattern is indicative of the tilt and pitch angles as well as reveals kinks or other irregular features. Other parameters that can be used for structural analysis are ^{13}C chemical shift or the ^{2}H quadrupolar splittings of methyl-deuterated alanines (9, 14), and the corresponding protocols can be found in a previous publication of this series (11).

Solid-state NMR spectroscopy is also used to monitor the macroscopic phase properties of phospholipids, membrane thinning, and formation of pore structures involving the phospholipids. First, ^{31}P solid-state NMR spectra have been used to monitor the macroscopic phase properties of phospholipid membranes as bilayers, hexagonal, cubic, and isotropic phases exhibit distinct spectral line shapes (15). Second, oriented ^{31}P and ^{2}H solid-state NMR spectra have been analyzed for membrane thinning and toroidal pore assemblies (16, 17), and the sample preparation and NMR measurements are presented in Subheadings 3.1 and 3.2.1, respectively.

Third, ^{2}H solid-state NMR spectroscopy of deuterium-labelled lipids allows one to monitor the order parameter of the fatty acyl chains in a segment-dependent manner and thereby also the membrane thickness, the protein penetration depth, and the resulting membrane disruptive properties of polypeptides (18). By mixing deuterated lipids with non-deuterated ones it is also possible to monitor interactions of the protein with specific membrane components using the same procedures (19). The protocols necessary for such investigations are provided in Subheading 3.2.4. Lastly, a similar approach can be used on head-group-deuterated lipids to monitor electrostatic interactions at the membrane interface (20, 21).

A major advantage of solid-state NMR spectroscopy is its capacity to work with bilayers in their native liquid-disordered state. However, the technique is also relatively insensitive when compared to other types of spectroscopy. Therefore it is worthwhile to consider the sensitivity of the measurements described here and the requirements for isotopic labelling. On the one hand the ^{31}P nucleus is present in every phospholipid at 100% natural abundance and exhibits a relatively high sensitivity. Therefore one of the most straightforward measurements is the analysis of the alignment and macroscopic phase properties of the membranes by ^{31}P solid-state NMR, and many of these measurements can be performed in time-scales of minutes (e.g., (16, 22)). On the other hand, recording the ^{15}N (or ^{2}H) solid-state NMR spectra of polypeptides reconstituted into membranes is more difficult as the proteins are diluted in the membrane and in water. In addition, in oriented membranes much of the space within the coil volume is occupied by glass plates. Furthermore, the spins typically observed (^{15}N, ^{2}H) are relatively insensitive when compared to ^{1}H or ^{31}P. Therefore, quantitative amounts of uniformly or selectively labelled proteins have to be prepared usually by a combination of molecular genetics, bacterial overexpression, and biochemical methods (23–25). A typical sample encompasses milligram amounts of proteins, labelled with ^{15}N in 100–200 mg of lipid, and even though signal enhancement techniques such as cross polarization are used the measurements can take several hours (one-dimensional experiments) to days (two-dimensional experiments). In this context it is noteworthy that dynamic nuclear polarization/solid-state NMR equipment has been developed and has recently become commercially available (26). This approach can boost the sensitivity of NMR spectroscopy by orders of magnitude, thereby reducing the acquisition times of spectra from days to minutes. Proof of principle has been provided that the technology also works for oriented membrane samples (27), and much progress in the field can be expected from this innovative technology.

2. Materials

2.1. Materials Needed for the Preparation of Samples of Membrane Proteins Reconstituted into Oriented Lipid Bilayers

1. Disposable glass tubes with conical ends ≥10 ml, e.g., ~15 mm × 100 mm; glass pipettes with long tip for organic solvents and standard glass pipettes for aqueous solutions; 250 μl Hamilton syringe; glass Petri dishes with lid (diameter about 10 cm); watchmaker forceps, one with pointed and one with flat tips (Outils Dumont SA, Montignez, Switzerland); Teflon tape; plastic foil that goes with an impulse sealer also used in kitchens (e.g., TEW Electrical Heating Equipment, Taipei, Taiwan).

2. Sterile filtered water. Solvents and reagents of analytical grade: Buffers, salts, sodium azide. To dissolve lipids dichloromethane or chloroform, methanol or other organic solvents (see Note 1).
3. Buffer stock solution used to dissolve the protein, e.g., 200 mM Tris/HCl, pH 7.4.
4. Protein sample in buffer.
5. Sodium azide (NaN_3) solution (0.02% w/v) for the resuspension of the protein/lipid pellet after ultracentrifugation.
6. 1-Palmitoyl-2-oleoyl-*sn*-glycero-3-phosphocholine (POPC) and depending on the desired lipid composition other phospholipids (Avanti Polar Lipids, Alabaster, AL, USA), cholesterol (Sigma-Aldrich) (see Note 2). In order to perform the ^2H solid-state NMR experiments described in Subheading 3.2.4 a number of different phospholipids deuterated at various head-group and/or fatty acyl chain positions are commercially available (e.g., Avanti Polar Lipids, Alabaster, AL, USA).
7. Lipid extruder (e.g., Avanti® Mini-Extruder, Avanti Polar Lipids, Alabaster, AL, USA), with polycarbonate filters (100 or 200 nm) and/or tip sonicator (e.g., Sonoplus, Bandelin electronic GmbH & Co. KG, Berlin, Germany).
8. Centrifuge (MiniSpin®, Eppendorf AG, Hamburg, Germany) for the removal of metal particles after sonication.
9. Dialysis chamber with molecular weight cutoff (MWCO) of 3.5–10 kDa and a sample volume of 3–12 or 0.5–3 ml (e.g., Slide-A-Lyzer Dialysis Cassette, Thermo Scientific, Rockford, IL) and syringes (e.g., 5 ml syringe, Terumo, Leuven, Belgium) equipped with an injection needle (0.8 mm × 4 mm Sterican, Braun, Melsungen, Germany, or 0.4 mm × 2 mm Neolus, Terumo, Leuven, Belgium) to insert the sample into the chamber in order to desalt the protein/lipid suspension.
10. Bio-Beads SM-2 (BioRad Laboratories, Richmond, CA) for the effective detergent removal from the protein sample after reconstitution.
11. Concentrifugal concentrator tubes (e.g., Macrosep® Centrifugal Devices, Pall, Port Washington, NY), molecular weight cutoff (MWCO) of typically 3 kDa [big enough for easy diffusion of buffer (and detergent), small enough to retain the protein].
12. To equilibrate the atmosphere at 93% r.h. a saturated solution of KNO_3 (for analysis min 99% Merck Darmstadt) is used in a closed chamber (see Note 3).
13. Nitrogen gas (bottled or prepared by evaporation of liquid nitrogen) with sensitive pressure gauge, connected to a silicon tube which finishes with a Pasteur pipette.

14. Ultrathin glass plates that fit your NMR coil (thickness 00 or 0 from Marienfeld, Lauda-Königshofen, Germany). Before preparing the sample test if a stack of glass plates sealed and wrapped in plastic film fits the coil (cf. Subheading "Formation of Oriented Lipid Bilayers Between Glass Plates"). Custom-made glass plates of dimensions 11 × 20, 9 × 20, or 8.5 mm × 20 mm have been used. If you have the choice 9 × 20 is about half the prize when compared to 8.5 × 20.
15. High vacuum with cold trap to remove traces of organic solvent (e.g., lyophilizer).
16. Ultracentrifuge to fractionize lipid-bound protein from unbound protein staying in solution (e.g., Optima L-100 XP, Beckman Coulter, Palo Alto, CA).

2.2. Materials Needed for NMR Spectroscopy

1. Access to a solid-state NMR spectrometer (at least two channels except for ^{2}H NMR measurements) with the corresponding probes: i.e., for the experiments described here static ^{31}P {^{1}H}, ^{15}N{^{1}H}, and/or ^{2}H. For better sensitivity during the investigation of peptides in membranes flat-coil NMR probes have been developed (28, 29) and are in the meantime commercially available from several manufacturers (e.g., Bruker, Rheinstetten, Germany; Agilent, Paolo Alto, CA; Doty Scientific, Columbia, SC). Preferably NMR probe heads that avoid strong electric fields at the interior of the coil are used as they assure less heat intake and thereby improved sample stability (see Note 4). For the investigation of lipids by ^{31}P{^{1}H} or ^{2}H solid-state NMR or for highly concentrated polypeptide samples by ^{15}N{^{1}H} (or ^{13}C{^{1}H}) solid-state NMR, where smaller overall quantities of sample are required (e.g., a ^{31}P NMR signal can already be obtained from 5 mg phospholipids oriented between a pair of glass plates), conventional solenoidal coil shapes have also been used.
2. Solid-state NMR references: ^{15}N-amonniumchloride or ^{15}N-amonniumsulfate for ^{15}N{^{1}H} experiments, 85% H_3PO_4 for ^{31}P{^{1}H} experiments, and water enriched with $^{2}H_2O$ for ^{2}H NMR.

3. Methods

3.1. Reconstitution of Proteins into Oriented Lipid Bilayers

An oriented solid-state NMR sample is typically composed of a few mg of protein in about 50–200 mg of lipid oriented between glass plates, the exact amounts depending on finding a compromise between a good "dilution" of the protein in the membrane and consequently good bilayer alignment, good signal intensity which is also correlated to the sensitivity (magnetic field strength) of the

NMR spectrometer, and the maximum sample size which is related to the volume of the NMR coil (typically in the range 10 mm × 8 mm × 3 mm). Also, to keep the protein functional or to test the effects of the membrane composition it may be desired that samples of variable lipid composition are prepared. Therefore it is notable that a large variety of bilayers with different composition have been oriented mechanically between glass slides (30), and as long as liquid-disordered bilayers form uniaxial alignment has been possible with many different lipid mixtures.

In the following a preparation scheme for different lipid mixtures is therefore provided. Whereas during the reconstitution procedure peptides can often be mixed with lipids in organic solvents (11, 31) this should in general be avoided for proteins. Therefore, the protocols presented here consist in preparing phospholipid vesicles (Subheading "Preparation of Lipid Vesicles"), followed by membrane reconstitution of the protein (Subheadings "Reconstitution of Water-Soluble Proteins into Lipid Vesicles and Reconstitution of Detergent-Solubilized Proteins into Lipid Vesicles"), removal of detergent/buffer exchange (Subheading "Detergent Removal Using Dialysis Tubes" and "Detergent Removal Using Micro Beads" or "Reducing the Volume of the Protein/Lipid Sample After Dialysis"), and orienting the resulting vesicles along the surface of small cover glasses (Subheading 3.1.2). The full protocol is summarized in the flowchart shown in Fig. 1.

3.1.1. Reconstitution of Proteins into Lipid Vesicles

Preparation of Lipid Vesicles

1. Prepare a homogenous mixture of the lipid organic solvents (at ~50 mg/ml) in glass tubes of approximate dimensions ~15 mm × 100 mm. Depending on the lipid composition chloroform, dichloromethane, or mixtures of one of the two solvents with methanol 2:1 v/v have been used (see Note 1).
2. The organic solution is gently evaporated in a hood using a stream of nitrogen in such a manner to create a film of the lipid on the inner surface of the test tube. Pasteur pipettes attached to a soft silicone tube can be used for this purpose and can be easily replaced by fresh ones to avoid cross-contamination of samples. Before exposing the solution to the stream of nitrogen carefully adjust the gas flow as turbulences may cause spray dispersion and the loss of sample. Warming the test tube in a small beaker of warm water helps in the steady evaporation of the solvent. If dichloromethane is used warming by the hand is sufficient.
3. Remove the remaining traces of organic solvents by high vacuum during a few hours or best over night. This can be achieved, e.g., in a lyophilizer.
4. Hydrate the lipid film with the buffer also used to dissolve the protein to obtain a lipid concentration ≤10 mM (7–8 mg/ml) (see Note 5), and close the tube with parafilm. Leave for about

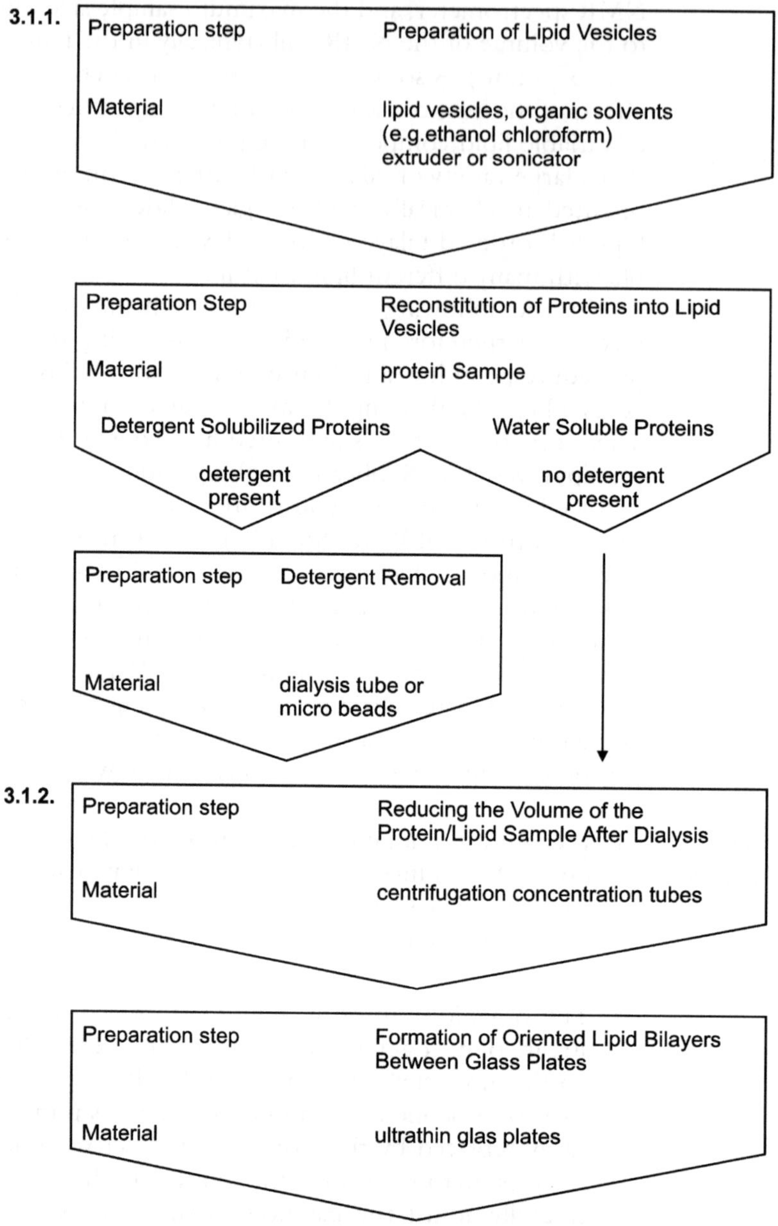

Fig. 1. Flowchart for the reconstitution of proteins into oriented lipid bilayers.

an hour or longer followed by vortexing. The addition of 3–4 glass beads of about 2 mm in diameter helps in transferring the film into a suspension of multilamellar vesicles.

5. To homogenize the samples submit to ≥3 freeze (using dry ice in isopropanol or a −20°C freezer/thaw (using lukewarm water)/vortex cycles (see Note 6).

Preparation of Unilamellar Vesicles by Extrusion

Unilamellar vesicles of a defined outer diameter in the range 100–1,000 nm can be produced by extrusion through a polycarbonate film that has a specific pore size. The extrusion device for a few milliliter amounts of solution typically consists of two syringes and a holder supporting the polycarbonate films.

1. Place two polycarbonate films in the holder onto a droplet of water, which has been positioned into the holder with the help of one syringe.
2. Wash the device with buffer. When the membranes are intact pressing the syringes requires certain strength.
3. Place the lipid suspension in one of the syringes, and press the suspension through the polycarbonate film for 21 times (see Note 7).

Preparation of Small Unilamellar Vesicles by Sonication

Alternatively, small unilamellar vesicles of a size that is typically <100 nm can be produced by sonication. This section thereby replaces Subheading "Preparation of Unilamellar Vesicles by Extrusion".

1. Position the sonicator tip centrally into the lipid suspension and the sample tube in an ice bath for cooling. Avoid direct contact of the sonication tip and the surface of the tube as both the tip and the glass tube may be damaged.
2. Sonication is performed at intervals that avoid sample heating until the suspension clears. The time varies between 10 and 60 min depending on the lipid composition and the salt concentration.
3. After sonication particles that tend to come off the metal tip of the sonicator are removed by centrifugation in tabletop centrifuge at 10,000 × *g* for 10 min (e.g., the highest speed on an Eppendorf-centrifuge, MiniSpin®) (see Note 8).

Protein Reconstitution

Reconstitution of Water-Soluble Proteins into Lipid Vesicles

In the simplest case the lipid vesicle suspension and the protein solution are mixed, and the protein spontaneously inserts into the lipid membranes. This is the case for membrane-active colicin domains, diphtheria T-domain, as well as several members of the Bcl-2 family of proteins, and the protocols described here follow our work on these proteins (23, 25, 32). The buffers involved in this step have to be adapted to the requirements of the protein and may, for example, involve triggering of membrane insertion by a decrease in pH (32, 33) (see Note 5).

Reconstitution of Detergent-Solubilized Proteins into Lipid Vesicles

Many membrane proteins need to be solubilized and stabilized by detergent during the extraction and purification steps. In these cases the protein/detergent solution and lipid bilayer suspension are mixed, and the detergent molecules are removed by one of the methods outlined below. Which detergent and which of the methods of its removal works best for a given protein has to be tested, or the information may be available in the literature (see Note 9). Two alternative procedures for removing the detergents after protein reconstitution into liposomes are given here in Subheadings “Detergent Removal Using Dialysis Tubes” and “Detergent Removal Using Micro Beads”.

Detergent Removal Using Dialysis Tubes

Detergent removal

1. The detergent molecules are removed in a dialysis chamber with pore size (MWCO) of 3.5–10 kDa with a sample volume of 3–12 or 0.5–3 ml. The pore size should be sufficiently small to avoid loss of the protein but large enough to allow efficient exchange of buffer and detergent. The sample (some milliliters) is placed into the dialysis chamber by a syringe equipped with an infusion needle following the instructions of the manufacturer.
2. The closed dialysis tube is placed into a large beaker (of volume >200 times the sample volume). The liquid in the beaker is constantly agitated by a magnetic stirrer and placed for 12–16 h typically at 4°C or at room temperature, respectively (depending on the phase transition temperature of the lipids and the requirements of the protein sample).
3. The buffer is replaced every 4 h.

Detergent Removal Using Micro Beads

The hydrophobic interaction of detergents like CHAPS and octyl-glucoside with Bio-Beads is an alternative procedure for the effective removal of detergents after reconstitution of proteins into model membranes (34).

1. Detergent-containing solutions are incubated with for instance Bio-Beads SM-2 (BioRad Laboratories, Richmond, CA) for 1 h with gentle shaking at room temperature and consequently eluted to get a detergent-free sample.
2. This procedure is repeated several times, as some less hydrophobic detergents like CHAPS associate less well with the beads (35).
3. The sample can be concentrated by a Macrosep Centrifugal Device.

Note 10 describes the usage of beads for the reconstitution of membrane proteins in a unique orientation.

Reducing the Volume of the Protein/Lipid Sample After Dialysis

In order to reduce the sample volume of the protein/lipid suspension the sample is ultracentrifuged after the dialysis step.

1. The suspension is placed into 25 mm × 89 mm polycarbonate tube (Beckman, Palo Alto, CA) and centrifuged at >100,000 × *g* for 5 h, 4°C.
2. Consequently, the supernatant is separated from the pellet and tested for protein remaining in solution, for instance by UV-absorption at 280 nm.
3. The pellet is washed and resuspended in sterile filtered water, complemented with 0.02% sodium azide (w/w) before being spread on glass plates (Subheading 3.1.2). This assures that only lipid-associated protein contributes to the solid-state NMR measurements and the salt concentration in the sample is reduced which is advantageous for the NMR spectroscopic performance (cf. Subheading 3.2).

3.1.2. Formation of Oriented Lipid Bilayers Between Glass Plates

A lipid vesicle suspension which is depleted of its bulk water and equilibrated at defined relative humidity spontaneously forms stacks of oriented bilayers on glass plates. However samples of proteins reconstituted in lipid vesicles are typically of several ml and contain salt, which would concentrate during the drying process and, therefore, should be removed before the sample is stacked. This can be achieved by dialysis and/or ultracentrifugation (cf. Subheadings "Detergent Removal Using Dialysis Tubes" and "Detergent Removal Using Micro Beads"). Alternatively, a concentration step followed by some washing steps can be used to reduce the salt concentration.

Concentration of the Samples

The samples can be easily concentrated by centrifugation devices where the solvent passes a filter of defined molecular size, but larger macromolecules or aggregates are retained. Different systems are available, and they usually come with instructions on how to use them. In our hands devices where the membrane runs parallel to the centrifugal forces work well as this arrangement avoids that the biological material precipitates onto and clogs the filter (e.g., Macrosep® Centrifugal Devices). These devices can also be used for washings and buffer exchange. However, it should be kept in mind that with each centrifugation (which takes several hours) only about 90% of the buffer has exchanged and that the volume inside the membranes remains inaccessible. The final volume should be in the range of 0.5 ml.

Formation of Oriented Lipid Bilayers Between Glass Plates

This part of the protocol is illustrated in ref. 11.

1. Carefully place about 30 clean ultrathin microscope cover glasses side by side on two or three glass Petri dishes. Use fine watchmaker forceps to manipulate the glass plates. Apply equal quantities as small elongated droplets onto the central areas of

all but one glass plates using a Hamilton syringe. Keep one glass plate on the side to be used during the stacking of the plates (step 3). Let the sample dry in air until most of the solvent has evaporated. Should dehydration be harmful to your protein perform this step in a more controlled fashion, for example in a chamber of defined relative humidity (see Note 3). Should the volume be too large to allow for the stable application onto the glass plates in a single round, it is possible to apply part of it, dry off some of the liquid, and reapply another droplet onto the same region of the glass plate.

2. After the sample has been equally distributed and the bulk solvent evaporated in air the samples are placed in a closed chamber of 93% r.h. (e.g., desiccator or a closed plastic box) and left to equilibrate for typically about 1–3 days. This humidity is established in contact with a saturated solution of KNO_3. Other hydration conditions are also possible (cf. Note 3).

3. Once the lipid films become translucent and soft (see Note 3) the glass plates are stacked on top of each other, with one empty glass plate used to cover the remaining surface. Membrane alignment can be improved by shearing with delicate pressure and small movements. The stack is stabilized by wrapping with Teflon tape and the whole sample sealed in a plastic wrapping using an impulse sealer. The volume of the sample should fit tightly into the NMR coil (cf. Subheading 3.2.1 step 6).

3.2. Solid-State NMR Measurements

In this paragraph the setup procedure for four types of NMR measurements is described. However, be aware of the precautions that have to be taken in the proximity of the high-field magnet and due to the high operating voltage of the electronic equipment. Because of these potential dangers the beginner should receive a detailed hands-on introduction into the use of a solid-state NMR spectrometer as well as into the language of the users interface. Furthermore, it is assumed that the pulse programmes have been established and tested. Some simple setup procedures and pulse sequences such as standard ^1H acquisition, Hartmann–Hahn matching, or cross polarization are also explained in an application-oriented manner in references (36). Furthermore, reference (11) describes some additional ^2H and $^{31}P\{^1H\}$ solid-state NMR measurements of oriented or non-oriented samples on membrane-associated peptides as well as magic angle sample spinning experiments, which may be also useful in this context.

Table 1
Typical parameters for proton-decoupled ^{31}P Hahn-echo experiments

Experiment	Proton-decoupled ^{31}P Hahn-echo experiment
Reference sample	85% Phosphoric acid
Parameters for optimization	^{31}P 90° and 180° pulses (power level, duration) ^{1}H decoupling field (power level)/corresponding ^{1}H 90° pulse ^{31}P carrier frequency, ^{1}H carrier frequency
Typical ^{31}P B_1 field for ^{31}P 90° and 180° pulses	50 kHz (corresponding to a 90° pulse of 5 μs)
Typical ^{1}H B_1 field for decoupling	>30 kHz (<8.5 μs)
Recycle delay	1.5 s for phospholipids in a fluid bilayer environment; 5 s for phosphoric acid
Interpulse delay (τ)	40 μs

3.2.1. Verification of Sample Alignment by Proton-Decoupled ^{31}P Solid-State NMR Spectroscopy

1. Insert a sample of 85% H_3PO_4 into the center of the NMR coil. Close probe head, and insert the probe into the NMR magnet. Connect the ^{31}P (X) and the ^{1}H electronic circuits of the probes to the respective electrical connections with incoming RF pulses from the console and outgoing signal from the sample via filters and the preamplifier that covers the ^{31}P Larmor frequency.
2. Tune and match the probe (^{31}P and ^{1}H electronic circuits).
3. Using a simple pulse-acquisition sequence vary the pulse length to determine the maximal and the zero intensities (90° and 180° pulses, respectively). During this setup procedure using an isotropic liquid the ^{1}H decoupling is unnecessary and can even heat and damage the sample and/or probe head as long acquisition times are used to fully record the free induction decay (FID). Adjust the carrier frequency to have the signal close to the center of your spectral window and all the other parameters according to Table 1. Adjust the receiver gain to avoid saturation and good sensitivity at the same time. Using a short pulse adjust and save the phases to have a purely positive signal intensity. Calibrate the chemical shift scale with the maximum of the phosphoric acid signal at 0 ppm. Adjust the power of the pulses to have a 90° pulse in the range of 5 μs. Once these parameters have been determined for a spectrometer and probe they provide a valuable starting point for future setup procedures, which should then require only minor readjustments.
4. Change the spectrometer settings to acquire the ^{1}H signal. Adjust the power levels of the ^{1}H channel to have a 90° pulse length ≤7 μs and the ^{1}H carrier frequency using a procedure

analogous to the previous step. This step is required only once in a while to measure or control the ^{1}H decoupling field strength and the ^{1}H frequency.

5. Using the ^{31}P and ^{1}H power levels determined in steps 3 and 4 adjust the durations of the 90°-τ–180°-τ-(acquisition with ^{1}H decoupling) pulse programme (37). Adjust the 90° and 180° pulses as determined and all the other parameters according to Table 1. Using the reference sample the performance of the pulse programme can be tested by adding a full phase cycle (16 scans). Check the FID for being additive from scan to scan, for the absence of artifacts in its beginning (probe ringing may require longer τ), or for spikes (reduction of the decoupling power and/or cleaning the coil may help).
6. Stably insert the sample into the center of the NMR coil (see Note 4). Reconnect the probe, tune, and match (steps 1 and 2) for each new sample and after equilibrating to a new temperature. Use the same parameters as before (Table 1) but reduce the recycle delay to 1.5–3 s. Readjust the receiver gain for the lipid sample for maximal sensitivity without saturating the receiver. For a sample containing 100 mg phospholipid the signal should become available with the first scans, a spectrum with good signal-to-noise ratio in a few minutes.

By inspecting (and simulating) the resulting proton-decoupled ^{31}P solid-state NMR experiments verify the morphology, alignment, and integrity of your lipid bilayer before and after the ^{15}N NMR measurements (Fig. 2).

3.2.2. Proton-Decoupled ^{15}N Cross-Polarization Solid-State NMR Spectroscopy of Membrane Samples

The setup procedure for a one-dimensional cross-polarization pulse sequence is described albeit with settings that already aim at being sufficiently close for the two-dimensional correlation experiment (separated local field spectroscopy) described in Subheading 3.2.3.

1. Insert a sample of $^{15}NH_4Cl$ (or $(^{15}NH_4)_2SO_4$) into the center of the NMR coil. Close probe head and insert the probe into the NMR magnet. Connect the ^{15}N (X) and the ^{1}H electronic circuits of the probes to the respective electrical connections with incoming radiofrequency pulses from the console and outgoing signal from the sample via filters and the preamplifier that covers the ^{15}N Larmor frequency.
2. Tune and match the probe (^{15}N and ^{1}H electronic circuits).
3. To acquire ^{1}H signals using a simple pulse-acquisition sequence use a short pulse (e.g., 2 μs) to adjust and save the phases to have a purely positive signal intensity. Adjust the receiver gain to avoid saturation and good sensitivity at the same time and

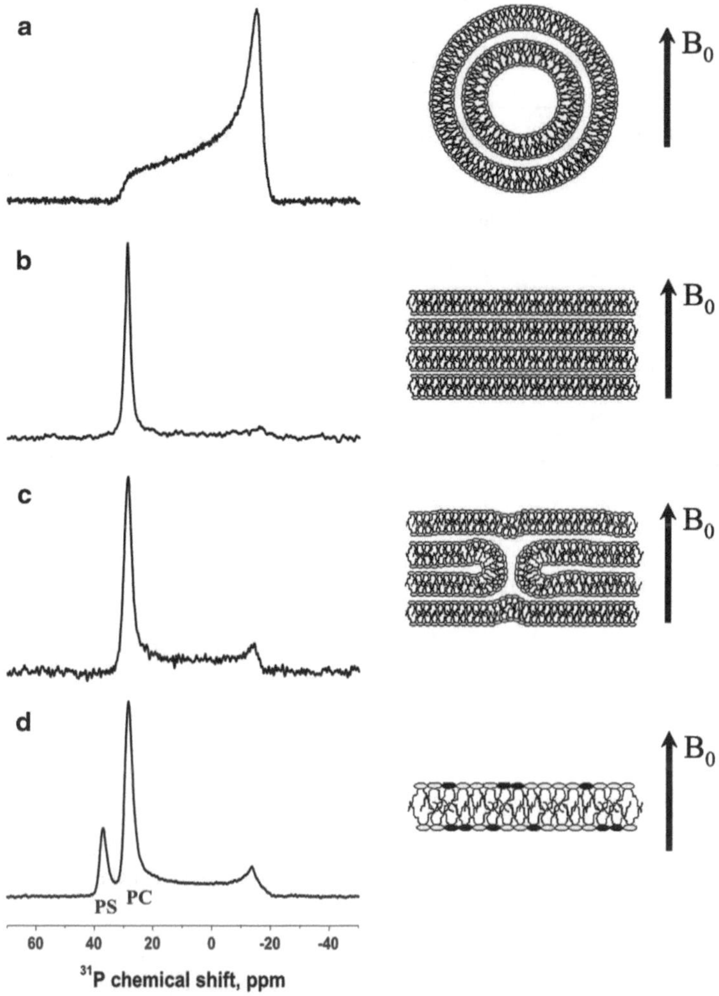

Fig. 2. Typical line shapes of proton-decoupled ^{31}P solid-state NMR experiments from liquid-disordered phosphatidylcholine bilayers. (**a**) Non-oriented sample where the powder pattern line shape reflects the functionality of the chemical shift interactions as well as the distribution of molecules relative to B_0. The chemical shift range of about 45 ppm is typical for liquid-disordered phosphatidylcholine bilayers. (**b**) Well-aligned sample where nearly all phospholipids oriented in a narrow range relative to B_0. (**c**) A small portion of non-oriented sample is superimposed on oriented phospholipids. (**d**) Mixture of phosphatidylcholine (PC) and phosphatidylserine (PS) supported by glass plates. The two main intensities around 30 and 40 ppm indicate that most lipids are well aligned albeit PC and PS exhibit different chemical shifts and chemical shift anisotropies. A small fraction of phospholipid resonates at values up to −20 ppm indicative of a limited amount of orientational or of conformational heterogeneity at the level of the phospholipid head groups. Possible morphologies as well as the dynamics explaining such line shapes are discussed in ref. 17.

the carrier frequency to have the signal close to the center of your spectral window. Keep a note of this frequency which should also be used later during cross polarization and decoupling. The spectrum is relatively broad, but the maximum gives

Table 2
Typical parameters for proton-decoupled ^{1}H-^{15}N cross-polarization experiments

Experiment	Standard cross-polarization (CP) experiment
Reference sample	^{15}N-ammoniumchloride or ^{15}N-ammoniumsulfate
Parameter for optimization	Proton 90° pulse (power level, duration) ^{15}N–^{1}H CP condition (power level ^{1}H and ^{15}N) ^{15}N carrier frequency, ^{1}H carrier frequency
Typical ^{1}H B_1 field for the initial 90° pulse	62.5 kHz (corresponding to a 90° pulse of 4 μs)
Typical ^{1}H B_1 field for CP	62.5 kHz (4 μs)
CP time for amides	800 μs
Recycle delay	3 s for amides (this time could in principle be shortened but for protecting the sample from excessive heating due to RF irradiation), >5 s for ammonium salts

a good initial value for the proton carrier frequency (see Note 11). After these adjustments verify the phase settings. Then vary the pulse length to determine the maximal and the zero intensities (90° and 180° pulses, respectively), and adjust the power of the pulses to have a 90° pulse in the range of 4 μs (Table 2). Once these parameters have been determined for a spectrometer and probe they provide a valuable starting point for future setup procedures, which should then require only minor readjustments (see Note 12).

4. Switch to the acquisition of the ^{15}N signal. It is recommended to use or programme a cross-polarization sequence (36) with the possibility to vary the power levels of the 90° ^{1}H pulse, the ^{1}H and ^{15}N contact pulses, and the ^{1}H decoupling independently (cf. the different CP pulse schemes shown in Fig. 3). Keep the ^{1}H power level fixed as determined in step 3 for all intervals (90° pulse, contact time, and decoupling during acquisition), set the length of the 90° ^{1}H pulse as determined, choose a contact time of ≤5 ms, and vary the power level of the ^{15}N channel to obtain the maximal intensity. Once you obtain the first signal correct and save the phases for a positive ^{15}N signal, calibrate the chemical shift scale with the maximum of the ammonium peak at 40 ppm, and set the ^{15}N carrier frequency close to 0 ppm (see Note 13). In case major corrections of the frequencies had to be made check and readjust the power levels for maximal signal intensity as well as the phases. At all times adjust the receiver gain to avoid saturation and good sensitivity at the same time.

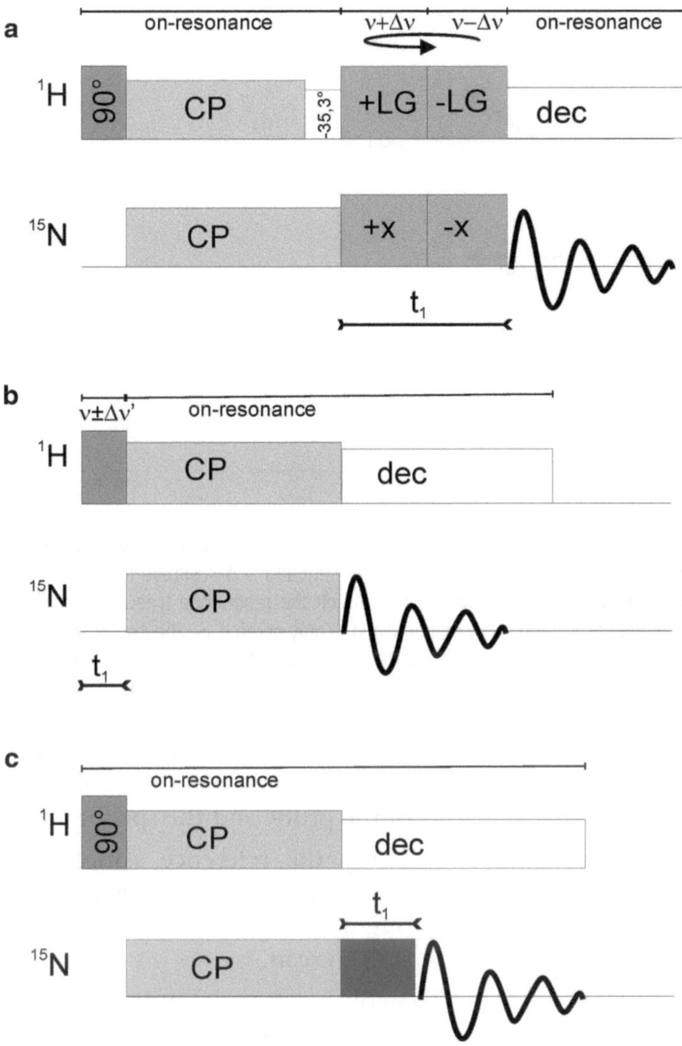

Fig. 3. (**a**) PISEMA pulse sequence. (**b**) Proton nutation pulse sequence with frequency offset of the ^{1}H pulse by $+\Delta\nu$ or $-\Delta\nu$, respectively. (**c**) ^{15}N nutation experiment. The cross-polarization (CP) and decoupling steps involve a frequency that is well centered with respect to the ^{1}H spectrum ("on-resonance"). The prime in panel (**b**) indicates that at this stage the offset is arbitrary, whereas during Lee-Goldburg decoupling it has a well-defined value [panel (**a**) and Fig. 4].

5. Once the cross-polarization conditions have been optimized verify that the 90° ^{1}H pulse provides indeed the maximal ^{15}N signal intensity within the cross-polarization pulse sequence. The ^{15}N signal should be zero when a 180° ^{1}H pulse is applied. Once all these parameters have been determined for a spectrometer and probe they provide a valuable starting point for future setup procedures, which should then require only minor corrections (see Note 12). By always comparing the noise level

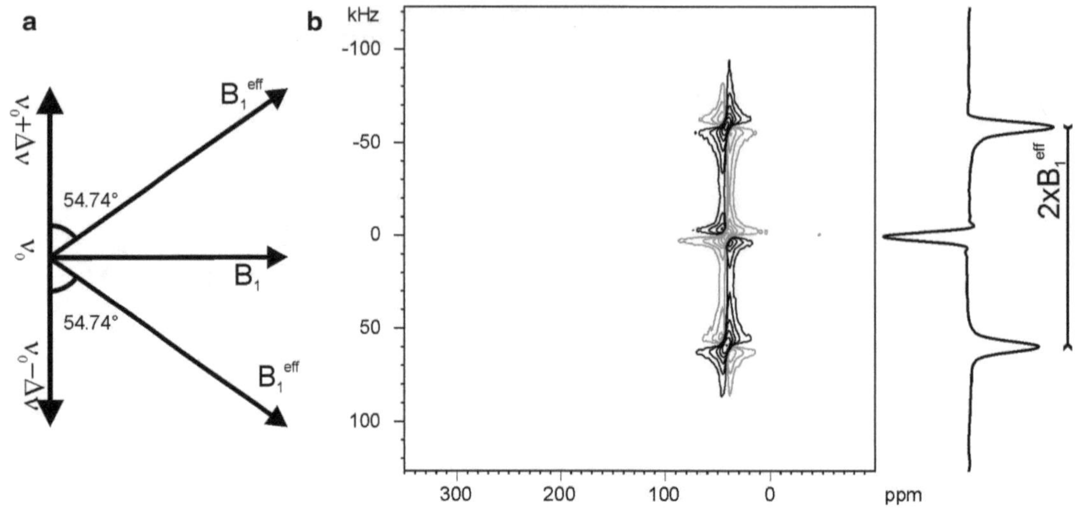

Fig. 4. (**a**) Calculation of the effective proton B_1 field (B_1^{eff}) during the SEMA pulse train. Whereas the B_1 field is controlled by the pulse power level the frequency offset $+\Delta\nu$ and $-\Delta\nu$ relative to the center frequency is generated by the RF pulse synthesizer. The center frequency has to match the resonance frequency of the nuclei. The symmetry for positive and negative offsets ($+\Delta\nu$ and $-\Delta\nu$ respectively) is only given if ν_0, the center frequency of the protons, is well adjusted. Note that here the "strength of the B_1^{eff} field" is indicated by the resulting nutation frequency (in Hz). Typical result of the two-dimensional nutation experiment for the proton channel (Fig. 3b).

and the signal intensity with previous settings using the same reference sample it is possible to keep track of the performance of the probe and the spectrometer.

6. Using the reference sample the performance of the pulse programme can be tested by adding a full phase cycle (eight scans). Check the growing FID for being additive from scan to scan.
7. Stably insert the sample into the center of the NMR coil, tune, and match the ^{15}N and ^{1}H channels for each new sample.
8. Switch to the acquisition of ^{1}H. Record the ^{1}H NMR spectrum of the sample, and adjust the ^{1}H power levels to obtain the same 90°/180° pulse length as determined in step 5. Depending on the content of water and salt a significant increase in power may be necessary (typically by 1–3 dB) to obtain the same probe performance (i.e., effective B_1 field) (see Note 4).

Switch to the cross-polarization pulse programme and the spectrometer to the acquisition of ^{15}N. Adjust the following parameters by keeping the previous ^{15}N but by using the new ^{1}H power levels (Table 2): 90° ^{1}H pulse as determined in steps 5 and 8, contact time 0.8–1 s, spectral width 40 kHz, and an acquisition time of 1–3 ms (long enough to avoid truncation of the FID, but short enough to avoid excessive sample heating during ^{1}H decoupling). Adjust the receiver gain to avoid saturation and good sensitivity at the same time. Use the chemical shift calibration of step 4. Check the first acquisitions (mostly noise) for the absence of artifacts in their beginning and spikes throughout the FID (reduction of the decoupling power and/or cleaning the coil may help). During acquisition the sample should be cooled with a stream of air, which can be humidified by bubbling through a gas wash bottle provided that acquisition is at greater than or equal to room temperature (if the temperature of the water bath is above that of the connecting lines and probe, water condensation will occur at the colder sites and affect the performance of the electronic circuits). It is recommended to check the tuning of the ^{1}H channel after a few hours of accumulation as water loss or redistribution within the sample can occur due to RF heating (see Note 4).

3.2.3. Two-Dimensional ^{15}N Solid-State NMR Spectra of Oriented Membrane Samples Correlating the ^{15}N Chemical Shift and the ^{15}N–^{1}H Dipolar Coupling

This family of experiments reveals from the labelled amide sites the ^{15}N chemical shift as well as the ^{15}N–^{1}H dipolar coupling (38), both parameters being orientation dependent (9). The two-dimensional experiment presented here has also been shown to provide additional spectral resolution when perlabelled polypeptides are investigated (Fig. 3a). It consists of two parts: a cross-polarization experiment (Subheading 3.2.2) and a SEMA pulse train during the t_1 evolution period (Subheading "Adjustment of the SEMA Pulse Train"). Following the flow of events described in Fig. 5, the corresponding settings are optimized step by step and should start with the simple one-dimensional cross-polarization setup described in Subheading 3.2.2 using, e.g., a ^{15}N-ammonium salt as a reference sample.

Adjustment of the SEMA Pulse Train

For setting up the SEMA pulse train the corresponding B_1 field has to be adjusted and calibrated (see Note 14). As the B_1^{eff} field determines the width of the spectral window in the t_1 dimension a minimum of about 50 kHz is required for the B_1^{eff} field to cover well the dipolar ^{15}N–^{1}H coupling in the amide bonds of peptides and proteins ($\leq$20 kHz; (12)) (see Note 15). Aiming at an efficient B_1 field of 50 to 65 kHz seems reasonable (Note 15), and this value will be adjusted during the following two-dimensional nutation experiments (Table 3, Figs. 3b and 4).

Determination of the Power Level for the Proton B_1^{eff} Field

1. Before starting this optimization step the nitrogen spectrum should be calibrated following Subheading 3.2.2 step 4, and the nitrogen carrier frequency should thereafter not be altered. Although the position of the proton carrier will be verified and adjusted during this experiment a good staring value will significantly accelerate the optimization steps (cf. Subheading 3.2.2 step 3). The CP conditions of the ^{1}H nutation experiment (Fig. 3b) are also taken from the previous settings (Subheading 3.2.2).
2. The effective B_1^{eff} field is measured from a nutation experiment (Fig. 3b) where the initial proton pulse is incremented in a

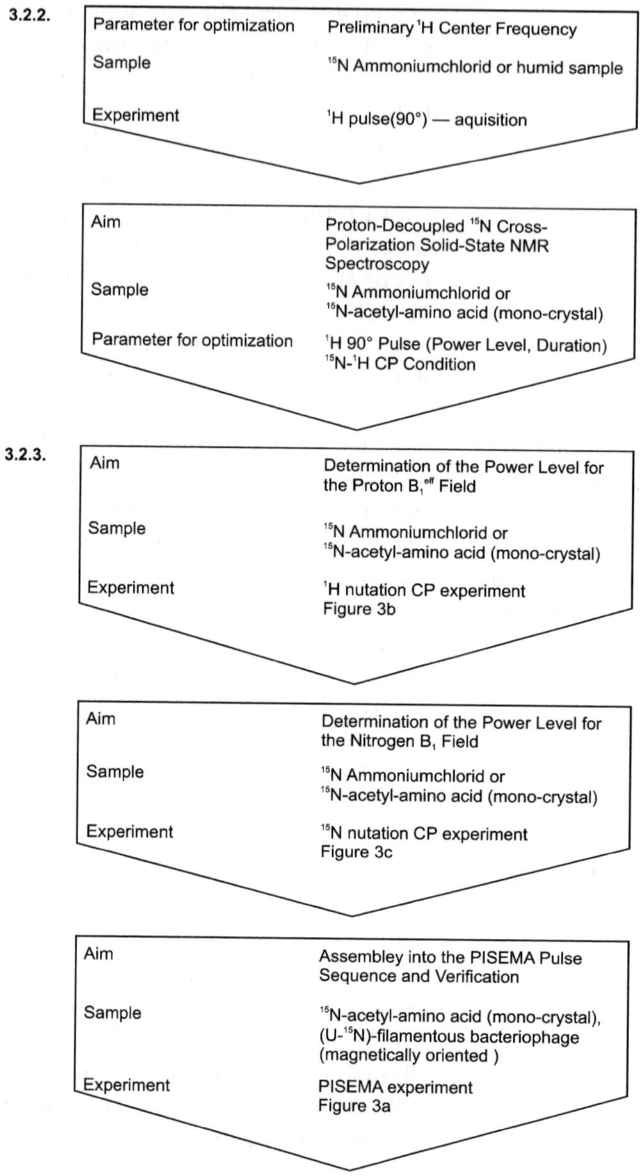

Fig. 5. Flowchart summarizing the setup protocol for the PISEMA experiment.

systematic manner during t_1 in steps of 4 μs for a spectral width of 250 kHz up to typically 0.25 ms for a resolution of 4 kHz. The carrier frequency of this initial pulse is fixed to either $\nu_0 + {B_1}^{\text{eff}} \times \cos(54.74°)$ or $\nu_0 - {B_1}^{\text{eff}} \times \cos(54.74°)$, respectively, where ν_0 is the center frequency and ${B_1}^{\text{eff}}$ the *intended* effective B_1 field (which should match the ^{15}N B_1 during t_1, Fig. 3a). The detection is performed on the ^{15}N channel after cross polarization to avoid the necessity for analysis of the broad ^{1}H signal of the protons. The resulting oscillation of

Table 3
Typical parameters for a ^{1}H nutation experiment with a frequency offset detected through ^{1}H–^{15}N cross polarization

Experiment	^{1}H nutation CP experiment (Fig. 3b)
Sample	^{15}N ammonium chloride or sulfate
Parameter for optimization	^{1}H power level for desired B_1^{eff} field ^{1}H center frequency (see Note 16)
Typical ^{1}H B_1^{eff} field	65 kHz (corresponding to a frequency offset of 37.528 kHz)
Typical ^{1}H B_1 field for CP	62.5 kHz (4 μs)
CP time	800 μs

the ^{15}N cross-polarization signal (Fig. 3b) reflects the nutation frequency of the ^{1}H magnetization around the magic angle during t_1 (Fig. 3a) and is easily measured after Fourier transformation in t_2 and t_1. The power level of this initial ^{1}H pulse is adjusted until the intended effective B_1 field is observed in the nutation experiment (Fig. 3b).

3. The proton center frequency is then verified and corrected by comparing a nutation experiment with *positive* and *negative* frequency offset of the initial proton irradiation period (both experiments should give similar B_1^{eff} if the center frequency is set correctly). If this is not the case modify the center frequency and go back to step 2. Perform steps 2 and 3 in an iterative manner where adjustments are made until a match is obtained. Note that the CP conditions should be verified after modifications of the center frequency (see Note 16) (Table 3).

Determination of Power Level for the Nitrogen B_1 Field

The B_1 field for nitrogen is measured by a ^{15}N-nutation experiment (Fig. 3c). It consists of a CP experiment followed by an incremented ^{15}N irradiation period. The experiment consists in adjusting the power level of the ^{15}N irradiation until the nutation of this nucleus matches B_1^{eff} of the ^{1}H (cf. "Determination of the Power Level for the Proton B_1^{eff} Field"); thus, Hartmann–Hahn matching is achieved during the SEMA pulse train (t_1 period in Fig. 3a) (Table 4).

Assembly into the PISEMA Pulse Sequence and Verification

To assemble the building blocks with the parameters adjusted in the previous sections the PISEMA pulse-sequence is first tested on a reference sample. Ideally, to this end a monocrystal of [^{15}N]-Ac-Leu, [^{15}N]-Ac-Gly, or another N-terminally protected amino acid is available. The FID of the ^{15}N signal should be clearly visible at least until 5 ms into the Lee-Goldburg decoupling period (i.e., the 325th 1D slice at B_1^{eff} = 65 kHz). This assures that the parameters

Table 4
Typical parameters for a ^{15}N nutation experiment detected through ^{1}H–^{15}N cross polarization

Experiment	^{15}N nutation CP experiment (Fig. 3c)
Sample	^{15}N ammonium chloride or sulfate
Parameter for optimization	^{15}N power level for requested B_1 filed
Typical ^{15}N B_1 field (matching the ^{1}H B_1^{eff} under off-resonance irradiation)	65 kHz
Typical ^{1}H B_1 field for CP	62.5 kHz (4 μs)
CP time	800 μs

for homonuclear decoupling were determined correctly. A new pulse sequence should be tested with a powder of a molecule representing a peptide bond, i.e., a N-terminally protected amino acid such as [^{15}N]-Ac-Leu and [^{15}N]-Ac-Gly. With such a compound one should obtain a well-known "butterfly" spectrum with correct dipolar couplings. This also allows verification of the correctness of the data processing protocol. The number of scans should be enough to unambiguously see the signal in nearly each 1D spectrum (typically 4,000 scans for membrane samples, which results in many days of acquisition). After acquisition of the first few 1D spectra oscillations of the signal intensities should become apparent. For membrane samples the ^{1}H power levels can be adjusted with higher precision by using one-pulse ^{1}H nutation experiment (Subheading "Determination of the Power Level for the Proton B_1^{eff} Field" and see Note 16). To compensate for B1 losses in wet samples it can be advantageous to test and fine-tune the PISEMA experiment on a fully hydrated biological sample such as magnetically oriented filamentous bacteriophage preparations. Note 4.

3.2.4. ^{2}H Solid-State NMR Spectra of Oriented or Non-oriented Samples

The experiments described here allow one to monitor a variety of physical properties of the membrane in the presence or in the absence of protein. These experiments require that during the preparation of membranes deuterated lipids are used and depending on the deuteration pattern different properties of the membranes can be revealed. For example, by incorporating fatty-acyl chain-deuterated phospholipids the lipid order parameter is monitored as a function of lipid composition, label position, temperature, buffer, and/or the presence of protein. The ^{2}H spectra therefore reveal membrane disruptive properties of proteins and peptides (17). By mixing deuterated and non-deuterated lipid species selective interactions of the polypeptides with one lipid over the other are detected (19). Finally, the ^{2}H spectra of lipids deuterated

Table 5
Typical parameters for ^{2}H-echo experiments

Experiment	^{2}H-echo experiment
Reference sample	Deuterium-enriched water
Parameters for optimization	^{2}H 90° (power level, duration) ^{2}H carrier frequency
Typical ^{1}H B_1 field for ^{2}H 90° pulses	>50 kHz (corresponding to a 90° pulse of <5 μs)
Recycle delay	0.3 s for phospholipids in a bilayer environment; 5 s for deuterated water
Interpulse delay (τ)	40 μs

at the lipid head groups have been calibrated and used for the investigations of electrostatic interactions at the level of the bilayer interface (20, 21). The measurements can be performed on non-oriented (usage of the pellet that results from Subheading 3.1.1) or oriented samples (procedures in Subheadings 3.1.1 and 3.1.2). The protocol is also provided and illustrated in (11).

1. Insert a sample of deuterated water into the center of the NMR coil. Close the probe head, and insert the probe into the NMR magnet. Connect the ^{2}H (X-channel) electronic circuit of the probe to the console. This line carries incoming radio frequency (RF) pulses from the console to the probe and outgoing signal from the sample via the preamplifier (that is tuned to the ^{2}H Larmor frequency) to the detector.
2. Tune and match the probe for the ^{2}H carrier frequency.
3. Using a simple pulse-acquisition sequence vary the pulse length to determine the maximal and the zero intensities (90° and 180° pulses, respectively). Adjust the carrier frequency to have the signal close to the center of your spectral window and all the other parameters according to Table 5. Adjust the receiver gain to avoid saturation and good sensitivity at the same time. Using a short pulse adjust and save the phases to have a purely positive signal intensity (Lorentzian). Calibrate the chemical shift scale by setting the maximum of the ^2H_2O signal to 0 Hz. Adjust the power of the pulses to have a 90° pulse in the range of ≤5 μs (Table 5). Once these parameters have been determined for a spectrometer and probe they provide a valuable starting point for future setup procedures, which should then require only minor readjustments.
4. Using the ^{2}H power levels determined in step 3 adjust the durations of a 90_x-τ–90_y-τ-acquisition pulse programme (39).

Adjust the 90° pulse as determined and all the other parameters according to Table 5. Using the reference sample the performance of the pulse programme can be tested by adding a full-phase cycle (eight scans). Check that the oscillations (FID) recorded from each repetition add to the overall signal intensity, for the absence of artifacts in its beginning (probe ringing may require longer τ) or of spikes (cleaning the coil may help). With some spectrometers an installed lock unit may interfere with the signal. This problem can be solved by reducing/turning off the lock power.

5. Stably insert the membrane sample into the center of the NMR coil. Reconnect the probe, tune, and match (steps 1 and 2) for each new sample and after equilibrating to a new temperature. Use the parameters determined before (Table 5). Readjust the receiver gain for the lipid sample to obtain the maximal sensitivity without saturating the receiver. The signal of deuterated lipids should become visible within minutes, a spectrum with reasonable signal-to-noise ratio in about 0.5–1 h. Note that the signal from natural abundance deuterium in water may show up first and can be reduced by using deuterium-depleted water for sample hydration.

4. Notes

1. The solubility of lipids in chloroform and dichloromethane is very similar. Therefore, preference should be given to the latter as it has less of an effect on the environment and is less toxic. To dissolve lipids and peptides solvents such as 2,2,2-trifluoroethanol, hexafluoroisopropanol, formic acid, and/or MilliQ water are commonly used (11). However, the use of such solvents to complex proteins can cause irreversible denaturation.
2. Often it is desired to test the effect of lipid composition on the membrane protein–lipid interactions. On the one hand POPC represents well the phospholipid composition of many biological membranes as the choline head group is abundant in the bilayers of higher organisms. The composition of these membranes is furthermore mirrored by a saturated fatty acyl chain at position 1 and an unsaturated chain at position 2, and POPC represents well their average hydrophobic thickness. Notably, whereas some polypeptides exhibit stable structural properties in a variety of different lipid membranes (30) others are quite sensitive to even small changes in lipid composition (40). Furthermore, the negative charges of eukaryotic membranes are often represented by phosphatidylserine, and these

membranes are also rich in cholesterol. On the other hand, a simple bacterial membrane mimic could be phosphatidylethanolamine/phosphatidylglycerol 3:1. Notably, many of these lipid components exhibit low solubility in alcohols. Therefore, dichloromethane (or chloroform) provides an alternative for most long-chain lipids (cf., e.g., the catalogue information by Avanti Polar Lipids, Alabaster, AL).

3. The equilibration period for hydration ranges from a few hours up to a few days. Hydration can be accelerated by passing through a vapor steam or by adding microliter amounts of water per glass plate. Diverse equilibration conditions are obtained by contact of the atmosphere with saturated solutions of LiCl (15% r.h), K_2CO_3 (44% r.h), NaBr (57% r.h.), NaCl (75% r.h.), KCl (86% r.h.), KNO_3 (93% r.h), or pure water (100% r.h.) (41). Extended equilibration and storage at 100% r.h. should be avoided as this promotes microbial growth. Hydration chambers can be easily prepared by filling a plastic box (e.g., lunch or freezer storage box) with salt and some water. An inverted Petri dish serves as a platform to avoid the direct contact of the salt solution with the sample.

4. Usually solid-state NMR experiments are set up with the help of standard samples with high signal intensity. By doing so it is assumed that the parameters do not change between the standard and the sample of interest. However, with the advent of high-field NMR spectrometers the proton B_1 fields tend to be attenuated in wet samples (i.e., hydrated lipid bilayers). This can be corrected by readjusting the proton parameters through their direct measurement on the sample itself (11). With linear amplifiers the correction increments (measured in dB) should be the same for all of the power levels. If this is not the case the initial pulse and the cross polarization need to be adjusted, and the decoupling power should remain sufficiently strong ($\geq$50 kHz). Depending on the instrumentation it may be easier to use the same power levels for all the periods of proton irradiation.

   Furthermore, conventional probes tend to heat the sample upon radio frequency irradiation, which leads to the loss of water and changes of the proton settings during long acquisitions. In such cases and in order to enhance sample stability the following modifications help (individually or in combination): reduction of the CP period, acquisition times and power levels of decoupling as short/small as possible (e.g., 1.5 ms, B_1 less than 50 kHz), and increased recycle delays. Furthermore, the recent development of probe heads where the coil center is depleted of the electric-field (e-free) has the advantage to reduce heating of the sample. In addition the proton-tuning and -parameters are barely dependent on the water content of the sample. This makes e-free probes an ideal tool for more sophisticated NMR experiments like SEMA, where a precise adjustment of the proton parameters is essential.

5. A wide variety of buffers can be used including Tris, HEPES, or citrate buffer. However, the interference of the buffer with the NMR experiments should be considered. For example, phosphate buffers result in large isotropic peaks hampering the analysis of ^{31}P NMR powder patterns, water contains a significant amount of deuterium (which can be reduced by using deuterium-depleted water; Sigma-Aldrich, St. Louis, MO, USA), or other buffers cause background signals in ^{13}C NMR experiments. Furthermore, high salt concentrations interfere with the tuning of NMR probes and the radio frequency irradiation during NMR spectroscopy (cf. Note 4). It should also be kept in mind that extreme pH values cause the hydrolysis of ester bonds and thereby lipid degradation. Long-term stability of lipids is best at pH 6–7 (42), reduced hydration, and/or storage in the freezer. When oriented samples are prepared at <100% r.h. it should further be considered that buffer and salt are concentrated into the small intermembrane space where the residual water molecules interact with the adjacent membranes. We have noted that the presence of buffer and salts often degrades the quality of the membrane alignment (e.g., (43)).
6. For freezing the lipid samples liquid nitrogen is also used. However, by doing so, we have observed the occurrence of isotropic signal intensities in ^{31}P NMR spectra of pure lipids which may be related to the speed of freezing and the corresponding ice/glass formation. Therefore, slow freezing in a −20°C freezer seems a better choice. The effect of freeze/thaw cycles on different membranes has been analyzed previously (44).
7. For extrusion the lipid concentration should not exceed a concentration of 10 mM (7–8 mg/ml). Furthermore, lipids with a phase transition temperature above room temperature require relatively high forces for extrusion.
8. The small diameter of the sonicated vesicle implies high curvature of the bilayers, which might influence the interaction to membrane proteins, and the small radius makes sedimentation by centrifugation very difficult.
9. Table 6 provides some basic physical parameters for a number of detergents used for the purification and/or structural analysis of membrane polypeptides. These data give some first indication of their suitability for the reconstitution protocol of a given protein.
10. Beads are also used to control the orientation of proteins reconstituted into lipid bilayers, like proton channels (45). Here, a recombinantly expressed protein contains a long histidine tag (His_{10}), which is used for immobilized metal affinity-

Table 6
Detergents used for membrane protein solubilization and their physical properties

Detergent	MW (g/mol)	Aggregation number	CMC (mM)	Charge	Dialyzable	Micelle MW
CHAPS	615	10	7.4–10	Zwitterionic	Yes	6,149
CHAPSO	631	11	8–10	Zwitterionic	Yes	6,940
DDM	511	110–140	0.18	Neutral	Yes	70,000
DPC	351	38–56	0.9–1.1	Zwitterionic	Yes	19,000
OG	292	27–100	23.2–25	Neutral	Yes	8,000–29,000
OTG	308	92–114	7.6	Neutral	Yes	18,000
SDS	288	63	1.4–8	Anionic	Yes	18,000

CHAPS: 3-[(3-Cholamidopropyl)dimethylammonio]-1-propane sulfonate (46–48); CHAPSO: 3-[(3-Cholamidopropyl)dimethylammonio]-2-hydroxy-1-propanesulfonate (48, 49); DDM: Dodecylmaltoside (50); DPC: Dodecylphosphocholine (51–53); OG: Octyl-β-glucopyranoside (54–56); OTG: Octyl-β-thio-glucopyranoside (57, 58); SDS: Sodium dodecylsulfate (59)

chromatography interactions with commercially available nickel-nitrilotriacetic acid (NTA) silica beads (Qiagen Inc., Valencia, CA, USA). Following the manufacturer's instructions the recombinant proteins are bound and immobilized on nickel beads via the decahistidine tag. Thereafter the protein-bead suspension is reconstituted into lipid vesicles by the detergent dialysis method. In doing so the part of the protein directly linked to the histidine tag faces the outer membrane of the double layer and is thereby available for interaction with the ligands in the bulk solution. In cases where further investigations of protein interactions are planned by NMR spectroscopy, the paramagnetic nickelions, leading to line broadening, should be exchanged with diamagnetic Zn^{2+} ions. To do so, nickel-beads are washed with a 0.3 mM EDTA solution to remove the nickel ion and consequently washed with water. Finally, incubation with 100 mM zinc sulfate solution recharges the NTA matrix, which then is used as described before.

11. If the ^{1}H line width of the ammonium salt hampers the accurate determination of the ^{1}H frequency a water or a humid membrane sample can be used giving sharper line width and allowing a more accurate frequency setting. The ^{1}H frequencies determined in Subheading 3.2.1 should be close to the one determined in this context.
12. Once the approximate settings of ^{1}H power and pulse length have been obtained with the probe and the same type of samples step 3 of 3.2.2 can be omitted as these parameters can be optimized in further experiments by first adjusting the Hartmann–Hahn matching conditions (step 4) and then by variation of the initial ^{1}H pulse (step 5). A pulse programme

that allows for the independent modification of the proton power levels during the various phases of the experiment (90° pulse, CP, decoupling) makes this task easier as the level of the 90° pulse can be changed without affecting the CP matching condition. At the Hartmann–Hahn matching condition of static samples the X and the ^{1}H spins precess at the same speed in the rotating frames defined by the respective B_1 fields (i.e., $\gamma_X B_{1X} = \gamma_H B_{1H}$), and the precession frequency is calculated from the 90°-pulse length (t_p) by $\nu_1 = 1/4t_p$.

13. The range of nearly constant excitation covers a spectral range of $1/4t_p$ where t_p is the pulse length. Therefore, the ^{15}N carrier frequency should be placed close to the chemical shift range of the amide (50 to 230 ppm) and amine (0 to 40 ppm) resonances. At the same time by placing the carrier outside the spectral range of interest possible artifacts at the position of the irradiation, that could potentially overlap with spectral resonances, are avoided.

14. The SEMA pulse provides a spin lock of the ^{1}H magnetization along the magic angle, thereby removing ^{1}H–^{1}H dipolar interactions. At the same time the ^{15}N power level should be adjusted to satisfy the Hartmann–Hahn matching condition. It allows the magnetization to travel between ^{15}N and surrounding protons but not within the ^{1}H network, thereby increasing the lifetime of the ^{15}N signal. The first condition is fulfilled by off-resonance ^{1}H irradiation ($\Delta\nu = B_{eff} \times \cos 54.7°$) to direct the vectorial addition of B_1 field and $\Delta\nu$ along the magic angle ($B_1^{eff} = 0.816 \times B_1$, which also provides the scaling factor for the dipolar interaction). The second condition consists in matching the ^{15}N B_1 field (i.e., the power level) to the B_1^{eff} field of the protons.

15. Generally it is considered that the spectral quality is ameliorated with higher B_1 fields although we have found, using ^{15}N,^{13}C-labelled crystals, that above a certain threshold artifacts tend to appear possibly due to limitations of the probe to take up the power properly in combination with the long duration of the SEMA pulse trains. Whereas 45 kHz are needed to cover the full spectral range it is desirable to reach higher B_1 fields. The maximal field strength that can be obtained depends on the fequencies, the coil geometry and to some extent also on the sample composition.

16. In real samples different types of protons exist, and these interact with each other and with the magnetic field in an anisotropic manner resulting in broad ^{1}H spectra (cf., e.g., Fig. 1A in ref. 10). During the SEMA sequence the protons which are in strong dipolar contact with the ^{15}N nuclei (e.g., the amide protons in the case of backbone-labelled polypeptides) are decoupled from the ^{1}H–^{1}H interaction network. The optimal

^{1}H center frequency is determined empirically using a reference sample (Subheading "Determination of the Power Level for the Proton B_1^{eff} Field") which ideally matches as closely as possible the situation of the oriented membrane samples. In practice, crystals of N-terminally protected amino acids and/or oriented bacteriophages are used (cf. also Note 4).

A precise measurement of the B_1 nutation frequency requires a relatively sharp nutation peak (Fig. 4). This is only the case when B_1^{eff} is close to the magic angle and the dipolar interactions are averaged. This also means that the nutation signal provides a good indicator for the Lee-Goldburg conditions.

If the two nutation frequencies at positive and negative offset (B_{1+}^{eff} and B_{1-}^{eff}, respectively) are different the correction term for the center frequency ($\delta \nu_0$) can be calculated from the respective B_{1+}^{eff} and B_{1-}^{eff} frequencies using the relationship $\delta \nu_0 = [(B_{1+}^{eff})^2 - (B_{1-}^{eff})^2]/4\Delta \nu$. Note that in this equation the "strength of the B_1^{eff} fields" is represented by the resulting nutation frequency (in Hz). An additional nutation experiment should be used to verify the correct setting, but the iterative approach described in Subheading "Determination of the Power Level for the Proton B_1^{eff} Field" (steps 2 and 3) may become unnecessary.

Acknowledgements

The help and comments by Jesus Raya and Philippe Bertani on the nutation and PISEMA experiments are gratefully acknowledged. We are grateful for the financial support by the Deutsche Forschungsgemeinschaft (postdoctoral grant to MM), the University of Strasbourg (PPF RMN), the CNRS (NMR equipment), and the Chemistry Institute (UMR7177).

References

1. Wallin E, von Heijne G (1998) Genome-wide analysis of integral membrane proteins from eubacterial, archaean, and eukaryotic organisms. Protein Sci 7:1029–1038
2. Boyd D, Schierle C, Beckwith J (1998) How many membrane proteins are there. Protein Sci 7:201–205
3. Wu BL, Chien EYT et al (2010) Structures of the CXCR4 chemokine GPCR with small-molecule and cyclic peptide antagonists. Science 330:1066–1071
4. Congreve M, Langmead C, Marshall FH (2011) The use of GPCR structures in drug design. Adv Pharamcol 62:1–36
5. Tapaneeyakorn S, Goddard AD et al (2011) Solution- and solid-state NMR studies of GPCRs and their ligands. Biochim Biophys Acta 1808:1462–1475
6. Ramamoorthy A, Kandasamy SK et al (2007) Structure, topology, and tilt of cell-signaling peptides containing nuclear localization sequences in membrane bilayers determined by solid-state NMR and molecular dynamics simulation studies. Biochemistry 46:965–975

7. Franks WT, Linden AH et al (2012) Solid-state magic-angle spinning NMR of membrane proteins and protein–ligand interactions. Eur J Cell Biol 91:340–348
8. Etzkorn M, Seidel K et al (2010) Complex formation and light activation in membrane-embedded sensory rhodopsin II as seen by solid-state NMR spectroscopy. Structure 18:293–300
9. Bechinger B, Resende JM, Aisenbrey C (2011) The structural and topological analysis of membrane-associated polypeptides by oriented solid-state NMR spectroscopy: established concepts and novel developments. Biophys Chem 153:115–125
10. Bechinger B, Kinder R et al (1999) Peptide structural analysis by solid-state NMR spectroscopy. Biopolymers 51:174–190
11. Aisenbrey C, Bertani P, Bechinger B (2010) Solid-state NMR investigations of membrane-associated antimicrobial peptides. In: Guiliani A, Rinaldi AC (eds) Antimicrobial peptides. Humana Press, Springer, N.Y., pp 209–233
12. Bechinger B, Sizun C (2003) Alignment and structural analysis of membrane polypeptides by 15N and 31P solid-state NMR spectroscopy. Concepts Magn Reson 18A:130–145
13. Salnikov E, Aisenbrey C et al (2010) Solid-state NMR approaches to measure topological equilibria and dynamics of membrane polypeptides. Biochim Biophys Acta 1798:258–265
14. Aisenbrey C, Bechinger B (2004) Tilt and rotational pitch angles of membrane-inserted polypeptides from combined 15N and 2H solid-state NMR spectroscopy. Biochemistry 43:10502–10512
15. Cullis PR, De Kruijff B (1979) Lipid polymorphism and the functional roles of lipids in biological membranes. Biochim Biophys Acta 559:399–420
16. Kim C, Spano J et al (2009) Evidence of pores and thinned lipid bilayers induced in oriented lipid membranes interacting with the antimicrobial peptides, magainin-2 and aurein-3.3. Biochim Biophys Acta 1788:1482–1496
17. Bechinger B, Salnikov ES (2012) The membrane interactions of antimicrobial peptides revealed by solid-state NMR spectroscopy. Chem Phys Lipids 165:282–301
18. Salnikov ES, Mason AJ, Bechinger B (2009) Membrane order perturbation in the presence of antimicrobial peptides by 2H solid-state NMR spectroscopy. Biochimie 91:743
19. Mason AJ, Martinez A et al (2006) The antibiotic and DNA-transfecting peptide LAH4 selectively associates with, and disorders, anionic lipids in mixed membranes. FASEB J 20:320–322
20. Seelig J, Macdonald PM, Scherer PG (1987) Phospholipid head groups as sensors of electric charge in membranes. Biochemistry 26:7535–7541
21. Bechinger B, Seelig J (1991) Interaction of electric dipoles with phospholipid head groups. A 2 H and 31 P NMR study of phloretin and phloretin analogues in phosphatidylcholine membranes. Biochemitry 30:3923–3929
22. Bechinger B (2005) Detergent-like properties of magainin antibiotic peptides: a 31P solid-state NMR study. Biochim Biophys Acta 1712:101–108
23. Aisenbrey C, Sudheendra US et al (2007) Helix orientations in membrane-associated Bcl-X L determined by 15 N solid-state NMR spectroscopy. Eur Biophys J 36:451–460
24. Nedelkina S, Gokce I et al (2008) High-yield expression and purification of soluble forms of the anti-apoptotic Bcl-x(L) and Bcl-2 as TolAIII-fusion proteins. Protein Expr Purif 29:1633–1644
25. Aisenbrey C, Cusan M et al (2008) Specific isotope labeling of colicin E1 and B channel domains for membrane topological analysis by oriented solid-state NMR spectroscopy. ChemBioChem 9:944–951
26. Rosay M, Tometich L et al (2010) Solid-state dynamic nuclear polarization at 263 GHz: spectrometer design and experimental results. Phys Chem Chem Phys 12:5850–5860
27. Salnikov E, Rosay M et al (2010) Solid-state NMR spectroscopy of oriented membrane polypeptides at 100 K with signal enhancement by dynamic nuclear polarization. J Am Chem Soc 132:5940–5941
28. Bechinger B, Opella SJ (1991) Flat-coil probe for NMR spectroscopy of oriented membrane samples. J Magn Reson 95:585–588
29. Nielsen NC, Daugaard P et al (1995) A flat-coil NMR probe with hydration control of oriented phospholipid-bilayer samples. J Biomol NMR 5:311–314
30. Bechinger B (2011) Insights into the mechanisms of action of host defence peptides from biophysical and structural investigations. J Pept Sci 17:306–314
31. Bechinger B, Bertani P et al (2010) The structural and topological analysis of membrane polypeptides by oriented solid-state NMR spectroscopy: sample preparation and theory. In: Castanho M (ed.) Membrane-active peptides: methods and results on structure and function. International University Line, La Jolla, California, USA, pp 196–215

32. Chenal A, Prongidi-Fix L et al (2009) Deciphering membrane insertion of the diphtheria toxin T domain by specular neutron reflectometry and solid-state NMR spectroscopy. J Mol Biol 391:872–883
33. Lakey JH, Slatin SL (2001) Pore-forming colicins and their relatives. Curr Top Microbiol Immunol 257:131–161
34. Raja M, Vales E (2011) Improved technique for reconstituting incredibly high and soluble amounts of tetrameric K(+) channel in natural membranes. J Membr Biol 241:141–144
35. Highsmith S (1990) On the mechanism of detergent modification of myosin structure and function. J Biochem Tokyo 107:554–558
36. Berger S, Braun S (2004) 200 And more basic NMR experiments: a practical course. Weinheim, Wiley-VCH Verlag
37. Rance M, Byrd RA (1983) Obtaining high-fidelity spin-1/2 powder spectra in anisotropic media: phase-cycled Hahn echo spectroscopy. J Magn Reson 52:221–240
38. Ramamoorthy A, Wei Y, Lee D (2004) PISEMA solid-state NMR spectroscopy. Ann Rep NMR Spec 52:1–52
39. Davis JH, Jeffrey KR et al (1976) Quadrupolar echo deuteron magnetic resonance spectroscopy in ordered hydrocarbon chains. Chem Phys Lett 42:390–394
40. Salnikov E, Bechinger B (2011) Lipid-mediated peptide–peptide interactions in bilayers: structural insights into the synergistic enhancement of the antimicrobial activities of PGLa and magainin 2. Biophys J 100:1473–1480
41. O'Brian FEM (1948) The control of humidity by saturated salt solutions. J Sci Instr 25:73–76
42. Ottiger M, Bax A (1998) Characterization of magnetically oriented phospholipid micelles for measurement of dipolar couplings in macromolecules. J Biomol NMR 12:361–372
43. Salnikov ES, Friedrich H et al (2009) Structure and alignment of the membrane-associated peptaibols ampullosporin A and alamethicin by oriented 15 N and 31 P solid-state NMR spectroscopy. Biophys J 96:86–100
44. Traikia M, Warschawski DE et al (2000) Formation of unilamellar vesicles by repetitive freeze-thaw cycles: characterization by electron microscopy and P-31-nuclear magnetic resonance. Eur Biophys J Biophys 29:184–195
45. Yokogawa M, Takeuchi K, Shimada I (2005) Bead-linked proteoliposomes: a reconstitution method for NMR analyses of membrane protein–ligand interactions. J Am Chem Soc 127:12021–12027
46. Hjelmeland LM, Nebert DW, Osborne JC (1983) Sulfobetaine derivatives of bile-acids – non-denaturing surfactants for membrane biochemistry. Anal Biochem 130:72–82
47. Womack MD, Kendall DA, Macdonald RC (1983) Detergent effects on enzyme-activity and solubilization of lipid bilayer-membranes. Biochim Biophys Acta 733:210–215
48. Banerjee P, Joo JB et al (1995) Differential solubilization of lipids along with membrane-proteins by different classes of detergents. Chem Phys Lipids 77:65–78
49. Hjelmeland LM (1980) A non-denaturing Zwitterionic detergent for membrane biochemistry – design and synthesis. Proc Natl Acad Sci USA 77:6368–6370
50. Vanaken T, Foxallvanaken S et al (1986) Alkyl glycoside detergents – synthesis and applications to the study of membrane-proteins. Methods Enzymol 125:27–35
51. Jacobs JJ, Anderson RA, Watson TR (1971) Interactions in phenol-sodium dodecyl sulphate-water systems. J Pharm Pharmacol 23:148–149
52. Lauterwein J, Bosch C et al (1979) Physicochemical studies of the protein–lipid interactions in melittin-containing micelles. Biochim Biophys Acta 556:244–264
53. Palladino P, Rossi F, Ragone R (2010) Effective critical Micellar concentration of a zwitterionic detergent: a fluorimetric study on n-dodecyl phosphocholine. J Fluoresc 20:191–196
54. Ross J, Olivier JP (1959) A new method for the determination of critical micelle concentrations of uncharged association colloids in aqueous or in non-aqueous solution. J Phys Chem 63:1671–1674
55. Dixon AM, Venable RM et al (2002) Micelle-bound conformation of a hairpin-forming peptide: combined NMR and molecular dynamics study. Biopolymers 65:284–298
56. Lorber B, Bishop JB, Delucas LJ (1990) Purification of octyl beta-D-glucopyranoside and reestimation of its Micellar size. Biochim Biophys Acta 1023:254–265
57. Hierrezuelo JM, Aguiar J, Ruiz CC (2005) Micellar properties of a mixed surfactant system constituted by n-octyl-beta-D-thioglucopyranoside and sodium dodecyl sulphate. Coll Surf A 264:29–36
58. Molina-Bolivar JA, Hierrezuelo JM, Ruiz CC (2006) Effect of NaCl on the self-aggregation of n-octyl beta-D-thioglucopyranoside in aqueous medium. J Phys Chem B 110:12089–12095
59. Coll H (1970) Study of ionic surfactants by membrane osmometry. J Phys Chem 74:520–528

Chapter 17

Solution NMR Spectroscopy for the Determination of Structures of Membrane Proteins in a Lipid Environment

Ashish Arora

Abstract

Several recent advancements have transformed solution NMR spectroscopy into a competitive, elegant, and eminently viable technique for determining the solution structures of membrane proteins at the level of atomic resolution. Once a good level of cell-based or cell-free expression and purification of a suitably sized membrane protein has been achieved, then NMR offers a combination of several versatile strategies, for example, choice of appropriate deuterated or non-deuterated detergents, temperature, and ionic strength; isotope labelling with ^{2}H, ^{13}C, ^{15}N, with or without protonation of Ile (δ1), Leu, and Val methyl protons; combinatorial labelling of specific amino acids; transverse relaxation-optimized NMR spectroscopy-based, Nonuniform sampling-based, and other NMR experiments; measurement of residual dipolar couplings using stretched polyacrylamide gels or DNA nanotubes; and spin-labelling and paramagnetic relaxation enhancements. Strategic combinations of these advancements together with availability of highly sensitive cryogenically cooled probes equipped high-field NMR spectrometers (up to 1 GHz ^{1}H frequency) have allowed the perseverant investigator to successfully overcome several of the conventional pitfalls associated with the NMR technique and membrane proteins, viz., low sensitivity, poor sample stability, spectral crowding, and a limited number of NOEs and other constraints for structure calculations. This has resulted in an unprecedented growth in the number of successfully determined NMR structures of large and complex membrane proteins, and this technique now holds great promise for the structure determination of an ever larger body of membrane proteins.

Key words: Solution NMR, Membrane proteins, Detergent micelles, Protein structures, β-Barrels, α-Helical bundles, Dynamics

1. Introduction

Integral membrane proteins are encoded by approximately 30% genes of any genome, and these proteins are involved in a wide range of functional activities that are essential for the cell (1, 2). Membrane proteins, especially the membrane-bound receptors and ion channels, are the known targets for more than half of all currently used drugs (3). Successful determination of almost 250 structures of membrane proteins has been achieved in various forms

Jörg H. Kleinschmidt (ed.), *Lipid-Protein Interactions: Methods and Protocols*, Methods in Molecular Biology, vol. 974, DOI 10.1007/978-1-62703-275-9_17,

such as crystals, micellar solutions, and membranes, by the use of X-ray crystallography, solution, and solid-state NMR spectroscopy, respectively (4). There still is a stark contrast between the number of protein structures studied and the functional significance of the membrane proteins in general. Determination of structures using NMR methods, in particular the high-resolution solution NMR methods, is very important because the technique provides information not only about the atomic coordinates of the structures but also dynamic motions of the backbone and side chains of membrane proteins solubilized in a large excess of mild detergents, such as DPC or DHPC, that structurally closely resemble membrane lipids and thus preserve the native structures and functions of membrane proteins in a biologically relevant environment. Membrane proteins fall into two families. Membrane proteins of the eukaryotic cell or plasma membrane and the inner (cytoplasmic) membrane of prokaryotes belong to the α-helical bundle family. Membrane proteins of the outer membranes of gram-negative bacteria, and of chloroplast and mitochondria, belong to the β-barrel family (5, 6). In recent times, solution NMR spectroscopy methods have been very successful for determining the solution structures of several large α-helical and β-barrel membrane proteins like OmpG, VDAC-1, UCP2, DsbB, DAGK, pSRII, KcsA, and proteorhodopsin (7–14). In addition to structure and dynamics, NMR technique also offers unique opportunities to study the interactions of the membrane proteins with several other interacting partners. This chapter aims to present the recent advances in the methodology of sample labelling and preparation and the development of newer NMR-based strategies and their successful applications. Needless to say that pushing the boundaries of existing techniques for structure determination of membrane proteins is an ongoing and continuous process.

2. Samples

2.1. Preparation of Isotopically Labeled Membrane Proteins

NMR solution studies on membrane proteins are based on multidimensional multinuclear NMR experiments on 0.5–1.5 mM proteins that are isotopically labeled with ^{15}N, or ^{13}C, ^{15}N, or ^{2}H, ^{13}C, ^{15}N isotopes. Based on the molecular weight, this can translate into 5–15 mg of membrane protein per sample at a concentration of 15–30 mg/ml. In most of the cases, the production of membrane protein has been achieved by bacterial expression using a suitable *E. coli* strain (15), although other available cellular expression systems are gram-positive bacteria (*Lactococcus lactis*), monocellular eukaryotes (*Pichia pastoris* and *Saccharomyces cerevisiae*), and cell lines from higher eukaryotes (insect cells and mammalian cells).

The gene encoding the desired protein is cloned in expression vectors and transformed into a bacterial host. Vectors containing the T7 and T5 viral promoters, especially the pET and pQE series, are quite commonly used. The common bacterial hosts are *E. coli* strains BL21(DE3), tuner, C43(DE3), and Rossetta(DE3). It is fairly common to express the membrane protein together with a cleavable poly-histidine tag at the N- or the C-terminal. In many cases, the cleavage site selected has been the one recognized by TEV protease. In the case of phospholamban, the 50 residue membrane peptide was expressed as MBP fusion protein, cleaved, and purified by reverse-phase gel chromatography (16).

2.1.1. Isotopic Labelling of Membrane Proteins

Assuming that bacterial expression and purification of unlabeled protein have been well established, the ^{2}H, ^{13}C, ^{15}N labelling process starts by making minimal media plates in D_2O in which deuterated glucose may be used if needed for acclimatization of the expression host. The competent cells of the selected host are freshly transformed with the plasmid construct, and the transformed cell suspension is plated on the MM-D_2O plates. The plates are incubated over 24- to 36-h period at set temperatures, and the colonies obtained are tested for contamination. A single colony is picked and grown in a small volume in a conical flask for 24–36 h. If the cell growth is healthy and no clumping is observed, then the overnight culture is used to inoculate a larger size (1:40–1:100 dilution factor) modified M9 medium prepared in D_2O containing 2–4 g/l ^{2}H (98%)- and ^{13}C (99%)-labeled D-glucose, and 1 g/l ^{15}N (98%)-labeled $(NH_4)_2SO_4$. Commercially available 10× media for various levels of deuterations (e.g., Bioexpress media from CIL) are available and can be prepared by simply diluting with D_2O under sterile conditions. The resulting media, although prohibitively costly, gives a good growth and yield of the expressed protein. A cheaper alternative is to spike the modified M9 minimal media described above with 1% of the 10× solution of CDN-Bioexpress-1000 medium for ^{2}H, ^{13}C, ^{15}N labelling. The growth of the cells is carefully monitored, and the cells are most commonly induced with standardized minimal amounts of isopropyl β-D-1-thiogalactopyranoside (IPTG). After letting the induced cells grow further for several hours, they are harvested and processed further (17).

For the production of uniformly ^{2}H, ^{13}C, ^{15}N protein samples with methyl protonation of Ile-δ1-[$^{13}CH_3$] and Val, Leu-[$^{13}CH_3$, $^{13}CH_3$], the CDN-Bioexpress-1000 is not added to the deuterated minimal medium described above, while 50 mg/l of 2-keto-3-d2-1,2,3,4-^{13}C-butyrate and 100 mg/l of 2-keto-3-methyl-^{13}C-3-d1-1,2,3,4-^{13}C-butyrate (18) are added 1 h before induction.

In addition to stable isotope labelling, it has become quite necessary for obtaining long-range constraints to strategically introduce spin-labels covalently bound at specific positions in the sequence. Structural restraints are derived from the paramagnetic relaxation enhancement (PRE) effect of the spin-label. For this strategy, normally a cys-less mutant of protein is prepared, and then several specific single-cystine mutants are generated and labeled with *S*-(2,2,5,5-tetramethyl-2,5-dihydro-1H-pyrrol-3-yl) methyl methanesulfonothioate (MTSL) (19).

2.2. Choice of Detergent, Temperature, Additives, and Other Considerations

Solution NMR studies are carried out by solubilizing the proteins in detergent micelles. Since this environment is topologically and structurally different from a typical lipid bilayer of a membrane, extensive validation of the detergent-solubilized membrane protein through biochemical and biophysical characterizations in detergent micelles is a prerequisite for initiating any serious structural work. For various proteins, this has to be taken up on a case-by-case basis, although, there are some general considerations which are described here.

For purification, it is often necessary to solubilize membrane proteins either in detergent micelles or detergent–phospholipid mixed micelles. β-Barrel membrane proteins are often purified from inclusion bodies that are dissolved in denaturing agents such as 8 M urea or 6 M guanidine hydrochloride (7, 8, 17, 20). For α-helical membrane proteins, more elaborate procedures are needed starting from solubilization and purification to final reconstitution. Shorter peptides corresponding to single to two transmembrane helices can be extracted in mixtures of organic solvents like methanol and chloroform. Solubilization in detergents, especially the nonionic detergents, is the mildest in terms of perturbation of protein structure in comparison to the alternatives mentioned above.

Wherever membrane proteins are reconstituted in detergent micelles for the purpose of solution NMR studies, the molar ratio of protein to detergent is always kept somewhere in between 1:100 and 1:600. This is done to minimize protein–protein contacts that would broaden and diminish the number of NMR peaks in the NMR spectra. Out of hundreds of seemingly suitable detergents for NMR studies, there are only handfuls that have been repeatedly used for most of the determined structures. Among these are dodecylphosphocholine (DPC), dihexanoyl-phosphatidylcholine (DHPC), lauryl dimethylamine oxide (LDAO), lyso-phosphatidylglycerol (LPPG), *N*-octyl-β-D-glucopyranoside (OG), *n*-dodecyl-β-D-maltoside (DDM), and sodium dodecyl sulfate (SDS).

DPC is most commonly used as a detergent for structural studies of membrane proteins by solution NMR (7, 10, 11, 16, 17, 20, 21). The CMC of DPC is 1.5 mM and typically in 50 mM NaCl, a micelle of DPC consists of 70–80 monomers at 25 C. This corresponds to an apparent molecular mass of ~26 kDa for the DPC micelle (22). Therefore, monomeric or oligomeric protein of total molecular mass up to 60–75 kDa can be reconstituted in DPC micelles for structural studies. The other detergents that have

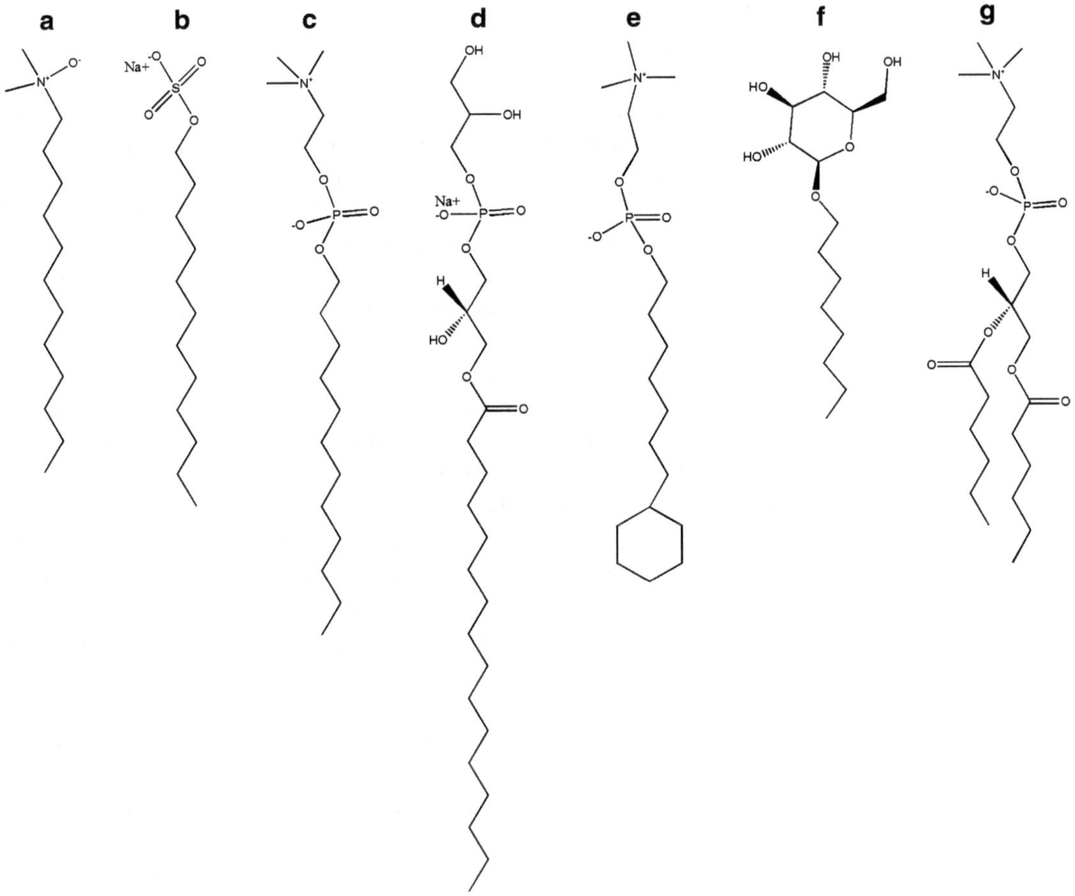

Fig. 1. Schematic representation of various detergents used for solubilization of membrane proteins for structure determination using NMR spectroscopy: (**a**) LDAO, (**b**) SDS, (**c**) DPC, (**d**) LPPG, (**e**) CYFOS-7, (**f**) β-OG, (**g**) DHPC.

been used are DHPC (23, 30), β-OG (20), SDS, lyso-myristoyl-PG (24), and LDAO (8, 25). Schematic representations of chemical structures of some commonly used detergents are shown in Fig. 1. For NMR structural studies of small α-helical membrane proteins of MW 4–10 kDa and having 2–4 transmembrane segments, LPPG has been found to be most suitable for obtaining samples that have a long lifetime and yield high-quality NMR spectra (26).

Detergent concentration needs to be standardized carefully by counting the number and also line widths at half height of the peaks in the HSQC spectra of proteins. For DPC, detergent concentrations from 150 to 600 mM have been used. In many of the reported studies, use of perdeuterated detergent has been made. However, based on the fast dynamics and exchange of individual detergent molecule, it would appear that perdeuteration of detergent is not necessary. The advantage of using non-deuterated over deuterated detergent is the huge difference in terms of the higher cost of the latter. However, where there is interference, such as in case of critical characterization of methyl–methyl NOEs, use of deuterated detergent is preferred. Most of the NMR studies have been conducted at temperatures between 30 and 50 C, as higher temperature improves the sensitivity of ^{13}C by reducing the line widths. However, this could also lead to higher conformational exchange, and this is sorted by recording some preliminary 2D and 3D experiments at two or more temperatures. In the presence of certain detergents, the NMR sample can solidify at ambient temperature, leading sometimes to irreversible changes in the sample state (26).

Detergents cooperatively assemble above their CMC in aqueous solutions to form micelles. Typically, detergents have CMCs in low millimolar range. The CMCs for SDS and DHPC are 2.2 and 15.2 mM, respectively. The aggregation numbers for SDS, DPC, and DHPC are ~70, ~80, and ~35, respectively, corresponding to aggregate molecular masses of ~25, ~28, and ~16 kDa, respectively. Detergent micelles are usually depicted as idealized spherical structures. Indeed, molecular dynamics (MD) simulations of pure DPC micelles with either 56 or 65 detergent molecules have revealed almost spherical structures with a ratio of the principal moments of inertia of 1.2:1.1:1 (27). However, when membrane proteins are solubilized in detergent micelles, the number of detergent molecules associated with the protein would be determined by the hydrophobic surface of the protein that needs to be coated with the hydrophobic alkyl chains of the detergent molecules (15) and the length of the alkyl chain and the head-group of the detergent. A variety of techniques, for example analytical ultracentrifugation, dynamic light scattering, gel exclusion chromatography, and NMR gradient spin-echo diffusion, have been used to determine the sizes of the detergent–protein aggregates. Indirect estimate of the size of the detergent–protein aggregate can also be derived from the rotational correlation time and global diffusion tensor by analysis of the amide ^{15}N T_1, T_2, and heteronuclear NOE relaxation parameters. The 8-stranded β-barrel proteins have all been found to have rotational correlation time of 20–22 ns corresponding to an apparent molecular weight of the detergent–protein aggregate of around 50–60 kDa (17, 20, 23, 28, 30). These values are in close agreement with the dynamic light scattering measurements and usually significantly lower than the molecular mass estimated from NMR gradient diffusion experiments (29).

There is a general consensus based on several experimental measurements like internuclear NOEs between membrane protein and detergent (OmpX/DHPC), rotational dynamics (OmpA/DPC), and molecular dynamics simulations (29) that the detergent molecules form an annular ring that coats the hydrophobic surface area of membrane proteins. The detergent/protein interface also appears to be slippery, such that the protein can rotate within the confines of the micelle. The lifetimes of detergent molecules in contact with the membrane protein could be of the order of 0.3 ns, which is based on the observation of negative intermolecular NOEs between the protein backbone amide protons and the protons on detergent molecules. Furthermore, the association of detergent molecules and protein is dynamic, and detergent molecules exchange between the protein–detergent complex, free monomeric detergent, and detergent-only micelles on a sub-millisecond timescale (29).

2.3. Sample Preparation with Some Specific Examples

2.3.1. β-Barrel Membrane Proteins

The β-barrel proteins for which NMR structures have been determined are all expressed into inclusion bodies. The inclusion bodies are solubilized into either 8 M urea (for OmpA) or 6 M guanidinium chloride (OmpX, PagP, and VDAC-1). For NMR studies, the proteins are refolded by diluting the urea or guanidinium chloride solution into buffer containing detergent micelles. Some of the specific cases are described below.

For the OmpA TM domain, refolding was carried out by slowly diluting protein solution in 15 mM Tris buffer, pH 8.5, containing 8 M urea into more than tenfold excess of 20 mM borate buffer, pH 10.0, containing 150 mM NaCl, 1 mM EDTA, and 20 mM DPC. After overnight refolding, the solution was concentrated through Amicon PM10 or YM1 membranes. The dilution–concentration step was repeated twice with NMR buffer (20 mM sodium phosphate, pH 6.3, 50 mM NaCl, 0.1% NaN_3) to give a final concentration of 1 mM protein and ~600 mM DPC (17).

PagP was refolded from 6 M GdmCl solution into phosphate buffer containing DPC, pH 6.0, using procedure similar to that described above for OmpA. For studying the interaction of PagP with Cyfos-7, a sample of PagP refolded in β-octyl glucoside (OG) was also prepared by dissolving precipitated PagP in 1 ml of 5% perdeuterated SDS and dialyzing for 5 days through membranes with MWCO of 3.5 kDa, against 50 mM sodium phosphate buffer, pH 6.0. Thirty-two milligram of OG was slowly dissolved, and then d5-ethanol was added to 1%. The sample was concentrated down to yield a 1 mM PagP NMR sample in a ~200 mM OG. Both SDS and ethanol were required for 100% refolding (20).

OmpX was solubilized in 6 M GdmCl in 20 mM Tris buffer (20 mM Tris and 5 mM EDTA), pH 8.5. Solubilization and reconstitution of OmpX were performed by slowly diluting the protein solution in 6 M GdmCl into a sixfold excess of refolding buffer containing 3% DHPC, 20 mM Tris-HCl, 5 mM EDTA, and 0.6 M arginine, pH 8.5, at 4 C. The OmpX/DHPC solution was then dialyzed against 2.5 l of 20 mM Tris buffer pH 8.5, having 5 mM EDTA and 100 mM NaCl, at 4 C for 20 min to remove residual GdmCl and L-arginine. The sample was concentrated to a volume of 300 μl. The final protein concentration was about 2 mM, and the detergent concentration was adjusted to 300 mM by adding solid DHPC to the solution (30).

The refolding of OmpG sample in 8 M urea was standardized in various detergents (7) like DHPC, Fos-cholines, β-OG, and DDM at various concentrations, pH values, and temperatures, and in presence of various additives like arginine, ethanol, glycerol, and glucose. Finally, the most suitable method for folding and preparation of NMR sample that was stable for a long time involved first refolding the protein in 70 mM β-OG at pH 9 at room temperature, followed by two steps of dilution and concentration, with 10× of 15 mM DPC in 25 mM bis-Tris buffer containing 50 mM NaCl and 0.05% NaN_3. The final NMR sample had OmpG concentration of 1.0–1.3 mM (7).

VDAC-1 was purified from inclusion bodies solubilized in 8 M urea, precipitated, and resolubilized in 6 M GdmHCl buffer (100 mM $NaPO_4$ pH 7.0, 100 mM NaCl, 6 M GdmHCl, 5 mM DTT, 1 mM EDTA). Refolding of VDAC-1 was carried out at 4 C by tenfold dilution in 1% LDAO-containing buffer, followed by further purification through cation-exchange chromatography, and a series of dialysis and dilution/concentration steps. The LDAO detergent concentration was monitored by 1D ^{1}H-NMR and adjusted by a series of dilution and concentration steps. The final sample conditions were 25 mM $NaPO_4$, 5 mM DTT, pH 6.8, 300–500 mM LDAO, and 0.5–1 mM VDAC-1 (8).

OprH from *Pseudomonas aeruginosa* was also purified from inclusion bodies and refolded from 8 M urea in DHPC micelles. Briefly, a sample containing 0.4 mM OprH in elution buffer was diluted tenfold into 20 mM Tris–HCl at pH 8.5, 5 mM EDTA, and 0.6 M L-arginine (refolding buffer) with 3% DHPC. The refolding solution was incubated for 72 h at 37 C before being dialyzed (20 min, room temperature) against 2.5 l of 20 mM Tris–HCl at pH 8.5, containing 5 mM EDTA, and 50 mM KCl. The solution was concentrated, and the buffer was exchanged against 25 mM $NaPO_4$ at pH 6.0, containing 50 mM KCl, 0.05% NaN_3, with 5% D_2O, by dilution/concentration. Final NMR samples were concentrated to 1.0–1.3 mM OprH and contained 150–175 mM DHPC as determined by ^{1}H NMR spectroscopy (23).

2.3.2. α-Helical Membrane Proteins

α-Helical membrane proteins differ in the number of transmembrane domains which can vary from a single transmembrane spanning α-helix to multiple monomeric or oligomeric α-helical bundles. Sample preparation protocols vary substantially, often depending on the number of transmembrane α-helices. Simpler membrane-spanning peptides can be synthesized through solid-phase peptide synthesis while more complex systems are overexpressed and purified in native form by solubilization into detergents. Several membrane peptides synthesized through solid-phase synthesis can be lyophilized and dissolved in buffer having the appropriate detergent. Lyophilized Pardaxin (33 residues) and phospholamban (50 residues) were dissolved in 20 mM phosphate-buffered saline, pH 6.5, containing 300 and 600 mM DPC, respectively (31, 32). Several small membrane peptides, for example, *E. coli* and *B. pseudofirmus* OF4 ATP synthase c subunits, have been expressed in bacterial cells and extracted with organic solvents. The peptides were reconstituted by mixing detergent and peptide in 3:1 chloroform:methanol and drying to a thin, clear film under a stream of argon for up to 12 h. Detergent/protein films were then dissolved in 1–50 mM aqueous potassium phosphate buffer, pH 4.5–8 for NMR experiments (26). For more complex α-helical membrane proteins that have to be purified in fully folded or partially unfolded form, more complex procedures have been followed as can be judged from the specific examples described below.

A specialized protocol of reconstitutive refolding was followed for the preparation of NMR sample for DAGK. The protein that was purified in DPC micelles was inactive. In order to get active protein, DAGK/DPC micelles were mixed with POPC to generate DAGK/DPC/POPC mixed micelles. DPC was removed by extensive dialysis to yield POPC lipid vesicles containing DAGK at 1:100 DAGK:POPC:mol:mol ratio, in refolding buffer (pH 7.8 buffer containing 10 mM imidazole and 0.05 mM EDTA). The membrane preparation was resolubilized in DPC micelles, and POPC was removed to give DAGK/DPC micelles having DAGK in active form (33).

Recently structure of a bacterial plasma membrane protein which plays a key role in disulfide bond formation, DsbB, was determined. Out of the six possible intermediate states representing steps of DsbB-catalyzed DsbA oxidation, the structure was determined for a Cys → Ser double mutant (DsbB(CSSC), C41S, C104S) termed as inter-loop disulfide bond state. DsbB(CSSC) was expressed in Rosetta(DE3) cells and purified by solubilizing the membranes with Tris-HCl/NaCl buffer containing 10 g/l DPC, followed by purification over Ni-NTA superflow column with Tris-HCl/NaCl buffer containing 1 g/l of DPC. For NMR experiments, the protein sample was concentrated and exchanged in 25 mM PBS, pH 6.2, containing 50 mM KCl, and 0.47 g/l DPC. The NMR experiments were carried out at 40 C. The DsbB [CSSC] sample was found to retain activity and displayed long-term stability under these conditions. The protein was found to have an apparent size in DPC micelles of 40 kDa, corresponding to a monomer (10).

Mouse UCP2 (residues 14–309 with a C-terminal His6 tag) was expressed using a pET-21 vector in *E. coli* Rosetta DE3 cells. Prior to purification, the lipid composition of the membrane fraction was adjusted with suitable amounts of DMPC, cardiolipin, and trace amounts of phytanoyl lipid, and these were pre-extracted with 10% OG at 4 C. UCP2 was extracted in 40 mM potassium phosphate (pH 8.0), 250 mM NaCl, 50 mM BME, 10 mM GDP, and 0.2% DPC and was purified by following a series of metal-affinity, ion-exchange, nucleotide-affinity, and gel-filtration chromatographic procedures. In the final step, UCP2 was eluted in 50 mM potassium phosphate (pH 6.5), 100 mM NaCl, and 5 mM DPC. The eluted UCP2 sample was supplemented with GDP, detergent, and lipids such that the final NMR sample contained 0.8 mM UCP2, 5 mM GDP, 150 mM DPC, 2 mM DMPC, 1 mM cardiolipin, 5 mM BME, 30 mM potassium phosphate (pH 6.5), and 80 mM NaCl (9).

Seven-helix transmembrane receptor sensory rhodopsin II (pSRII, 1–241) from *Natronomonas pharaonis* plus a sequence encoding a C-terminal hexahistidine-tag was expressed from pET-28b(+) vector into the *E. coli* expression strain BL21 Tuner (DE3). Protein expression was carried out in the presence of 10 μM all-*trans*-retinal. Crude membranes were extracted with 1.5% (w/v) of detergent DDM, and solubilized pSRII was purified with nickel-nitrilotriacetic affinity beads. The protein was eluted in 50 mM Tris–HCl, pH 7.0, 300 mM NaCl, and 0.1% DDM containing 150 mM imidazole, and the solution was exchanged into phosphate buffer by repeated rounds of concentration and dilution with a final volume of 0.5 ml. For the preparation of NMR samples, the protein was mixed with nickel-nitrilotriacetic acid beads as before and, after removal of the flow-through, washed with 50 column volumes of 20 mM sodium phosphate buffer, pH 7.0, 50 mM NaCl containing 0.6% (w/v) 1,2-diheptanoyl-sn-glycero-3-phosphocholine (diC7PC). The protein was eluted in the same buffer containing 300 mM imidazole, and the eluted solution was repeatedly concentrated and diluted in a concentrator (10-kDa cutoff) with 60 ml sodium phosphate, pH 5.9, and 50 mM NaCl containing 0.06% (w/v) DHPC to a final volume of 0.45 ml (12).

Proteorhodopsin (PR) has been expressed and isotopically labeled through the cell-free expression system. The green-absorbing variant of PR was cloned without the signal peptide into the pIVEX2.3d vector optimized for cell-free expression. The construct yielded 229 residues from PR (amino acids 21–249), a five residue linker, and a hexa-histidine tag, resulting in a final size of 26 kDa corresponding to 241 amino acids. For assignment of the characteristic H75 a Strep-tag (WSHPQFEK) was cloned at the C-terminal end in place of the His-tag. Continuous-exchange cell-free expression was based on an S30 extract and followed standard protocols. PR was expressed in the presence of 0.6 mM all-*trans* retinal in the detergent mode (DCF) using 0.4% digitonin mixed with diC7PC in a 4:1 molar ratio. Ni-affinity purification was conducted with 0.05% DDM in the loading and washing buffer, and the detergent was exchanged into 0.1% diC7PC prior to elution. Strep-tagged PR was purified via a *Strep*-Tactin matrix with 0.1% diC7PC in the washing and elution buffer. Buffer exchange was achieved by either centrifugal concentration devices (10,000 MWCO) or gravity flow PD10 Desalting columns. NMR samples were buffered with 25 mM NaOAc, pH 5, and 2 mM DTT. The diC7PC concentration was brought up to 2%. For side chain assignment and ^{13}C-edited NOESY experiments, deuterated acetate and diC7PC were used in the sample. For preparation of deuterated protein samples, cell-free expression was conducted in completely deuterated solutions (except RNAse inhibitor and T7 Polymerase). Selective labelling was achieved by replacing single types of amino acid in the unlabeled amino acid mixture used for cell-free expression with the labeled equivalent. In addition to the overlap and relaxation-optimized (oro)-SAIL amino acid, the remaining 19 amino acids were ^{2}H labeled. Side chain assignment of oro-SAIL Leu and oro-SAIL Val was achieved with information from ^{1}H-methyl-detected out-and-back experiments, 4D ^{15}N,^{13}C-separated NOESY, as well as 3D ^{13}C-separated NOESY. The 11, 20-^{13}C-labeled retinal was co-translationally bound to K231 by replacing the unlabeled cofactor in the cell-free expression system (14).

3. NMR Methodology

3.1. NMR Methods

Structure determination of membrane proteins has benefited tremendously from the advent of newer NMR techniques, especially transverse relaxation-optimized NMR spectroscopy (TROSY (34)). TROSY results in a substantial reduction of the ^{15}N relaxation rates of the ^{15}N–^{1}H moieties during coherence transfer steps by constructive use of the interference between the ^{15}N and ^{1}H dipolar--dipolar coupling and ^{15}N-chemical shift anisotropy. The TROSY pulse scheme selects the narrowest of the four component signals of the doubly decoupled ^{15}N–^{1}H correlations in a ^{15}N–^{1}H-HSQC spectrum. The TROSY pulse scheme has already been implemented in many of the standard and specialized 3D and 4D triple-resonance experiments for sequential assignments of proteins, in 3D ^{15}N-edited NOESY experiments, and in ^{15}N T_1, T_2, and {^{1}H}^{15}N-NOE relaxation experiments. In all such cases, the TROSY-type selection improves the sensitivity and the resolution of experiments starting on ^{1}H^{N} and ^{15}N of the amide residues. Further, the TROSY effect is most efficient for deuterated proteins and at high magnetic fields (between 900 and 1,100 MHz), both of which are increasingly amenable (34–36).

3.1.1. NMR Assignments

Sequential assignment is easier to obtain for the β-barrel membrane proteins because of the relatively better chemical shift dispersion of resonances in β-sheets and the alternating nature of hydrophilic and hydrophobic residues in this class of proteins. In addition, a good number of long-range NOEs are available which, together with theoretical hydrogen bonds, allow calculation of the backbone fold to high accuracy. The sequential assignments are harder for α-helical membrane proteins because of the smaller dispersion of the shifts of many consecutive aliphatic and other apolar residues that typically occur in this class of proteins. Although sequential H_N–H_N NOEs in the α-helix can be helpful to some extent, it is ordinarily difficult to find enough long-range NOEs that are required to pack helices against each other and thus fold the tertiary structure.

The triple-resonance suite of experiments that have been utilized for the sequential backbone assignments include the 3D TROSY versions of HNCA, 3D ct-TROSY-HNCA, HN(CO)CA, HN(CA)CB, HNCACB, HN(CO)CACB, HN(COCA)CB, HN (CA)CO, and HNCO (35). In case of OmpG, a new, just-in-time (JIT) version of HN(CA)CO (7) was found to be critical for obtaining the backbone CO assignments. Assignments of NOEs, while being critical for generating distance constraints for structure calculations, are also useful for aiding and removing ambiguities in the backbone assignments. NOE assignments are made by recording 3D and 4D ^{15}N-edited ^{1}H–^{1}H NOESY experiments in combination with HMQC, HSQC, and TROSY. For OmpG, NOEs were determined from 3D ^{15}N–^{15}N–^{1}H TROSY–NOESY–TROSY and [^{1}H, ^{1}H]-NOESY-^{15}N-TROSY experiments (7). For UCP2, the 3D ($^{1}H_N$, $^{1}H_N$)-HMQC-NOESY-TROSY experiment was utilized (9). For human VDAC-1, up to 600 NOEs were identified from 3D [^{1}H, ^{1}H]-NOESY-^{15}N-TROSY and [^{1}H, ^{1}H]-NOESY-^{13}C-HMQC experiments in combination with a set of 4D experiments employing multidimensional decomposition (MDD) NOESY with nonuniform sampling (NUS), viz., 4D NUS-MDD-^{13}C-HMQC-[^{1}H, ^{1}H]-NOESY-^{13}C-HMQC and 4D NUS-MDD-^{15}N-HMQC-[^{1}H, ^{1}H]-NOESY-^{13}C-HMQC, at a field strength of 900 MHz (8).

For DsbB, the Ile, Leu, and Val methyl assignments were achieved by using a HMCM[CG]CBCA experiment and 3D [^{13}C-*F1*, ^{13}C-*F2*]-edited NOESY (200 ms mixing time) (18) on a 1.5 mM U-[^{2}H]-, Ile-δ1-[$^{13}CH_3$]-, and Val, Leu-[$^{13}CH_3$, $^{13}CH_3$]-labeled DsbB[CSSC] sample in deuterated DPC micelles. Using these experiments, methyl groups in all 8 Ile(δ1), 14 Val, and 14 out of 28 Leu side chains of DsbB[CSSC] were assigned. In general, for α-helical membrane proteins, the assignment of the chemical shifts of methyl protons can often not be completed due to the high degeneracy of chemical shifts of ^{13}Cα/^{13}Cβ pairs and $^{13}CH_3$ proton and carbon chemical shifts of the methyl groups. However, even a few unambiguous methyl–methyl long-range NOEs are sufficient to determine the tertiary fold as in the case of DsbB in DPC micelles (10).

3.1.2. NMR Constraints

Backbone dihedral angle constraints are usually determined from chemical shift-based predictions by TALOS or TALOS+ software program which utilizes database mining against fragments of known structure (37). For a deuterated protein, the Cα, Cβ, and CO chemical shifts are corrected for deuteration effects, and the H_N and N chemical shifts are corrected for the TROSY–HSQC difference (38). The mainstay of structure determination by solution-state NMR spectroscopy is to assign short-, medium-, and long-range ^{1}H–^{1}H NOEs which may fall into various categories like intra-residue, sequential, inter-residue, intra- and inter-β-strand, intra- and inter-α-helical, and also inter-monomer NOEs in case of oligomeric proteins. These are often combined with hydrogen bonds which are particularly useful in determining the 3D-fold of closed β-barrel proteins. For β-barrel proteins, measurement of H_N–H_N NOEs is amenable through ^{15}N-edited [^{15}N,^{1}H]-TROSY-[^{1}H-^{1}H]-NOESY-[^{15}N-^{1}H]-TROSY or its HSQC versions (39). For smaller β-barrel systems, H_N–H_α NOEs can be assigned through partial deuteration. These NOEs in combination with theoretical hydrogen bonds and dihedral angle constraints are sufficient to give good RMSD for the β-strands of small closed β-barrels. Additional NOE information can be obtained through I,L,V-methyl protonation and measurement of H_N-methyl and methyl–methyl NOEs. This approach has contributed towards structure determination and refinement of several membrane proteins (8, 10, 40–43). The inclusion of even modest number of inter-helical NOEs in conjunction with additional structural restraints, such as RDCs and PREs as described below, offers the possibility of significantly improving the quality of membrane protein structure.

3.1.3. Residual Dipolar Couplings

Residual dipolar couplings (RDCs) in the range of roughly −27 to +25 Hz for single-bond-coupled heteronuclear spin pairs have offered an alternative route for structure calculation and refinement of membrane proteins (44–47). RDCs measured for membrane proteins involve directly bonded spin pairs: H_N–N, H_α–C_α, N–CO, and CO–C_α. These are obtained from IPAP-HSQC and TROSY-HNCO experiments. RDCs provide angular constraints based on the orientation of internuclear vectors relative to the magnetic field, which can be used for improving structural accuracy. However, for the measurement of RDCs, weak alignment of the protein molecules is required. Weak alignment in membrane protein samples has been achieved through either steric or electrostatic interactions with an alignment medium such as bicelles, filamentous phage, strained or charged polyacrylamide gel matrices, and DNA nanotubes. Usually, DMPC/DHPC or DMPC/CHAPS bicelles are not suitable for alignment of integral membrane proteins. Filamentous phage pf1 is also not stable in the presence of detergents especially at the high experiment temperature.

The rehydrated compressed or stretched polyacrylamide gels for alignment have been successfully used for OmpA, phospholamban, and PR (14, 48, 49). Useful degree of alignment was achieved for OmpA in DPC micelles in negatively and positively charged copolymer gels (48). Negatively charged copolymer gels were prepared from a 1:1 mixture of acrylamide and either acrylic acid or 2-acrylamido-2-methyl-1-propanesulfonic acid (AMPS). Positively charged copolymer gel was prepared from a mixture of 1:1 acrylamide and *N*-(2-acryloamidoethyl)triethylammonium chloride (APTMAC) (50, 51). Compressed gels in Shigemi tubes having copolymer concentration of 3–4% gave alignment of OmpA characterized by $^{1}D_{HN}$ in the range between −22 and +25 Hz, −10 and +20 Hz, and −10 and +25 Hz, for AMPS, acrylic acid, and APTMAC copolymers, respectively. In all cases, very good agreement, with a quality factor of ~24%, was obtained between the experimental RDCs and those calculated from the 1.65 Å crystal structure of OmpA TM domain (48, 52).

For PR, best spectral quality and sufficiently large dipolar couplings were obtained at an acrylamide:*N*,*N′*-methylene-bisacrylamide ratio of 150:1. For sample preparation, a commercially available gel system (New Era Enterprises, Inc.) was used. Gels were cast with a diameter of 6 mm, dialyzed against an excessive amount of H_2O overnight, and dried at room temperature. Dried gels were incubated with the PR NMR samples and allowed to swell to its original diameter within 24 h. Afterwards, the gels were transferred into NMR tubes with an inner diameter of 4.2 mm. To reduce peak overlap, three separate combination of selectively labeled PR were prepared as follows: sample 1—Ala, Gly, Leu, Phe, and Ser; sample 2—Ala, Met, Ile; and sample 3—Phe, Thr, Trp, Tyr, and Val (14).

For alignment of UCP2, use of DNA nanotubes was made. 260 μl of 0.5 mM (^{2}H, ^{13}C, ^{15}N) UCP2 sample containing 5 mM GDP, 100 mM DPC, 2 mM DMPC, 1 mM cardiolipin, 5 mM BME, 30 mM potassium phosphate (pH 6.5), and 80 mM NaCl was mixed with 260 μl of 20 mg/ml preformed DNA nanotubes. The mixture was then concentrated down to 260 μl using a Centricon. The final aligned sample gave ^{2}H quadrupolar splitting of 5.0 Hz. A homogeneous suspension of DNA nanotubes was prepared by mixing folded monomers of staple-strand oligodeoxyribonucleotides, followed by precipitation with PEG8000 and centrifugation. The recovered nanotubes were resuspended and centrifuged once again, before concentrating to 30 mg/ml with Centricon-100 concentrator by centrifugation (9).

3.1.4. Paramagnetic Relaxation Enhancement Measurements

The PRE approach has been quite useful in enabling long-range distance constraints that are very important for getting the correct tertiary structure. It exploits the distance-dependent line broadening of NMR resonances caused by the interaction of an NMR-active nucleus with an unpaired electron. This approach has been successfully used to complement the ^{1}H–^{1}H NOEs long-range distance information and the other experimental constraints for determining the structures of several α-helix and β-barrel membrane proteins. The starting point for PRE is preparation of cysteine-less protein and then rationally introducing single Cys mutations usually at rigid positions where spin-labelling is not expected to spatially or functionally affect the protein being studied. The single Cys residues are most commonly covalently linked with the spin-label MTSL. Due care must, however, be taken to check for incomplete tagging. Once a probe has been successfully incorporated, the effects of the PRE are typically measured by comparing peak line widths and intensities between spectra acquired under paramagnetic and diamagnetic conditions. The PREs induced by the spin-label can be measured by comparing the labeled sample either with the unlabeled samples or with the same sample after reduction of the spin-label with excess of ascorbic acid. A specific protocol is described below for UCP2 (9).

To introduce a single paramagnetic site for PRE measurement, the five cysteines of UCP2 were all mutated to alanine or serine (Cys 25 Ala, Cys 191 Ser, Cys 217 Ser, Cys 227 Ser, Cys 256 Ser); the cysteine-free UCP2 had GDP binding properties similar to those of wild-type protein. A single cysteine was introduced into the protein on the basis of the known secondary structures from molecular fragment replacement (MFR) segments. MTSL was then attached at the cysteine position by adding sevenfold excess label to 10 mM UCP2 in the NMR buffer at pH 8.0 and incubating at 25 C for 4 h. Excess label was removed to avoid nonspecific broadening. The pH was changed back to 6.5 for NMR measurements. To quantify residue-specific broadening of backbone $^{1}H_{N}$, two TROSY-HNCO spectra were recorded, one after nitroxide labelling and the other after reduction of the nitroxide-free electron with 5 M excess of ascorbic acid. On the basis of observable PREs, distance constraints were created as follows: For resonances that were strongly affected by the spin-label, distances of less than 13 Å were assumed. For resonances showing measurable change in intensities and line widths, distances ranging from 13 to 20 Å were converted into upper and lower limits with a ±4 Å error margin. While for resonances that did not display a detectable PRE effect, a lower distance limit of 20 Å was used (9).

3.2. Structure Calculations and Refinement

The recipe for structure calculations and refinement by most of the common programs like CYANA, CNS, and XPLOR (53–55) is to analyze, calculate, and iteratively refine both the calculated structures and the experimental information to a high level of consistency. The input parameters for structure calculations are the dihedral angle restraints, NOE-derived distance restraints, hydrogen bonds, PRE-derived long-distance restraints, RDC restraints, radius of gyration restraint, database potentials, and restraints from biochemical experiments and/or those involving ligand/cofactor. Some specific examples are described below. Overall, the methodologies described have proven to be of great value in calculation and refinement of structures of large membrane proteins and in particular α-helical proteins having many transmembrane segments.

In the simplest of cases as for PagP, backbone dihedral angles were calculated with TALOS using chemical shift assignments corrected for deuterium isotope shifts. ^{1}H–^{1}H NOEs were obtained from 4D ^{15}N-edited NOESY spectra and were separated into strong, medium, and weak, corresponding to distance restraints with upper limits of 3.5, 5.0, and 6.0 Å, respectively. Hydrogen bonding partners were selected based on NOE patterns, and two distance restraints were used to simulate a hydrogen bond. NOE- and hydrogen bond-based distance restraints and backbone dihedral angle restraints were used as input for the simulated annealing protocol of CNS v. 1.0 with 2,000 high-temperature steps, 6,000 torsion angle dynamics cooling steps, and 5,000 Cartesian coordinate dynamics cooling steps. Two hundred and fifty structures were generated, and 20 lowest energy structures were selected for representing the structures for PagP-DPC and PagP-β-OG. The selected structures had a backbone RMSD for well-defined regions of 0.90 ± 0.19 and 0.80 ± 0.13 Å, respectively (20).

For OprH, distance constraints were calibrated using Cyana version 2.1, and backbone dihedral angle constraints were determined using TALOS. Hydrogen bond constraints derived from $^{1}H/^{2}H$ exchange experiments were set at 2.5 and 3.5 Å for H_N–O and N–O distances, respectively. Structure calculations were performed using CNS version 1.2 with 4,000 high-temperature, 8,000 torsion slow-cool, and 8,000 Cartesian slow-cool annealing steps. A total of 200 structures were calculated, and the 20 lowest energy structures were selected for ensemble analysis. The mean global backbone RMSD over the 20 structures for β-sheet residues was 0.85 ± 0.20 Å (23).

Structure of OmpX was calculated in a similar way and was refined/recalculated by including additional hydrogen bonds and NOEs between the protons of the amide group and the methyl groups of I,L,V-methyl protonated and otherwise uniformly ^{2}H-, ^{13}C-, ^{15}N-labeled OmpX (40), which resulted in significant improvement in the quality over the structures that were determined initially (30).

For high-resolution structure determination of a four-helix bundle membrane integrating protein Mistic (110 residues, 13 kDa) in LDAO micelles, long-range distance constraints, necessary for the determination of three-dimensional fold, were obtained from PREs by incorporating site-directed spin-labels individually at five different positions along the sequence of Mistic. The long-range distance constraints aided in the determination of a preliminary scaffold, which was refined through an iterative process by collecting long-range and medium-range NOEs and refining calibration of the spin-label restraints. The final structure calculation was performed with 573 NOE distance restraints, 346 angle restraints from chemical shifts and NOEs, and 478 distance restraints from the spin-label experiments (25).

The 3D structure of pSRII was calculated with the program ARIA27 interfaced to CNS on the basis of 5,564 distance restraints (4,055 unambiguous, 1,509 ambiguous), 190 pairs of backbone torsion angle restraints, and 132 hydrogen bond restraints. In this case, the large number of NOEs assigned is quite noteworthy. The final structures were well defined, with average pair-wise RMSD of 0.48 Å for the backbone residues 1–221 over 30 structures. The overall structure was in excellent agreement with the previously determined X-ray structures (14).

For calculation of PR structures, PRE was assigned into three categories of long-range distance restraints. The experimental RDCs, covering the range from −26.9 to −2.4 Hz, were used to determine the initial alignment tensor magnitude and rhombicity via the histogram method incorporated in CYANA, which were subsequently refined by fitting against the structure. CYANA structure calculations with torsion angle dynamics were performed with 200 conformers. The 20 conformers with the lowest target function values were superimposed with an RMSD to the mean coordinates of 0.81 Å within the transmembrane region. To validate the structure, complete sets of PRE restraints for each cysteine mutant were excluded one by one from the structure calculation. Comparison of [^{15}N–^{1}H]-correlation spectra was made with deuterated PR in deuterated diC7PC/DMPC bicelle which further established that the structure in micelle faithfully represented that in the bilayer membrane environment (14).

A novel approach has been used very recently for the determination of structure of mouse UCP2. This approach utilized the RDC-based MFR that had been validated for structure determination of ubiquitin using molecular fragments fit to RDCs. RDCs were measured from protein samples that were weakly aligned in a DNA nanotube liquid crystal as described above. The local and secondary structures of the protein were determined by constructing a database containing 320,000 seven-residue fragments from structures. For each seven-residue stretch along UCP2, the corresponding RDCs were fitted to all fragments using singular value decomposition incorporated in the program PALES (56). The candidate fragments were selected on the basis of quality of fit and were used to determine the local backbone structure by following a three-step protocol of fragment assignment, gap filling, and end extension. The RDCs also determined the relative orientation of the secondary structural segments. In addition to RDCs, semiquantitative distance restraints were obtained from PRE measurements, and these restraints provided the spatial arrangement of the transmembrane helices in the tertiary fold. Structures were

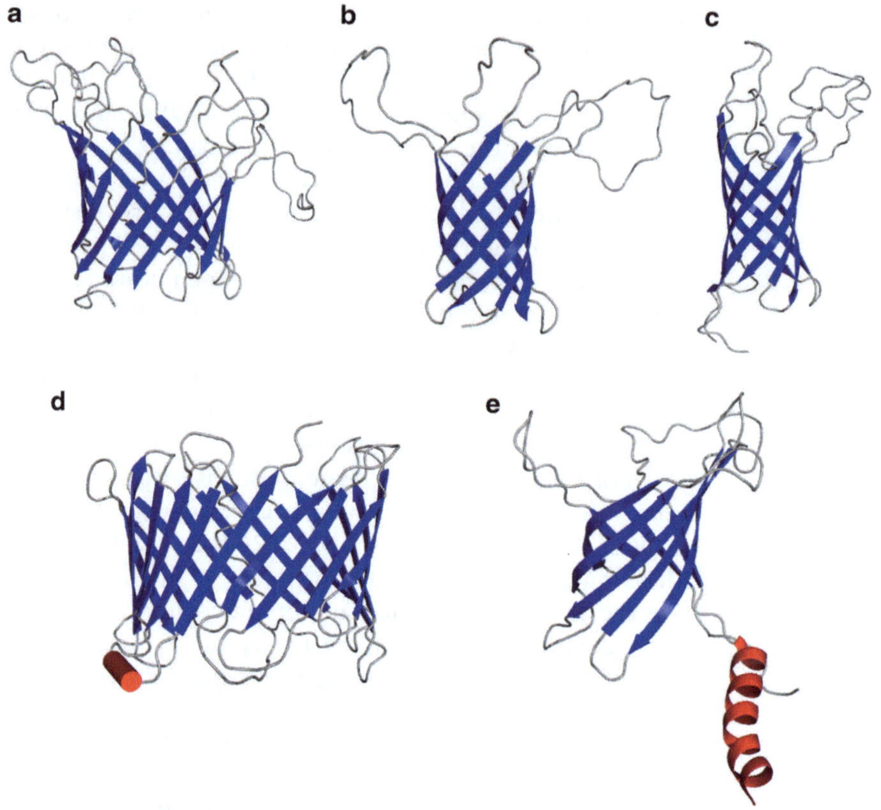

Fig. 2. Solution NMR structures of selected β-barrel membrane proteins: (**a**) OmpG, PDB ID: 2JQY ; (**b**) OprH, PDB ID: 2LHF; (**c**) outer membrane protein OmpA transmembrane domain, PDB ID: 1G90; (**d**) VDAC-1, PDB ID: 2K4T; (**e**) PagP, PDB ID: 1MM4.

calculated using XPLOR-NIH with the assigned structured segments, RDCs and PREs. The ϕ and ψ values of the segments were strongly enforced by a harmonic potential with force constant ramped from 10 to 1,000 kcal/mol/rad^2. All RDCs used for determining the segments were applied, and the RDC force constant was ramped from 0.01 to 1.5 kcal/mol/Hz^2. PRE restraints were enforced with flat-well harmonic potentials, with the force constant ramped from 1 to 40 kcal/mol/A^2. In addition to experimental restraints, a weak database-derived 'Rama' potential function was ramped from 0.02 to 0.2 for the general treatment of side-chain rotamers. A total of 30 monomer structures were calculated using a simulated annealing protocol in which the bath temperature was cooled from 2,000 to 200 K. Fifteen low-energy structures were selected as the structural ensemble. The average backbone pair-wise RMSD for well-structured regions over 15 structures was 1.32 Å (9).

The structures determined by solution NMR spectroscopy of selected β-barrel and α-helical bundle membrane proteins are shown in Figs. 2 and 3, respectively. The experimental restraints

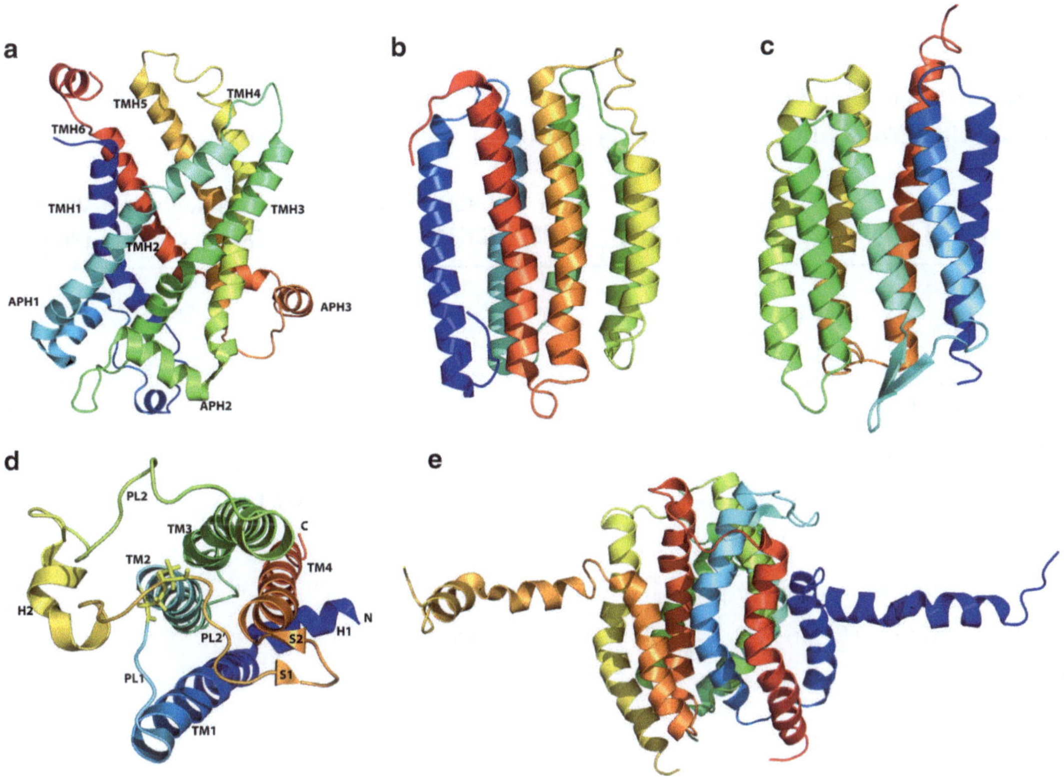

Fig. 3. Solution NMR structures of selected α-helical bundle membrane proteins: (**a**) UCP2, PDB ID: 2LCK; (**b**) proteorhodopsin, PDB ID: 2L6X; (**c**) pSRII, PDB ID: 2KSY; (**d**) DsbB, PDB ID: 2K74; (**e**) DAGK, PDB ID: 2KDC.

and structural statistics for the final ensemble of structures for selected membrane proteins are given in Table 1.

3.3. Protein–Detergent Interactions Monitored by NMR Spectroscopy

The interaction between the membrane protein and the detergents in the surrounding micelle can be studied by a variety of NMR methods and is quite useful for structural validation. Exposure of backbone amides to solvent can simply be studied by solvent exchange experiments like the SEA-TROSY (57). Based on these experiments, the picture that usually emerges is that the detergent is "wrapped" as a belt around the hydrophobic core of the membrane protein with the loops and turns extending into the solvent. Internuclear NOEs between distinct group of protons of the detergent and the assigned backbone and side-chain protons of the protein in protein–detergent micelle can be measured directly or through designing suitable saturation transfer difference (STD) NMR experiments. In many cases, internuclear NOEs have been measured between the detergent and the protonated methyl groups of an I,L,V-methyl protonated ^{2}H-, ^{13}C-, ^{15}N-labeled protein. Generally, intermolecular NOEs between the I,L,V-methyl groups and detergent alkyl groups are strongest for side chains in the hydrophobic core that are pointing outwards, while for regions

Table 1
Experimental restraints and structural statistics for the final ensemble of solution structures of selected β-barrel and α-helical bundle membrane proteins determined with NMR spectroscopy

Protein (PDB ID)	OmpA (2GE4)	OmpG (2JQY)	OprH (2LHF)	DsbB (2K74)	UCP2 (2LCK)	Proteorhodopsin (2L6X)	VDAC-1 (2K4T)
Distance restraints	90	316	199	487	452	376	612
Hydrogen bonds	68	55	134	295	–	133	139
Dihedral angle constraints (ϕ and φ)	142	262	188	97	454	392	316
RMSD values (Å)							
All backbone atoms	0.48	2.38	0.85	0.8	1.32	0.81	2.9
All heavy atoms	1.86	7.37	2.07	1.75	1.84	1.42	3.6
Ramachandran plot statistics							
Most favored region (%)	–	66.7	66	86.9	95.4	–	77.1
Allowed region (%)	–	27.1	29.3	10	3.2	–	21.6
Additionally allowed region (%)	–	4.2	3.4	3.1	1.3	–	0.8
Disallowed region (%)	–	1.9	1.3	0	0.2	–	0.4
Reference	(64)	(7)	(23)	(10)	(9)	(14)	(8)

exposed to the aqueous medium, like the extracellular loops and turns, such NOEs are weaker. This also suggests that the orientation of detergent molecules is in the form of an annular ring which is generally perpendicular to the long axis of the solubilized membrane protein (29).

The organization of detergent molecules around membrane proteins can also be studied by monitoring the differential relaxation effects of water-soluble and lipophilic paramagnetic relaxation agents on the resonances of the detergent and the protein. The paramagnetic relaxation agents that have been used are Gd(DOTA) (Gd3+ chelated with 1,4,7,10-tetraazocyclododecane-*N*,*N*′,*N*″,*N*‴-tetraaceticacid), 16-doxylstearc acid (16-DSA), and 5-doxylstearic acid (5-DSA). Gd(DOTA) is water soluble and does not penetrate into the micelle interior. On the other hand, 16-DSA and 5-DSA are incorporated into the detergent micelles. The latter two spin-labels can help to further differentiate between the residues in the central hydrophobic belt and those in the interfacial region. Such studies have been used for characterization of protein–detergent micelle as well as for validation of protein structure for several membrane proteins (8, 12, 14, 25, 57).

3.4. Conformational Transitions of Membrane Proteins in Detergent Micelles

Experimental measurement of molecular dynamics is a major advantage that NMR has over X-ray crystallography. Detailed NMR dynamics measurements are based on the analysis of amide ^{15}N longitudinal (T_1) and transverse (T_2) relaxation times and heteronuclear nuclear Overhauser enhancement ({^{1}H}^{15}N-NOE) which can characterize motions of the protein on timescales ranging from picoseconds to milliseconds. The relaxation parameters measured at one or more fields can be analyzed using model-free formalisms to yield dynamics information on picosecond to millisecond timescales (58, 59). Further direct characterization of exchange processes, in which a protein locally exchanges between two or multiple states, can be carried out through relaxation–dispersion experiments. The relaxation–dispersion experiments yield kinetic rate constants for the exchange, and these can be coupled to functional aspects of the protein (59–61).

As a general feature, membrane protein dynamics are characterized by a relatively rigid transmembrane region combined with flexible loops and turns that connect the TM β-strands of α-helices. However, some finer dynamic elements have also been observed in many cases. For example, existence of a flexibility gradient was first determined from {^{1}H}^{15}N-NOE measurements on OmpA in DPC micelles (17). The central portion of the β-barrel was found to be very rigid, with flexibility increasing towards the two ends. Similar results were obtained for OmpX and PagP in detergent micelles, suggesting that this may be a common feature of membrane proteins.

Backbone amide ^{15}N NMR dynamics measurements of OmpA have been performed at three magnetic fields. A total of nine relaxation data sets were globally analyzed using an extended model-free formalism. The diffusion tensor was found to be prolate axially symmetric with an axial ratio of 5.75, indicating that the protein rotated free from the micelle along its long axis, while along the perpendicular axis, it rotated along with the annular ring of the micelle. The generalized order parameters gradually decreased from the mid-plane towards the two ends of the barrel establishing the flexibility gradient described above. This protein dynamics gradient possibly counteracts the dynamics gradient of the lipids in a bilayer. The extracellular loops displayed large-scale internal motions on the ns timescale, and it was suggested that they were likely to undergo concerted ("sea anemone"-like) motions emanating from their anchoring points on the barrel (62).

For PagP, very interesting dynamics measurements have been made in detergent CYFOS-7. PagP is an 8-stranded β-barrel outer membrane enzyme that catalyzes the transfer of a palmitate chain from phospholipids to the lipid A moiety of lipopolysaccharide. The putative catalytic site of PagP is located on the highly mobile extracellular L1 loop that connects strands A and B. PagP is functionally inactive in LDAO, DPC, and β-OG, perhaps because the detergent molecules occupy the binding site and compete with the phospholipid substrate. PagP is active in CYFOS-7 which has a cyclohexyl ring at the end of its alkyl chain and therefore cannot penetrate into the β-barrel. Two distinct set of peaks for the amide residues are seen in the HSQC spectrum of PagP in CYFOS-7 detergent at 25 C. Based on the chemical shifts of the L1 loop region, the major conformer has been designated to represent the relaxed or "R" form, while the minor conformer has been assigned to represent the tense or "T" form. Structural differences between the two forms are localized mainly to the L1 loop, the β-bulge of strand A, and the adjacent region on strand H. The R → T and T → R conversion rates at 25 C has been found to be $2.8 \pm 0.5/s$ and $6.5 \pm 0.9/s$, respectively, from $^{15}N_{ZZ}$ exchange spectroscopy, based on which the proportion of the minor conformer was determined to be ~30%. The choice of detergent and temperature are both important for the observation of the minor component. Existence of the two conformers and interchange between them in CYFOS-7 are thought to reflect the functioning of this protein in its native membrane environment with the R form facilitating substrate entry and the T form facilitating substrate catalysis (63).

Role of local dynamics in the channel function (13) of the voltage-gated potassium channel from *Streptomyces lividans* (KcsA) has been studied by dynamics measurements in DPC micelles at pH 4 (open state) and pH 7 (closed state), which roughly corresponds to the pH-dependent gating. The functional form of KcsA is a tetramer. Each monomer features two transmembrane helices and an additional, shorter helix in the transmembrane region that forms a selectivity filter for ions, which in terms of size corresponds to the hydration shell of K+ ions in water. Several recent dynamics studies have indicated that two or more conformational states can exist for the selectivity filter and the C-terminal of second transmembrane helix, with rapid interconversion between the two states in ion-conducting open state of the channel. The structural changes observed at different pH values are correlated with the pH-dependent open probability profile of the channel measured by functional tests.

The membrane protein dynamics of the seven transmembrane helical protein pSRII have also been investigated. Amide ^{15}N relaxation rate constants indicate that the helices of this protein are well ordered and pack to a regular bundle. Some of the connecting loop regions are highly flexible and experience motions on the picosecond-to-nanosecond timescale. In the protein core, a limited amount of motion on the microsecond-to-millisecond timescale was observed (12).

4. Conclusion

The belief in high potential of NMR spectroscopy for structural and functional characterization of membrane proteins is being consolidated with every newer and more challenging problem that is being addressed by the courageous investigators in this field. Advanced expression and labelling methods, methyl-protonation of I, L, V, use of RDC and PRE, molecular fragment searches, use of SAXS and SANS data, and functional measurements are likely to become the norm for studying membrane proteins with this technique. The value of NMR spectroscopy would lie in high-fidelity correlation of protein dynamics and mapping of structural interactions to obtain mechanistic information for hundreds of functionally relevant membrane proteins that are being discovered continuously. As all aspects associated with NMR spectroscopy, like the high-field magnets, newer pulse techniques, high-sensitivity multichannel cryogenically cooled probes, and labelling strategies, are undergoing spontaneous development, the technique, therefore, holds a great promise for studying membrane proteins of far greater complexity in the future.

References

1. Wallin E, von Heijne G (1998) Genome-wide analysis of integral membrane proteins from eubacterial, archaean, and eukaryotic organisms. Protein Sci 7:1029–1038
2. Kim H et al (2006) A global topology map of the *Saccharomyces cerevisiae* membrane proteome. Proc Natl Acad Sci USA 103:11142–11147
3. Terstappen GC, Reggiani A (2001) In silico reseach in drug discovery. Trends Pharmacol Sci 22:23–26
4. http://www.drorlist.com/nmr/MPNMR.html
5. Lomize MA et al (2006) OPM: orientations of proteins in membranes database. Bioinformatics 22:623–625
6. Tusnady GE, Dosztanyi Z, Simon I (2004) Transmembrane proteins in protein data bank: identification and classification. Bioinformatics 20:2964–2972
7. Liang B, Tamm LK (2007) Structure of outer membrane protein G by solution NMR spectroscopy. Proc Natl Acad Sci USA 104:16140–16145
8. Hiller S et al (2008) Solution structure of the integral human membrane protein VDAC-1 in detergent micelles. Science 321:1206–1210
9. Berardi MJ et al (2011) Mitochondrial uncoupling protein 2 structure determined by NMR molecular fragment searching. Nature 476:109–113
10. Zhou Y et al (2008) NMR solution structure of the integral membrane enzyme DsbB: functional insights into DsbB catalyzed disulfide bond formation. Mol Cell 31:896–908
11. Van Horn WD et al (2009) Solution nuclear magnetic resonance structure of membrane-integral diacylglycerol kinase. Science 324:1726–1729
12. Gautier A et al (2010) Structure determination of the seven-helix transmembrane receptor sensory rhodopsin II by solution NMR spectroscopy. Nat Struct Mol Biol 17:768–774
13. Baker KA et al (2007) Conformational dynamics of the KcsA potassium channel governs gating properties. Nat Struct Mol Biol 14:1089–1095
14. Reckel S et al (2011) Solution NMR structure of proteorhodopsin. Angew Chem Int Ed 50:11942–11946
15. Kim HJ et al (2009) Recent advances in the application of solution NMR spectroscopy to multispan integral membrane proteins. Prog Nucl Magn Reson Spectrosc 55:335–360

16. Oxenoid K, Chou JJ (2005) The structure of phospholamban pentamer reveals a channel-like architecture in membranes. Proc Natl Acad Sci USA 102:10870–10875
17. Arora A et al (2001) Structure of outer membrane protein A transmembrane domain by NMR spectroscopy. Nat Struct Biol 83:34–338
18. Tugarinov V, Hwang PM, Kay LE (2004) Nuclear magnetic resonance spectroscopy of high-molecular-weight proteins. Annu Rev Biochem 73:107–146
19. Battiste JL, Wagner G (2000) Utilization of site-directed spin labelling and high resolution heteronuclear nuclear magnetic resonance for global fold determination of large proteins with limited nuclear Overhauser effect data. Biochemistry 39:5355–5365
20. Hwang PM et al (2002) Solution structure and dynamics of the outer membrane enzyme PagP by NMR. Proc Natl Acad Sci USA 99:13560–13565
21. MacKenzie KR, Prestegard JH, Engelman DM (1997) A transmembrane helix dimer: structure and implications. Science 276:131–133
22. Arora A, Tamm LK (2001) Biophysical approaches to membrane protein structure determination. Curr Opin Struct Biol 11:540–547
23. Edrington TC et al (2011) Structural basis for the interaction of lipopolysaccharide with outer membrane protein H (OprH) from *Pseudomonas aeruginosa*. J Biol Chem 286:39211–39223
24. Williams KA et al (1996) Structure and dynamics of bacteriophage IKe major coat protein in MPG micelles by solution NMR. Biochemistry 35:5145–5157
25. Roosild TP et al (2005) NMR structure of Mistic, a membrane integrating protein for membrane protein expression. Science 307:1317–1321
26. Krueger-Koplin RD et al (2004) An evaluation of detergents for NMR structural studies of membrane proteins. J Biomol NMR 28:43–57
27. Tieleman DP, van der Spoel D, Berendsen HJ (2000) Molecular dynamics simulations of dodecylphosphocholine micelles at three different aggregate sizes: micellar structure and chain relaxation. J Phys Chem B 104:6380–6388
28. Tamm LK et al (2003) Structure, dynamics and function of the outer membrane protein A (OmpA) and influenza hemagglutinin fusion domain in detergent micelles by solution NMR. FEBS Lett 555:139–143
29. Fernandez C et al (2002) Lipid-protein interactions in DHPC micelles containing the integral membrane protein OmpX investigated by NMR spectroscopy. Proc Natl Acad Sci USA 99:13533–13537
30. Fernandez C, Adeishvili K, Wuthrich K (2001) Transverse relaxation-optimized NMR spectroscopy with the outer membrane protein OmpX in dihexanoyl phosphatidylcholine micelles. Proc Natl Acad Sci USA 98:2358–2363
31. Porcelli F et al (2004) Structure and orientation of pardaxin determined by NMR experiments in model membranes. J Biol Chem 279:45815–45823
32. Zamoon J et al (2003) NMR solution structure and topological orientation of monomeric phospholamban in dodecylphosphocholine micelles. Biophys J 85:2589–2598
33. Gorzelle BM et al (1999) Reconstitutive refolding of diacylglycerol kinase, an integral membrane protein. Biochemistry 38:16373–16382
34. Pervushin K et al (1997) Attenuated T2 relaxation by mutual cancellation of dipole-dipole coupling and chemical shift anisotropy indicates an avenue to NMR structures of very large biological macromolecules in solution. Proc Natl Acad Sci USA 94:12366–12371
35. Fernandez C, Wider G (2003) TROSY in NMR studies of the structure and function of large biological macromolecules. Curr Opin Struct Biol 13:570–580
36. Tugarinov V et al (2002) Four-dimensional NMR spectroscopy of a 723-residue protein: chemical shift assignments and secondary structure of malate synthase g. J Am Chem Soc 124:10025–10035
37. Shen Y et al (2009) TALOS+: a hybrid method for predicting protein backbone torsion angles from NMR chemical shifts. J Biomol NMR 44:213–223
38. Gardner KH, Rosen MK, Kay LE (1997) Global folds of highly deuterated, methyl-protonated proteins by multidimensional NMR. Biochemistry 36:1389–1401
39. Ikura M et al (1990) Detection of nuclear Overhauser effects between degenerate amide proton resonances by heteronuclear three-dimensional NMR spectroscopy. J Am Chem Soc 112:9020–9022
40. Fernandez C et al (2004) NMR structure of the integral membrane protein OmpX. J Mol Biol 336:1211–1221
41. Renault M et al (2009) Solution state NMR structure and dynamics of KpOmpA, a 210 residue transmembrane domain possessing a high potential for immunological applications. J Mol Biol 385:117–130

42. Lau TL et al (2009) The structure of the integrin alphaIIbbeta3 transmembrane complex explains integrin transmembrane signalling. EMBO J 28:1351–1361
43. Yu L et al (2005) Nuclear magnetic resonance structural studies of a potassium channel–charybdotoxin complex. Biochemistry 44:15834–15841
44. Prestegard JH, al-Hashimi HM, Tolman JR (2000) NMR structures of biomolecules using field oriented media and residual dipolar couplings. Q Rev Biophys 33:371–424
45. Bax A (2003) Weak alignment offers new NMR opportunities to study protein structure and dynamics. Protein Sci 12:1–16
46. de Alba E, Tjandra N (2004) Residual dipolar couplings in protein structure determination. Methods Mol Biol 278:89–106
47. Bax A, Grishaev A (2005) Weak alignment NMR: a hawk-eyed view of biomolecular structure. Curr Opin Struct Biol 15:563–570
48. Cierpicki T, Bushweller JH (2004) Charged gels as orienting media for measurement of residual dipolar couplings in soluble and integral membrane proteins. J Am Chem Soc 126:16259–16266
49. Mascioni A, Eggimann BL, Veglia G (2004) Determination of helical membrane protein topology using residual dipolar couplings and exhaustive search algorithm: application to phospholamban. Chem Phys Lipids 132:133–144
50. Meier S, Haussinger D, Grzesiek S (2002) Charged acrylamide copolymer gels as media for weak alignment. J Biomol NMR 24:351–356
51. Ulmer TS et al (2003) Evaluation of backbone proton positions and dynamics in a small protein by liquid crystal NMR spectroscopy. J Am Chem Soc 125:9179–9191
52. Pautsch A, Schulz GE (2000) High-resolution structure of the OmpA membrane domain. J Mol Biol 298:273–282
53. Guntert P, Mumenthaler C, Wuthrich K (1997) Torsion angle dynamics for NMR structure calculation with the new program DYANA. J Mol Biol 271:283–298
54. Brunger AT et al (1998) Crystallography & NMR system: a new software suite for macromolecular structure determination. Acta Crystallogr 54:905–921
55. Schwieters CD et al (2003) The Xplore-NIH NMR molecular structure determination package. J Magn Reson 160:65–73
56. Zweckstetter M, Bax A (2000) Prediction of sterically induced alignment in a dilute liquid crystalline phase: aid to protein structure determination by NMR. J Am Chem Soc 122:3791–3792
57. Hilty C et al (2004) Membrane protein-lipid interactions in mixed micelles studied by NMR spectroscopy with the use of paramagnetic reagents. Chembiochem 5:467–473
58. Lipari G, Szabo A (1982) Model-free approach to the interpretation of nuclear magnetic-resonance relaxation in macromolecules I. Theory and range of validity. J Am Chem Soc 104:4546–4559
59. Lipari G, Szabo A (1982) Model-free approach to the interpretation of nuclear magnetic-resonance relaxation in macromolecules II. Analysis of experimental results. J Am Chem Soc 104:4559–4570
60. Palmer AG, Kroenke CD, Loria JP (2001) Nuclear magnetic resonance methods for quantifying microsecond-to-millisecond motions in biological macromolecules. Methods Enzymol 339:204–238
61. Wang C, Rance M, Palmer AGIII (2003) Mapping chemical exchange in proteins with MW > 50 kD. J Am Chem Soc 125:8968–8969
62. Liang B, Arora A, Tamm LK (2010) Fast-time scale dynamics of outer membrane protein A by extended model-free analysis of NMR relaxation data. Biochim Biophys Acta 1798:68–76
63. Hwang PM, Bishop RE, Kay LE (2004) The integral membrane enzyme PagP alternates between two dynamically distinct states. Proc Natl Acad Sci USA 101:9618–9623
64. Cierpicki T et al (2006) Increasing the accuracy of solution NMR structures of membrane proteins by application of residual dipolar couplings. High-resolution structure of outer membrane protein A. J Am Chem Soc 128:6947–6951

Chapter 18

Nanodiscs as a New Tool to Examine Lipid–Protein Interactions

Mary A. Schuler, Ilia G. Denisov, and Stephen G. Sligar

Abstract

Nanodiscs are self-assembled discoidal fragments of lipid bilayers 8–16 nm in diameter, stabilized in solution by two amphipathic helical scaffold proteins. As stable and highly soluble membrane mimetics with controlled lipid composition and ability to add affinity tags to the scaffold protein, nanodiscs represent an attractive model system for solubilization, isolation, purification, and biophysical and biochemical studies of membrane proteins. In this chapter we overview various approaches to structural and functional studies of different classes of integral membrane proteins such as ion channels, transporters, GPCR and other receptors, membrane enzymes, and blood coagulation cascade proteins which have been incorporated into nanodiscs. We outline the advantages provided by homogeneity, ability to control oligomerization state of the target protein and lipid composition of the bilayer. Special attention is paid to the opportunities afforded by nanodisc system for the detailed studies of the role of different lipid properties and protein–lipid interactions in the functional behavior of membrane proteins.

Key words: Nanodiscs, Lipoproteins, Lipid–protein interactions, GPCR, Cytochrome P450, Tissue factor, Membrane protein solubilization

1. Introduction

The interactions between lipids and proteins are one of the most fundamental processes in living organisms, responsible for critical cellular events ranging from replication, cell division, signaling, and movement. Forming the central coupling responsible for maintaining the functionality of the breadth of proteins, receptors, and enzymes that find their natural home in biological membranes, the fundamental mechanisms of recognition of protein for lipid, and vice versa has been a focal point of biochemical and biophysical investigations for many decades. Complexes of lipids and proteins, such as the various lipoprotein factions, play central roles in the trafficking of important proteins, small molecules, and metabolites

Jörg H. Kleinschmidt (ed.), *Lipid-Protein Interactions: Methods and Protocols*, Methods in Molecular Biology, vol. 974,
DOI 10.1007/978-1-62703-275-9_18,

such as cholesterol and are often implicated in disease states. Recently an engineered lipoprotein particle, termed the nanodisc, a modified form of the human high-density lipoprotein fraction, has served as a membrane miminic for the investigation of membrane proteins and studies of lipid–protein interactions. In this review, we summarize the current knowledge regarding this self-assembling lipid–protein complex and provide examples for its utility in the investigation of a large number of biological systems.

1.1. Development of the Nanodisc System

By the late 1990s, we had acquired one of the new atomic force microscopes (AFM) that could operate under water and had modified it to execute a form of force-volume imaging for soft biological samples. Looking for good candidates for AFM structural studies, we were struck by the pioneering work of Professor Ana Jonas at the University of Illinois in which she described the atherosclerotic process and the role of lipoproteins in reverse cholesterol transport. In this clinical process, the predominant circulating forms of high-density lipoproteins (HDL) are balls of variable size containing cholesterol esters, lipids, and proteins, although a transient form is roughly discoidal and stabilized by the amphipathic apolipoprotein A-I (ApoA-I). Because methods had been developed for reconstituting the larger HDL particles by the removal of detergents from a lipid-enriched ApoA-I fraction isolated from human blood (1, 2), we were able to image these HDL particles on atomically flat mica and show that they were extremely heterogeneous with peaks in the size distribution that corresponded to the lengths of the repeating ApoA-I protein (3). Having previously discovered that many eukaryotic proteins could be expressed in *Escherichia coli* from synthetic genes (4), we undertook the synthesis of a recombinant ApoA-I coding sequence suitable for expression in this bacterial host and the modification of reconstitution procedures for production of uniformly sized nanoparticles. The result of this genetic engineering exercise was a set of "membrane scaffold proteins" (MSPs) that were capable of self-assembly into discoidal phospholipid bilayers (termed nanodiscs) wrapped with an amphipathic helical belt surrounding the alkyl chains on the phospholipids.

The importance of these soluble nanodiscs to studies of membrane proteins was recognized by others in our laboratory struggling with the challenge of making precise biophysical measurements on detergent-solubilized membrane proteins. The wild idea emerged that perhaps if these integral membrane proteins were included in the detergent-solubilized mixture of lipid and MSP, and the same process of self-assembly were initiated to form nanodiscs, the membrane protein target would find its natural home in the bilayer. Thus, the discoidal particle would become a model system wherein the membrane protein would think it was in a native-like membrane environment, maintaining its naturalistic presentation in a monomeric state with full functionality and yet rendered soluble in aqueous solutions via its encircling MSP belt. We tried this multicomponent assembly idea with the simplest membrane proteins that we had on hand, mammalian cytochrome b5 and the cytochrome P450 reductase, which both have large soluble domains but simple helical anchors in the membrane. Working exceedingly well, subsequent fractionations allowed us to separate "free nanodiscs" from "protein-containing nanodiscs" and to quantitate the topology of surface protrusions relative to the nanodisc surface (5, 6). Further experiments led to the development of a general method for assembling membrane proteins into nanodiscs (7–9). While the public response to this bio-assembly system was initially somewhat muted due to concerns that deeply embedded and totally integral membrane proteins would not readily be incorporated, subsequent efforts over the next decade (including that by many independent laboratories) demonstrated the broad applicability of nanodiscs for studying the breadth and complexity of membrane protein complexes. The individual sections of this review highlight current information on the structure and assembly process of nanodiscs as well as the range of proteins and methods accessible with these technologies.

2. The Nanodisc Self-Assembly Process

The nanodisc is composed of one or more lipids and two copies of the amphipathic helical MSP. An initial goal of our research was to provide a structural determination of the lipid and protein factors that make up the nanodisc. Various lengths of the MSP were generated by recombinant manipulation of the MSP sequence and its subsequent expression in *E. coli.* By using a combination of small-angle X-ray scattering (SAXS), size-exclusion chromatography, and direct analysis of the lipid-to-protein ratios, the nanodisc structure was verified (7, 10). Together with the solid-state NMR (11), these results proved that the earlier suggestions that the amphipathic helices of the MSP were directed parallel to the bilayer normal in a "picket fence" arrangement (12) were in error. Rather, the MSP encircles the discoidal membrane fragment in a "double belt" configuration as initially suggested by Wlodawer et al. (13). Data obtained for nanodiscs of different sizes, generated by using varying length MSPs, provided us with overall structures similar to those proposed for nascent discoidal HDL particles. But, importantly, the different length MSP belts, and exact control of the assembly conditions, generated homogeneous and monodisperse populations with a size uniformity not evident in previous biological assemblies made with full-length ApoA-I. In the final derived nanodisc structure, a circular fragment of the phospholipid

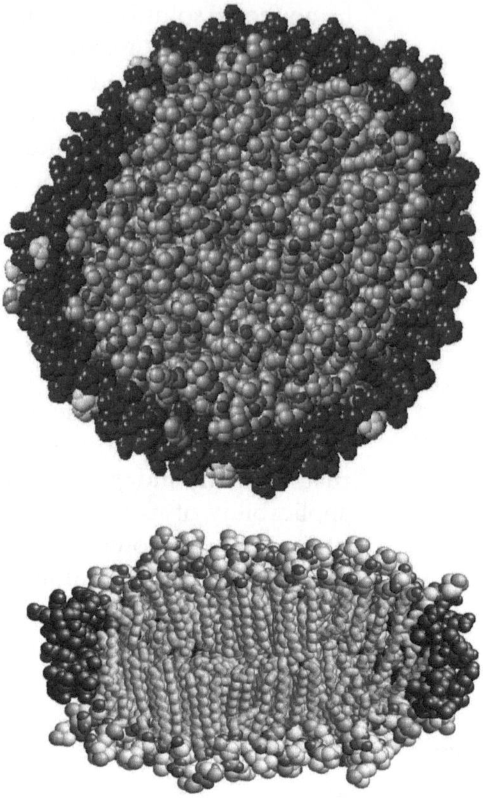

Fig. 1. Results of MD simulations of nanodiscs assembled with DPPC molecules and scaffold protein MSP1D1 (14). Alignment of prolines shown in *yellow* can be seen at the left part of the top view (*top*), while the different conformations of the lipid acyl tails in the middle of nanodisc and at the protein–lipid interface can be seen at the *bottom*.

bilayer is circumscribed by two copies of the MSP unit (7, 10), as illustrated in Fig. 1.

Detailed reviews of these structural and biological investigations using native and altered MSP sequences wherein the number of amphipathic helices punctuated by prolines and glycines is varied have been published (7, 10, 15). This set of MSPs that allow nanodiscs to be assembled with varying sizes is summarized in Table 1. Analogous constructs incorporating hexahistidine or FLAG tags at the amino- or carboxy-terminus of MSP were created, allowing for convenient purification of the MSP protein from *E. coli* lysates or the assembled nanodiscs following incorporation of membrane proteins. An engineered TEV protease or Factor X protease cleavage site was also incorporated, enabling facile removal of the affinity tag should this be desired, as is in cases where the membrane protein target carries the affinity label. This large collection of MSPs of varying composition and the optimization of lipid–protein ratios for the self-assembly process have enabled the generation of uniformly sized nanodiscs ranging in diameter from 9.8 to 17 nm (15).

Further examination of the effect of varying MSP unit lengths, by determinations of the amounts of radiolabeled lipid and protein in the nanodisc fractions on size-exclusion columns, illustrated an interesting trend first described by Denisov et al. (10) which was later expanded (15). It was found that insertion of extra helices in the central portion of the MSP unit (MSP1E1, MSP1E2, MSP1E3) resulted in nanodiscs of increasing size, but deletions of the N-terminal hexahistidine tag and the subsequent 22 amino acids of the MSP protein did not significantly decrease the size of the nanodisc formed. This suggested that the first 22 amino acids are

Table 1
Labels and amino acid sequences of membrane scaffold proteins used for self-assembly of nanodiscs (adapted with permission from (15))

Abbreviation	Description	Amino acid sequence
H1	Helix 1	LKLLDNWDSVTSTFSKLREQLG
H1Δ(1–11)	Truncated helix 1	STFSKLREQLG
H1Δ(1–17)	Truncated helix 1	REQLG
H2	Helix 2	PVTQEFWDNLEKETEGLRQEMS
H3	Helix 3	KDLEEVKAKVQ
H4	Helix 4	PYLDDFQKKWQEEMELYRQKVE
H5	Helix 5	PLRAELQEGARQKLHELQEKLS
H6	Helix 6	PLGEEMRDRARAHVDALRTHLA
H7	Helix 7	PYSDELRQRLAARLEALKENGG
H8	Helix 8	ARLAEYHAKATEHLSTLSEKAK
H9	Helix 9	PALEDLRQGLL
H10	Helix 10	PVLESFKVSFLSALEEYTKKLNTQ
FX	Original N-terminus	MGHHHHHHIEGR
TEV	Modified N-terminus	MGHHHHHHH DYDIPTTENLYFQG

Scaffold protein variant	Abbreviated name[a]	Composition
1	MSP1	FX-H1-H2-H3-H4-H5-H6-H7-H8-H9-H10
2	MSP2	FX-H1-H2-H3-H4-H5-H6-H7-H8-H9-H10-GT-H1-H2-H3-H4-H5-H6-H7-H8-H9-H10
3	MSP1E1	FX-H1-H2-H3-H4-H4-H5-H6-H7-H8-H9-H10
4	MSP1E2	FX-H1-H2-H3-H4-H5-H4-H5-H6-H7-H8-H9-H10
5	MSP1E3	FX-H1-H2-H3-H4-H5-H6-H4-H5-H6-H7-H8-H9-H10
6	MSP1D1	TEV-H1Δ(1–11)-H2-H3-H4-H5-H6-H7-H8-H9-H10

(continued)

Table 1 (continued)

Scaffold protein variant	Abbreviated name[a]	Composition
7	MSP1D2	TEV-H2-H3-H4-H5-H6-H7-H8-H9-H10
8	MSP2N1	TEV-H1Δ(1–11)-H2-H3-H4-H5-H6-H7-H8-H9-H10-GT-H1Δ(1–11)-H2-H3-H4-H5-H6-H7-H8-H9-H10
9	MSP2N2	TEV-H1Δ(1–11)-H2-H3-H4-H5-H6-H7-H8-H9-H10-GT-H2-H3-H4-H5-H6-H7-H8-H9-H10
10	MSP2N3	TEV-H1Δ(1–11)-H2-H3-H4-H5-H6-H7-H8-H9-H10-GT-H1Δ(1–17)-H2-H3-H4-H5-H6-H7-H8-H9-H10

Design of the membrane scaffold proteins. [a]Abbreviated names: (**1**) MSP1, membrane scaffold protein 1 with N-terminal hexahistidine tag, followed by Factor X recognition site.
(**2**) MSP2, fusion of two MSP1 molecules with short glycine–threonine linker and N-terminal hexahistidine tag, followed by Factor X recognition site.
(**3–5**) MSP1E1, MSP1E2, MSP1E3, extended membrane scaffold proteins, obtained via insertion of one (residues 56–77 of original MSP1), two (residues 56–99), or three (residues 56–121) extra 22-mer amphipathic helices after the residue Q55 into original MSP1 (**1**) sequence, as described (10). All have N-terminal hexahistidine tag and Factor X recognition site.
(**6**) MSP1D1, MSP1 with the first 11 N-terminal amino acids removed which contains the N-terminal heptahistidine tag and TEV protease recognition site with linker.
(**7**) MSP1D2, MSP1 with the first 22 N-terminal amino acids removed contains the N-terminal heptahistidine tag and TEV protease recognition site with linker.
(**8–10**) New MSP2 proteins based on MSP1 deletion mutants rather than original MSP1. All three have the same first MSP1D1 molecule, but vary in the extent of truncation in the second MSP.

marginally, if at all, involved in the self-assembly process and the resultant stabilization of the discoidal nanoparticle. Additional truncations from the N-terminal end beyond the first 22 amino acids led to a gradual decrease in lipid–protein ratio accompanied by an increase in the heterogeneity of the assembled nanodiscs and a corresponding appearance of aggregated fractions. Further systematic studies of the lipid–protein ratio in nanodiscs assembled from different MSP constructs have shown that the number of lipids per nanodisc (N) and the number of amino acids in the scaffold protein (M) can be described a simple relationship (10, 15):

$$N_L S = (0.423M - 9.75)^2,$$

where S represents the mean surface area (measured in Å^2) per lipid used to form the nanodisc. The quadratic relationship between the number of lipid molecules per nanodisc and the length of the MSP unit provides a confirmation of the flat two-dimensional morphology of nanodisc particles. The similar sizes of nanodiscs formed using the same MSP unit and varying lipids provide a clear indication that the length of the protein's amphipathic helix is the sole determinant of the diameter of nanodiscs with the different optimized lipid–protein stoichiometry reflecting the different surface areas occupied by individual lipids.

2.1. Molecular Dynamics Simulations of Nanodisc Structure

One structural question relates to the azimuthal alignment of the two MSP belts. The involvement of the N-terminal segment of the MSP protein in binding the lipid bilayer and the alignment of the two MSP units were examined by Molecular Dynamics (MD) simulations of nanodiscs generated from various MSP units listed in Table 1 (14). Simulations of nanodisc formed with MSP1Δ (1–11) and MSP1Δ(1–22) scaffold proteins, which lacking the first 11 or 22 N-terminal residues, respectively, resulted in discoidal structures with little deformation of either the lipid bilayer or the scaffold proteins. Notably, the two MSP units and their interspersed proline residues retain alignment with respect to each other with virtually no misalignment (14). Inspection of the hydrophobic tail groups of the DPPC lipids showed them effectively covered by the encircling MSP1Δ(1–11) or MSP1Δ(1–22) and in all other modified MSP analyzed. Within each amphipathic MSP unit, the majority of the hydrophobic residues are on the interior side of the helices, making contact with the lipid hydrophobic tail and with the hydrophilic residues are oriented outwardly, effectively shielding water from the hydrophobic interior of the nanodisc. These simulations also indicated that the presence of a hydrophilic solvent, such as water, is needed to maintain the discoidal nanodisc structure. It was also noted that the diffusion of sodium ions was unhindered, with no obvious adhesion to the scaffold protein or lipid head groups. Additional MD simulations of nanodiscs with bacteriorhodopsin (bR), an integral membrane protein that contains seven transmembrane helices, indicated that nanodisc structure does not assume any out-of-plane deformation with the addition of bR (14). However, the bR-containing nanodisc adopts a slightly rectangular shape that is likely due to perturbations of the lipid bilayer caused by protein–lipid interactions. These initial models and subsequent improved models have been used to design MSP units optimal for individual membrane proteins and to plan experiments with membrane-embedded or membrane-associated proteins.

The results of these MD simulations also allow determining the average number of *gauche* isomers per lipid molecule in nanodiscs for both the bulk, at the center, as well as the boundary lipids interacting with the MSP (14). Because the initial configuration of the lipid bilayer corresponds to the liquid crystalline state, it includes a high number of *gauche* isomers, the same for the bulk and boundary lipids. As the temperature of simulation is 27 °C (i.e., below the phase transition for DPPC), the bilayer is equilibrated in the gel state as seen by a significant decrease in the *gauche* isomer fraction. But this decrease is much less pronounced for the boundary lipids (i.e., those within 15 Å from the nearest heavy atom of MSP). In this new equilibrium, the number of *gauche* isomers per lipid molecule is two for the bulk lipids and almost five for the boundary lipids with both numbers being in general agreement with those estimated from spectroscopic measurements for gel and liquid crystal phases (16) and from FTIR spectroscopy in reconstituted lipoproteins (17). This statistical analysis of MD simulations does not depend on the assumed thickness of boundary lipid layer, between 13 and 18 Å (where 15 Å is the average thickness of two lipid layers), if the mean area per lipid is 56 $Å^2$. Yet, as can be seen in Fig. 1, the thickness of the bilayer is higher in the center of the nanodisc. While the strict quantitative comparison of all structural parameters derived from MD simulation with experimental results is a challenge, this analysis suggests a specific mode of structural perturbation of the lipid bilayer at the protein–lipid interface in nanodiscs. The main reason for such perturbation is the hydrophobic mismatch between the average thickness of the protein helix of 10–11 Å and the hydrophobic thickness of the lipid bilayer, 14–19 Å for DMPC and DPPC, depending on the phase state (18, 19). An important aspect of such perturbation is also the difference in hydrophobic mismatch below and above the main transition temperature and the thermodynamic perturbation of lipid bilayer phase state by the scaffold protein.

2.2. Nanodisc Bilayer Phase Transition and MSP Interactions

The properties of the lipid bilayer nanoscale fragment containing 120–650 lipids have been studied further by observing the main thermotropic phase transition in nanodiscs formed with DPPC and DMPC (15, 20, 21). For this purpose, nanodiscs provide a unique system for the evaluation of fundamental physical properties of nanoscale lipid bilayers surrounded by protein. This is important as the protein weight percentage in biological membranes varies from 20 to 80 % suggesting that a significant fraction of the lipid is in direct contact with protein (22). Such interfacial, or boundary, lipids are often perturbed by interactions with proteins, and the character and magnitude of this perturbation depend on the protein and lipid structures as well as the bilayer phase state (23, 24). Additionally, biological membranes often contain segregated domains in the submicron size range, which are usually formed with different lipid and protein contents. The properties of these discrete submicron domains are the subject of intensive experimental and theoretical studies (25, 26) that are difficult to interpret because of heterogeneities in natural biological membranes.

To understand the organization of subdomains in monodisperse nanodiscs, we characterized the thermotropic properties of lipid bilayers assembled with the MSP1D1 and MSP1E3 scaffold proteins which correspond to ~140–340 lipids per disc (20, 21). Measurements of SAXS at different temperatures allowed monitoring the size changes of nanodiscs due to the linear thermal expansion of DMPC and DPPC bilayers in nanodiscs. The coefficient of lateral thermal expansion coefficients, $C_T = (1/A)(dA/dT)$ measured for DPPC and DMPC below phase transition, is 0.002–0.003 K^{-1}, almost twice lower than that reported for giant unilamellar vesicles (27). Such differences are attributed partly to the confinement provided by the scaffold protein and partly to the inability of the boundary lipids to undergo the same phase changes as the unrestricted lipids in the center of nanodisc. Consistent with this confinement, the temperatures of the phase transition in nanodiscs are higher than those observed for liposomes by 2–4 K (20). In addition, differential scanning calorimetry (DSC) shows significantly lower values of melting enthalpy and broader peaks for nanoscale lipid bilayers in nanodiscs than those observed for the pure DMPC and DPPC. Comparison of DSC data and also results of Laurdan fluorescence measurements in nanodiscs of different sizes allowed to suggest that the layer of 1 or 2 lipids directly interacting with the scaffold protein at the periphery of nanodisc is virtually excluded from the main thermotropic transition, in agreement with other observations cited in the original study (20), as well as with the MD results described earlier in this review.

In larger nanodiscs, such as the ~17 nm version assembled with an MSP fusion containing 16 amphipathic helices in the scaffold protein and ~650 lipids per disc, the main phase transition from gel to liquid crystalline state of DPPC is indicated as a sharp increase of nanodisc size at 39 °C (15), which is close to the phase transition documented in multilamellar vesicles (28). The lateral thermal expansion coefficient C_T for DPPC in large nanodiscs is 0.005 K^{-1} for the liquid crystalline phase and 0.007 K^{-1} above phase transition, both values slightly higher than reported for smaller nanodiscs (20), but similar to those measured for DMPC liposomes (27). This suggests that the high elasticity and flexibility of longer scaffold proteins in nanodiscs allow for almost unrestricted expansion of the lipid bilayer with increasing temperature. It was also noted that the thermally induced phase transition is reversible in nanodiscs if the temperature of the sample did not exceed 55 °C for more than an hour. In this case the SAXS curves measured for the control (sample kept at 22 °C) and the experimental sample were virtually identical to those measured before heating. This confirms the overall stability of these nanodiscs in solution and provides additional support to the reported method of generation of large lipoprotein particles suitable for solubilization of functional complexes of membrane proteins.

3. Nanodiscs for Analysis of Lipid–Protein Interactions

As reviewed recently by Bayburt and Sligar (29), nanodiscs have provided unprecedented opportunities for studying a wide spectrum of membrane proteins that have been self-assembled into nanodiscs, including many investigations that probe specific lipid

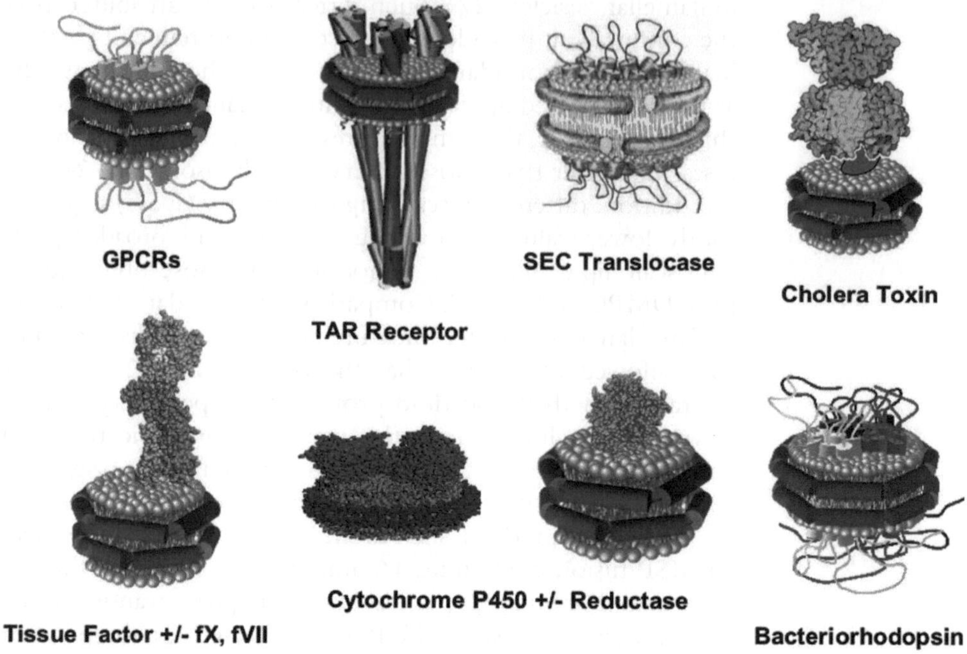

Fig. 2. Representative examples of integral membrane proteins successfully self-assembled into nanodiscs. This include GPCRs (29, 30, 33, 34, 53, 54), multi-subunit chemotactic receptors (48, 49, 55), complex assemblies such as the SEC translocase (40, 41, 56), many different cytochromes P450 both with and without co-incorporated reductase (9, 57–63), human tissue factor and the enzymes of the blood coagulation cascade (64–67), active trimeric bacteriorhodopsin (68), and cholera toxin binding to gangliosides (69, 70).

requirements for activity or the lipid–protein interface. Included in this collection are integral membrane proteins, such as G-protein-coupled receptors (GPCRs) (30–35), ion channels (36–39), transporters (40–45), pores and toxins (46, 47), receptors (48–51), and cytochrome c oxidase (52); each is incorporated into nanodiscs with precisely controlled compositions. A pictorial "Rogues Gallery" of these embedded proteins is represented in Fig. 2.

The large size range of those successfully self-assembled into nanodiscs includes some with only one transmembrane helix and others with as many as 24 transmembrane helices. These assemblies have allowed for the detailed investigation of the effect of bilayer properties (structure, charge, etc.) on the physical, chemical, and biochemical properties in well-defined oligomeric states associated with the functioning of these proteins in biological systems. They have also provided multiple opportunities for studying interactions of proteins with their surrounding lipids (33, 43, 54, 71–73) as well as interactions of membrane lipids and embedded proteins with soluble interacting proteins (i.e., blood coagulation factors) (64, 66, 74). Nanodiscs without any incorporated proteins have been also used as nanoscale mimics of the biomembrane surface and its interactions with soluble proteins and peptides (75).

3.1. Use of Nanodiscs as a Controlled Membrane Surface

Many biochemical processes occur on the surface of a membrane and require specific lipid compositions for full functionality. One example is the blood coagulation cascade that operates to inhibit blood loss following traumatic injury and depends on the recruitment of various serine proteases in the circulatory system to membrane surfaces. The mechanism of molecular recognition for these proteases involves both protein–protein interactions, in which the integral membrane protein tissue factor (TF) provides a surface for association, and protein–lipid interactions, in which the gamma-carboxyglutamate domain (Gla) interacts through bridging calcium ions with negatively charged lipids (45). In the past, difficulties in precisely quantitating the stoichiometry and affinity for these associations arose because typical liposomal preparations loaded with TF and anionic lipids, such as phosphatidylserine (PS), displayed significant mobility in their lipids and proteins that complicated analysis. In more recent analyses, nanodiscs have provided the controlled membrane surface with precise lipid compositions needed to mechanistically appreciate these interactions. The ability to derivatize nanodiscs with various anchors, biotin, nucleic acid, or hexahistidine, presented the means for affixing nanodiscs to sensor surfaces as needed to monitor associations with blood coagulation factors. In one particular study (66), nanodiscs loaded with varying levels of PS in a pure phosphatidylcholine (PC) background were attached to sensor chips suitable for surface plasmon resonance readout. The binding affinity and stoichiometry for human factor VII and factor X as a function of the PS/(PC + PS) mole percent were readily determined. In another study (67) more complex lipid mixtures were accommodated in the nanodisc platform to provide for the molecular analysis of the synergies of particular lipids in the coagulation factor activation cascade. This work led to the ABC "anything but choline" hypothesis describing the ability of smaller head group lipids, such as PG, to dramatically accelerate the initiation of the clotting process. Nanodiscs provide a stable supported lipid bilayer platform and hence have provided an excellent vehicle for structural investigations. For example, they have been used extensively in both solid-state and solution NMR spectroscopy. This has provided a detailed picture of how the Gla domains, calcium, and the anionic lipid head groups of phosphoserine come together to form a high-affinity complex (65, 76, 77).

3.2. Membrane Protein Libraries Using Nanodiscs

The most direct method for incorporating integral membrane proteins into the nanodisc bilayer is to self-assemble it from a detergent-solubilized mixture of MSP and lipid, maintaining the correct overall stoichiometry of the component parts. Problematically, this necessitates the isolation and purification of the target protein, often a process that inactivates membrane proteins as they are separated from their native environment. To circumvent this problem, we demonstrated some years ago that one could detergent-solubilize a native membrane preparation in the presence of excess MSP and lipid whereupon detergent removal would distribute the starting membrane protein population into individual nanodiscs, thus forming a soluble library of the components of the starting preparation (9, 59). In collaboration with Professor W. Kline and Dr. K. Wilcox at Northwestern University, Mr. Marty in our laboratory has advanced this concept to develop a protocol for generating Soluble Membrane Protein Libraries (SMPL) and pioneered their use in experiments seeking a potential intervention targets for Alzheimer's disease. These experiments have focused on identifying the synthetic receptor proteins, which interact with the soluble amyloid beta oligomers (known as amyloid beta-derived diffusible ligands, ADDLs) that are potent neurotoxins involved in Alzheimer's disease (78). In previous studies, the identification of synaptic receptor protein(s) has proven difficult and very controversial (79) since this integral membrane protein is restricted to synaptic junctions. Extending our earlier studies by inserting either purified membrane proteins or mixtures of membrane proteins from over-expression systems (9, 59) into nanodiscs, it is now possible to array the population of membrane proteins from synaptic vesicles into nanodiscs. We are using this concept to isolate and identify ADDL-binding proteins using co-immunoprecipitation capture and Difference Imaging Gel Electrophoresis (DIGE) high-throughput proteomics. This same approach can also be used to isolate the protein–protein binding partners present in cell–cell and cell–viral recognition events and to further define the critical players in cellular communication and viral entry.

A first example using nanodiscs to define the associations between membrane components and soluble interacting proteins (termed the interactome) has been published recently for the membrane-embedded bacterial SecYEG channel, maltose transporter MalFGK$_2$, and membrane integrase YidC (56). Modifications of the lipid composition in the nanodisc during the reconstitution process have proved to be a critical parameter for the successful isolation of SecA as a bona fide SecYEG interactor. This suggests that the nanodisc system may be an excellent tool to test whether specific lipidic environments are critical for protein binding.

3.3. Protein–Protein Interactions Using Nanodiscs

Interactions of human hepatic P450s and their electron transfer partners operate in the complex environment of the endoplasmic reticulum membrane. One important question with these various P450s is the potential role of specific phospholipids in controlling their metabolic activities. Some early investigations with membrane reconstitution systems suggested that trace amounts of phospholipids were required for optimal activity (80), while others suggested that the presence of specific phospholipids might stabilize some kinetically distinguishable conformations of P450s, such as the major drug-metabolizing isoform CYP3A4, and directly modulate its affinity for particular substrates (81, 82). Subsequent recent investigations (71) attempted to define the role of specific phospholipids in maintaining high catalytic turnover.

With the nanodisc system representing a controllable membrane environment, we have an ideal platform for mechanistically assessing the role of PL and other anionic lipids in hepatic drug metabolism. Many of our previous nanodisc assemblies have routinely used mixed lipids ranging in composition from 100 % POPC to 100 % POPS (66), and others have used mixtures of PA, PS, and PE head groups in PC backgrounds to evaluate the role of these in controlling X, Y, and Z (64, 66, 74). Extending these previous experiments to the analysis of P450 systems, we co-assembled human CYP3A4 together with its P450 reductase and varying lipids in nanodiscs and quantitated NADPH-driven testosterone hydroxylation. Our initial analysis found that the steady-state rate of testosterone hydroxylation increases by 40 % in nanodiscs containing 30 % POPS/70 % POPC compared to nanodiscs containing pure POPC and that the overall coupling (defined as the fraction of oxygen used for product formation) increases by 60 %. Even larger effects of the anionic lipid POPS were observed for the CYP3A4 hydroxylation of bromocriptine with the product formation rate and overall coupling increasing by 80 % and 100 %, respectively (Grinkova et al., unpublished results). The molecular origin of this effect is in the process of being evaluated via the careful measurement of the functional properties (spin state, rate of autoxidation, uncoupling, and product formation) of various CYP3A4 states as a function of substrate loading and various anionic lipid compositions. It is possible that there exists a synergy between phospholipids manifested through the increased accessibility of the proteins to the negatively charged head group, as was previously described for the case of blood coagulation factors binding to nanodisc surfaces (66, 67).

3.4. Protein–Lipid Interaction Studies Using Nanodiscs

In addition to these previously mentioned studies, interactions of phosphatidylinositol (PI)-binding proteins with membrane bilayers have been assessed in nanodiscs generated predominantly with DOPC and minor amounts (1.25 %) of PI (75). Using pull-down assays, NMR, and fluorescence polarization methods, dissociation constants for the PI-binding proteins from the nanodisc surface have been measured to be $K_d \sim 30$ nM, a tighter binding to PI in bilayers than previously defined in earlier studies with DPC micelles ($K_d \sim 100$ nM). Control experiments performed with nanodiscs containing no PI show negligible binding, confirming the stoichiometric 1:1 interaction of PI-binding proteins with single molecules of PI incorporated into the lipid bilayer.

Modulations of proteorhodopsin (PR) by different lipid composition have also been assessed in nanodiscs constructed with DMPC or POPC (54). Controlled assembly of PR monomers has provided the opportunity to compare its photophysical properties with the known properties of oligomeric PR in native membranes. Kinetic measurements have shown that the M and N states, which are key intermediates in the photocycle, have longer lifetimes in DMPC bilayers and shorter lifetimes in POPC bilayers. These results have been attributed to the better modeling of physiological conditions in *E. coli* membrane by the unsaturated longer hydrophobic tails of POPC. The authors also suggest that the difference in hydrophobic mismatch between the bilayer thickness and the length of the PR transmembrane helices also plays an important role in the observed variations of the photophysical properties of this protein. It thus appears that the composition of the lipid bilayer is a vital component of the mechanism that controls the photophysical properties of PR and that nanodiscs will be a very useful tool for the further analysis in this field.

Possible effect of anionic lipids on the yield of photoinitiated phosphorylation of monomeric rhodopsin incorporated in POPC/POPS nanodiscs has been studied by Bayburt et al. (33). The results of this work show that the presence of POPS is not necessary for the efficient phosphorylation of rhodopsin by the rhodopsin kinase GRK1. However, arrestin binding was significantly stronger in the presence of 20–50 % POPS and could be substantially inhibited by increasing ionic strength. These results suggest the dominant role of electrostatic repulsion between positively charged arrestin and POPC head groups which could be efficiently decreased by the presence of negatively charged POPS lipids in the nanodisc bilayer or by high ionic strength. Thus, incorporation of monomeric rhodopsin into nanodiscs with different lipid composition allowed to directly prove that the rhodopsin monomers are as functional as oligomers with respect to arrestin binding and phosphorylation and to evaluate the role of anionic lipids in the functional properties of this prototype GPCR.

Further extending the analysis of bR and membrane interactions, the effect of lipid composition on bR unfolding has recently been studied by single-molecule atomic force spectroscopy (73). Unfolding of single bR molecules as a function of applied strength from the AFM tip attached to the terminal amino acid was compared for the native purple membrane fragments and for single bR molecules incorporated into nanodiscs with DMPC. Somewhat unexpectedly, there was no difference between unfolding intermediates, consistently with the similar optical properties observed for bR in nanodiscs and in purple membranes. This pioneering experiment opens a new venue for the direct straightforward measurements of the strength and stability of protein–lipid interactions for the integral membrane proteins.

Using epitope-tagged MSP proteins as a means of immunoprecipitating nanodiscs containing embedded glycosphingolipid ganglioside (GM1) and immobilizing them on sensor chips, tightly bound proteins associated with GM1 have been evaluated by a number of physical means (69, 70). Initial studies with surface plasmon resonance identified several tightly binding proteins in *E. coli* extracts. Subsequent mass-spectroscopic analysis identified one of these as the heat-labile enterotoxin LTB (69). These results demonstrate the effectiveness of approaches combining nanodisc immunoprecipitations with mass-spectroscopy for clarification of proteins interacting with components in membrane surface. Extensions of this can be widely used for studies of protein–protein, protein–small molecule, and protein–lipid interactions (83).

Reconstitution of an ABC transporter MsbA into nanodiscs of different size and composition revealed the details of the functional interactions of the enzyme with the membrane bilayer (43). Enzymatic activity of MsbA varied up to fourfold in nanodiscs assembled with different lipids and also depended on the size of the nanodisc particles. The highest rate was observed in DMPC discs of smaller size, while incorporation into the negatively charged DMPG discs resulted in the lowest ATPAse activity.

Effect of the membrane bilayer on the midpoint potential of human cytochrome P450 CYP3A4 has been measured by potentiometric titration of this heme enzyme incorporated into nanodiscs assembled with zwitterionic POPC lipids (84). For both substrate-free and substrate-bound proteins, the measured absolute values of midpoint potentials turned out to be higher by 50–100 mV than for soluble isozymes CYP101 and CYP102. The same effect of the lipid bilayer has been observed later for the flavoprotein cytochrome P450 reductase (85). In addition, the presence of the negatively charged lipid POPS shifted midpoint potentials of both flavins to the more negative values, thus favoring the more efficient electron transfer to the cytochrome P450.

Application of nanodiscs for the studies of the membrane-active peptides by solution NMR has been recently described for the peptide antibiotic Aam-I from fungi (72). This hydrophobic and poorly soluble in water ^{15}N-labeled hexadecapeptide has been incorporated into nanodiscs formed with various lipids by stepwise titration using stock solution in methanol up to 1:30 peptide-to-lipid molar stoichiometry. Comparison of 2D ^{1}H–^{15}N correlation spectra of Aam-I in nanodiscs assembled with different lipids reveals broadened signals with the estimated rotational correlation time ~40 ns, lower than expected for nanodiscs at the same conditions, i. e., 55–60 ns, indicating the presence of additional mobility of the peptide incorporated into bilayer. At the same time, many signals undergo the selective broadening and disappear from the 2D spectra in the samples made with DMPC and POPC lipids. The detailed comparison of these signals allows for the site-selective assignment of the mobile fragments of Aam-I in nanodiscs with different composition and suggests the presence of several conformational states at equilibrium.

4. Summary and Future Directions

In this brief review we have attempted to provide examples that speak to the utility of the nanodisc system for the investigations of membrane proteins and their interaction with lipids. While many membrane mimetic models can provide critical biochemical, biophysical, and structural input (86–88), the nanodisc system has been shown to be a robust means for stabilizing and investigating protein and lipid interactions. Detailed experiments have allowed one to study the interactions of the membrane proteins with a membrane, the mechanism of action of membrane proteins, and the three-dimensional structure in a native lipid membrane environment. No other experimental system currently allows for such precise control over the environment of membrane proteins. Since nanodiscs contain a small patch of membrane of known composition, they are also ideally suited for studying the properties of bilayers such as lipid phase transitions and the binding and transport of small molecules such as cholesterol. As the critical MSP components are now readily available, many laboratories are now independently using this technology. The future promises to bring numerous additional advances in the use of these lipoprotein particles in structure determination, formulation and delivery of therapeutics, and elucidation of mechanism.

Acknowledgments

Our work is supported by grants from the National Institutes of Health GM33775 and GM31756.

References

1. Jonas A, Kezdy KE, Wald JH (1989) Defined apolipoprotein A-I conformations in reconstituted high density lipoprotein discs. J Biol Chem 264:4818–4824
2. Wald JH, Krul ES, Jonas A (1990) Structure of apolipoprotein A-I in three homogeneous, reconstituted high density lipoprotein particles. J Biol Chem 265:20037–20043
3. Carlson JW, Jonas A, Sligar SG (1997) Imaging and manipulation of high-density lipoproteins. Biophys J 73:1184–1189
4. Beck von Bodman S et al (1986) Synthesis, bacterial expression, and mutagenesis of the gene coding for mammalian cytochrome b5. Proc Natl Acad Sci U S A 83:9443–9447
5. Bayburt TH, Carlson JW, Sligar SG (2000) Single molecule height measurements on a membrane protein in nanometer-scale phospholipid bilayer disks. Langmuir 16:5993–5997
6. Bayburt TH, Sligar SG (2002) Single-molecule height measurements on microsomal cytochrome P450 in nanometer-scale phospholipid bilayer disks. Proc Natl Acad Sci U S A 99:6725–6730
7. Bayburt TH, Grinkova YV, Sligar SG (2002) Self-assembly of discoidal phospholipid bilayer

nanoparticles with membrane scaffold proteins. Nano Lett 2:853–856
8. Bayburt TH, Sligar SG (2003) Self-assembly of single integral membrane proteins into soluble nanoscale phospholipid bilayers. Protein Sci 12:2476–2481
9. Civjan NR et al (2003) Direct solubilization of heterologously expressed membrane proteins by incorporation into nanoscale lipid bilayers. Biotechniques 35:556–558, 560, 562–563
10. Denisov IG et al (2004) Directed Self-assembly of monodisperse phospholipid bilayer Nanodiscs with controlled size. J Am Chem Soc 126:3477–3487
11. Li Y et al (2006) Structural analysis of nanoscale self-assembled discoidal lipid bilayers by solid-state NMR spectroscopy. Biophys J 91:3819–3828
12. Phillips JC et al (1997) Predicting the structure of apolipoprotein A-I in reconstituted high-density lipoprotein disks. Biophys J 73:2337–2346
13. Wlodawer A et al (1979) High-density lipoprotein recombinants: evidence for a bicycle tire micelle structure obtained by neutron scattering and electron microscopy. FEBS Lett 104:231–235
14. Shih AY et al (2005) Molecular dynamics simulations of discoidal bilayers assembled from truncated human lipoproteins. Biophys J 88:548–556
15. Grinkova YV, Denisov IG, Sligar SG (2010) Engineering extended membrane scaffold proteins for self-assembly of soluble nanoscale lipid bilayers. Protein Eng Des Sel 23:843–848
16. Mendelsohn R et al (1989) Quantitative determination of conformational disorder in the acyl chains of phospholipid bilayers by infrared spectroscopy. Biochemistry 28:8934–8939
17. Lins L et al (1993) Structure of the apolipoprotein A-IV/lipid discoidal complexes: an attenuated total reflection polarized Fourier transform infrared spectroscopy study. Biochim Biophys Acta 1149:267–277
18. Nagle JF, Tristram-Nagle S (2000) Structure of lipid bilayers. Biochim Biophys Acta 1469:159–195
19. Tristram-Nagle S, Nagle JF (2004) Lipid bilayers: thermodynamics, structure, fluctuations, and interactions. Chem Phys Lipids 127:3–14
20. Denisov IG et al (2005) Thermotropic phase transition in soluble nanoscale lipid bilayers. J Phys Chem B 109:15580–15588
21. Shaw AW, McLean MA, Sligar SG (2004) Phospholipid phase transitions in homogeneous nanometer scale bilayer discs. FEBS Lett 556:260–264
22. Gennis RB (1989) Biomembranes. Molecular structure and function. Springer, New York
23. Marsh D (2008) Protein modulation of lipids, and vice-versa, in membranes. Biochim Biophys Acta 1778:1545–1575
24. Marsh D (2010) Electron spin resonance in membrane research: protein-lipid interactions from challenging beginnings to state of the art. Eur Biophys J 39:513–525
25. Simons K, Gerl MJ (2010) Revitalizing membrane rafts: new tools and insights. Nat Rev Mol Cell Biol 11:688–699
26. He HT, Marguet D (2011) Detecting nanodomains in living cell membrane by fluorescence correlation spectroscopy. Annu Rev Phys Chem 62:417–436
27. Needham D, Evans E (1988) Structure and mechanical properties of giant lipid (DMPC) vesicle bilayers from 20 degrees C below to 10 degrees C above the liquid crystal-crystalline phase transition at 24 degrees C. Biochemistry 27:8261–8269
28. Cevc G (ed) (1993) Phospholipids handbook. Marcel Dekker, Inc, New York
29. Bayburt TH, Sligar SG (2010) Membrane protein assembly into nanodisks. FEBS Lett 584:1721–1727
30. Leitz AJ et al (2006) Functional reconstitution of β2-adrenergic receptors utilizing self-assembling nanodisc technology. Biotechniques 40:601–602, 604, 606, 608, 610, 612
31. Whorton MR et al (2007) A monomeric G protein-coupled receptor isolated in a high-density lipoprotein particle efficiently activates its G protein. Proc Natl Acad Sci U S A 104:7682–7687
32. Whorton MR et al (2008) Efficient coupling of transducin to monomeric rhodopsin in a phospholipid bilayer. J Biol Chem 283:4387–4394
33. Bayburt TH et al (2011) Monomeric rhodopsin is sufficient for normal rhodopsin kinase (GRK1) phosphorylation and arrestin-1 binding. J Biol Chem 286:1420–1428
34. Knepp AM et al (2011) Direct measurement of thermal stability of expressed CCR5 and stabilization by small molecule ligands. Biochemistry 50:502–511
35. Goldsmith BR et al (2011) Biomimetic chemical sensors using nanoelectronic readout of olfactory receptor proteins. ACS Nano 5:5408–5416
36. Raschle T et al (2009) Structural and functional characterization of the integral membrane

protein VDAC-1 in lipid bilayer Nanodiscs. J Am Chem Soc 131:17777–17779

37. Yu T-Y et al (2012) Solution NMR spectroscopic characterization of human VDAC-2 in detergent micelles and lipid bilayer nanodiscs. Biochim Biophys Acta. doi:10.1016/j.bbamem.2011.11.012
38. Shenkarev ZO et al (2010) Lipid-protein nanodiscs as reference medium in detergent screening for high-resolution NMR studies of integral membrane proteins. J Am Chem Soc 132:5628–5629
39. Shenkarev ZO et al (2010) NMR structural and dynamical investigation of the isolated voltage-sensing domain of the potassium channel KvAP: implications for voltage gating. J Am Chem Soc 132:5630–5637
40. Alami M et al (2007) Nanodiscs unravel the interaction between the SecYEG channel and its cytosolic partner SecA. EMBO J 26:1995–2004
41. Dalal K, Duong F (2010) Reconstitution of the SecY translocon in nanodiscs. Methods Mol Biol 619:145–156
42. Dalal K et al (2009) Structure, binding, and activity of Syd, a SecY-interacting protein. J Biol Chem 284:7897–7902
43. Kawai T et al (2011) Catalytic activity of MsbA reconstituted in nanodisc particles is modulated by remote interactions with the bilayer. FEBS Lett 585:3533–3537
44. Ritchie TK, Kwon H, Atkins WM (2011) Conformational analysis of human ATP-binding cassette transporter ABCB1 in lipid nanodiscs and inhibition by the antibodies MRK16 and UIC2. J Biol Chem 286:39489–39496
45. Alvarez FJ, Orelle C, Davidson AL (2010) Functional reconstitution of an ABC transporter in nanodiscs for use in electron paramagnetic resonance spectroscopy. J Am Chem Soc 132:9513–9515
46. Dalal K, Duong F (2011) The SecY complex: conducting the orchestra of protein translocation. Trends Cell Biol 21:506–514
47. Katayama H et al (2010) Three-dimensional structure of the anthrax toxin pore inserted into lipid nanodiscs and lipid vesicles. Proc Natl Acad Sci U S A 107:3453–3457, S3453/3451-S3453/3453
48. Boldog T et al (2006) Nanodiscs separate chemoreceptor oligomeric states and reveal their signaling properties. Proc Natl Acad Sci U S A 103:11509–11514
49. Li M, Hazelbauer GL (2011) Core unit of chemotaxis signaling complexes. Proc Natl Acad Sci U S A 108:9390–9395
50. Glueck JM, Koenig BW, Willbold D (2011) Nanodiscs allow the use of integral membrane proteins as analytes in surface plasmon resonance studies. Anal Biochem 408:46–52
51. Mi L-Z et al (2008) Functional and structural stability of the epidermal growth factor receptor in detergent micelles and phospholipid nanodiscs. Biochemistry 47:10314–10323
52. Näsvik Öjemyr L et al (2012) Reconstitution of respiratory oxidases in membrane nanodiscs for investigation of proton-coupled electron transfer. FEBS Lett 586(5):640–645
53. Bayburt TH et al (2007) Transducin activation by nanoscale lipid bilayers containing one and two rhodopsins. J Biol Chem 282:14875–14881
54. Ranaghan MJ et al (2011) Green proteorhodopsin reconstituted into nanoscale phospholipid bilayers (nanodiscs) as photoactive monomers. J Am Chem Soc 133:18318–18327
55. Boldog T, Li M, Hazelbauer GL (2007) Using nanodiscs to create water-soluble transmembrane chemoreceptors inserted in lipid bilayers. Methods Enzymol 423:317–335
56. Zhang XX et al (2012) Nanodiscs and SILAC-based mass spectrometry to identify a membrane protein interactome. J Proteome Res 11:1454–1459
57. Baas BJ, Denisov IG, Sligar SG (2004) Homotropic cooperativity of monomeric cytochrome P450 3A4 in a nanoscale native bilayer environment. Arch Biochem Biophys 430:218–228
58. Denisov IG, Sligar SG (2011) Cytochromes P450 in nanodiscs. Biochim Biophys Acta 1814:223–229
59. Duan H et al (2004) Co-incorporation of heterologously expressed Arabidopsis cytochrome P450 and P450 reductase into soluble nanoscale lipid bilayers. Arch Biochem Biophys 424:141–153
60. Duan H, Schuler MA (2006) Heterologous expression and strategies for encapsulation of membrane-localized plant P450s. Phytochem Rev 5:507–523
61. Frank DJ, Denisov IG, Sligar SG (2011) Analysis of heterotropic cooperativity in cytochrome P450 3A4 using α-naphthoflavone and testosterone. J Biol Chem 286:5540–5545
62. Grinkova YV, Denisov IG, Sligar SG (2010) Functional reconstitution of monomeric CYP3A4 with multiple cytochrome P450 reductase molecules in nanodiscs. Biochem Biophys Res Commun 398:194–198
63. Grinkova YV et al (2008) The ferrous-oxy complex of human aromatase. Biochem Biophys Res Commun 372:379–382

64. Morrissey JH et al (2008) Blood clotting reactions on nanoscale phospholipid bilayers. Thromb Res 122:S23–S26
65. Jonas A, Krajnovich DJ (1978) Effect of cholesterol on the formation of micellar complexes between bovine A-I apolipoprotein and L-alpha-dimyristoyl-phosphatidylcholine. J Biol Chem 253:5758–5763
66. Shaw AW et al (2007) The local phospholipid environment modulates the activation of blood clotting. J Biol Chem 282:6556–6563
67. Jonas A, Maine GT (1979) Kinetics and mechanism of phosphatidylcholine and cholesterol exchange between single bilayer vesicles and bovine serum high-density lipoprotein. Biochemistry 18:1722–1728
68. Bayburt TH, Grinkova YV, Sligar SG (2006) Assembly of single bacteriorhodopsin trimers in bilayer nanodiscs. Arch Biochem Biophys 450:215–222
69. Borch J, Roepstorff P, Moeller-Jensen J (2011) Nanodisc-based co-immunoprecipitation for mass spectrometric identification of membrane-interacting proteins. Mol Cell Proteomics 10:O110 006775, 006779
70. Borch J et al (2008) Nanodiscs for immobilization of lipid bilayers and membrane receptors: kinetic analysis of cholera toxin binding to a glycolipid receptor. Anal Chem 80:6245–6252
71. Jonas A (1976) Microviscosity of lipid domains in human serum lipoproteins. Biochim Biophys Acta 486:10–22
72. Shenkarev ZO et al (2009) Lipid-protein nanodiscs: possible application in high-resolution NMR investigations of membrane proteins and membrane-active peptides. Biochemistry (Mosc) 74:756–765
73. Zocher M et al (2012) Single-molecule force spectroscopy from nanodiscs: an assay to quantify folding, stability, and interactions of native membrane proteins. ACS Nano 6:961–971
74. Morrissey JH et al (2009) Protein-membrane interactions: blood clotting on nanoscale bilayers. J Thromb Haemost 7:169–172
75. Kobashigawa Y et al (2011) Phosphoinositide-incorporated lipid-protein nanodiscs: a tool for studying protein-lipid interactions. Anal Biochem 410:77–83
76. Boettcher JM et al (2011) Atomic view of calcium-induced clustering of phosphatidylserine in mixed lipid bilayers. Biochemistry 50:2264–2273
77. Morrissey JH et al (2010) Protein-phospholipid interactions in blood clotting. Thromb Res 125:S23–S25
78. Lambert MP et al (1998) Diffusible, nonfibrillar ligands derived from Abeta42 are potent central nervous system neurotoxins. Proc Natl Acad Sci U S A 95:6448–6453
79. Wilcox K et al (2011) Abeta oligomer-induced synapse degeneration in Alzheimer's disease. Cell Mol Neurobiol 31:939–948
80. Jonas A (1986) Reconstitution of high-density lipoproteins. Methods Enzymol 128:553–582
81. Jonas A, Jung RW (1975) Fluidity of the lipid phase of bovine serum high density lipoprotein from fluorescence polarization measurements. Biochem Biophys Res Commun 66:651–657
82. Matz CE, Jonas A (1982) Micellar complexes of human apolipoprotein A-I with phosphatidylcholines and cholesterol prepared from cholate-lipid dispersions. J Biol Chem 257:4535–4540
83. Marty MT, Das A, Sligar SG (2012) Ultra-thin layer MALDI mass spectrometry of membrane proteins in nanodiscs. Anal Bioanal Chem 402:721–729
84. Das A, Grinkova YV, Sligar SG (2007) Redox potential control by drug binding to cytochrome P 450 3A4. J Am Chem Soc 129:13778–13779
85. Das A, Sligar SG (2009) Modulation of the cytochrome P450 reductase redox potential by the phospholipid bilayer. Biochemistry 48:12104–12112
86. Jonas A, Hesterberg LK, Drengler SM (1978) Incorporation of excess cholesterol by high density serum lipoproteins. Biochim Biophys Acta 528:47–57
87. Raschle T et al (2010) Nonmicellar systems for solution NMR spectroscopy of membrane proteins. Curr Opin Struct Biol 20:471–479
88. Serebryany E, Zhu GA, Yan ECY (2012) Artificial membrane-like environments for in vitro studies of purified G-protein coupled receptors. Biochim Biophys Acta 1818:225–233

Chapter 19

The Simulation Approach to Lipid–Protein Interactions

Teresa Paramo, Diana Garzón, Daniel A. Holdbrook, Syma Khalid, and Peter J. Bond

Abstract

The interactions between lipids and proteins are crucial for a range of biological processes, from the folding and stability of membrane proteins to signaling and metabolism facilitated by lipid-binding proteins. However, high-resolution structural details concerning functional lipid/protein interactions are scarce due to barriers in both experimental isolation of native lipid-bound complexes and subsequent biophysical characterization. The molecular dynamics (MD) simulation approach provides a means to complement available structural data, yielding dynamic, structural, and thermodynamic data for a protein embedded within a physiologically realistic, modelled lipid environment. In this chapter, we provide a guide to current methods for setting up and running simulations of membrane proteins and soluble, lipid-binding proteins, using standard atomistically detailed representations, as well as simplified, coarse-grained models. In addition, we outline recent studies that illustrate the power of the simulation approach in the context of biologically relevant lipid/protein interactions.

Key words: Molecular dynamics, Simulation, Molecular modelling, Lipid-binding proteins, Membrane proteins, Lipids, Coarse-grained models

1. Introduction

Understanding the role of interactions between lipids and proteins, and predicting the sites of such interactions, is a key challenge in modern structural biology. Membrane proteins constitute ~20–30% of most genomes (1), and it has been estimated that ~50% of future drug targets are membrane proteins (2). They are of great biological and biomedical importance, as reflected by their involvement in a variety of physiological functions, including transport, energy transduction, cellular signaling, etc. It is therefore essential that we understand the nature of interactions between membrane proteins and their surrounding biological membrane milieu; lipid–protein interactions are important in stabilizing the structure and assembly of membrane proteins in their native

Jörg H. Kleinschmidt (ed.), *Lipid-Protein Interactions: Methods and Protocols*, Methods in Molecular Biology, vol. 974, DOI 10.1007/978-1-62703-275-9_19, © Springer Science+Business Media New York 2013

environment (3), whilst specific lipid–protein interactions have been shown to be essential for their function (4). In addition, such interactions are vital for the function of lipid-binding protein domains, which are specialized for the recognition of specific lipid molecules and serve to carry out a broad range of transfer, metabolism, and signaling processes (5, 6).

Our knowledge of the native interactions between membrane proteins and the surrounding lipid environment is limited at the atomic level (7), with only ~300 unique structures currently known (see, e.g., http://blanco.biomol.uci.edu/mpstruc/listAll/list). This small number (relative to their genomic representation and biomedical importance) is largely due to difficulties with expression, purification, and crystallization (8). For both soluble and membrane proteins of known structure, it is frequently difficult to observe bound lipids due to their poorly resolved electron density (4). Moreover, exposure to nonnative detergent or precipitant molecules prior to crystallization means that native lipid molecules are often lost from exposed protein surfaces or cavities (9).

Because of the difficulties outlined above associated with experimental characterization of protein–lipid interactions, computational simulation-based methods are now routinely utilized to supplement our interpretation of data from structural biology. In particular, molecular dynamics (MD) uses a physics-based framework to describe the dynamics of a biomolecular system in a model of a physiologically relevant environment (10–12). The typical starting point is a time- and space-averaged experimental protein structure obtained under in vitro conditions. Such simulations, in which all atoms, including protein, lipid, and solvent, are represented in atomistic or near-atomistic detail, allow one to extract the maximum possible information from the limited structural data available for lipid–protein interactions, as well as to pose questions concerning the role of modelled interactions in functional mechanisms that may be subsequently tested experimentally (13–21). In addition, coarse-grained modelling approaches to biomolecular systems, in which the number of degrees of freedom in the system is reduced, have recently gained popularity (22). Coarse-grained simulations enable longer time- and length-scales to be accessed, at the cost of reduced accuracy, and have proven useful, e.g., for the spontaneous self-assembly of protein/lipid complexes. Reverse-mapping of such models back to the atomistic level of detail then allows refinement of the interactions by including the relevant chemical moieties. Thus molecular dynamics simulations provide an ideal complement to biophysical experimental methods for the study of protein–lipid interactions.

With the growing, worldwide accessibility to large-scale, high-performance computing facilities, along with continuing advances in simulation algorithms highly optimized for parallelized computation, it is now relatively straightforward to perform simulations of systems with dimensions of tens of nanometers (representing, e.g., a patch of membrane containing a protein) over biologically relevant timescales of fractions of microseconds or beyond. In this chapter, we discuss the major considerations for running an MD simulation of a protein/lipid system. We introduce the theoretical framework of the simulation method, before describing the common approaches for setting up atomistic and coarse-grained models of both membrane proteins and soluble lipid-binding protein domains. We also detail recent applications which demonstrate the power of using MD simulations to further our knowledge of lipid–protein interactions.

2. Materials

1. To carry out a simulation of a protein/lipid system, an initial configuration of atomic coordinates must be generated. This requires consideration of the protein structure, bound ligand molecules, as well as the nature of the desired local lipid/solvent environment. Protein coordinates may be obtained from the Protein Data Bank (http://www.rcsb.org), derived from X-ray crystallographic or NMR sources, or alternatively may be modelled on the basis of homologous structures, using, e.g., Modeller (see Note 1).
2. When using experimentally derived coordinates, it may be necessary to "repair" parts of the molecule, for example, to replace the wild-type coordinates of experimentally introduced mutations or to introduce crystallographically unresolved atoms or entire amino acids due to mobility of, e.g., unstructured loops. For small modifications of the protein structure, the simulation package itself (or a standard molecular modelling program) will likely be sufficient. For more complicated problems, homology modelling approaches and/or ab initio loop building may be necessary (see Note 1). In the case of missing protein termini, it may suffice to use "capping" groups which mimic the adjacent, missing residues. Finally, it may be of interest to retain experimentally observed, nonprotein molecules, such as crystallographic water molecules and ions, or ligands such as lipids that could be functionally or structurally important. This, of course, must be adjudged by the user on a case-by-case basis.
3. For protein/membrane systems, initial coordinates for a lipid bilayer are required. A number of groups have made available online previously built and equilibrated membrane models for this purpose, covering a range of lipid compositions and topologies for different force fields (see Note 2). In addition, the

CHARMM membrane builder server may be used to create mixed lipid systems (http://www.charmm-gui.org/?doc=input/membrane). Downloaded equilibrated membrane coordinates may be directly utilized for setting up a membrane/protein system, assuming that they are of sufficient size, or as the starting point for building and equilibrating a new lipid bilayer with different dimensions. In this context, it is important to bear in mind that MD simulations typically utilize periodic boundaries to avoid "edge effects"; therefore the membrane must be large enough to avoid significant nonbonded interactions between the protein and its "periodic image."

4. Once a particular system configuration has been constructed from the initial protein/lipid coordinates (described in Subheading 3), it is necessary to calculate the potential energy (and from the negative gradient, the force) on all atoms in the system. In the classical simulation approach, each atom in a system is treated as a van der Waals sphere with a point charge. The potential energy of the system may then be described as a series of pairwise equations that approximate the bonded and nonbonded chemical interactions. The former is typically composed of harmonic springs approximating bond length and bond angle terms, along with sinusoidal functions to describe dihedral angles. The latter normally consists of the Lennard-Jones 6–12 potential to represent van der Waals interactions, which includes terms for both attractive dispersion forces and for the unfavorable energy of atomic "overlap" at short distances, and electrostatic interactions between partial charges centered on each atom, described using Coulomb's law. This set of equations is fairly standard and efficiently implemented in a number of commonly used simulation packages (see step 7).
5. The potential energy function of the system is completed by the force field, which consists of a series of reference parameters (equilibrium molecular properties and associated force constants) that have been derived in a self-consistent manner. This latter point is key: there are a number of ways of parameterizing a force field, but typically one force field may not be arbitrarily "mixed" with another. Historically, individual simulation force fields have tended to be chosen for their accuracy in representing specific groups of biomolecules. However, continuing refinements have led to a number of highly accurate force fields applicable to a whole range of molecular systems. In the context of biomolecular systems, some of the main force fields utilized include versions of the GROMOS (23–26), CHARMM (27–29), OPLS (30), and AMBER parameter sets (31). For membrane simulations, the Berger force field has been particularly widely used (32) due to its reproduction of

experimentally determined bilayer properties (such as bilayer normal density profiles, area per lipid headgroup, and order parameters) and its capacity to be combined with a number of commonly used protein parameter sets. Alternatively, CHARMM lipid parameter sets have also been widely utilized (27, 29); a major advantage here is that they have been developed in a consistent manner with the parameters for proteins and other biomolecules in the force field. In some simulation studies, the original CHARMM27 lipid parameters led to incorrect lipid surface area values and the erroneous formation of gel-phase membranes, necessitating the application of an artificial surface tension term (see Note 3), but improvements in the more recent CHARMM36 force field have corrected these issues. Moreover, the continuous CHARMM force field optimization (including, e.g., for carbohydrates and small, drug-like molecules (28)) by the MacKerell group means that a variety of heterogeneous biomolecular systems may be accurately modelled (see Note 4).

6. Instead of atomistic approaches, coarse-grained (CG) simulation models may be used, in which sets of atoms are grouped together and adopt the collective physiochemical properties of their constituent atoms. At the cost of detail, the reduced degrees of freedom and softer potentials involved in CG models lead to orders of magnitude increase in the time- and length-scales that may typically be accessed (18, 22, 33). CG models may be considered an extreme form of the "united atom" force fields already in common use (26), which group together nonpolar hydrogens with their associated heavy atoms (i.e., a methyl group is approximated as a single particle) and lead to significant reduction in system size with little loss of accuracy. A simple lipid CG model developed by Marrink and coworkers has become popular and in general groups together approximately four heavy atoms into single, coarse-grained particles (34, 35). This model enables the rapid calculation of lipid simulations spanning microsecond timescales. Based on the model, several compatible extensions to CG peptides and proteins have been developed (36–40). A similar chemical "philosophy" to that used for the lipid model has generally been adopted; thus, the protein backbone is normally represented as one CG particle, with side chains described by between one and three particles. In common with all these models, by the nature of the "coarseness" of the representation, higher-order structure may not be reliably reproduced without the use of structural restraints (e.g., via distance restraints or elastic networks), thus limiting the degree to which CG simulations may be used to investigate protein conformational changes.

7. Once a suitable force field has been selected, an energy minimization (EM) protocol (typically a standard mathematical optimization such as the method of steepest descents) is required to relax the system and remove "bad contacts" in the context of the chosen force field. Subsequently, an MD simulation trajectory may be generated. In MD, the forces on all atoms in the system are calculated, and Newton's laws of motion are subsequently used to calculate new accelerations, velocities, and positions; this is repeated over small time steps to generate a trajectory (see Note 5). To perform both EM and MD, a number of highly efficient, freely available, and well-supported codes are available (41–44). The choice of which code to use is to a certain extent dependent upon personal preference; factors to consider include user-friendliness, code scalability, and requirement for algorithms beyond standard simulation methods, such as biased sampling or free-energy calculations. Traditionally, MD packages have been compatible with only one or a few parameter sets, but more recently, popular packages such as GROMACS, NAMD, and AMBER provide the capability to choose from several popular force fields. Various other issues must be considered when performing MD simulations including, for example, treatment of long-range electrostatics (see Note 6) and appropriate simulation ensembles (see Note 3).
8. In principle, a simulation trajectory may be calculated on a desktop computer. However, most common simulation packages enable one to efficiently parallelize a simulation by splitting it across multiple processors, meaning that the maximum available computational power may be utilized. This may, for example, take the form of widely distributed computing networks, although the significant inter-processor communication necessary for efficiently parallelized MD calculations represents a major bottleneck in speedup. This may be at least partially circumvented by the use of in-house compute clusters or large-scale, high-performance, remote supercomputing facilities. Furthermore, recent advances in algorithms and hardware provide further routes to speedup, as illustrated by the repurposing of graphics processing units (45) for simulation and the development by D. E. Shaw Research of software/supercomputer solutions specialized for high-performance MD calculation (46).

3. Methods

3.1. Lipid/Protein System Setup

Once the coordinates for the protein model have been prepared, it is necessary to embed them in an appropriate lipid environment, prior to solvation (see Note 7). The slow dynamics associated with the exchange of lipids (47, 48) means that care must be taken to correctly position a protein within a model membrane or mimetic environment, to ensure that biologically meaningful results are obtained from the subsequent production simulation. Several approaches are outlined below for building biologically or experimentally relevant systems.

3.1.1. Insertion of a Membrane Protein Within a Lipid Bilayer

The basic principle of constructing a membrane system is to overlay the protein onto previously built bilayer coordinates, remove overlapping lipid molecules, and subsequently equilibrate the system (though some earlier techniques involved assembling lipid molecules around the protein instead (49)).

1. Choosing the initial location for the protein may be performed "by eye," using as a guide the match between the physiochemical properties of the protein surface and the membrane environment (in some cases aided by the presence of any bound lipid molecules co-crystallized with the protein). The pattern of hydrophobicity and hydrophilicity along the transmembrane protein axis may be matched up to the different parts of the bilayer, facilitated in particular by the characteristic, amphipathic "girdles" of aromatic amino acids which are thought to "anchor" themselves to the membrane lipid/headgroup interfacial region (50, 51).
2. Alternatively, empirical approaches may also be used to facilitate positioning, such as the Orientations of Proteins in Membranes (OPM) database (52), which predicts the location based on the transfer energy of accessible amino acids between water and the membrane core (http://opm.phar.umich.edu/).
3. Once an initial protein/bilayer system has been prepared, it is likely that the interaction interface between the transmembrane protein surface and the surrounding layer of lipids is suboptimal, particularly in the case of membrane proteins with complex and irregular shapes. Various approaches are available to "relax" this interaction interface whilst minimizing the degree to which the membrane is disrupted. In the method of Faraldo-Gómez et al. (53), weak repulsive forces are applied to nearby lipid molecules perpendicular to the solvent-accessible protein surface during multiple short MD simulations, resulting in a volume adapted to the protein surface with only minimal perturbation of the existing bilayer structure; thus, a better match between the protein surface and membrane cavity is formed in comparison with similar methods which only form a cylindrical cavity (54). The method has recently been extended to incorporate chemical specificity at the protein/lipid interface (via use

of an implicit protein grid-based force field), and the code may be used alongside any existing MD package (55).

4. An alternative, straightforward approach for membrane/protein embedding developed by Wolf et al. involves growing an initially "contracted" protein back to its correct size in the membrane plane whilst pushing lipids away during a short MD simulation (56), and yet more methods are based on the use of a predefined grid (57).
5. Finally, although a number of methods are available for fixing a poorly packed protein/lipid interface, with increasing computational power and faster algorithms, it is also now sometimes possible to simply use an extended "equilibration period." In this approach, positional restraints are applied to the protein coordinates to prevent structural drift during an extended MD simulation, during which local lipid and solvent molecules can relax freely around the protein surface. Subsequently, protein restraints may be gradually released, prior to production MD. In fact, it is recommended that at least a few nanoseconds of position restrained MD (or at least a "thermalization" protocol) is performed prior to any biomolecular simulation, irrespective of the method of setup, to ensure that the system is equilibrated at the temperature of interest.
6. Equilibration of a membrane system during restrained MD may be adjudged by following the evolution of bilayer structural properties such as the area per lipid within the membrane plane. The importance of careful equilibration is illustrated by a recent study of minor pseudopilin membrane proteins associated with secretion (58). Extensive equilibration revealed that the trimeric pseudopilin complex was associated with significant local deformation around one of the partially membrane-extracted transmembrane helices, whereas equilibration of individual chains led to relaxation of the membrane defect (Fig. 1). Supported by experimental data, this suggested a mechanism by which self-assembly of the pseudopilins may be thermodynamically coupled to the nucleation of the pseudopilus, prior to protein secretion.

3.1.2. Other Approaches to Insertion and Assembly of Membrane Proteins

Because membrane proteins are solubilized in membrane-mimetic environments (9) prior to structure determination (8), it may be of interest to study these non-physiological conditions via simulation. To achieve useful results, it is essential to reproduce as closely as possible the experimental conditions in the initial configuration of the simulation system. Moreover, CG simulations of membrane proteins have become useful, providing an alternative approach to assemble lipid/protein systems ab initio.

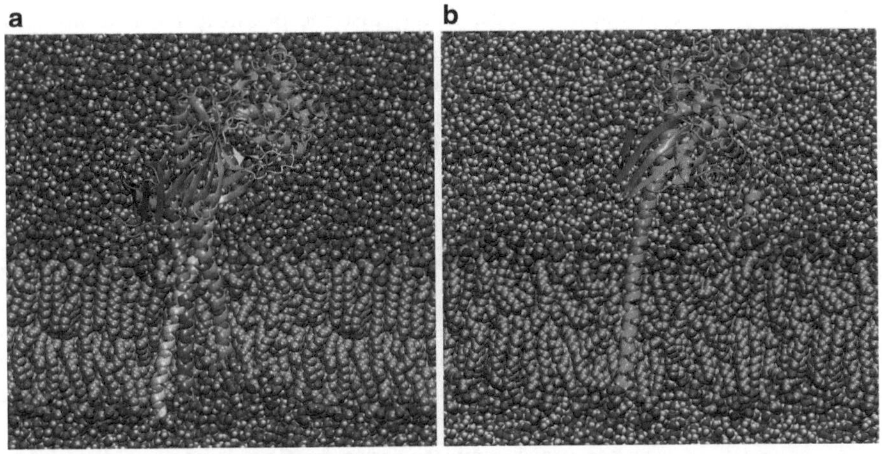

Fig. 1. Snapshots from extensively equilibrated simulation systems of minor pseudopilin components within a lipid bilayer. In (**a**) a significant membrane defect is associated with one of the protein chain termini. In (**b**) the isolated protein chain is shown in its "relaxed" state, which leads to the loss of the local membrane defect. In both figures, the membrane lipids and water are shown in spacefill format and the protein chains have a cartoon representation. This figure was adapted from work performed in ref. 58.

1. The information provided within the PDB of a protein structure solved by X-ray diffraction is sufficient to reconstruct the crystallographic unit cell. This provides a means to compare the simulated protein dynamics to those from experimental temperature-factors. A simulation of four membrane protein chains in a crystallographic unit cell, beginning from the precise protein/detergent arrangement observed in the original X-ray lattice (59), revealed that the detergent molecules form a dynamic, membrane-mimicking environment on the surface of the crystallized proteins. Moreover, a high degree of structural stability in the protein/detergent complex confirmed the accuracy of the simulation force field and protocol.
2. The dynamics of detergent/protein complexes are also of interest, since they are utilized during biophysical characterization and solution NMR studies. The setup of such systems is nontrivial but may be guided by experimental parameters such as the critical micelle concentration and detergent aggregation number, as well as higher resolution information such as measurements of intermolecular NOEs between the protein surface and detergent molecules (60). Although the detailed geometry of a protein/detergent complex is difficult to define, some attempts have been made to model the detergent micelle "jacket" around simple membrane protein architectures, including for inner-membrane α-helical dimers (61) and bundles (62, 63) and β-barrel outer membrane proteins (64, 65), in general revealing that the protein conformational dynamics are increased in comparison with the membrane environment.

3. In some cases, it may also be possible to simulate in atomistic detail the spontaneous self-assembly of protein/detergent systems, avoiding the need for user intervention (66–68). Unfortunately, the compute times for performing such simulations are rather costly. Coarse-grained approaches can help to solve this problem.
4. The CG simulation approach may be applied to spontaneously assemble proteins into membranes or other non-bilayer phases (69). Such a process (beginning from a random mixture of protein, lipid, and solvent) would be expected to take on the order of a few hundred nanoseconds of simulation time, which translates to around a day of compute time on a desktop computer for a typical system. The successful assembly, and subsequent accurate reproduction of protein/lipid interactions observed in all-atom models (70), results from the fact that the CG protein/lipid mixing parameters have been derived on the basis of thermodynamic partitioning behavior between polar and nonpolar phases (36, 38).
5. Once a particular configuration of a protein/lipid system has been obtained via CG approaches, it may be of interest to convert back to atomistic resolution, in order to perform all-atom simulations and refine the protein structure, study conformational changes, or analyze specific lipid–protein interactions, the details of which are inaccessible to CG models. This may be achieved by aligning atomistic lipid fragments and/or amino acids to their CG counterparts, either by least-squares fitting or mapping procedures or by using a restrained, simulated annealing procedure to optimize the positions of atoms within each CG particle (71–82). It should also be noted that dual-resolution approaches, in which molecules are represented at both atomistic and CG resolution within a single simulation system, are showing great promise (83–90).

3.2. Optimization and Prediction of Lipid-Binding Modes in Protein Cavities via Simulation

1. As well as investigating the likely orientation of membrane proteins with a lipid environment, the converse problem may also be tackled from a simulation perspective, i.e., prediction of the position of lipid ligands within the cavity of soluble proteins. A good example of this is illustrated by recent work on the immunological, MHC-like CD1 proteins, which present lipid antigens to the T-cell receptor (TCR). CD1-binding domains contain a deep hydrophobic cavity, and variations in the shape and size of the cavity in different isoforms define their capacity to anchor a diverse array of lipids. In the absence of available crystallographic data, prediction of the structure of the CD1c isoform enabled molecular modelling of possible lipid-binding modes, which could then be tested via simulation, and provided potential clues for the rational design of

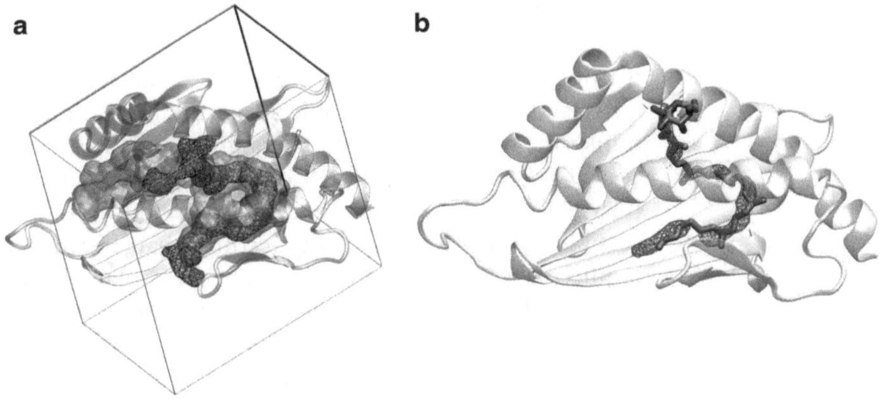

Fig. 2. Use of the MDFF algorithm to flexibly fit an MPM molecule into the CD1c crystallographic density. In (**a**) identified electron density map (grid surface) is overlaid onto the internal cavity (smooth molecular surface) of the CD1c-binding domain (transparent cartoon). In (**b**) the final MDFF refined conformation of MPM (wireframe) is overlaid over part of the observable electron density (grid surface) of the CD1c-binding domain (transparent cartoon). This figure was adapted from work performed in ref. 93.

therapeutic molecules (91). Moreover, the effects of naturally occurring sequence variations in the TCR on the experimentally measured binding affinity of CD1-presented lipid molecules could be rationalized via in silico "mutation" combined with standard, all-atom MD simulation approaches in the presence of lipid antigen molecules (92).

2. In some cases, MD simulations may be used to integrate available low- or high-resolution structural data from experiments to understand lipid-binding modes. For example, the structure of CD1c was recently solved via X-ray crystallography (93), bound to mannosyl-β1-phosphomycoketide (MPM), a single-chain, methyl-branched glycolipid from *Mycobacterium tuberculosis* whose interaction with CD1c could guide vaccine development. As the electron density for the bound lipid was poorly resolved, the MD "flexible fitting" (MDFF) algorithm (94) was utilized within the NAMD simulation framework (43) to apply forces on MPM (in addition to those from the standard MD potential) proportional to the gradient of an overlaid density map (Fig. 2). In this way, candidate lipid conformations could be flexibly docked into the available electron density. Based on conformations of these biased simulations, further rounds of crystallographic refinement led to the identification of a plausible MPM conformation within the CD1c-binding groove and revealed that the phosphomannose headgroup is stabilized at the cavity entrance by a number of high-occupancy water-mediated hydrogen bonds.
3. Simulations may also be used to investigate the *mechanisms* of lipid binding. For example, atomistic MD simulations have been extensively carried out for members of the fatty acid-binding

protein (FABP) family, typically over timescales of tens or hundreds of nanoseconds. FABPs limit the potential membrane-dissolving, "detergent-like" activity of free fatty acids and other hydrophobic molecules and also aid in lipid transport, metabolism, and signaling. Their structure is highly conserved, consisting of a β-barrel with an intercalated helix-turn-helix motif (1). The binding site is an internal cavity inaccessible from the bulk, suggesting that ligand absorption requires a conformational change. To explain how the fatty acid is able to reach the unexposed binding cavity, a dynamic "portal" site has been proposed (95–97), and the application of MD simulations to various FABPs has revealed the spontaneous absorption of fatty acids at this site (98, 99). On the other hand, MD simulation studies on various other FABPs revealed a possible alternative site of entry ("anti-portal"), located in an evolutionarily well-conserved region close to the amino-terminus (100, 101). Simulations of the interaction of liver basic FABP with preformed micelles revealed that the anti-portal was the preferred lipid-binding site (102). In fact, this may be the more physiologically relevant environment for fatty acid extraction by FABPs, since local fatty acid concentrations normally exceed their critical micelle concentration during absorption. MD simulations have also been applied to identify possible water/lipid exchange pathways within FABPs (103, 104) (105–107).

3.3. Towards More Biologically Realistic Membrane Simulations

1. With ever-increasing timescales accessible to MD simulations, it has become possible to observe the folding and membrane insertion of simple peptide models in atomistically detailed systems. Long-timescale simulations combined with protocols designed to improve sampling, such as replica-exchange (108) or elevated temperature MD (109), have yielded observations of the spontaneous folding and insertion of the thermostable WALP model peptides (110) and combined with the use of refined force fields reproduce experimentally validated folding pathways in terms of both kinetics and thermodynamics (111).
2. Whilst simulations of spontaneous peptide insertion will likely become more frequent in the future, biased potential of mean force (PMF) simulation approaches have become popular in recent years (112) and may be used to quantitatively calculate the thermodynamics of individual amino acid partitioning within membranes, providing a simple means to compare in silico observations with in vitro or in vivo measurements of lipid–protein interactions whilst yielding unparalleled spatial resolution. In the typical umbrella sampling approach, a series of simulations are performed with the center of mass of a molecule harmonically restrained at different positions along

the bilayer normal, and subsequent "unbiasing" of the simulated distributions yields the PMF. Although a single PMF requires extensive simulation sampling, the nature of umbrella sampling makes it "embarrassingly parallel." PMFs calculated for amino acid side-chain analogues (113–115) have successfully reproduced probability distributions along the bilayer normal expected from bioinformatics analyses (50) and correlate well with hydrophobicity scales (116).

3. Some quantitative differences between simulated and experimental partitioning thermodynamics have nevertheless been observed. In particular, poor agreement in the *absolute* magnitude of the experimental "biological hydrophobicity scale" (117) (based on translocon-mediated insertion into the endoplasmic reticulum) with scales calculated from simulated PMFs has been reported. However, simulations of more biologically realistic systems have gone some way to explaining this disparity (118–120). Recent MD simulations of amino acid partitioning in the presence of the SecY machinery indicate that the insertion penalty for, e.g., arginine is significantly reduced in the presence of the protein environment of the translocon (121).

4. Significant progress has also been made in the simulation of more biologically realistic membrane lipid compositions. For example, the dynamics of rhodopsin in lipid membrane compositions representing those in mammalian rod cells have been studied (122, 123). Extensive atomistic simulation sampling, consisting of 26 sets of 100 ns simulations, was utilized to ensure that sufficient independent configurations of mixed lipid bilayers were sampled. These enabled the identification of eight sites on the protein surface that interacted with unusual polyunsaturated lipids, along with less populated clusters of stearoyl lipids and cholesterol. The observation of such specific lipid–protein interactions helps to rationalize the observation that the amounts of these lipids are tightly controlled within the native environment, since they have significant effects on the functional activity of rhodopsin. Moreover, multi-scale simulations of the interactions of the transmembrane domain of fukutin, a protein implicated in muscular dystrophy, with membrane compositions representing the Golgi Apparatus have been reported (124). The short 36-residue transmembrane fukutin fragment was modelled as a perfect α-helix, guided by CD spectroscopy data. Having established the stability of the helical conformation at the atomistic level, the authors performed CG simulations of the peptide in a range of more realistic membrane mixtures (Fig. 3). Initially, the effect of the nature of the lipid tails and headgroups on the orientation of the protein was investigated using

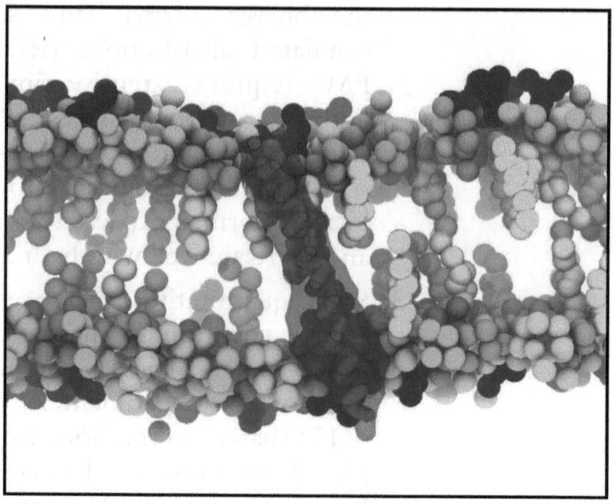

Fig. 3. A coarse-grained simulation snapshot of the alpha helical membrane domain of Fukutin (tube format with overlaid transparent molecular surface) in a phospholipid membrane composed of DPPC, DPPE, PIP2, sphingomyelin, DPPS, and cholesterol (all shown in spacefill format). The membrane composition closely reflects the in vivo concentrations of lipids within the Golgi apparatus. The water, lipid acyl tails, and ions have been omitted for clarity. This figure was adapted from work performed in ref. 124.

dilauroylphosphatidylcholine (DLPC), dipalmitoylphosphatidylcholine (DPPC), and palmitoleoylphosphatidylcholine (POPC) bilayers. Small tilt angles were observed for longer lipid tail membranes, whereas in the shorter DLPC lipids, substantial transmembrane tilting occurred to minimize hydrophobic mismatch. The protein was subsequently simulated in a mixture of lipids reflecting the Golgi membrane, including sphingomyelin, cholesterol, DPPC, dipalmitoylphosphatidylethanolamine (DPPE), dipalmitoylphosphatidylserine (DPPS), and phosphatidylinositol 4,5-bisphosphate (PIP_2). A marked difference in the protein–lipid interactions at the amino- and carboxy-termini of the fukutin transmembrane fragment was observed, with local clustering of the PIP_2 lipids around the amino-terminal domain, and the authors suggested that the interaction of the amino-terminal domain with the Golgi membrane may induce the formation of microdomains.

4. Notes

1. The homology modelling approach, in which one or more homologous "template" sequences of proteins (whose structures are known) are identified, aligned to the unknown "target" structure, and subsequently used to build a model, has

become a useful approach in recent years (sometimes combined with simulation (125)) to, e.g., interpret low-resolution structural data and to help design novel experiments, for both soluble and membrane-bound proteins (126). Normally, the sequences of the target and template(s) must share at least ~30% identity to produce a useful model, and known experimental structural information can be used to constrain the model and hence improve its quality. Additionally, conserved topological features such as inverted repeats may be used to model different conformational states of membrane protein structures (127). Accurate homology modelling has benefited greatly in recent years from the development of powerful programs such as Modeller (128) and the availability of automated online servers such as SWISS-PROT (http://swissmodel.expasy.org/).

2. Web sites that have made available coordinates and/or topologies for lipids and/or lipid bilayers include (with respective contributor's research groups in parentheses) the following: http://moose.bio.ucalgary.ca/index.php?page=Structures_and_Topologies (Tieleman); http://lipidbook.bioch.ox.ac.uk/; (Sansom); http://terpconnect.umd.edu/~jbklauda/research/download.html (Klauda); http://www.apmaths.uwo.ca/~mkarttu/downloads.shtml (Karttunen); http://www.chms.ucdavis.edu/research/web/faller/ (Faller); http://www.charmm-gui.org/?doc=archive&lib=lipid_pure (Im); http://mackerell.umaryland.edu/CHARMM_ff_params.html (MacKerell); and http://compbio.biosci.uq.edu.au/atb/ (Mark).

3. Generally speaking, the so-called isothermal–isobaric or NPT ensemble (constant number of particles, constant temperature, and constant pressure, with the latter two coupled to an appropriate barostat and thermostat within the simulation code) is most appropriate for biological systems. In this context, the exact details of the method used for controlling the constant-pressure conditions must be considered carefully (129). Isotropic pressure coupling may be reasonably used for approximately spherical globular proteins (or micelles) in solvent, with all three axes of the simulation box coupled together. However, for membrane simulations, this does not allow fluctuations to occur independently in the surface area, which is a key property of the lipid bilayer. An alternative approach is to constrain the box in the membrane plane to enforce an appropriate area per lipid and/or to apply a surface tension term, as in the case of CHARMM27 lipid simulations. Assuming the chosen force field is sufficiently accurate, however, the recommended approach for membranes is normally to use semi-isotropic pressure coupling, in which the dimension parallel to the bilayer normal fluctuates independently, but the two

dimensions perpendicular to the bilayer normal are coupled. This also enables one to assess the reliability of a membrane simulation, by comparing the equilibrated mean area per lipid to experimental values.

4. Sometimes parameters are not available within the force field being used for “nonstandard” molecules of interest, e.g., a ligand or a cofactor. In this case, it may be possible to build a topology by “analogy,” i.e., by finding fragments of chemical groups within the force field and combining them to describe the molecule of interest. Alternatively, a number of Web servers are now available for the automatic parameterization of molecular groups, although the user should always carefully check the resultant topology and the behavior of the molecule in preliminary test simulations, modifying parameters where appropriate using the same approaches used to derive the original force field. Popular servers include ParamChem (https://www.paramchem.org/), SwissParam (http://swissparam.ch/), and PRODRG (http://davapc1.bioch.dundee.ac.uk/prodrg/).
5. Because of the many-body nature of typical, charged molecular systems (i.e., the force on each atom depends on the position of every other atom), it is not possible to calculate a simulation analytically. Therefore, numerical finite difference methods are used to integrate the equations of motion over small time steps. The chosen time step is limited by the highest-frequency vibration within the system of interest, which is typically ~10^{-15} s. Thus, this process must be repeated many millions of times to gradually propagate a simulation and hence obtain a trajectory over biologically relevant timescales.
6. A correct treatment of electrostatic interactions during MD simulations, particularly in highly charged biomolecular systems, is essential. Their calculation tends to be extremely costly, because electrostatic interactions are significant even at large distances (the Coulomb potential decays as $1/r$). One way to lower the computational cost is to use a “cutoff,” whereby only particles within a certain distance (typically ~1.5–2 nm) of each other interact. Because the truncation of nonbonded potential interactions is rather artificial, the potential may be modified with a smoothing function or by including the response of a dielectric medium at long range with a “reaction-field.” However, in general it is desirable to include all long-range electrostatic interactions explicitly, to ensure an accurate representation of the system in question. Thus, methods based on Ewald summation may be used; this was first developed for the calculation of electrostatics in an infinite crystal lattice. Modern, computationally efficient Ewald-based algorithms such as Particle Mesh Ewald (PME) are normally the preferred choice for treatment of electrostatics in biomolecular

simulations, despite concerns that they may introduce artificial periodicity into the system (130).

7. The solvation of simulation systems is relatively straightforward and generally entails an automated process of superimposition of previously equilibrated water boxes onto the coordinates of the protein/lipid complex, followed by deletion of overlapping water molecules. Addition of ions may also be reasonably carried out via replacement of water molecules at random, to a final concentration approximating physiological conditions (typically a ~0.15 M NaCl solution). In the case of membrane simulations, a few other details must be considered; in particular, it is normally undesirable to place water molecules within the hydrophobic bilayer core. Water molecules within the membrane may be manually removed using molecular graphics tools (see ref. 57 for a description of such approaches), or alternatively, a protein-restrained equilibration simulation lasting a few nanoseconds should lead to the spontaneous exit of any water from the lipid tail phase.

References

1. Russ AP, Lampel S (2005) The druggable genome: an update. Drug Discov Today 10:1607–1610
2. Overington JP, Al-Lazikani B, Hopkins AL (2006) How many drug targets are there? Nat Rev Drug Discov 5:993–996
3. Phillips R et al (2009) Emerging roles for lipids in shaping membrane-protein function. Nature 459:379–385
4. Lee AG (2011) Biological membranes: the importance of molecular detail. Trends Biochem Sci 36:493–500
5. Bernlohr DA et al (1997) Intracellular lipid-binding proteins and their genes. Annu Rev Nutr 17:277–303
6. De Libero G, Mori L (2005) Recognition of lipid antigens by T cells. Nat Rev Immunol 5:485–496
7. Fleishman SJ, Unger VM, Ben-Tal N (2006) Transmembrane protein structures without X-rays. Trends Biochem Sci 31:106–113
8. Arora A, Tamm LK (2001) Biophysical approaches to membrane protein structure determination. Curr Opin Struct Biol 11:540–547
9. Garavito R, Ferguson-Miller S (2001) Detergents as tools in membrane biochemistry. J Biol Chem 276:32403–32406
10. Karplus M, McCammon JA (2002) Molecular dynamics simulations of biomolecules. Nat Struct Biol 9:646–652
11. Durrant JD, McCammon JA (2011) Molecular dynamics simulations and drug discovery. BMC Biol 9:71
12. Dror RO et al (2010) Exploring atomic resolution physiology on a femtosecond to millisecond timescale using molecular dynamics simulations. J Gen Physiol 135:555–562
13. Domene C, Bond PJ, Sansom MSP (2003) Membrane protein simulations: ion channels and bacterial outer membrane proteins. Adv Protein Chem 66:159–193
14. Sansom MSP et al (2002) Water in ion channels and pores—simulation studies. Novartis Found Symp 245:66–78
15. Sansom MSP et al (2005) Molecular simulations and lipid-protein interactions: potassium channels and other membrane proteins. Biochem Soc Trans 33:916–920
16. Shaw DE et al (2010) Atomic-level characterization of the structural dynamics of proteins. Science 330:341–346
17. Ash WL et al (2004) Computer simulations of membrane proteins. Biochim Biophys Acta 1666:158–189
18. Bond PJ, Khalid S (2010) Antimicrobial and cell-penetrating peptides: structure, assembly and mechanisms of membrane lysis via atomistic and coarse-grained molecular dynamics simulations. Protein Pept Lett 17:1313–1327

19. Gumbart J et al (2005) Molecular dynamics simulations of proteins in lipid bilayers. Curr Opin Struct Biol 15:423–431
20. Beckstein O et al (2003) Ion channel gating: insights via molecular simulation. FEBS Lett 555:85–90
21. Bond PJ, Sansom MS (2004) The simulation approach to bacterial outer membrane proteins. Mol Membr Biol 21:151–161
22. Nielsen SO et al (2004) Coarse grain models and the computer simulation of soft materials. J Phys Condense Matter 16:R481–R512
23. Chandrasekhar I et al (2003) A consistent potential energy parameter set for lipids: dipalmitoylphosphatidylcholine as a benchmark of the GROMOS96 45A3 force field. Eur Biophys J 32:67–77
24. Chiu SW et al (2009) An improved united atom force field for simulation of mixed lipid bilayers. J Phys Chem B 113:2748–2763
25. Poger D, Van Gunsteren WF, Mark AE (2010) A new force field for simulating phosphatidylcholine bilayers. J Comput Chem 31:1117–1125
26. Oostenbrink C et al (2004) A biomolecular force field based on the free enthalpy of hydration and solvation: the GROMOS force-field parameter sets 53A5 and 53A6. J Comput Chem 25:1656–1676
27. Klauda JB et al (2010) Update of the CHARMM all-atom additive force field for lipids: validation on six lipid types. J Phys Chem B 114:7830–7843
28. Vanommeslaeghe K et al (2010) CHARMM general force field: a force field for drug-like molecules compatible with the CHARMM all-atom additive biological force fields. J Comput Chem 31:671–690
29. MacKerell AD Jr et al (1998) CHARMM: the energy function and its parameterization with an overview of the program. Wiley, Chichester, pp 271–277
30. Kaminski GA et al (2001) Evaluation and reparametrization of the OPLS-AA force field for proteins via comparison with accurate quantum chemical calculations on peptides. J Phys Chem B 105:6474–6487
31. Lindorff-Larsen K et al (2010) Improved side-chain torsion potentials for the Amber ff99SB protein force field. Proteins 78:1950–1958
32. Berger O, Edholm O, Jahnig F (1997) Molecular dynamics simulations of a fluid bilayer of dipalmitoylphosphatidylcholine at full hydration, constant pressure, and constant temperature. Biophys J 72:2002–2013
33. Venturoli M, Smit B, Sperotto MM (2005) Simulation studies of protein-induced bilayer deformations, and lipid-induced protein tilting, on a mesoscopic model for lipid bilayers with embedded proteins. Biophys J 88:1778–1798
34. Marrink SJ, de Vries AH, Mark AE (2004) Coarse grained model for semiquantitative lipid simulations. J Phys Chem B 108:750–760
35. Marrink SJ et al (2007) The MARTINI force field: coarse grained model for biomolecular simulations. J Phys Chem B 111:7812–7824
36. Monticelli L et al (2008) The MARTINI coarse-grained force field: extension to proteins. J Chem Theory Comput 4:819–834
37. Bond PJ, Sansom MSP (2006) Insertion and assembly of membrane proteins via simulation. J Am Chem Soc 128:2697–2704
38. Bond PJ, Wee CL, Sansom MS (2008) Coarse-grained molecular dynamics simulations of the energetics of helix insertion into a lipid bilayer. Biochemistry 47:11321–11331
39. Bond PJ, Sansom MSP (2007) Bilayer deformation by the Kv channel volage sensor domain revealed by self-assembly simulations. Proc Natl Acad Sci U S A 104:2631–2636
40. Shih AY et al (2006) Coarse grained protein-lipid model with application to lipoprotein particles. J Phys Chem B 110:3674–3684
41. Van Der Spoel D et al (2005) GROMACS: fast, flexible, and free. J Comput Chem 26:1701–1718
42. Hess B et al (2008) GROMACS 4: algorithms for highly efficient, load-balanced, and scalable molecular simulation. J Chem Theory Comput 4:435–447
43. Phillips JC et al (2005) Scalable molecular dynamics with NAMD. J Comput Chem 26:1781–1802
44. Brooks BR et al (2009) CHARMM: the biomolecular simulation program. J Comput Chem 30:1545–1614
45. Friedrichs MS et al (2009) Accelerating molecular dynamic simulation on graphics processing units. J Comput Chem 30:864–872
46. Klepeis JL et al (2009) Long-timescale molecular dynamics simulations of protein structure and function. Curr Opin Struct Biol 19:120–127
47. Deol SS et al (2004) Lipid-protein interactions of integral membrane proteins: a comparative simulation study. Biophys J 87:1–13
48. Domene C et al (2003) Lipid/protein interactions and the membrane/water interfacial region. J Am Chem Soc 125:14966–14967
49. Woolf TB, Roux B (1996) Structure, energetics, and dynamics of lipid-protein interactions:

a molecular dynamics study of the gramicidin A channel in a DMPC bilayer. Proteins 24:92–114

50. Ulmschneider MB, Sansom MS (2001) Amino acid distributions in integral membrane protein structures. Biochim Biophys Acta 1512:1–14
51. Ulmschneider MB, Sansom MS, Di Nola A (2005) Properties of integral membrane protein structures: derivation of an implicit membrane potential. Proteins 59:252–265
52. Lomize MA et al (2006) OPM: orientations of proteins in membranes database. Bioinformatics 22:623–625
53. Faraldo-Gómez JD, Smith GR, Sansom MS (2002) Setting up and optimization of membrane protein simulations. Eur Biophys J 31:217–227
54. Tieleman DP, Berendsen HJ (1998) A molecular dynamics study of the pores formed by Escherichia coli OmpF porin in a fully hydrated palmitoyloleoylphosphatidylcholine bilayer. Biophys J 74:2786–2801
55. Staritzbichler R et al (2011) GRIFFIN: a versatile methodology for optimization of protein-lipid interfaces for membrane protein simulations. J Chem Theory Comput 7:1167–1176
56. Wolf MG et al (2010) g_membed: efficient insertion of a membrane protein into an equilibrated lipid bilayer with minimal perturbation. J Comput Chem 31:2169–2174
57. Kandt C, Ash WL, Tieleman DP (2007) Setting up and running molecular dynamics simulations of membrane proteins. Methods 41:475–488
58. Cisneros DA et al (2011) Minor pseudopilin self-assembly primes type II secretion pseudopilus elongation. EMBO J 31(4):1041–1053
59. Bond PJ et al (2006) Membrane protein dynamics and detergent interactions within a crystal: a simulation study of OmpA. Proc Natl Acad Sci U S A 103:9518–9523
60. Fernandez C et al (2002) Lipid-protein interactions in DHPC micelles containing the integral membrane protein OmpX investigated by NMR spectroscopy. Proc Natl Acad Sci U S A 99:13533–13537
61. Cuthbertson JM, Bond PJ, Sansom MSP (2006) Transmembrane helix–helix interactions: comparative simulations of the glycophorin A dimer. Biochemistry 45:14298–14310
62. Sands ZA, Sansom MS (2007) How does a voltage sensor interact with a lipid bilayer? Simulations of a potassium channel domain. Structure 15:235–244
63. Patargias G et al (2005) Molecular dynamics simulations of GlpF in a micelle vs. in a bilayer: conformational dynamics of a membrane protein as a function of environment. J Phys Chem B 109:575–582
64. Bond PJ, Sansom MSP (2003) Membrane protein dynamics versus environment: simulations of OmpA in a micelle and in a bilayer. J Mol Biol 329:1035–1053
65. Cox K, Sansom MS (2009) One membrane protein, two structures and six environments: a comparative molecular dynamics simulation study of the bacterial outer membrane protein PagP. Mol Membr Biol 26:205–214
66. Bond PJ et al (2004) MD simulations of spontaneous membrane protein/detergent micelle formation. J Am Chem Soc 126:15948–15949
67. Braun R, Engelman DM, Schulten K (2004) Molecular dynamics simulations of micelle formation around dimeric Glycophorin A transmembrane helices. Biophys J 87:754–763
68. Bockmann RA, Caflisch A (2005) Spontaneous formation of detergent micelles around the outer membrane protein OmpX. Biophys J 88:3191–3204
69. Scott KA et al (2008) Coarse-grained MD simulations of membrane protein-bilayer self-assembly. Structure 16:621–630
70. Bond PJ et al (2007) Coarse-grained molecular dynamics simulations of membrane proteins and peptides. J Struct Biol 157:593–605
71. Rouse SL et al (2009) Simulations of the BM2 proton channel transmembrane domain from influenza virus B. Biochemistry 48:9949–9951
72. Carpenter T et al (2008) Self-assembly of a simple membrane protein: coarse-grained molecular dynamics simulations of the influenza M2 channel. Biophys J 95:3790–3801
73. Shih AY et al (2007) Disassembly of nanodiscs with cholate. Nano Lett 7:1692–1696
74. Shih AY, Sligar SG, Schulten K (2009) Maturation of high-density lipoproteins. J R Soc Interface 6:863–871
75. Stansfeld PJ et al (2009) PIP(2)-binding site in Kir channels: definition by multiscale biomolecular simulations. Biochemistry 48:10926–10933
76. Rzepiela AJ et al (2010) Reconstruction of atomistic details from coarse-grained structures. J Comput Chem 31:1333–1343
77. Rzepiela AJ et al (2010) Membrane poration by antimicrobial peptides combining atomistic and coarse-grained descriptions. Faraday Discuss 144:431–443, discussion 445–481

78. Cojocaru V et al (2011) Structure and dynamics of the membrane-bound cytochrome P450 2C9. PLoS Comput Biol 7:e1002152
79. Sengupta D et al (2008) Toroidal pores formed by antimicrobial peptides show significant disorder. Biochim Biophys Acta 1778:2308–2317
80. Leontiadou H, Mark AE, Marrink SJ (2006) Antimicrobial peptides in action. J Am Chem Soc 128:12156–12161
81. Chen R, Mark AE (2011) The effect of membrane curvature on the conformation of antimicrobial peptides: implications for binding and the mechanism of action. Eur Biophys J 40:545–553
82. Bond PJ et al (2008) Coarse-grained simulations of the membrane-active antimicrobial peptide maculatin 1.1. Biophys J 95:3802–3815
83. Chang R, Ayton GS, Voth GA (2005) Multiscale coupling of mesoscopic- and atomistic-level lipid bilayer simulations. J Chem Phys 122:244716
84. Christen M, van Gunsteren WF (2006) Multigraining: an algorithm for simultaneous fine-grained and coarse-grained simulation of molecular systems. J Chem Phys 124:154106
85. Izvekov S, Voth GA (2005) A multiscale coarse-graining method for biomolecular systems. J Phys Chem B 109:2469–2473
86. Michel J, Orsi M, Essex JW (2008) Prediction of partition coefficients by multiscale hybrid atomic-level/coarse-grain simulations. J Phys Chem B 112:657–660
87. Orsi M, Noro MG, Essex JW (2011) Dual-resolution molecular dynamics simulation of antimicrobials in biomembranes. J R Soc Interface 8:826–841
88. Orsi M, Sanderson WE, Essex JW (2009) Permeability of small molecules through a lipid bilayer: a multiscale simulation study. J Phys Chem B 113:12019–12029
89. Shi Q, Izvekov S, Voth GA (2006) Mixed atomistic and coarse-grained molecular dynamics: simulation of a membrane-bound ion channel. J Phys Chem B 110:15045–15048
90. Rzepiela AJ et al (2011) Hybrid simulations: combining atomistic and coarse-grained force fields using virtual sites. Phys Chem Chem Phys 13:10437–10448
91. Garzón D, Bond PJ, Faraldo-Gómez JD (2009) Predicted structural basis for CD1c presentation of mycobacterial branched polyketides and long lipopeptide antigens. Mol Immunol 47:253–260
92. Sanderson JP et al (2012) Natural variations at position 93 of the invariant Valpha24-Jalpha18 alpha chain of human iNKT-cell TCRs strongly impact on CD1d binding. Eur J Immunol 42:248–255
93. Scharf L et al (2010) The 2.5 Å structure of CD1c in complex with a mycobacterial lipid reveals an open groove ideally suited for diverse antigen presentation. Immunity 33:853–862
94. Trabuco LG et al (2011) Applications of the molecular dynamics flexible fitting method. J Struct Biol 173:420–427
95. Sacchettini JC, Gordon JI, Banaszak LJ (1989) Crystal structure of rat intestinal fatty-acid-binding protein. Refinement and analysis of the Escherichia coli-derived protein with bound palmitate. J Mol Biol 208:327–339
96. Hodsdon ME, Cistola DP (1997) Discrete backbone disorder in the nuclear magnetic resonance structure of apo intestinal fatty acid-binding protein: implications for the mechanism of ligand entry. Biochemistry 36:1450–1460
97. Zhang F et al (1997) Solution structure of human intestinal fatty acid binding protein: implications for ligand entry and exit. J Biomol NMR 9:213–228
98. Friedman R, Nachliel E, Gutman M (2006) Fatty acid binding proteins: same structure but different binding mechanisms? Molecular dynamics simulations of intestinal fatty acid binding protein. Biophys J 90:1535–1545
99. van Aalten DM et al (1995) Essential dynamics of the cellular retinol-binding protein–evidence for ligand-induced conformational changes. Protein Eng 8:1129–1135
100. Friedman R, Nachliel E, Gutman M (2005) Molecular dynamics simulations of the adipocyte lipid binding protein reveal a novel entry site for the ligand. Biochemistry 44:4275–4283
101. Levin LB et al (2010) Insight into the interaction sites between fatty acid binding proteins and their ligands. J Mol Model 16:929–938
102. Levin LB et al (2009) Molecular dynamics study of the interaction between fatty acid binding proteins with palmitate mini-micelles. Mol Cell Biochem 326:29–33
103. Likic VA, Prendergast FG (2001) Dynamics of internal water in fatty acid binding protein: computer simulations and comparison with experiments. Proteins 43:65–72
104. Bakowies D, Van Gunsteren WF (2002) Simulations of apo and holo-fatty acid binding protein: structure and dynamics of protein, ligand and internal water. J Mol Biol 315:713–736

105. Woolf TB, Grossfield A, Tychko M (2000) Differences between apo and three holo forms of the intestinal fatty acid binding protein seen by molecular dynamics computer calculations. Biophys J 78:608–625
106. Eberini I et al (2008) Conformational and dynamics changes induced by bile acids binding to chicken liver bile acid binding protein. Proteins 71:1889–1898
107. Ricchiuto P et al (2008) Structural and dynamic roles of permanent water molecules in ligand molecular recognition by chicken liver bile acid binding protein. J Mol Recognit 21:348–354
108. Nymeyer H, Woolf TB, Garcia AE (2005) Folding is not required for bilayer insertion: replica exchange simulations of an alpha-helical peptide with an explicit lipid bilayer. Proteins 59:783–790
109. Ulmschneider MB, Ulmschneider JP (2008) Folding peptides into lipid bilayer membranes. J Chem Theory Comput 4:1807–1809
110. Killian JA, von Heijne G (2000) How proteins adapt to a membrane-water interface. Trends Biochem Sci 25:429–434
111. Ulmschneider MB et al (2010) Mechanism and kinetics of peptide partitioning into membranes from all-atom simulations of thermostable peptides. J Am Chem Soc 132:3452–3460
112. Deng Y, Roux B (2009) Computations of standard binding free energies with molecular dynamics simulations. J Phys Chem B 113:2234–2246
113. Johansson AC, Lindahl E (2008) Position-resolved free energy of solvation for amino acids in lipid membranes from molecular dynamics simulations. Proteins 70:1332–1344
114. MacCallum JL, Bennett WF, Tieleman DP (2007) Partitioning of amino acid side chains into lipid bilayers: results from computer simulations and comparison to experiment. J Gen Physiol 129:371–377
115. MacCallum JL, Bennett WF, Tieleman DP (2008) Distribution of amino acids in a lipid bilayer from computer simulations. Biophys J 94:3393–3404
116. MacCallum JL, Tieleman DP (2011) Hydrophobicity scales: a thermodynamic looking glass into lipid–protein interactions. Trends Biochem Sci 36:653–662
117. Hessa T et al (2007) Molecular code for transmembrane-helix recognition by the Sec61 translocon. Nature 450:1026–1030
118. Li L, Vorobyov I, Allen TW (2008) Potential of mean force and pKa profile calculation for a lipid membrane-exposed arginine side chain. J Phys Chem B 112:9574–9587
119. Johansson AC, Lindahl E (2009) Protein contents in biological membranes can explain abnormal solvation of charged and polar residues. Proc Natl Acad Sci U S A 106:15684–15689
120. Schow EV et al (2011) Arginine in membranes: the connection between molecular dynamics simulations and translocon-mediated insertion experiments. J Membr Biol 239:35–48
121. Gumbart J, Chipot C, Schulten K (2011) Free-energy cost for translocon-assisted insertion of membrane proteins. Proc Natl Acad Sci U S A 108:3596–3601
122. Pitman MC et al (2005) Role of cholesterol and polyunsaturated chains in lipid–protein interactions: molecular dynamics simulation of rhodopsin in a realistic membrane environment. J Am Chem Soc 127:4576–4577
123. Grossfield A, Feller SE, Pitman MC (2006) A role for direct interactions in the modulation of rhodopsin by omega-3 polyunsaturated lipids. Proc Natl Acad Sci U S A 103:4888–4893
124. Holdbrook DA et al (2010) Stability and membrane orientation of the fukutin transmembrane domain: a combined multiscale molecular dynamics and circular dichroism study. Biochemistry 49:10796–10802
125. Law RJ et al (2005) Membrane protein structure quality in molecular dynamics simulation. J Mol Graph Model 24:157–165
126. Punta M et al (2007) Membrane protein prediction methods. Methods 41:460–474
127. Crisman TJ et al (2009) Inward-facing conformation of glutamate transporters as revealed by their inverted-topology structural repeats. Proc Natl Acad Sci U S A 106:20752–20757
128. Fiser A, Sali A (2003) Modeller: generation and refinement of homology-based protein structure models. Methods Enzymol 374:461–491
129. Anezo C et al (2003) Methodological issues in lipid bilayer simulations. J Phys Chem B 107:9424–9433
130. Patra M et al (2003) Molecular dynamics simulations of lipid bilayers: major artifacts due to truncating electrostatic interactions. Biophys J 84:3636–3645

INDEX

Jörg H. Kleinschmidt (ed.), *Lipid-Protein Interactions: Methods and Protocols*, Methods in Molecular Biology, vol. 974, DOI 10.1007/978-1-62703-275-9, © Springer Science+Business Media New York 2013

M

N

O

P

Q

R

S

MIX
Papier aus verantwortungsvollen Quellen
Paper from responsible sources
FSC® C105338

If you have any concerns about our products,
you can contact us on
ProductSafety@springernature.com

In case Publisher is established outside the EU,
the EU authorized representative is:
Springer Nature Customer Service Center GmbH
Europaplatz 3, 69115 Heidelberg, Germany

Printed by Libri Plureos GmbH
in Hamburg, Germany